Basalts and Phase Diagrams

Stearns A. Morse

Basalts and Phase Diagrams

An Introduction to the Quantitative Use of Phase Diagrams in Igneous Petrology

Second Edition

Stearns A. Morse
Department of Geosciences
University of Massachusetts
Amherst, MA, USA

ISBN 978-3-030-97880-8 ISBN 978-3-030-97882-2 (eBook)
https://doi.org/10.1007/978-3-030-97882-2

This Springer imprint is published by the registered company Springer Nature Switzerland AG
The registered company address is: Gewerbestrasse 11, 6330 Cham, Switzerland

For Dorothy

Preface to the Second Edition

The original edition of *Basalts and Phase Diagrams* contained, in more than half the book, just the story of how the principles of physical chemistry work for rocks. New and old readers will discover that the same part of the book will look the same as before, because that's the idea. However, in the Preamble and later parts of this second edition, the news is good: there are wonderful new discoveries and applications that place geochemistry and petrology among the richest of the earth sciences. When it comes to finding natural ice_{VII} in a very deep part of the Earth's mantle, or reaching out to a one-in-a-million chance of finding a natural example of the most abundant rock in the whole Earth (in a meteorite), you will begin to see how some earth sciences are the most fun. Now we are able, 23 years into a new century, to correctly list ALL the mineral species from the tip of your tongue to the core–mantle boundary. And the real colour of freshly split olivine is not yellow but emerald green. Fortunately, for the challenges of science, there are matters of the inner core that seem not yet quite clear. But not for lack of trying! And it has not been that long since a core–mantle disturbance has been tracked all the way up in a plume to the Western USA. Discoveries in geophysics ramify to our interpretations of the rocks themselves.

Amherst, MA, USA
2023

Stearns A. Morse

Preamble: A Remarkable Experiment

The experimental petrologist leads a charmed life: Every experiment yields a result never seen before by man or beast. If she mixes and pressures up a sample every morning, she then opens up a previous day's (or year's) sample and sees a unique result. It may be dull or useless, but good or bad, it is unique. And often it is a happy surprise. This note is from an experimentalist's last experiment, or "run" as we call it.

With a young colleague, Deb Banks, and Professor John Brady as director of the Five College Experimental Laboratory at Smith College, we had finished a study of the thermal history of the 1.307-Ga-old Kiglapait Layered Intrusion in Labrador. This study had reached 14 kbar pressure and 1300°C temperature with highly aluminous minerals augite, orthopyroxene and garnet owing to their unusual compositions to high pressure and temperature. With regard to previous studies, we had extrapolated a mantle source of the parent liquid to occur at 1450°C and 15 kbar before emplacement at 1250°C and 5–6 kbar.

The definitive, multiple-saturated experiments had occurred at about 13 kbar and 1300°C. But I was curious to know what the mineralogy might look like at the liquidus. So I built a final, off-the-wall experiment with a natural olivine-rich composition. As always, this sample was prepared by grinding the natural rock in a mullite mortar under acetone to maintain a dry condition. Then the micron-scale powder was packed into a hole in graphite to create a high surface energy and easy melting, to exclude any existing macroscopic crystals. The sample was then taken up to a pressure of 13 kbar and an extreme temperature of 1450°C. This high temperature was not expected to survive, and indeed after a small interval, it began to sag.

I watched the temperature closely and wrote it down along with the pressure and time. The temperature sagged at a rate of 1.9°C per minute until it reached a final temperature at 1375°C, and then remained there. I kept watching for 3.5 h and then quenched the sample, opened it, ground off the graphite and placed the sample in a 1-in. epoxy mould overnight. In the morning, I retrieved the sample, ground and polished it, and held it under mild pressure until the next day.

This was a "unique discovery" day to cap all the others. When examined in the scanning electron microscope and then in the analytical electron microprobe, it showed only one single, enormous, euhedral crystal of olivine (see the Figure). By "enormous", I mean a rhombic section 60 × 100 microns (micrometres), which is unique in size of all the thousands of experimental crystals I have previously made. Not only is the crystal large, but it is uniform in composition, with classical orthorhombic symmetry. It has three major parallel sides, and at each corner, a tiny growth spurt which signifies how the crystal grew: the corners show the crystal reaching out for food (as it were) which when gained brings each crystal edge out at constant composition. This is a classic example of how silicate crystals grow from a melt.

But that is not all, by a long shot. The quenched matrix outside the crystal shows local dendrites that look like feathers: they are quench crystals formed during the very fast cooling of the experiment. One corner of the crystal shows rapid external growth in a rectangle. And of greatest interest, the centre of the crystal is a negative crystal form, with edges essentially parallel to those at the outside crystal. Within this pocket is a crystalline product made by the quenching of trapped melt. This pocket, called a melt inclusion, is far more fractionated than the initial bulk composition. Its normative composition includes 68% feldspar, 26% nepheline, a small amount of magnetite, a trace of apatite, and no olivine. The presence of nepheline shows the low fugacity of silica, which has been used to make the olivine.

Melt inclusions can be found in many coarse-grained intrusive rocks, and it is tempting to many observers to infer that these inclusions have trapped a sample of the parent magma. Such an inference must be made with great caution, however. If a melt inclusion can fractionate to a highly evolved composition in 3.5 h, it is not likely to retain a parent liquid composition on a geologic time scale.

The thermal history of this experiment contains a gold-mine of information. What happened during the 75° and 39 min of cooling to 1375°C? Nothing except loss of heat to the surroundings. What then stopped this loss? Crystallization! The experiment is a classic thermal history of the fractional latent heat of crystallizing magmas. We recognize two types of heat loss in melt systems: the so-called sensible heat that we can feel, and, only when crystals form, the latent heat of crystallization. It turns out that the latent heat that must be removed to allow crystal growth is a very large fraction of the total heat loss of a cooling crystalline system. The so-called fractional latent heat can amount to 80% of the total heat loss, and at the end of crystallization, it is 100% when the last bit of crystal growth stops.

Although the principles of latent heat loss on crystallization have been well-known for ages in the sciences, they have not routinely been invoked in observational petrology until the seminal work of Marian Holness at Cambridge. That work becomes the focus of a chapter on Layered Intrusions in this book.

Further studies of this experiment can be found in the article "Solidification of Trapped Liquid in Rocks and Minerals" American Mineralogist 99, pp888–896, 2013.

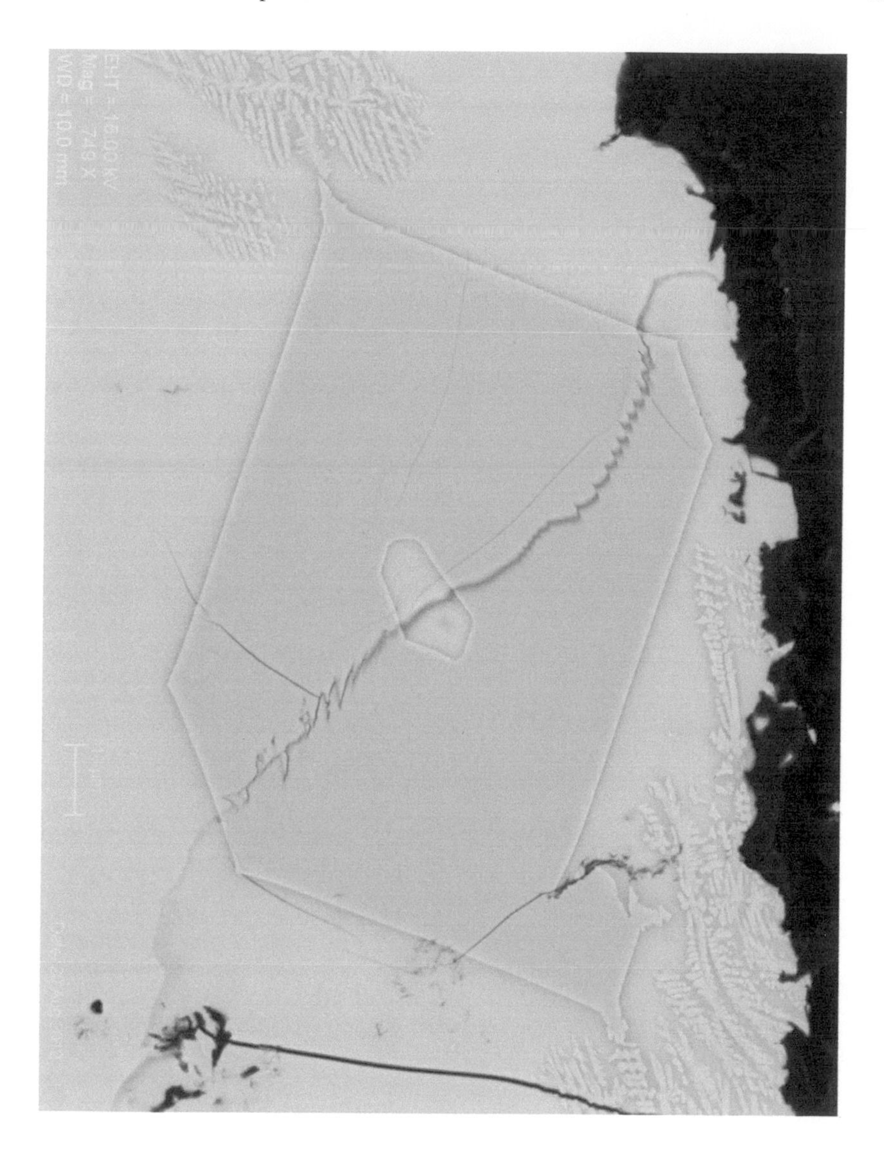
Mag = 749 X
WD = 10.0 mm

Acknowledgments

My first encounter with phase equilibria was from my undergraduate professor John Lyons of Dartmouth College, whose introduction to this rigorous science sparked my interest. In due course this led to a joint paper on the Gibbs free energy in garnet-biotite equilibria. Professor Richard Stoiber taught me optical mineralogy and we wrote a book on it. In graduate school at McGill University, visits of Harry Hess and Hat Yoder inspired us particularly. I ran the daily bulletin board and it was my great sorrow to post the death of N. L. Bowen, whom I had hoped to meet. Instead, I discovered Don Lindsley, whose depth of understanding has taught us new tricks and produced wise thoughts over many years. I have been stimulated and edified by the remarkable insights of my Carnegie office-mate Ikuo Kushiro, who shares the third most references in this book after N. L. Bowen and J. F. Schairer. He and I (among others) share an appreciation of the structure of silicate liquids. Among other supporters who inspire the work of this book have been Gary Ernst, Paul Asimow, Ed Stolper, Lew Ashwal, Dave Stevenson, Ron (the Lord Baron) Oxburgh, Sir Keith Onions, Mark Ghiorso, and Dean Presnall. Many worthy others are cited in an Appendix to this book. Finally, boundless appreciation to John Brady for over two decades of high-pressure collaboration, in petrology and in publication.

Contents

Chapter 1
Introduction

Basalt Problems

Basalt is the most abundant rock on Earth. It forms the ocean ridges, slides off onto the ocean floor, and eventually dives back down. Basalt is the lava which pours out of Kilauea in Hawaii, and maybe crosses a tar road without melting it (we love to play that ace). Basalt is the rock that formed the island of Surtsey, off Iceland, as people watched. It appears in Japan and, in the US, in both New York and Washington State. The name is applied to the solid rock, the lava, and to the magma which, on eruption, becomes lava.

What do we know of these materials? We may know the composition of the rock, and even of the lava at various stages of eruption, from chemical analyses. Seismic evidence tells us that earthquakes from 50 to 60 km deep precede Hawaiian eruptions; this is far beneath the Mohorovičić discontinuity, and we can reasonably infer that the earthquakes are related to movement of magma in the upper mantle. The source of magma is therefore evidently the mantle. The temperature of the lava at eruption can be measured, and commonly falls in the range 1150–1250°C. We know from physical chemistry that hot solid rock on ascending in the mantle can partially melt to yield basalt. We know that three major types of peridotite (plagioclase, spinel, and garnet) can occur in the mantle and act as the source rocks for basalt magma. It would seem that we know a great deal about basalt.

But many matters remain uncertain about basalt. We do not really know the depth of magma generation in specific cases, for the seismic events may only represent the passage of magma. We do not know the exact composition or temperature of generation, since both of these are altered on passage to the surface. We do not know precisely what is melted to form the original magma. We do not often know what crystals form in transit, and of these, which are left behind and which are floated upward. We do not always know what forces impel the magma to the surface, although buoyancy forces would seem most important. We do not know in detail what processes of melting or crystallization may produce the variety of basalt

S. A. Morse, *Basalts and Phase Diagrams*,
https://doi.org/10.1007/978-3-030-97882-2_1

magma compositions which we see. We do not understand the ultimate sources of heat required to generate basalt magma, although we know that some of it must come from the decay of radioactive elements and some from primordial heat left in the deep earth. In particular, we do not understand very well how heat is concentrated within the mantle. We do not even know very well the chemical and physical changes which take place during crystallization at the surface, although studies in Hawaii have greatly improved our knowledge of at least one specific batch of magma. In short, we still have only a general idea of the origin of basalt magma, the progenitor of the most common igneous rock type at the surface of the crust. These problems are multiplied further in the case of plutonic (gabbroic) rocks which we can never observe in the process of formation.

What we do know of basalt genesis comes from a wide variety of studies: geochemical, geophysical, and petrological. Of petrological questions, the chief ones are: what rock is melted or partly melted in the mantle to form the relatively uniform magma which we classify as basalt, and what processes can work on such a magma to produce the suites of differentiated rocks which are associated with basalt? The role of experimental petrology has been to place limits on the answers to these questions, to identify parents and processes which are capable of producing what we see, and to shelve those which appear not to work.

There is no single, sharply defined basalt problem. There is a range of problems having to do, first, with the source of magma, and second, with what the magma may become. These two threads of basalt study are inextricably entwined, for the question of the fate of magmas may or may not be resolved at the instant of origin. In this book, we shall keep both melting within the mantle and crystallization of the resulting magmas in view, and we shall find, as we go, a number of working hypotheses which deserve to be considered concurrently.

The study of basalts can be helpfully framed in terms of the dichotomy between alkali basalt and tholeiite (defined and discussed in the next chapter). The ocean floor is made of tholeiite, erupted at ocean ridges and carried away from them by flowing mantle beneath. Tholeiite also occurs abundantly in continental flood basalts. It is therefore the most abundant and commonly made rock in the earth's crust. Alkali basalts occur on many ocean islands and at hotspots, in lesser quantities, but with chemical characteristics that are important clues to the structure and history of the earth's mantle.

A particular basalt problem lies in the eruption, often from the same vent, of tholeiitic basalts at one time and alkali basalts at another time. Alkali basalt tends to have a low silica content along with high alkalies. Tholeiite tends to have more silica and lesser alkalies. Tholeiitic basalts most commonly carry interstitial quartz or micropegmatite, or both, when well-crystallized. It is also commonly observed that silica-saturated rocks such as dacite are found associated with tholeiitic basalts. Since the interstitial materials on the one hand and dacitic or rhyolitic rocks on the other hand have much the same composition, a genetic sequence tholeiite → rhyolite may reasonably be inferred. Similarly, alkali basalts never carry interstitial quartz or micropegmatite, but may carry interstitial microsyenite or an undersaturated mineral such as a feldspathoid, commonly nepheline. They are commonly associated in the field with such silica-undersaturated rocks as trachyte and phonolite. Hence no great intuitive leap is required to deduce the genetic sequence alkali basalt → phonolite.

A genetic sequence might be implemented in a variety of ways; we shall commonly be concerned in this book with both partial melting at the source and later fractional crystallization as competing means of achieving a genetic sequence. The fractional crystallization mechanism is one suggested by the internal evidence of the rocks, for if, as textures indicate, an interstitial residuum (of, say, micropegmatite or microsyenite) represents a late-solidified fraction of liquid (i.e. magma), it is but a small step to infer that separation of all the other crystals from this small fraction of magma would yield an appropriately advanced rock, such as rhyolite or trachyte.

Basalts other than alkali basalt and tholeiite are well known. Among these are high-alumina basalts, which occur in island arcs and continental margins. High-alumina basalts in general appear to be high in the components of plagioclase feldspar, even when glassy, hence apparently not mechanically enriched in plagioclase crystals. There is also a class of basalts intermediate between alkali basalt and tholeiite, as defined above, which are neither oversaturated nor undersaturated with respect to silica, and which belong to a genetic sequence culminating in trachyte, itself neither oversaturated nor undersaturated. But for present purposes we may treat these other types of basalt as bothersome diversions from a simple model, and illustrate many valid points in terms of arbitrarily defined alkali basalt and tholeiite.

The implied genetic sequences of these latter two types pose a fundamental question: do the two magma types have a common or diverse origin? If common, is there a common parent magma, or a common parental mantle rock? If diverse, what are the diverse parent magmas or rocks? In this book, we shall not attempt to arrive at a firm decision in favour of a single one of these possibilities, but instead we shall inquire how to evaluate the various propositions by means of phase diagrams.

Other important methods of evaluation, such as studies of minor element distribution, are largely beyond the scope of this book, but in regard to minor elements it should be pointed out that the principles of phase diagrams usually apply to minor as well as major elements. The experimental study of basaltic minerals and their phase relations is, in any event, a fundamental and powerful means of evaluating hypotheses of basalt origin.

Phase Diagrams as Models of Rocks

At one time there was quite an uproar over the application of phase diagrams to natural rocks, some claiming that since the natural rocks are obviously more complicated than synthetic systems, the phase diagrams are simplistic and misleading. We may now cheerfully relegate the uproar to the history and philosophy of science, and say we are sophisticated enough to have a word which should long ago have disarmed the combatants: phase diagrams are *models* of natural rocks. We should not hesitate to confess that they are simplistic, but if we are conscious of using them as models, they need not be misleading. As models they can help us to understand the ground rules of the game, the purpose of which is to discover rock-forming processes which are inaccessible to direct view.

In order for phase diagrams to serve as instructive models, they must deal initially with relatively few chemical elements, each of which is present in a state of high purity; in other words, the models with which we work should be chemically simple. If this were not so, we should quickly be presented with experimental results, due to impurities or complexities, which we could not interpret in any straightforward way. This is simply to say that we wish our experiments to be controlled, so that we may change one variable at a time and discover the result of changing that variable. Once a few fundamental principles are discovered, we may make more and more complicated models, till they approach natural rocks themselves.

Experiments on natural rocks may provide even better models than those on synthetic systems, but even these experiments generate models rather than duplicating natural processes, because we cannot *a priori* define, let alone control, all the variables of the natural process. If we could, our work would be done. The interpretation of experiments on natural rocks is risky, and impossible without the principles furnished by synthetic systems in which all the variables may be controlled.

This book treats a limited number of experimental systems, and shows how they may be brought to bear on the problems of basalts. The emphasis is on the application of phase diagrams to both petrologic problems and experimental problems. We shall see that they may be used both qualitatively and quantitatively, as guides to research and guides to rock genesis. Although the theme is basalt genesis, the principles encountered are applicable over the field of igneous petrology in general.

We shall be concerned, in each system studied, with both crystallization and melting phenomena, since basalt magmas must be approached both from the standpoint of what they may become (residual liquids produced by fractional crystallization) and from the standpoint of how they originate (partial or complete melting of mantle rocks).

We begin with a brief discussion of basalts, in Chap. 2, followed by discussions of some pertinent physical chemistry in Chaps. 3 and 4. With Chap. 5 there begins a systematic exposition of phase diagrams, their experimental background, and their interpretation. Towards the end of the book, we shall encounter some welcome new treatments of phase equilibria.

Chapter 2
Mineralogy and Chemistry of Basalts

Mineralogy

When basaltic lava is abruptly quenched, it solidifies to glass. When slowly cooled, it crystallizes, more or less completely, to an assemblage of minerals. We may speak of the mineralogy of basalts in terms of these minerals (*modal* composition), and of the chemistry of basalts both in terms of elements (or oxides) and the *potential* minerals that might form if the lava were slowly cooled (*normative* composition). The calculation of normative minerals is discussed below, but first we may consider the actual minerals found in well-crystallized basalts.

The two essential minerals of greatest importance are plagioclase and pyroxene, since these make up perhaps 80% of many basalts. Indeed, for a starting model, we should not be far off base to consider plagioclase and pyroxene alone for ideas of basalt crystallization; this is precisely what we shall do in Chaps. 5–8. The plagioclase involved is an intermediate member of the An-Ab series, and Bowen (1928) showed that a composition very near An_{50} occurs most frequently in basalts. The pyroxene is chiefly calcic, that is, a member of the augite series, usually more magnesian than ferrous-rich ($Mg/(Mg + Fe) > 0.5$), and commonly not very far from diopside, although it ranges to subcalcic augite. Calcium-poor pyroxene, generally hypersthene, may or may not be present, depending on the abundance of silica relative to other constituents. Ca-poor pyroxene is a critical mineral in basalt classification, its presence signifying tholeiitic basalt, and its absence usually implying alkali basalt.

Olivine is another critical mineral in basalt classification. In alkali basalt, it is commonly present instead of Ca-poor pyroxene. In olivine tholeiites, it is present in company with Ca-poor pyroxene.

By adding olivine, our basalt model has grown to four minerals: plagioclase, two pyroxenes, and olivine. These four minerals make a model that is indeed a close analogue of natural basalts. The only ubiquitous mineral that we have left out so far is Fe-Ti oxide, usually titaniferous magnetite, perhaps in company with ilmenite,

S. A. Morse, *Basalts and Phase Diagrams*,
https://doi.org/10.1007/978-3-030-97882-2_2

although more oxidized varieties occur. We shall find that consideration of these non-silicates of cations which can easily be oxidized requires a special type of experimental and theoretical treatment, in which the availability of oxygen must be controlled or specified in some way.

Other minerals that occur in basalts, and that play an important role in classification when they occur, are quartz and nepheline. Quartz commonly occurs in company with Ca-poor pyroxene. Olivine may occur, but if so it is usually mantled by a more siliceous mineral, commonly Ca-poor pyroxene, for magnesian olivine is not stable in the presence of excess silica, because of the reaction

$$\underset{(Fo)}{Mg_2SiO_4} + \underset{(Q)}{SiO_2} = \underset{(En)}{2MgSiO_3} \tag{2.1}$$

Rocks that carry quartz are *silica-oversaturated* rocks, and rocks that carry Ca-poor pyroxene instead of quartz could be described as *silica-saturated* rocks.

Nepheline occurs in some rocks lacking Ca-poor pyroxene; its presence is due to a deficiency of silica required to form feldspar. Were such silica available, nepheline would react to form the albite component of plagioclase by the reaction

$$\underset{(Ne)}{NaAlSiO_4} + \underset{(Q)}{2SiO_2} = \underset{(Ab)}{NaAlSi_3O_8} \tag{2.2}$$

The presence of nepheline (or another feldspathoid) therefore clearly signifies a state of *silica-undersaturation*, i.e. an amount of silica too small for the continual formation of feldspar throughout the crystallization history of the rocks. Nepheline is accompanied by olivine, this being the most silica-poor Fe-Mg silicate in common basaltic rocks. Nepheline-bearing basalts lie at the opposite end of a spectrum from quartz-bearing basalts; the spectrum consists of the sequence

Basanite (Ne-bearing):	critically undersaturated
Alkali basalt:	critically undersaturated
Olivine tholeiite:	undersaturated
Nepheline basalt:	saturated
Tholeiite (Q-bearing):	oversaturated

Chemistry

Table 2.1 lists the average chemical composition of crustal rocks, estimated from Rudnick and Gao (2005, Table 10) who give all Fe as FeO. The compositions are expressed as weight, atom, and volume per cent. This table leaves no doubt that the earth's crust is dominated by oxygen, and that in terms of weight of atoms, silicon runs a strong second. The volume per cent column reflects the small ionic radius of silicon and the large radii of Ca, Na, and K relative to Si.

Table 2.1 Chemical composition of the earth's crust (oceanic and continental), as estimated from Rudnick and Gao (2005)

		As atoms				
As oxides	Wt%	Atoms	Wt %	Atom%	Radius[a]	Vol %
		O	46.6	60.48	1.32	78.45
SiO_2	60.6	Si	28.3	22.23	0.34	7.43
TiO_2	0.7	Ti	0.4	0.19	0.69	0.13
Al_2O_3	15.9	Al	8.4	6.88	0.56	3.78
FeO	6.7	Fe	5.2	2.06	0.86	1.74
MgO	4.7	Mg	2.8	2.57	0.80	2.02
CaO	6.4	Ca	4.6	2.52	1.15	2.84
Na_2O	3.1	Na	2.3	2.21	1.10	2.38
K_2O	1.8	K	1.5	0.84	1.46	1.21
P_2O_5	0.1	P	0.0	0.03	0.25	0.01
SUM	100.0		100.0	100.00		100.00

[a]After Whittaker and Muntus (1970); in Å (0.001 picometers)

Table 2.2 Average basalt

	wt. %
SiO_2	47.65
TiO_2	2.14
Al_2O_3	15.28
Fe_2O_3	3.57
FeO	7.54
MgO	7.52
CaO	9.91
Na_2O	2.98
K_2O	1.23
P_2O_5	0.44
H_2O	1.51
Total	99.77

Table 2.2 gives the chemical composition of the average basalt, as estimated from the data of Chayes (1972), and stated in the usual format of weight per cent of oxides. Again, silicon and oxygen dominate the analysis (as SiO_2, near 50% in all basalts); Al_2O_3 is next in abundance, and CaO, MgO, and FeO are closely similar, averaging a little less than 10%. Other oxides are almost always below the 5% level, and until one has inspected many basalt analyses, they all tend to look about the same, despite important differences that do exist. There is a way of treating chemical analyses, however, in such a way as to bring out important differences that are not manifest on simple inspection. This treatment is the norm calculation.

The CIPW Norm

The norm calculation is an arbitrary formula for casting a chemical analysis of a rock into potential minerals that could form if the rock crystallized completely under idealized conditions. Many different formulas have been proposed, some with a view to enabling further calculations (e.g. the molecular norms of Niggli 1936 and Eskola 1954), and some with a view to representing more exactly the actual mineralogy of the rock (metamorphic norms: see Barth 1962, pp. 337–343). For many purposes, however, the CIPW[1] norm is quite adequate,[2] and valuable because of the large catalogue of CIPW norms available for comparison, e.g. Washington. The calculation has great utility in discussing fine-grained or glassy rocks whose actual mineralogy is obscure or non-existent, and it provides an important link between natural rocks and experimental systems that are defined by their chemistry alone.

The CIPW norm is a weight norm, that is, the final result is expressed as weight per cent of the various normative minerals, or potential minerals implied by the chemical analysis. With common basaltic rocks, these minerals are the feldspars, pyroxenes, olivine, ilmenite, magnetite, apatite, and sometimes nepheline or quartz. The first three groups, which are solid solutions, are calculated in terms of their end members, so that one may recalculate the normative feldspar composition as percentages of An,[3] Ab, and Or, or the normative pyroxene compositions in terms of Wo, En, and Fs, or the normative olivine composition in terms of Fo and Fa (Table 2.3).

One may rant on at some length about the method and purpose of the norm calculation, but no amount of ranting is so instructive as the performance of a few calculations. The calculation is easily done with a pocket calculator, and three or four examples should suffice to show what the norm calculation is all about.[4] A simplified calculation procedure, adequate for most basaltic rocks, is given in Appendix I, along with a shorthand form that facilitates the calculation. A single example is given below to suggest the general outline of the calculation, using the Chayes average basalt from Table 2.2.

In order to calculate mineral formulae from a weight per cent analysis, the analysis must first be recast into molecular proportions (Tables 2.4 and 2.5). This is done by dividing the weight per cent of each oxide by the rounded molecular weight of that oxide. The molecular proportions are then used up in a standard order, first by assigning all P_2O_5 to apatite, using an appropriate amount of CaO to yield an

[1]Named after its inventors, Whitman Cross, Joseph Iddings, L.V. Pirsson, and Henry S. Washington (1917).

[2]Or superior; see Chayes (1963).

[3]A list of common normative minerals and their abbreviations is given in Table 2.3.

[4]The norm calculation is best done digitally, and there are many varieties available. However, for historical reasons it is useful to retain a formal format with decimals rounded to a common value of two places. There are also other types of norms, to be discussed.

Table 2.3 Common normative mineral molecules in the CIPW norm

	Mineral	Formula
	Salic group	
	Quartz (*q*)	SiO_2
	Orthoclase (*or*)	$K_2 \cdot Al_2O_3 \cdot 6SiO_2$
	Albite (*ab*)	$Na_2O \cdot Al_2O_3 \cdot 6SiO_2$
	Anorthite (*an*)	$CaO \cdot Al_2O_3 \cdot 2SiO_2$
	Leucite (*lc*)	$K_2O \cdot Al_2O_3 \cdot 4SiO_2$
	Nepheline (*ne*)	$Na_2O \cdot Al_2O_3 \cdot 2SiO_2$
	Kalsilite (*ks*)	$K_2O \cdot Al_2O_3 \cdot 2SiO_2$
	Femic group	
	Diopside (*di*)	$CaO \cdot (Mg,Fe)O \cdot 2SiO_2$
	Wollastonite (*wo*)	$CaO \cdot SiO_2$
Hypersthene (*hy*)	Enstatite (*en*)	$MgO \cdot SiO_2$
	Ferrosilite (*fs*)	$FeO \cdot SiO_2$
	Olivine (*ol*)	$2(Mg,Fe)O \cdot SiO_2$
	Forsterite (*fo*)	$2MgO \cdot SiO_2$
	Fayalite (*fa*)	$2FeO \cdot SiO_2$
	Magnetite (*mt*)	$FeO \cdot Fe_2O_3$
	Ilmenite (*il*)	$FeO \cdot TiO_2$
	Apatite (*ap*)	$3(3CaO \cdot P_2O_5) \cdot CaF_2$

Note: Diopside may be calculated into its *wo*, *en*, *fs* components. Hypersthene, if present, is always calculated in addition to these

Table 2.4 Example of norm calculation, using the average basalt of Table 2.2

Oxide	Wt. %	Mol. Wt.	Mol. No.[a]	Subtractions	Calculations	
SiO_2	*47.65*	60	*794* – 794 = 0		*Mg:Fe*	
Al_2O_3	*15.28*	102	*149* – 13 – 48 = 88 – 88 = 0		188 + 56 = 244	
Fe_2O_3	*3.57*	160	22 – 22 = 0		188/2.44 = 77%	
FeO	*7.54*	72	*105* – 22 – 27 = 56 – 18 = 38 – 38 – 0		*hy, ol*	
MnO	–	71	–		2*S* = 188	*M* = 165
MgO	*7.52*	40	*188* – 61 = 127 – 127 = 0		*x* = 23	*y* = 142
CaO	*9.91*	56	*177* – 10 – 88 = 79 – 79 = 0			
Na_2O	*2.98*	62	*48* – 48 = 0		*ab, ne*	
K_2O	*1.23*	94	*13* – 13 = 0		2*N*=	*x*=
P_2O_5	*0.44*	142	*3* – 3 = 0		*S*=	*y*=
TiO_2	*2.14*	80	*27* – 27 = 0			
Rest	*1.51*	("Rest" = H_2O etc)			Analysis sum *99.77*	

[a] 10^3 Wt.%/Mol. Wt.

arbitrary apatite formula. Then TiO_2 is used to form ilmenite, $FeO \cdot TiO_2$, with an equal amount of FeO, and so on. A debit ledger enables one to keep track of the amounts of CaO, FeO, etc. used. The calculation proceeds in routine fashion through hypersthene (Hy), at which point a summation is made of all silica used. If the

Table 2.5 List of normative minerals from the calculation above

Normative minerals											
I						II SiO_2 used		III Mol. Wt./1000		III Weight norm[a]	
ap	(P + [CaO = 3.33P])		3	10				0.336		*ap*	1.00
mt	(Fe''' + Fe'')		22	22				0.232		*mt*	5.10
il	(Ti + Fe'')		27	27				0.152		*il*	4.10
or	(K + Al)		13	13	(6K)	78		0.556		*or*	7.23
ab	(Na + Al)		48	48	(6Na)	288		0.524		*ab*	25.15
an	(Al + Ca)		88	88	(2Ca)	176		0.278		*an*	24.46
	Σ MgO + FeO left = 244, %MgO = 77									Total feldspar	56.84
										An/Plag =	49.3
di	(Ca + MgFe)	79	*wo*	79	(2Ca)	158	For (Σ *di*)	0.116	9.16		
			en	61				0.100	6.10		
			fs	18				0.132	2.38	*di*	17.64
					Σ SiO_2 =	700					
hy	(Ca + MgFe)	23	*en*	18	(1 Mg)	18		0.100	1.8		
			fs	5	(1 Fe)	5		0.132	0.66	*hy*	2.46
ol	(Ca + MgFe)	142	*fo*	109	(0.5 Mg)	54	Use half	0.140	7.56		
			fa	33	(0.5 Fe)	17	MgO, FeO	0.204	3.47	*ol*	11.03
ne	(Na + Al)				(2 Na)			0.284		*ne*	–
q	(Si)				(1 Si)			0.060		*q*	–
										Rest	1.51
										Total	99.68

[a] Mol.No. of first oxide in Col.I × Col.III

original molecular proportion of silica is in excess of that used, the rock is oversaturated, and the excess silica is reported as quartz in the norm. If the original amount is less than that used, some or all of the MgO, FeO assigned to hypersthene must be recalculated to the less siliceous mineral olivine. If, when all hypersthene is thus converted to olivine, too much silica has still been used, some or all of the Na_2O, Al_2O_3 assigned to albite must be recalculated to the less siliceous mineral nepheline, and the rock is silica-undersaturated. If, as rarely happens, too much silica has still been used, orthoclase must be converted to leucite. When the appropriate silica balance has been achieved, the molecular amounts for each normative mineral are multiplied by the molecular weight of the mineral to achieve a weight per cent norm. H_2O and other volatiles are ignored in the calculation, and simply added on at the end to see if the summation approximates that of the original chemical analysis, which it should if no errors have been made.

A few trial norm calculations should be made by hand in order to appreciate the chemical mineralogy that underlies the norm scheme. It is best to begin with an oversaturated tholeiite, which is the simplest calculation, then proceed to an olivine-hypersthene tholeiite, and finally to a nepheline-normative alkali basalt. There are further teaching and learning approaches available in the exercise procedures of Brady (http://www.science.smith.edu/~jbrady/petrology/igrocks-topics/norms/norms-page01.php).

Historical Note: CIPW

The turn of the century at 1900 marked a peak in the connection between field and laboratory studies of rocks. By then the United States Geological Survey was already 20 years old with an analytical branch in Colorado. Young scientists were travelling to Germany and bringing back methods and examples of analysis. Among these were C. Whitman Cross, Joseph P. Iddings, Louis V. Pirsson, and the bearded Henry Stephens Washington. In and out of the Survey, this group of four generated the CIPW classification system that is still in use. There is so much of interest in this process that it helped to generate a history of Petrology in a 686-page book "Mind over Magma" by Davis A. Young (2003, Princeton). The portrait of Washington stood on the wall over my desk at the old Geophysical Laboratory, and there was an old anecdote that he was quoted as saying "If the alkalies are low, smoke another cigar." Professor Young also edited a newly discovered and wonderful autobiography of Joseph Iddings: "Recollections of a Petrologist." GSA Special Paper 512 (2015). He also wrote a study of Bowen: "Crystallization and Differentiation: The evolution of a Theory." (Mineralogical Society of America Monograph Series #4, 1998).

The Basalt Tetrahedron: A Model for Basalt Study

Yoder and Tilley (1962) devised a scheme for classifying basaltic rocks by their CIPW norms, using the concept of a "basalt tetrahedron." This tetrahedron (Figs. 2.1 and 2.2) is simply the quaternary system forsterite-diopside-nepheline-quartz, a group of minerals that serves, to a first approximation, as a model of basalts. Plagioclase is represented by albite, which lies on the line nepheline (Ne)-quartz (Q) by virtue of reaction (2.2). Pyroxene is represented by diopside, at one corner of the tetrahedron, and enstatite, which lies between forsterite (Fo) and quartz by virtue of reaction (2.1).

Olivine is represented by the forsterite corner, quartz by the quartz corner, and feldspathoid by the nepheline corner of the tetrahedron. As sketched in Fig. 2.1, the tetrahedron is cut by two planes, one a critical plane of silica undersaturation (Fo-Di-Ab) which separates the region of critically undersaturated rocks towards nepheline from the region of undersaturated and saturated rocks towards quartz. Those rocks falling exactly in the critical plane Fo-Di-Ab could be called critically undersaturated rocks in terms of their CIPW norm. The other plane of importance is the plane of silica saturation (En-Di-Ab), which separates the region of silica-undersaturated

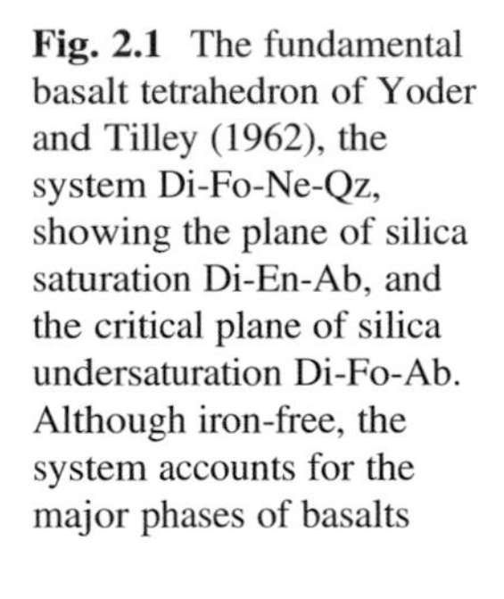

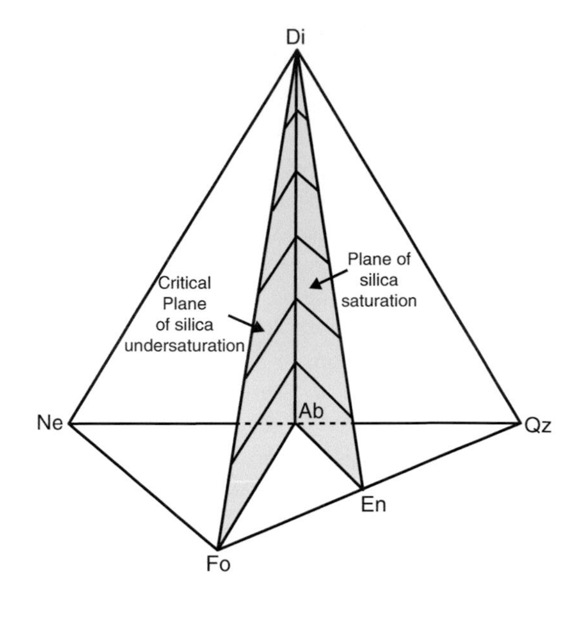

Fig. 2.1 The fundamental basalt tetrahedron of Yoder and Tilley (1962), the system Di-Fo-Ne-Qz, showing the plane of silica saturation Di-En-Ab, and the critical plane of silica undersaturation Di-Fo-Ab. Although iron-free, the system accounts for the major phases of basalts

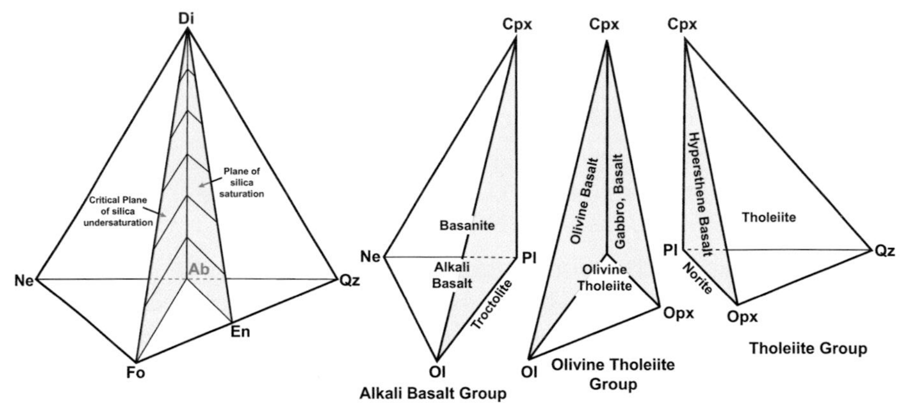

Fig. 2.2 The basalt tetrahedron generalized to admit iron members of Di and Fo corners (i.e. olivine in general instead of the end member Fo, etc.), and An, and exploded to show nomenclature of major basalt types based on their normative composition

rocks towards the critical plane from the region of silica-oversaturated rocks towards quartz.

Figure 2.2 is the generalized basalt tetrahedron of Yoder and Tilley, so called because plagioclase, olivine, augite, and hypersthene in general are substituted for the pure end members albite, forsterite, diopside, and hypersthene. This implies consideration of FeO in addition to the oxides Na_2O, Al_2O_3, CaO, MgO, and SiO_2 required to form the fundamental tetrahedron of Fig. 2.1, and also requires CaO and $A1_2O_3$ to be projected in such a way as to furnish anorthite. The generalized tetrahedron offers a convenient means of summarizing the essential characteristics of basaltic rocks in terms of their chemistry. The positions of the rock names inserted in the expanded tetrahedron of Fig. 2.2 reflect the salient mineral and chemical features of the rocks, which can be summarized as follows:

1. *Tholeiite* (oversaturated): normative *quartz* and *hypersthene* (region to the right of the plane of silica saturation).
2. *Hypersthene basalt* (also tholeiite, saturated): normative *hypersthene* (on plane of silica-saturation—has just enough SiO_2 to make hypersthene, but not enough for free quartz).
3. *Olivine tholeiite* (undersaturated): normative *hypersthene* and *olivine* (region to the left of the plane of silica saturation).
4. *Olivine basalt*: normative *olivine* (on the critical plane of silica undersaturation—not enough SiO_2 to make hypersthene).
5. *Alkali basalt*: normative *olivine* and *hypersthene* (region to the left of the critical plane of silica undersaturation.

Note that saturation with respect to silica does not depend on silica percentage alone, but on the ratio of silica to alkalies, etc. In calculating the norm of a rock high in alkalies, silica is used to make *or* and *ab*, so there may not be enough left over for *hy* or even olivine, in which case the norm shows *ne* and represents alkali basalt. The same silica percentage with low alkalies could be an olivine tholeiite or even tholeiite.

Similar nomenclature applies to plutonic rocks if the word *gabbro* is substituted for basalt.

Petrographic Characteristics

1. Tholeiite: *Augite* (or subcalcic augite), *plagioclase* (commonly labradorite), and *iron oxides* are essential minerals. Olivine is minor or absent. *Interstitial acid residuum* (glass or quartz-feldspar intergrowth) is *characteristic*. Hypersthene is present or "occult" in the augite. Olivine, if present, generally shows reaction to Hy where in contact with groundmass (= liquid).

These remarks apply to all the tholeiites (Groups 1–3 above); the olivine percentage generally rises towards the critical plane of silica undersaturation.

2. Alkali basalt: *Nepheline* is present well within the volume, but generally absent near the critical plane. Pyroxene is high in CaO. No nepheline occurs. *Plagioclase* near An_{50}, commonly zoned to oligoclase. Olivine both as phenocrysts *and* in groundmass. *Characteristically holocrystalline* without the interstitial residuum common in tholeiite.

The basalt tetrahedron is a model, and it does not satisfy all the characteristics of natural basalts. In particular, it ignores apatite and Fe-Ti oxides. Moreover, it formalizes the silicate minerals by means of the artificial CIPW norm. In spite of this, it is a useful conceptual model for beginning the study of basalt, and many of its apparent shortcomings can be systematically eliminated by expanding the model after the fashion of Fig. 2.2. We shall find in particular that the region around the critical plane is very instructive, since it generates some important hypotheses about the relationships of alkali basalt and tholeiite.

The basalt tetrahedron will be used in this book as a framework on which to hang ideas about basalt, and as a checklist with which to keep track of a systematic examination of phase diagrams. The phase diagrams that we will examine will concern those two- and three-mineral assemblages in the tetrahedron, and a few others (such as An-Ab) that lie contiguous to the tetrahedron.

Other models can of course be constructed for the examination of basalt. The basalt tetrahedron of Yoder and Tilley is a part of the 5-component system Na_2O-CaO-Al_2O_3-MgO-SiO_2. The omission of Na_2O from this system permits a tetrahedral representation of the remainder, "CMAS", which contains the basaltic minerals diopside, enstatite, and anorthite, as well as important related compounds such as wollastonite, quartz, spinel, and the aluminous, pyroxene-like compound known as calcium-Tschermak's molecule. We shall find occasion to refer to this system in later chapters, where its utility will be demonstrated.

Other Types of Norm Calculation

The CIPW norm is stated in weight units, and is convenient for comparing, in general terms, rock analyses and phase diagrams that are reported in weight units. The latter are inappropriate, however, when strictly quantitative comparisons are desired, because the iron end members of femic molecules are heavier than the magnesian end members, and therefore do not plot at the same point in a diagram expressed in weight units. The problem is overcome by the use of another type of norm, such as the cation or oxygen norm. Each of these uses the same combining conventions as the CIPW norm, but each is expressed in a different way.

The cation norm, sometimes called the molecular norm, is calculated from oxides stated in terms of one cation. Thus, for example, Na_2O is calculated as $2NaO_{0.5}$, and the weight percentage is divided by 31 rather than 62. Similarly, the oxygen norm is calculated from oxides stated in terms of one oxygen. Thus Na_2O is divided by 62, but SiO_2 is divided by 30 rather than 60, to yield units of $Si_{0.5}O$. The symbols of the cation norm are rendered "Ap", etc., to distinguish them from the CIPW "*ap*", etc.

Cation norms are useful for comparing chemical analyses of natural rocks to phase diagrams plotted in cation units. Oxygen norms are convenient to use and diagrams plotted in oxygen units are especially useful for plotting modal (volume) analyses of rocks, since oxygen accounts for most of the volume of rock-forming silicate minerals. Therefore, oxygen norms have a special place in petrology. They are especially useful in plotting experimental results that can usefully be compared with modal data.

Table 2.6 Comparison of CIPW, cation, and oxygen norms, Gough Island Alkali Olivine Basalt of Barth (1962: Table III-9)

Oxide	Wt%		NORM			
			CIPW		Cation	Oxygen
SiO_2	49.10	*ap*	1.34	**Ap**	1.18	1.17
TiO_2	3.59	*mt*	4.18	**Mt**	3.04	2.62
Al_2O_3	16.21	*il*	6.84	**Il**	5.06	4.92
Fe_2O_3	2.87	*or*	16.12	**Or**	16.59	16.9
FeO	6.84	*ab*	25.68	**Ab**	27.00	28.56
MnO	0.05	*an*	20.29	**An**	20.42	21.28
MgO	5.04	*di*	16.25	**Di**	16.42	15.96
CaO	8.90	*ol*	6.60	**Ol**	7.26	6.27
Na_2O	3.53	*ne*	2.27	**Ne**	3.04	2.33
K_2O	2.76	Rest	0.47			
P_2O_5	0.54	Sum	100.04		100.01	100.01
H_2O	0.47	XAn	0.44		0.43	0.43
SUM	99.90	Fsp	62.09	%	64.01	65.8

The qualitative characteristics (such as the presence of *ne* and Ne) of CIPW, cation, and oxygen norms will always be the same, because the same CIPW combining conventions are used in each. The CIPW, cation, and oxygen norms of an alkali olivine basalt are compared in Table 2.6.

Chapter 3
Crystals and Melts

Introduction

It might be supposed that crystals and melts are so familiar that their properties would be self-evident. Our familiarity is, however, based largely on the crystal ice and the melt water, and the substance H_2O has such unusual properties that we would be seriously misled if we used this as a model for silicates (except quartz!), as we shall see below. The phenomena of melting and crystallization are unfamiliar to many readers, and the atomic organization and bonding of liquids turns out to have much importance for the theory and practice of melting.

The classic treatise on crystals and melts is the 1903 work of Tammann, *Kristallisieren and Schmelzen* (*Crystallization and Melting*). This remarkable work is still of fundamental importance to students of the subject, and one finds reference to it in many research papers in the literature. Tammann's work was developed without consideration of the atomic arrangement of matter which we call structure, and the modern treatise which gives the structure of crystals and liquids their appropriate emphasis is the classic book *Melting and Crystal Structure* by Ubbelohde (1965, 1979), a fascinating and readable work to which the serious student is commended. A special source of wisdom on crystallization is Tiller's (1991) comprehensive study of microscopic interfacial phenomena.

Definitions

A *crystal* is a region of matter characterized by periodicity. More precisely, it is a regular arrangement of atoms which occupy sites in the mathematical construct called a *lattice*. A crystal is said to have *structure*, a term which describes the combination of lattice sites, which are imaginary, with atoms, which are real. Most of us are accustomed to ascribing further characteristics to crystals. Among these are

S. A. Morse, *Basalts and Phase Diagrams*,
https://doi.org/10.1007/978-3-030-97882-2_3

a definite chemical composition, which is to say that the crystal has a homogeneous distribution of atom species, ideally in a fixed proportion throughout the crystal, so that once a *unit cell* is defined as the smallest and simplest repeat unit required to reproduce the crystal by the operations of symmetry, we can suppose that every other unit cell is identical to the first. As a concept, this is helpful, but we must recall that it will not do for zoned crystals. In any event, the concept of definite chemical composition rightly demands that there be severe constraints on the kind and variety of atom species present in the crystal.

Under the term crystal we also comprehend a morphological meaning: a polyhedral form bounded by plane faces, or what is called a euhedral crystal. We shall ordinarily be concerned in this book with the structural definition of crystals and not the morphological one. The general designation solid (S) will be taken to imply the crystalline state, structurally defined.

A *liquid* is not so easy to define as a crystal, surprising as it may seem. Without doubt it is a substance of very low rigidity compared to the crystalline form of the same substance: it is a fluid. Also without doubt its component atoms are more highly disordered than those of a crystal. The atom positions in a liquid are not likely to be random. Short-range order is common, although disorder increases with temperature. The chemical composition of a liquid is subject to far fewer constraints than that of a crystal, hence wide compositional variability is a characteristic of liquids in general, although liquid composition may be constrained artificially in experiments. In terms of compressibility, liquids lie between gases and solids, separated from each of these by orders of magnitude under certain conditions of temperature and pressure.

It is convenient to describe a liquid near its freezing point (or the beginning of its freezing range) as a *melt*. Since in this book we are concerned largely with equilibria between crystals and liquids, our liquids are best considered melts, whether natural magmas or synthetic liquids. Melts, since they lie by our definition near the freezing point, are those liquids whose structure is most ordered, and about whose structure we are most concerned. Melts and crystals are both referred to as *condensed phases*, in contrast to the vapour state.

As melts are superheated (i.e. heated above their equilibrium with crystals) they pass into liquids in the more general sense, and finally pass into the *vapour* (or *gaseous*) state. A vapour has the unique property of filling its container, hence is in principle easier to define than a liquid. It is, like liquid, a fluid, but one in which the atoms are more dispersed and more disordered. It is more compressible, less rigid, and less dense than liquid. At temperatures and pressures below a certain value called the *critical point*, vapour and liquid form an interface with each other. At a given pressure below the critical point, a liquid passes into a vapour (boils) at a fixed value of temperature. Above the pressure of the critical point, liquid passes continuously into vapour with rising temperature, by becoming more and more disordered and less and less dense, so that the distinction between liquid and vapour fades above the critical point. Critical points involving the merged identity of liquid and vapour are of no importance[1] in petrologic systems without a volatile component, because the vapour pressure of silicate liquids is very small. Such critical points may be of interest, however, when a volatile component such as water is present.

In principle, critical phenomena between solid and liquid states of matter are possible, but they have not been conclusively demonstrated to exist,[2] and they need not concern us further.

Water: A Familiar But Misleading Example

It is pertinent to discuss briefly the states of matter of the compound H_2O, and to point out how unusual this compound is relative to the silicates with which we shall be chiefly concerned.

The most obvious difference between H_2O and silicates is that the former is liquid at room temperature, and is relatively close to its freezing point over the earth's surface. A more subtle difference, and one which is the source of much initial confusion among students of silicates, is that ice shows a negative change of volume on melting under normal conditions, so that ice floats in the sea and in cocktails, and the milk left too long on the winter doorstep extrudes its creamy top and cap from the neck of the bottle. That is, ice is less dense than water. This is true of remarkably few substances, among which are the elements Ga, Sb, and Bi. For all known rock-forming materials, the reverse is true, crystals being denser than their pure liquids, although there is some doubt about the relative densities of plagioclase feldspar and the complex liquids which are basaltic magmas.

Liquid water has a very low viscosity, resulting in hydrodynamic properties familiar to all. Silicate liquids, on the other hand, are very viscous; the melts used in glassmaking are more fluid than most silicate liquids, and many melts, for example feldspar liquids, cannot be poured because they are so viscous. Natural basalt lava may flow rather freely, of course, as motion pictures of Kilauea and other eruptions show, but much of the hydrodynamic behaviour of erupting lava is determined by small quantities of water vapour and other volatiles which diminish viscosity. The stiffness of silicate liquids and the fluidity of water are directly related to the differences in ordering of the two liquid structures. Water consists almost totally of individual molecules of H_2O, which, having only weak polarity and attraction for each other, are rarely hooked together. Silicate liquids, on the other hand, are dominated by $(SiO_4)^{-4}$ tetrahedral groups, which have a relatively strong affinity for each other (by sharing corners) and for cations, so that there is a rather high degree of polymerization in silicate liquids which has a profound effect on their physical properties and their ability to nucleate on freezing.

---

[1] Extraterrestrial petrology is excluded from this remark.

[2] See Ubbelohde (1965, pp. 19–24) for discussion.

Water, like most metals, freezes readily into crystals, and in ordinary circumstances cannot be quenched to a glass, because the molecules are so mobile that they very quickly arrange themselves in the pattern of the crystal once nucleation has begun. Silicate liquids, on the other hand, tend to be easily quenched to the glassy state, their molecules being so tangled up in such a sticky medium that they may require a relatively long time (hours, days, or months) to form crystal nuclei from which crystal growth may proceed. Moreover, crystal growth is slow once begun, because of slow diffusion rates in viscous silicate liquids. The reluctance of some silicate liquids to nucleate in the laboratory is probably not accurately indicative of most natural events, since a natural magma is usually under a flow regime rather than a static regime, and a flow regime greatly increases the number and variety of sites favourable for nucleation. In static laboratory experiments, the ability to quench silicate liquids to glass is of fundamental utility, since this provides a *post facto* test for the presence of liquid at the temperature of the experiment, and allows interpretation of textural relations between crystals and "liquid" (now glass).

In summary, water is very different from silicate liquids in its low freezing point, its negative change of volume on melting, its viscosity, its degree of polymerization, its ease of nucleation and its unquenchability. Water, however, is capable of dissolving in silicate liquids to an appreciable degree, and it is not surprising, in view of its unusual and contrasting properties, that even small amounts of water produce drastic changes in the behaviour of silicate melts, as discussed in Chap. 19.

It is appropriate at this point to introduce the phase diagram for water, which summarizes some of the things we have said about volume change on melting and the critical point, and which contains some surprises at high pressure which ought to be common knowledge among students of the Earth. Figure 3.1 is the phase diagram for water, drawn approximately to scale, for low pressures. In it, we ignore for present purposes the finer points of phase diagrams which occupy most of this book. We can view Fig. 3.1 simply as a map, in two-dimensional pressure-temperature space, of the regions where ice, liquid, and vapour are stable. Ice is stable in the left portion of the diagram, at low temperatures, and its stability field is bounded by a line which is the *melting curve,* denoting equilibrium with liquid, and another line, the *sublimation curve,* denoting equilibrium with vapour. To the right of the melting curve, the field of liquid extends, at low pressures, to the *boiling point curve,* which denotes the equilibrium of water and vapour. To the right of the sublimation curve, the field of vapour extends outward indefinitely and upward to the boiling point curve. The three curves separating regions of different states of matter intersect in a triple point, a unique point in *P-T* space (i.e. having a single value of pressure and a single value of temperature), where all three states of matter coexist. Since we shall rarely be concerned with all three states of matter for single compounds of petrologic interest, we shall rarely encounter this sort of triple point, although triple points involving solid and liquid states of matter (three solids, or two solids and one liquid) are common and important in petrology.

The melting curve, or *S-L* curve, is of special interest, as it has a negative slope, which results from the decrease of volume on melting discussed above. A decrease

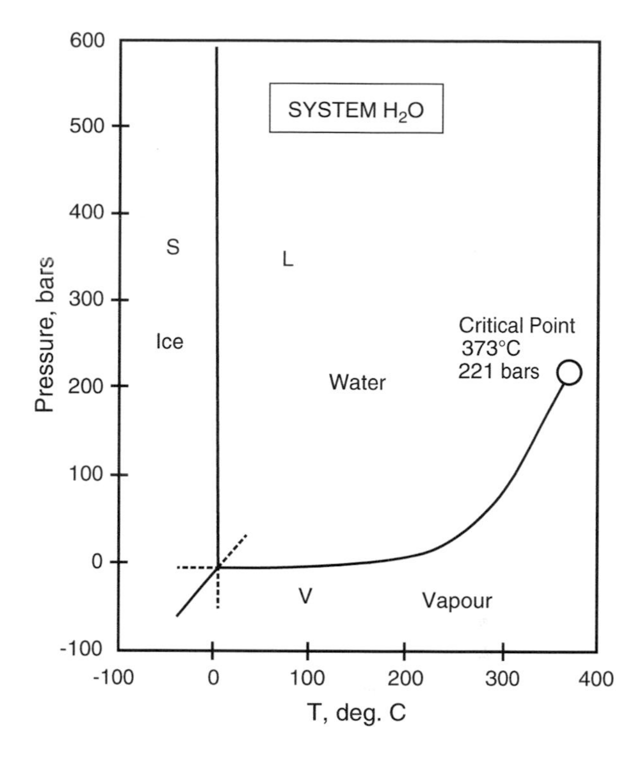

Fig. 3.1 Part of the phase diagram for water, approximately to scale. The zero point of the Celsius temperature scale is on the *S/L* curve at 1 atm in air (system FLO-air)

of volume on melting implies that pressure favours the liquid state (it being the denser), hence the negative slope.

Another feature of interest in Fig. 3.1 is the critical point at 221 bars,[3] 373°C, where the separate identities of liquid and vapour merge and become indistinguishable. The liquid near this temperature is quite different in its properties from the melt near the melting curve, even though one may be changed continuously into the other.

Figure 3.2 shows the phase diagram for water at high pressures as well as low. This reveals seven different crystalline modifications (polymorphs) of ice, each of which has its own field of stability in *P-T* space. The negative melting curve of ice_I terminates at a triple point with ice_{III} and liquid, and no other form of ice has a negative melting curve, so this unusual aspect of water disappears above about 2 kbar pressure. Since pressure favours the denser phase, and the higher polymorphs of ice melt with an increase in volume, the solid phase is favoured, and higher and higher temperatures are required to melt ice at very high pressures. Note that at about 23 kbar, the melting point of ice_{VII} is the same as the *boiling* point of water at atmospheric pressure!

---

[3] One bar is close to 1 atm of pressure (0.987 atm).

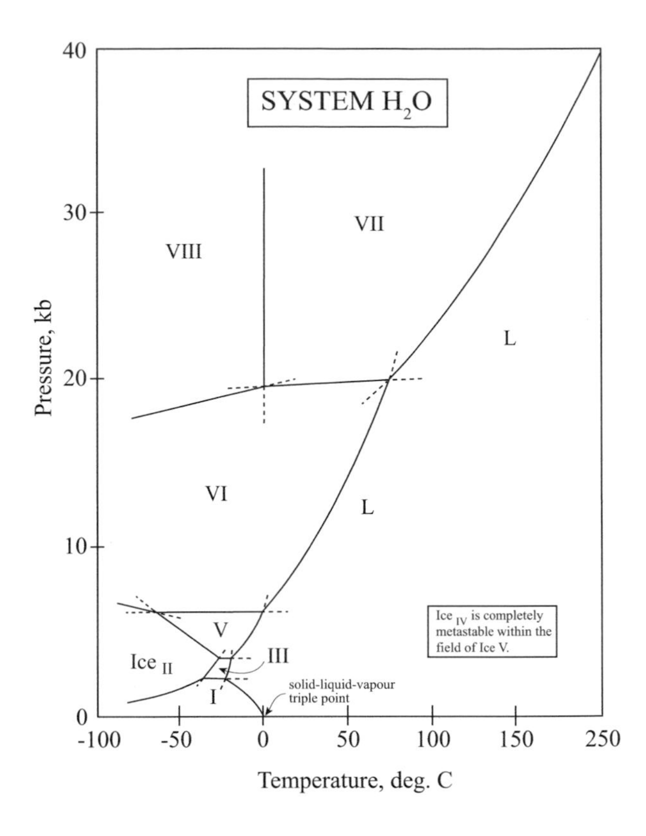

Fig. 3.2 Phase diagram for water at high pressures (Chaplin 2016). Ice_{IV} is a metastable phase occurring in the *P-T* region of Ice_{V}

The Phenomena of Melting and Crystallization

The change from the solid to the liquid state involves a large jump in the randomness of atom positions, from somewhere near non-random (ordered) in the crystal to highly random (disordered) in the liquid. The ordering of atoms or ions into specific sites in a crystal is, however, perfect only at 0°K, and from absolute zero to the melting temperature, disorder increases, sometimes in discrete jumps implying phase transformations, and sometimes smoothly. The underlying cause of disorder is the thermal motion of atoms, which increases smoothly with temperature above 0°K. The phenomenon of melting, then, is one of a discontinuous jump in disorder accompanying a change from solid to liquid state. The "disorder" of which we speak is of several kinds, both in crystal and liquid states. Among these are positional disorder, which relates simply to the mutual positions of atoms, as in crystal structure sites or in the liquid, and orientational disorder, which relates to the orientation of units of the structure such as molecules or polymers.[4]

[4]For a discussion of disorder and melting, see Ubbelohde (1965), Chap. 5.

That melts retain some degree of order, i.e. structure, was shown directly by X-ray diffraction studies, which in some cases demonstrated a quasi-crystalline structure in the melt, even to the degree of implying the existence of unit cells much like those of the crystal (Ubbelohde 1965, p. 91). Further effects related to silicate melt structure are discussed in some detail in Chap. 6.

The changes which occur in the vicinity of the melting point in both crystals and melts are collectively referred to as *premonitory phenomena*.[5] In crystals, these phenomena can be related to vibrational disorder and to defects in the crystal structure. In melts, pre-freezing effects have been described by Ubbelohde (1965) as involving the formation of aggregates or clusters (perhaps "polymers" would be appropriate in silicates) which may serve as nuclei for spontaneous crystallization. The sum of these effects is such as to diminish the difference between solid and liquid states near the melting point.

Supercooling and Superheating

It is common knowledge that water can be supercooled, by slow abstraction of heat in the absence of impurities or motion, to temperatures well below the freezing point. The same is probably true of most silicate liquids, and supercooling is a quite general phenomenon in melts. Supercooling results from the fact that the nucleation of a crystal is a random process, requiring the accidental arrangement, in the melt, of atoms in the pattern of the crystal structure, after which a crystal nucleus may grow at a great rate. The presence of foreign particles or dislocations due to motion in the melt merely increases the probability of nucleation by affording a sort of substrate upon which atoms may become ordered. If such *heteronuclei*[6] are present in experimental work, supercooling is extremely difficult to achieve. With careful experimental control, however, it is found that a wide range of substances, including pure metals, halides, and molecular liquids, can be supercooled to about 80% of their Kelvin melting temperature. The recurrence of values around 80% suggests that this represents a fundamental limit of supercooling. Supercooling of this sort of magnitude is probably extremely rare in natural silicate melts, because of the presence of heteronuclei and shearing motion in the magma, and in fact it is a fair question whether any appreciable supercooling ever occurs in magmas. In experimental work, however, especially with viscous melts, supercooling may be a serious hindrance to experimental success.

Superheating of crystals above their melting temperature is a rare phenomenon, and usually amounts to only a fraction of a degree when it occurs. However, in crystals which melt to very viscous liquids, such as quartz or albite, superheating of as much as 300°C has been reported. This emphasizes the fact that melting

[5] Ubbelohde (1965) p. 4.

[6] Ubbelohde (1965), p. 284.

and crystallization are rate processes, and when a viscous melt with a high degree of structure is involved, rates of disordering and diffusion may become infinitesimally slow.

Melting as a Function of Pressure

Most solid compounds, with the notable exception of ice$_{I}$ and a few others, have positive volume changes on melting, and therefore have positive melting curves in P-T space, by the Clausius Clapeyron equation

$$\frac{\overline{S}_{\mathrm{f}}}{\left(\overline{V}_{\mathrm{L}}-\overline{V}_{\mathrm{S}}\right)}=\frac{\mathrm{d}P}{\mathrm{d}T} \tag{3.1}$$

in which $\overline{S}_{\mathrm{f}}$ is the molar entropy of fusion, always positive, and $\overline{V}_{\mathrm{L}}$ and $\overline{V}_{\mathrm{s}}$ are the molar volumes of liquid and solid, respectively, their difference being usually positive. It is then the normal expectation for silicate minerals to have positive melting curves, which they do in known cases, in the absence of volatile components, at low to intermediate pressures.

The introduction of water to anhydrous silicate systems produces negative melting curves, because water dissolves in the melt to a greater extent than in the crystal, stabilizing the melt at a lower temperature as more pressure is applied to the vapour phase, by LeChatelier's principle. This is a most important effect for petrology because of the general availability of H_2O in the earth's crust, and parts of the mantle, and because of the very large temperature differences between dry and hydrous melting curves for the same crystal. The effect on a number of silicate systems is discussed in Chap. 19.

Solutions: Solid and Liquid

It is common experience that liquids are likely to be solutions, and natural magmas are no exception; they are in fact complex solutions of large numbers of components, of which only about ten are major. Most of the melts of interest to our present purpose will be solutions.

Most naturally occurring minerals are also solutions, called solid or crystalline solutions, of two or more end members which are stable in their pure state only in the absence of suitable contaminants, or at 0 °K. The extent of solid solution is very limited in many minerals, and complete between end members in many others, for example, in plagioclase feldspar at high temperatures. Except in rare cases, we shall be dealing with natural and synthetic crystals which show solid solution toward other compositions to a degree which should not be ignored, although we shall find it helpful to ignore this phenomenon temporarily, in some cases, to simplify the presentation of chemical principles.

The Structure of Melts

The science of crystallization and melting was long dominated by the classical field of metallurgy, in which there is not much evidence for structure in molten metal. This trade grew alongside the field of ceramics, in which the structure of the melt is of great importance and perhaps familiar. The oldest evidence of humans intentionally melting silicate rocks elicits surprise and admiration: it appears that an ancient culture in the Mesopotamian Middle East discovered about 4000 years ago how to make artificial basalt by melting silt (Stone et al. 1998). The secrets of their science are obscure. Substances created in the advancing field of polymer science may be ambiguous between melts and crystals.

In petrology, our recent understanding of melt structure owes much to the pioneering studies of Mysen, Virgo, and Seifert (1982), Kushiro and Walter (1998), Toplis (2005), Mysen and Richet (2005), and Mysen (2007). In particular, Mysen developed the concept of quantifying the likelihood and style of polymerization in silicate liquids by the measure of what is called NBO/*T*.

In this algorithm, NBO stands for the number of Non-Bridging Oxygens, and *T* stands for the number of Tetrahedral cations. Think of the crystal structures of silicate minerals: independent tetrahedra in orthosilicates like olivine, chains in inosilicates like pyroxenes, sheets in phyllosilicates like mica, and frameworks in quartz and feldspars. The olivine structure has only non-bridging oxygens, four to a tetrahedron, with one central tetrahedral cation, silicon: $NBO = 4$, $T = 1$; $NBO/T = 4$. The mineral lawsonite has double tetrahedra with three non-bridging oxygens each for two silicons: $NBO/T = 3$. The single-chain silicate enstatite has two tetrahedra linked by one oxygen to the next, so each tetrahedron has two non-bridging oxygens and one tetrahedral cation: $NBO/T = 2$. The sheet silicate Tschermakite has six Si and two Al in eight tetrahedra: $NBO/T = 1$. The sheet silicate phlogopite has a ring of six tetrahedra in which three oxygens are bridging in each tetrahedron, leaving one not bridging: $NBO/T = 1$. In quartz and feldspar, there are no non-bridging oxygens, hence $NBO/T = 0$.

These results emerge from the bulk chemistry of any silicate melt by means of a systematic calculation that is best done in a spreadsheet. One of these kindly contributed by Bjorn Mysen is accessible as Deposit Item AM-15-55056, Calculation Table, at the Mineralogical Society of America at www.minsocam.org or ammin.geoscienceworld.org.

A Final Note About Freezing

The freezing of ice on a pond is a special phenomenon because the more ice that forms, the more the impediment to the growth rate. That effect occurs because the pond water must cool through its upper surface, because its sides and bottom are warmer than the overlying atmosphere. This famous wisdom is called the Stefan effect; the latent heat of crystallization must pass through the ice, so the problem gets worse the thicker the ice (see Turcotte and Schubert 2002). In nature, the upper and lower boundaries of magma bodies that crystallize at both floor and roof are Stefan problems. This principle has nothing to do with phase diagrams, but it has everything to do with intrusive magmas and should be on our minds when we address intrusive igneous rocks in the field or in theory.

Chapter 4
The Phase Rule as a Tool

Introduction

Phase diagrams are both constructed and interpreted in accordance with certain principles of physical chemistry embodied in what is known as the Gibbs phase rule. A working knowledge of the phase rule is essential for the effective use of phase diagrams. Such a working knowledge is very simple to acquire, since the phase rule is a very simple statement about the relationships of three variables. Like many rules, the phase rule conceals in its simplicity a very subtle, brilliant, and powerful chain of logical reasoning, and it is possible, perhaps even desirable for advanced students, to analyse phase diagrams from first principles without using the rule as a crutch. We shall use the phase rule as a tool, however, and the purpose of this chapter is to provide some fundamental definitions and a very rudimentary sketch of the principles upon which the phase rule rests. As we shall develop and use it, the phase rule is a part of what is called classical thermodynamics, the counterpart of which is statistical mechanics.

Systems, Components, Phases, Variables, Equilibrium

A *system* may be defined as an assemblage of material bodies interacting among themselves. As generally used, the term implies that matter cannot be exchanged with the system's surroundings, a condition which is often explicitly stated by the use of the term *closed system*. Heat can be exchanged with the surroundings, so the temperature of the system may be externally controlled. Pressure may be applied to the system, and its volume thereby adjusted, again without the exchange of matter. A system may also be described as an arbitrarily chosen small part of the universe. Most of our systems will be made up from and described in terms of a few selected chemical compounds, such as, for example, diopside and anorthite, but the term

S. A. Morse, *Basalts and Phase Diagrams*,
https://doi.org/10.1007/978-3-030-97882-2_4

system can also apply to a small region of the mantle undergoing melting, or to a magma flowing through a fissure, or any chosen part of such a magma.

An open system is a system which may undergo interchange of matter with its surroundings, usually limited interchange, so that the amount of one component, such as water vapour, may be controlled externally while the amounts of all other components remain fixed at their initial values. A container sealed by an osmotic membrane is a good example of such an open system, and a sealed capsule of platinum at high temperatures is a good example of such a membrane with respect to hydrogen, which diffuses easily through hot platinum. As a counter example, a similarly filled gold capsule under the same conditions is essentially a closed system, gold being impermeable to hydrogen, or very nearly so for most practical purposes.

A *component* of a system is a chemical constituent whose quantity may be independently and arbitrarily varied. In a closed system, the amounts of the various components are chosen at some initial time, and left unchanged during the process of interest. The simplest sort of component is an atom of a given chemical species, and the components of a system may, if necessary or convenient, be described and controlled in terms of masses of elements. If the expected chemical reactions so warrant, more complicated constituents, such as the metal oxides usually reported in a chemical analysis, may be chosen, or even constituents such as mineral formulae. Thus one and the same system could be described in terms of the four components Ca, Mg, Si, O, or the three components CaO, MgO, SiO_2, or the two components $CaSiO_3$ and $MgSiO_3$ or the single component $CaMgSi_2O_6$. Systems are described by the number (c) and identity of their components, for example, as follows:

The unary ($c = 1$) system $CaMgSi_2O_6$
The binary ($c = 2$) system $CaSiO_3 – MgSiO_3$
The ternary ($c = 3$) system $CaO – MgO – SiO_2$
The quaternary ($c = 4$) system Ca-Mg-Si-O

The choice of components is a matter of experience. It is mandatory to choose the smallest number, hence the most complicated identity, of components until experimental or theoretical experience dictates otherwise. It develops, for example, that choice of the single component $CaA1_2Si_2O_8$ is often adequate for discussing the melting of pure anorthite, whereas in discussing accurately the melting of mixtures of diopside and anorthite the two components diopside and anorthite will not suffice, because diopside is not pure $CaMgSi_2O_6$ in the presence of Al, and both crystals and melts must be discussed in terms of at least a ternary system, as discussed in Chap. 5.

A *phase* may be described as a homogeneous, mechanically separable part of a system. Phases are separated from one another by interfaces. Each mineral or crystal species or glass is a phase (S), each magma or liquid or melt is a phase (L), each vapour is a phase (V). In a system of many phases, many of these may be solid, but commonly only one (or zero) is a liquid, and only one (or zero) a gas. A system may be partly described by the number (ϕ) and identity of its phases. In the phase diagram for water at low pressure (Fig. 3.1) each phase is a different state of matter, but as Fig. 3.2 illustrates, there are numerous phases of ice which are all members of the solid state. The mistake of equating phases with states of matter must be avoided.

The criterion of mechanical separability is a potentially troublesome part of the definition of phase. The criterion means separability in principle, such as could be ascertained with a microscope. For example, it is quite feasible to separate the pyroxene and plagioclase from one another in a coarse gabbro, and in principle it is also possible to separate them in a very fine-grained basalt, although one would prefer to avoid such a task. But how good a microscope should one use to deduce separability? One may see fine exsolution lamellae in pyroxene, and conclude that these should be counted as a phase separate from the host pyroxene, but how about the different types of layers in a mixed-layer silicate (Zen 1963), or the different domains in a feldspar crystal? Such subdivisions may reach such a small scale that chemical reactions with the external environment "ignore" the separate parts and respond only to the bulk crystal, although chemical reactions between or among the various subdivisions may still be imagined. In rare cases, therefore, the definition of a phase must be determined by experience, or at least theory, and tailored to the scale of the questions being asked.

Variables or *parameters* of state are quantities used to describe the state of a system. These fall into two classes, extensive and intensive. *Extensive* variables are those which depend on the size (mass) of the system or its parts, such as

Masses of phases (or of the system)
Volume of the system
Entropy (S) of the system
Enthalpy (H) of the system

Intensive variables are those which are *in*dependent of mass, such as

Temperature (T)
Pressure (P)
Concentrations of the components (X)

The object of the phase rule is to discover the minimum number of independent statements which must be made about intensive and extensive variables in order to describe the system completely. The phase rule applies only to systems in *chemical equilibrium*, which may be described as a state of no observable change in the masses of the various reactants and products, or a state in which chemical reactions proceed with equal rate in either direction. Chemical equilibrium also implies thermal and pressure equilibrium, so we should expect the system to be isothermal and isobaric at equilibrium. This definition excludes *local equilibrium*, which is a useful concept for dealing with systems in a chemical or thermal gradient, for example. The zoning of a crystal of feldspar represents a kind of local equilibrium.

It is useful to identify two kinds of chemical equilibrium. We shall most often be concerned with *stable* equilibrium, a state which represents the lowest free energy of the system under the specified P and T. Another kind, *metastable* equilibrium, represents a supposedly transient state of affairs wherein the system has not achieved the lowest free energy state, but looks and acts as though it were at equilibrium over the time scale of observation. This time scale may be indefinitely long, and for example, all igneous and metamorphic hand specimens of rocks are at metastable equilibrium, at room temperature, because the chemical reactions which should continue in cooling simply do not proceed at a finite rate near room temperature. If it were not for this fact, we could never observe the properties of minerals at high temperature which in fact are quenched in by prohibitively slow reaction rates. Other examples of metastable equilibrium are supercooled melts, and glass quenched from the liquid state. In most phase diagrams, solid lines denote stable equilibrium, and when produced beyond an intersection such as a triple point, are dashed to denote metastable equilibrium. A few examples of such *metastable extensions* are shown in the upper part of Fig. 3.2, and at the triple point in Fig. 3.1. The metastable extension denotes the fact that, at least in principle, reactions can be made to run just outside the true field of stability of the reactants. For example, supercooled liquid and vapour can coexist at metastable equilibrium in the field of $\mathrm{Ice_I}$ near the triple point. We shall not be greatly concerned with such phenomena in the context of this book, but the metastable state is important in much experimental work, and in the interpretation of natural rocks.

The Phase Rule and Variance

Statements about components, phases, and variables may be marshalled into a formula known as the phase rule which gives the number of additional statements that must be made to define the system completely. It can be shown that, in a system at equilibrium, the number of variables which can be *independently* varied (without disturbing the equilibrium) is equal to the number of components in the system, plus two. The reasoning goes briefly like this. An equation describing the internal energy E of a system can be written

$$\mathrm{d}E = T\mathrm{d}S - P\mathrm{d}V + \mu_{\mathrm{a}}\mathrm{d}n_{\mathrm{a}} + \cdots + \mu_{\mathrm{k}}\mathrm{d}n_{\mathrm{k}} \qquad (4.1)$$

where T is absolute temperature, S is entropy, P is pressure, V is volume, μ is the chemical potential, or the partial free energy per mole of a component, n is the number of moles (i.e. amount), and a…k are each components of the system. Readers not familiar with chemical thermodynamics should not throw up their hands in horror, for what we propose to show does not require an understanding of all these terms. At equilibrium, by definition, entropy, volume, and amounts of components may not change, so their differentials equal zero: $\mathrm{d}S = 0$, $\mathrm{d}V = 0$, $\mathrm{d}n_{\mathrm{i}} = 0$. Therefore at equilibrium, Eq. (4.1) integrates to

$$E = TS - PV + \mu_a n_a + \cdots + \mu_k n_k \qquad (4.2)$$

We see now, merely as a book-keeping exercise, that the energy of the system depends upon two terms, *TS* and *PV*, and *as many other terms* $\mu_{\mathrm{i}}\, n_{\mathrm{i}}$ as there are components. Hence the energy of the system, which is to say the state of the system, is completely described by $c + 2$ independent variables.

These $c + 2$ variables could all be extensive, such as the masses of phases, or partly extensive and partly intensive, such as T and P. If they all happened to be extensive, then specifying the number and identity of each phase present would serve to describe the system, or in other words we should need to make ϕ statements, one for each phase, in order to describe the system. This condition of description can be written

$$\phi - c + 2 \tag{4.3}$$

and interpreted as meaning "when the number of phases equals the number of components plus two, the system is uniquely defined in terms of its intensive variables." In other words, there is only one temperature, one pressure, and one concentration of components in some phase when $\phi = c + 2$ identifiable phases are present in the system. A corollary of this is that $c + 2$ gives the maximum number of phases which can coexist at equilibrium.

If by chance some of the $c + 2$ variables needed to describe the system were intensive variables, then of course $\phi < c + 2$, i.e. the number of phases present is less than $c + 2$, and a certain number of intensive variables must be specified to describe the system completely. Let W be this number of intensive variables to be specified, and then

$$W = c + 2 - \phi \tag{4.4}$$

which is the Gibbs phase rule. W is an integer commonly called the variance, and is also commonly denoted F, for *degrees of freedom*.

We can now see that for completeness, Eq. (4.3) should read

$$\phi \leq c + 2 \tag{4.5}$$

with the variance W giving the difference when the inequality applies.

We identify with W the intensive variables P, T and X. As an example, suppose $W = c + 2 - \phi = 1$; the system is said to be univariant, and a single statement, either about pressure, or about temperature, or about the concentrations of components, suffices to describe the system completely. This is true whether $c = 1$ and $\phi = 2$, or whether $c = 19$ and $\phi = 20$. Suppose $W = 2$; the system is *divariant*, and two statements, one each about two of the variables P, T, X, are required to describe the system completely, and so on.

The example chosen in Eq. (4.3) is one in which $W = 0$, and such a system with $\phi = c + 2$ is invariant, a state which is very informative because for any particular set of ϕ phases, it is unique.

The phase rule is thus a shorthand way of assessing the maximum number of intensive parameters which can change independently of each other without decreasing the number of coexisting phases in the system. The nature and utility of the phase rule becomes clearer with its actual application in the succeeding chapters.

Exchange Reactions

It is convenient to mention this species of reaction here, because it is fundamental to physical chemistry and not used here until later in the book. Suppose phase "A" is physically similar enough to phase "B" that the two may be interchanged within a crystal according to *T*, *P*, conditions within the system. We may then write the exchange A $\leftrightarrows$ B more simply as AB_{-1}.

As an example, consider crystals of olivine in the presence of a magma. They have a general composition $(Mg, Fe)_2$ SiO_4. If crystallization has occurred more rapidly than the supply of fresh magma, we may write the activity at the crystal as $FeMg_{-1}$ meaning that the crystal is gaining Fe at the expense of Mg. Similarly if the opposite occurred and the fresh magma was richer than the growing crystal in Mg, the reaction would be written $MgFe_{-1}$.

Chapter 5
Diopside and Anorthite: Supposedly a Binary Eutectic System

Introduction

In this chapter, we introduce the study of phase equilibrium diagrams as models of natural rocks. The system to be described is simple, yet it furnishes a neat example of what may go wrong with a model when a mineral chooses not to have the idealized composition you so hopefully expected it would. This complication, fortunately, increases rather than decreases the relevance of the model system to natural basalts.

This is a longish chapter, because it must cope with the basic principles of phase diagrams in the course of describing a single example. If it occasionally seems that we are straying rather far from basalts, it need only be recalled that a flanking manoeuvre is sometimes more productive than a frontal assault.

Preparation and Melting of Diopside (Di)

Diopside is one of the easiest minerals to synthesize accurately, i.e. with correct composition, so easy in fact that the melting point of pure synthetic diopside is used as a calibration point (1391.5°C) in the Geophysical Laboratory temperature scale.[1] The synthesis procedures described in this book are limited to those in which the components are mixed as the oxides. Another very common technique is to mix the materials as an aqueous gel, which is then fired to an anhydrous state.

The ideal formula for diopside is $CaMgSi_2O_6$; stated in oxides, convenient for experimental work, $CaO{\cdot}MgO{\cdot}2SiO_2$. Each of these oxides may be obtained in a state of high purity. One uses reagent grade $CaCO_3$ (calcite) as a source of CaO.

[1] Temperature scales in silicate research are discussed by Sosman (1952). Despite the use of diopside as a calibrant, it actually has a small melting range (Kushiro 1972), and the figure 1391.5°C refers to the upper limit of the range.

S. A. Morse, *Basalts and Phase Diagrams*,
https://doi.org/10.1007/978-3-030-97882-2_5

MgO may be obtained commercially in high purity, although Mg metal may be even purer, and this may be oxidized in acid to form MgO, which is the mineral periclase. Many schemes have been used to obtain pure SiO_2. One of the simplest, probably as good as any, is to crush and mill optical grade quartz crystal in steel, and remove the contaminating steel magnetically and with acid. This quartz contains fluid inclusions, and is not yet pure SiO_2 in its most reactive form, so one converts it to cristobalite by heating in a platinum crucible at 1500°C or higher for a number of hours (the tridymite/cristobalite transformation occurs at about 1470°C). Complete conversion to cristobalite is seldom achieved short of several days, as found by Yoder and Schairer in 1968 (verbal communication), but the material is probably adequate for most purposes after a few hours of heating. One must temper the desire for complete conversion with the reality of heating platinum crucibles at these temperatures: whiskers of platinum will form in time on the surface of the charge, and one will be weighing a very heavy contaminant into one's silicate mixture. Schairer was also fond of pointing out that the quartz must lie *loosely* in the crucible (don't breathe on it!), or the volume changes on heating through the high-low quartz and high quartz-tridymite transformations will deform the crucible severely. Pure silica can also be obtained commercially from some glass companies as lump cullet.

Given the suitable, desiccated starting materials, one simply mixes in a platinum crucible appropriate weights of $CaCO_3$, MgO, and SiO_2 to form diopside (a 10-g yield is convenient). Enough $CaCO_3$ must be weighed in to yield the appropriate amount of CaO; the CO_2 will escape on heating. The crucible with its charge is placed in a ceramic-lined, platinum-wound furnace and heated to around 1500°C for several hours, overnight for example. The charge is quenched by setting the crucible in a pan of water, the resultant glass is broken and knocked out onto a paper by hammering the crucible,[2] the glass is crushed in a steel mortar to mix it, tramp steel is removed with a magnet, and the charge is returned to the crucible and melted again. A few fusions suffice to produce a homogeneous glass. Homogeneity is checked by examining an immersion mount of a bit of the crushed glass under the microscope; the glass particles should all have the same refractive index and be free of any inclusions of undissolved material.

Once a pure, homogeneous diopside glass is obtained, it can be crystallized readily by holding it at 1200–1300°C overnight, or several times overnight if desired for better crystallinity. The resulting crystals are extremely small (usually much less than 10 μm in length), but give sharp X-ray peaks. They can be used in melting experiments by the quenching method (or by differential thermal analysis, which measures a discontinuity in temperature rise with constant application of heat, thus indicating the latent heat absorbed on melting). The quenching method[3] utilizes the fact that most silicate liquids can be converted to a glass by abrupt cooling (quenching). In the quenching method, a small amount (such as 10 mg) of crystalline diopside is wrapped in an envelope of platinum foil and suspended in a vertical-tube

[2] Hammering is not necessary if $Pt_{95}Au_5$ is used, as the melt does not wet this alloy.

[3] A description of the quenching method is given by Osborn and Schairer (1941).

furnace by a fine platinum wire. A thermocouple is suspended beside the envelope, furnishing the temperature of the experiment. After heating at constant temperature for an hour or so, an electric current is applied to melt the platinum suspension wire, and the platinum envelope with its diopside charge falls into a dish of mercury, cooling to nearly room temperature in a few seconds. When examined under the microscope, the charge will be seen to consist of diopside crystals if the temperature was below the melting temperature, or of glass if above. With careful measurement of temperatures, runs at 1390°C yield crystals, and runs at 1393°C yield glass, indicating a melting point of 1391.5 ± 1.5°C. Few silicate melting points can be determined so closely by this method; an uncertainty of ±3°C to ±10°C is more normal. Diopside has given consistent results and its melting point is now used routinely (along with NaCl, 800.5°C) for the calibration of thermocouples used in experimental work.

Preparation and Melting of Anorthite (An)

Anorthite, $CaAl_2SiO_8$ or $CaO{\cdot}Al_2O_3{\cdot}2SiO_2$, can also be synthesized readily. The sources of CaO and SiO_2 have already been discussed in connection with diopside. Alumina, Al_2O_3, can be obtained commercially in high purity. Once again, the CaO is weighed in as $CaCO_3$ with allowance for the escape of CO_2. The problems of synthesis are somewhat greater in this case, since Al_2O_3, better known as the mineral corundum or the gems ruby and sapphire, is highly refractory and doesn't dissolve easily in a silicate melt. The possibility of melting the Al_2O_3 itself is precluded in ordinary work by its high melting point (2072°C), which is in excess of that of the platinum (1770°C) used as both container and furnace winding. At least three fusions at about 1600°C are made, each separated by a quench and rough crush of the glass, before the mixture is examined for homogeneity. After this, fusions are separated by finer crushing in a steel mortar, with magnetic removal of tramp steel, and are continued until no corundum grains remain and the glass has a homogeneous refractive index. The melting point of anorthite, determined by the quenching method, is taken as 1553°C, since it demonstrably lies between 1550 and 1555°C. There is some doubt that this compound melts (congruently) to a liquid of its own composition at a single value of temperature. Corundum has been observed in the glass quenched from temperatures near the assumed melting point, and anorthite may in fact melt to corundum and a more lime-silicate-rich liquid over a small interval. The difficulty of thermal control and measurement at these high temperatures renders a precise answer to this question elusive.

Preparation of Di-An Mixtures

It might be supposed that homogeneous glasses with compositions between pure Di and pure An would be made by melting weighed mixtures of the pure end-member glasses. For tactical reasons, however, a better approach is to make each composition from scratch, in 10 g batches, using the same starting materials as for the pure compounds, i.e. $CaCO_3$, MgO, Al_2O_3, and SiO_2. The principal reason for this is that the addition of almost anything (MgO in this case) facilitates the dissolution of Al_2O_3 into the silicate melt, so it is easier to make a homogeneous glass of $Di_{10}An_{90}$ composition than a glass of An_{100} composition. The ease of dissolution of Al_2O_3 can be ascribed to the lower viscosity of the liquid when MgO is present. Pure diopside liquid, rich in network modifiers, is so runny that it can be poured, while silicate liquids rich in alumina, a network former, are stiff as a board. It is to be presumed that liquids of lower viscosity tend to make better solvents, since they can carry away the solute components more rapidly.

Most silicate glasses are in like manner made directly to their composition from oxide mixes rather than from end-member compounds, for similar reasons. Alkalies, however, are usually added as disilicates to minimize alkali loss by volatilization.

For the study of a system such as Diopside-Anorthite, it is customary to prepare compositions at 10% intervals by weight. Additional compositions may be prepared at a later time if more precise experimental data are needed.

T-X Plots: Liquidus and Solidus

The experimental study of a system such as Di-An proceeds by a determination of the beginning of melting (first appearance of glass in quenched runs) and end of melting (last appearance of crystals in quenched runs) for a series of compositions, determined by the quenching method described above under *Preparation and melting of diopside*. The information obtained from a series of experiments is best appreciated in graphical form, such as a graph of temperature versus composition, or *T-X* plot. The ordinate is conventionally temperature in degrees Celsius, and the abscissa, composition. In our examples, *X* will denote composition in weight per cent, simply because weights are the experimentally convenient measures of amounts. Composition could be stated in other terms, for example, molecular per cent or mole fraction, cation units, or oxygen units.

A typical *T-X* plot (see Fig. 5.1) of a binary (two-component) silicate system consists of symbols or labels depicting solid, solid plus liquid, or liquid states at the termination of each run, and a collection of curves marking the boundaries between

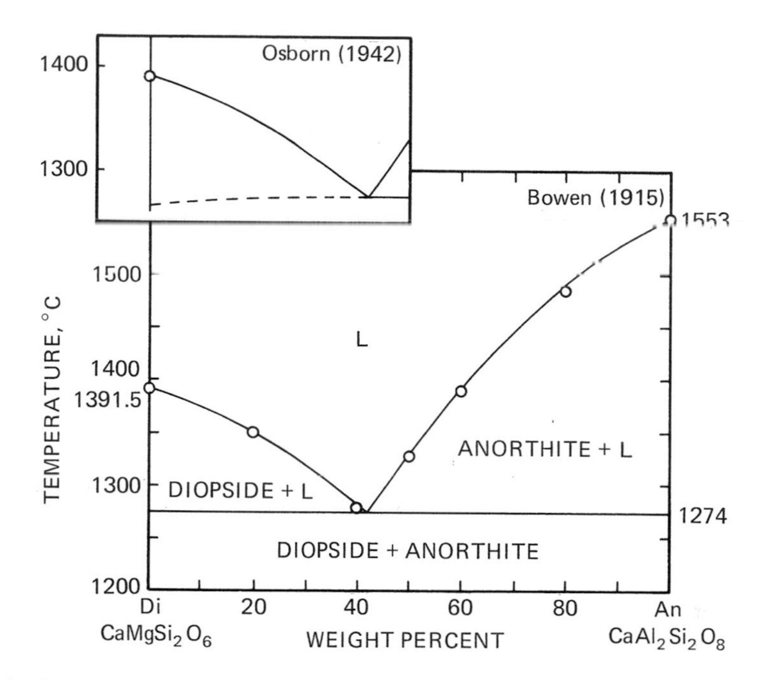

Fig. 5.1 Phase equilibrium diagram of the system diopside-anorthite, after Bowen (1915) with inset after Osborn (1942). Experimental points defining the liquidus curves are circled. The Bowen diagram assumes a binary eutectic system. The Osborn diagram reflects the discovery that the system is not binary for diopside-rich compositions. This is because Al enters diopside crystals, as discussed later in the chapter

regions of solid (*S*), solid plus liquid, and liquid (*L*).[4] These curves have special names, and properties which are very useful. The curve separating the liquid region from the solid plus liquid region is termed the *liquidus*, best remembered as the *curve which at any temperature gives the composition of the liquid in equilibrium with a specified solid*. The curve separating the solid plus liquid region from the solid region is termed the *solidus* or beginning-of-melting curve. If the solid consists strictly of crystals of end-member composition, such as pure Di and pure An, the solidus is a *straight, isothermal line*, and the compositions of crystals lie at either end of the line. If solid solution occurs, and the system is strictly binary, the solidus is a *curve which at any temperature gives the composition of the crystals in equilibrium with a specified liquid*. The case where solid solution occurs is discussed more fully in Chap. 6.

[4] A complete plot would also show a vapor (*V*) region. However, the vapor pressure of silicates and silicate liquids is so low that the vapor state is usually ignored. *T-X* diagrams are assumed to represent a pressure of 1 atm unless otherwise stated.

Diopside-Anorthite as a Binary System

On Earth, the minerals diopside and anorthite occur with compositions very close to the ideal end members almost solely in metamorphic rocks. In igneous rocks, formed at higher temperatures, they contain various "impurities", or more realistically, show solid solution towards other components. It should not then surprise us if the crystals formed in high-temperature laboratory experiments showed compositions other than end-member compositions. This is the case with the system Di-An. There is evidence that the system is not strictly binary, in other words that solid solution does not involve only the two components, and we shall consider this problem at the end of this chapter. For the moment, let us pretend that the system is binary, and that all crystals are pure diopside or pure anorthite. In this way, we can illustrate some important principles of phase diagrams and igneous rocks.

The system diopside-anorthite has been studied at 1-atm pressure by Bowen (1915) and Osborn (1942). The *T-X* plot, or equilibrium diagram, or phase diagram, essentially as determined by Bowen, is shown in Fig. 5.1. Small circles are shown at the melting points of the end members, diopside and anorthite, and at the final melting points of Di-An mixtures at several intervals along the *X* axis. These circles represent the midpoint between the two experimental runs bracketing the final melting point. For example, if a run quenched from 1352°C shows glass only, and a run quenched from 1348°C shows glass with a small amount of diopside crystals, the final melting point for the mixture (in this case $Di_{80}An_{20}$) is plotted at 1350°C. The collection of small circles showing the final melting points defines the liquidus curves. Above these curves the system consists only of liquid. It is clear from the shape of the liquidus curves that they will intersect somewhere near a composition of $Di_{58}An_{42}$ and a temperature somewhere near 1274°C. The lowering of liquidus temperatures away from end members is a common phenomenon in silicate systems and one familiar to chemists under the title of *freezing point depression.* Either end member may be considered an "impurity" which, when added to the other, causes lowering of the final melting point (or first freezing point, depending on the point of view).

It is to be noted that none of the mixtures studied *within* the system melts at a single temperature, but that all show a melting *range* beginning (in this case) somewhere near 1274°C and terminating at the liquidus curve. The same is true of virtually all rocks, which crystallize or melt completely over a range of temperatures rather than at a point. This apparently innocent fact is one of fundamental importance in petrology, for it accounts for the possibility of magmas changing their composition by removal of crystals, or fractional crystallization. Conversely, the melting interval requires that natural rocks will begin melting to a magma quite different from the rock composition. We shall have much to say about both these processes of generating diversity in magmas.

The horizontal line at 1274°C lies through the beginning of melting of each composition studied. In Bowen's work, these all fell at the same temperature, within experimental error. The horizontal line is an isotherm, and it is also the solidus for the system in the sense that it locates the boundary between the fields of solid (S) and solid + liquid ($S + L$).

The point marked by the intersection of the two liquidus curves and the solidus ($Di_{58}An_{42}$; 1274°C) is termed the *eutectic*[5]; it lies at the one and only composition within the system which has a true melting point rather than a melting interval. The diagram of Fig. 5.1 is one of a *binary eutectic* system. The eutectic is an important phenomenon which we shall consider in more detail below.

Variance in Di-An

We continue to assume that Di and An are pure compounds, and that the system is binary.

The phase rule may be stated as

$$W = c + 2 - \phi \qquad (5.1)$$

We recall that the 2 signifies intensive parameters not dependent on the masses of things in the system; we may identify among these parameters temperature (T), and pressure (P). The phase rule reminds us that if we wish to describe the system completely, we shall need to make $c + 2 - \phi$ statements about independent parameters. However, as long as we are considering experiments made at constant pressure (isobaric conditions), we are holding one variable (P) fixed, and we need make only $c + 1 - \phi$ statements to describe the system. Any other arbitrary restriction, such as on T (an isothermal restriction), would have the same effect. We may then rewrite the phase rule for isobaric conditions:

$$W_p = c + 1 - \phi \qquad (5.2)$$

in which the subscript p reminds us that the variance is reduced by one because of an isobaric restriction. For present purposes, the constant pressure is 1 atm, since the experiments were performed at atmospheric pressure, i.e. with no pressure control.[6]

We now need to evaluate c, and then ϕ, in order to evaluate the variance. We have already decided to call the system Di-An binary ($c = 2$), but let us consider some

[5]Guthrie (1884) coined this term to signify the lowest-melting point in a system. Strictly speaking, this point in this system implies the coexistence of Di, An, L, and vapor which occurs at a single value of pressure well below 1 atm. Because of the low vapor pressure of silicate liquids, this strict definition is usually ignored.

[6]The atmosphere is frowned upon as a unit of pressure in these days of organized science, but this is precisely the context in which it is useful, denoting an experiment made at ambient (and slightly variable) conditions of pressure. Authors attempting to describe this condition as 0.1 MPa are at fault in implying a strict pressure constraint that does not exist at atmospheric pressure. “A foolish consistency is the hobgoblin of little minds.”

alternatives for the moment. The system contains only the elements Ca, Al, Mg, Si, and O, so at worst it would have to be described in five components (quinary system). If we believe that the cations are almost always bound to oxygens in melts and crystals, we could use as components CaO, Al_2O_3, MgO, and SiO_2 ($c = 4$, quaternary system). If we believed that the liquids and crystals can be described as mixtures of $CaSiO_3$, $MgSiO_3$, and Al_2SiO_5 (an unlikely eventuality), we could describe the system as ternary ($c = 3$). Another ternary alternative will be described at the end of this chapter. Finally, if we believe that the crystals and liquids can be described in terms of pure Di ($CaMgSi_2O_6$), pure An ($CaAl_2Si_2O_8$) and liquid solutions or solid mixtures of these, we are justified in calling the system binary, as we shall now do. Hence we shall use the phase rule statement $W_p = 3 - \phi$.

As for ϕ, the experiments, summarized in the phase diagram (Fig. 5.1), describe the phases for us. As the figure suggests, above the liquidus curves, there is but a single phase, L, and $W_p = 3 - 1 = 2$; this is a *divariant region.*[7] Such a region is often described as having *two degrees of freedom*, and it is interesting to note that, under the restriction of constant pressure, this region is a *two-dimensional field* in the phase diagram. Saying that $W_p = 2$ is equivalent to saying that we must make two statements, one for each dimension, to know where we are in that part of the phase diagram labelled L. We must specify the temperature and the chosen bulk composition[8] of the system in order to plot a point in the L field, for example, 1500°C and $Di_{50}An_{50}$.

On either of the liquidus curves in Fig. 5.1, two phases are stable, either Di + L or An + L, depending on the curve. In this case $W_p = 3 - 2 = 1$, and the system is *univariant* for any point on either of the liquidus curves. It is interesting to note that, with the restriction of constant pressure, univariance corresponds to a *one*-dimensional line. Saying that $W_p = 1$ is equivalent to saying that we need make only one statement (about one dimension) to plot a point in the $S - L$ part of the phase diagram. For example, given the identity of the phases as An + L, one need only specify the temperature as 1328°C to be informed from the phase diagram that a liquid of composition $Di_{50}An_{50}$ coexists with crystals of pure An. (One learns nothing of the *bulk* composition, which could lie anywhere between $Di_{50}An_{50}$ and Di_0An_{100}, but that is immaterial as far as chemical reactions at equilibrium are concerned.) Alternatively one could specify the composition of the liquid in equilibrium with crystals of pure An as $Di_{50}An_{50}$, and learn from the diagram that the temperature for such a situation is 1328°C.

---

[7] Strictly speaking, it is an isobarically divariant region. If P is allowed to vary, $W = c + 2 - \phi = 4 - \phi$ and the region of $\phi = 1$ is truly trivariant. Whenever W_p is written, it must be remembered that the variance is under *restriction* and is always one less than the unrestricted variance W. This understanding will be assumed in the following pages.

[8] "Bulk composition" is a term used to describe any chosen mixture within a system for purposes of discussion: it is so called to help distinguish the total composition from the individual compositions of crystals or liquids. It is very important to remember that the system is defined by the bulk composition chosen, which may not change during any single process or experiment.

There is a unique point, the eutectic, in the diagram of Fig. 5.1 at which the two liquidus curves intersect at the solidus. Since the liquidus curves specify equilibrium between Di + L on the one hand and An + L on the other hand, the eutectic apparently involves an equilibrium among all three, Di + An + L. Here and only here, $\phi = 3$, and $W_p = 3 - 3 = 0$; the eutectic is an *invariant* point. Notice again the correspondence between variance and dimensions, both being zero in this case.[9] Saying that the variance is zero is a most informative statement, since nothing further need be said to learn from the phase diagram that the temperature is 1274°C, and that the liquid has the composition $Di_{58}An_{42}$ and coexists with pure Di crystals and pure An crystals.

The regions between the liquidus and solidus curves are convenient places for labels, and for plotting bulk compositions at specified temperatures, but no phases lie within them. They may each contain an infinite number of horizontal (isothermal) *tie lines* connecting points on the liquidus with the appropriate ordinate, in other words, tying liquids of appropriate composition to the (unchanging) compositions of the crystals with which they coexist. On any of these tie lines there can be plotted a bulk composition. The tie lines and bulk compositions are usually omitted in phase diagrams except during graphical analysis, which is described below.

Below the solidus is a region where crystals of Di and crystals of An coexist (in proportions dependent on the bulk composition). Here $\phi = 2$, and it is clear that $W_p = 3 - 2 = 1$. That simply means that there is only one intensive variable to be specified, and it is temperature. The state of the system is not responsive to the bulk composition.

The Lever Rule

When crystals coexist with liquid in a binary system, it is possible (except when liquid is at a eutectic) to know the relative percentages of each (S and L) if the temperature and bulk composition are known. This important fact permits the quantitative use of phase diagrams in the analysis of crystallization and melting. The treatment of relative percentages is a simple graphical procedure based on the *lever rule* or *rule of moments*. This rule is best described by a sketch, Fig. 5.2, of a lever sitting on a fulcrum. The fulcrum is labelled BC for "bulk composition" and the ends of the lever are labelled S for the position of the solid (or crystal) and L for the position of the liquid. Lever lengths are labelled s and l. In the illustration it is clear that l outweighs s because of occupying a longer arm of the lever. Quantitatively, the fraction of the total mass of the system represented by L is $l/(s + l)$; similarly, the fraction of S is $s/(s + l)$, and the sum of s and l is 1. To convert fraction to percentage only requires multiplication of the fraction by 100. For example, if $l = 9$ (in arbitrary

[9]The true eutectic (see footnote 5 above) involves Di + An + L + V; $\phi = 4$, and this equilibrium is truly invariant, i.e. $W = 4 - 4 = 0$.

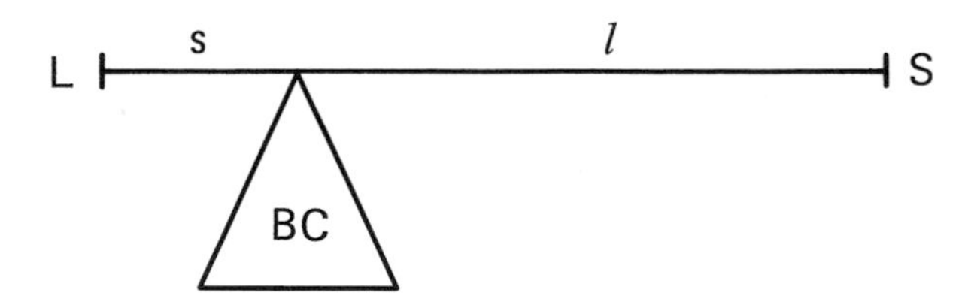

Fig. 5.2 Principle of the lever rule

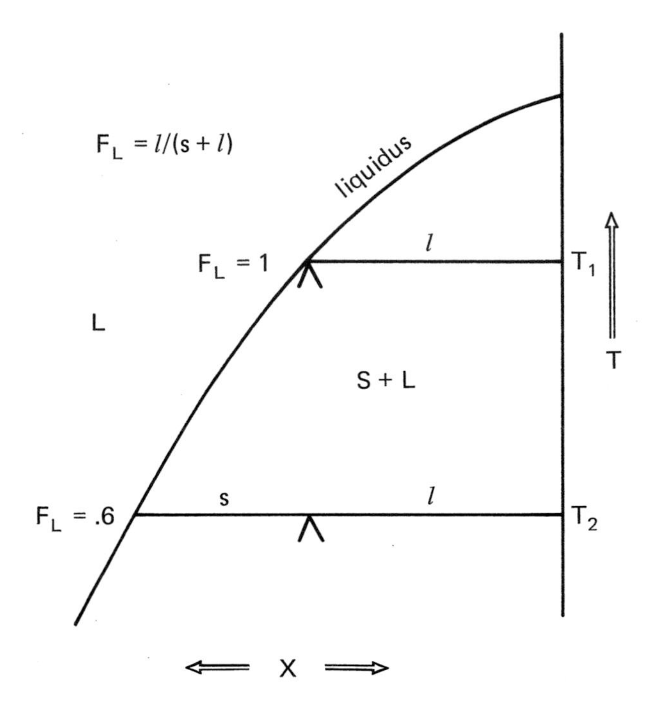

Fig. 5.3 Example of the lever rule

units of length—millimetres are convenient), and $s = 3$, the fraction of L is $9/12 = 0.75$, or 75%, and the fraction of S is $3/12 = 0.25$, or 25%. We shall frequently use the symbol F_L for *fraction of liquid remaining (on crystallization), or produced (on melting), or present in the system (without regard to process).*

The only problem with the lever rule lies in knowing which side of the bulk composition fulcrum to label l and which to label s. This problem is easily solved by a commonsense analysis. If the fulcrum lies *at* the liquidus (see Fig. 5.3) at a certain temperature T_1, the bulk composition is practically in the liquid field and the entire lever is represented by l (to the right of the fulcrum), and $s = 0$. At a lower temperature (T_2 in Fig. 5.3), the proportion of crystals has, logically, increased, and the end of the lever labelled s has grown while l remains the same length. The

l end of the lever is always the end *away* from the liquidus curve with respect to the fulcrum, and the s end is always away from the crystal composition (the ordinate in this case).

Nature of the Eutectic: Isothermal Melting and Crystallization

If one pursues the lever rule of Fig. 5.3 in the context of the An + L field of Di-An (Fig. 5.1), one quickly discovers that, since l has fixed length, there is some liquid left at a temperature of 1274°C, when the liquid has just reached the eutectic composition, no matter where the original bulk composition lay. We also know, from the fact that $W_p = 0$ at the eutectic, that neither temperature, nor the composition of the liquid, nor the compositions of crystals of Di or crystals of An may change as long as the liquid remains at E. Furthermore, we comprehend from the diagrams that a minute decrement of temperature below 1274°C results in solid Di + An only, without liquid. It is apparent that an unusual event occurs at E, namely the *isothermal crystallization* of liquid to yield a completely solid mass of Di and An crystals. The lever rule is of no use while the liquid is at E, the ratio of crystals to liquid being no longer a function of T, but only of time, dictated by the rate of heat loss from the system. The reaction at E is $L = S$ + cals (calories) on crystallization and S + cals $= L$ on melting. Furthermore, since the liquid does not change composition, it must produce (on crystallization) Di and An in exactly its own ratio, i.e. 58:42, so the isothermal, eutectic crystallization of 1 g of liquid yields 0.58 g of Di crystals and 0.42 g of An crystals. Melting consumes crystals in these proportions to yield the eutectic liquid.

A very interesting property of the eutectic, then, is that it is a *thermal buffer*, just as is the melting point of a single pure phase such as ice or diopside. If we make a sketch of temperature versus time, we shall find that the temperature of the eutectic is occupied by a flat, isothermal line representing a *thermal arrest*. In fact, such a thermal arrest furnishes one way of determining a melting point or a eutectic experimentally. Most of us have more than once made such an experiment with ice and water and a thermometer. The lucky ones have repeated the experiment many times with a hand-cranked ice cream freezer surrounded by ice and rock salt.

Although the isothermal crystallization of rocks is probably rare in nature, nearly isothermal melting may occur at the sources of magmas in the mantle. In either event, the process should be kept in mind, since the heat to be removed (or added) is only the latent heat of crystallization (or fusion), and not the heat required to change the temperature of the body as well.

Analysis of Crystallization and Melting

We are now sufficiently prepared with concepts and jargon to proceed to the denouement of any experimental system, a complete analysis of the course of crystallization (or melting) of any selected bulk composition. We begin with crystallization, using Fig. 5.4, which is again the system diopside-anorthite with some graphical constructions and a doubled scale of temperature.

Choosing $Di_{40}An_{60}$ as a bulk composition, we imagine a liquid of this composition at temperature T_1. There is one phase, L, and $W_p = 3 - 1 = 2$. As the liquid is cooled, nothing happens except for some ordering of atoms in the liquid until the An + L liquidus curve is reached at T_2. At this moment, the first few crystals of An begin to form; $\phi = 2$, and $W_p = 3 - 2$ equals 1. After an infinitesimal drop in

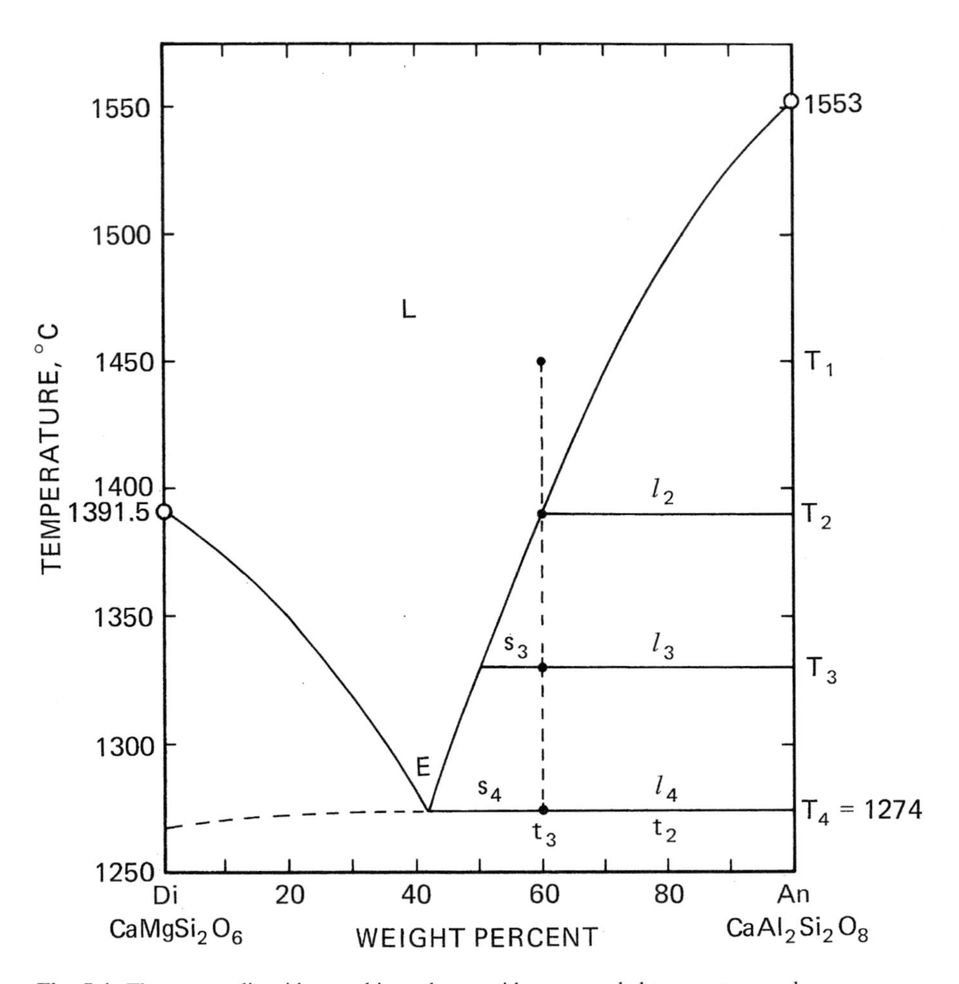

Fig. 5.4 The system diopside-anorthite redrawn with an expanded temperature scale

temperature, the liquid has a composition of nearly $Di_{40}An_{60}$, but is shifted slightly to the left of this, by an amount corresponding to the mass of An crystals which have formed. Cooling is continued, and more crystals of An form as a result. If we take stock of the situation at T_3, we deduce from the diagram that the weight of An crystals formed by this time is given by $s_3/(s_3 + l_3)$ or about 20%. At this point the liquid, having yielded so much pure An, is correspondingly poorer in An and richer in Di, with a composition of about $Di_{50}An_{50}$. Continued cooling, as before, enriches the liquid in Di by the crystallization of An, and as before, further crystallization of An requires both heat loss and temperature drop. The slope of the liquidus at any point is the resultant of a horizontal vector away from An and a vertical vector downward.

Eventually, cooling of the liquid with its charge of An crystals, accompanied by further crystallization of An, brings the liquid to T_4 and the eutectic composition. The instant before the temperature reaches T_4, the amount of An which has crystallized is given by $s_4/(s_4 + l_4)$ or about 31%. At the instant when the liquid reaches the eutectic E and temperature T_4, crystals of Di begin to form for the first time. With continued removal of heat, the temperature remains at T_4 while the liquid of composition E produces crystals of Di and An in the ratio of E (58:42) until it is used up, at which moment the entire mass is solid, and the temperature drops with further removal of heat.

We now turn our attention to the history of the solids produced by this entire process. The bulk composition having been chosen as $Di_{40}An_{60}$, it is clear that the entire crystallization process should yield 40% Di, 60% An by the time the liquid is used up. During the cooling from T_2 to T_4, the *total solid composition* (TSC) lay on the An ordinate. While the liquid lay at the eutectic, at temperature T_4, crystallization of Di as well as An gradually moved the TSC away from An, along the isotherm T_4, as a function of time (t). There was a time t_2 when the TSC was $Di_{20}An_{80}$, for example, although we cannot know when that was from the phase diagram. Eventually there was a time t_3 when the TSC reached $Di_{40}An_{60}$, the bulk composition, and it was at this moment that the last of the liquid disappeared.

The same principles apply to crystallization initially in the Di field, as for example that of a bulk composition $Di_{80}An_{20}$. An initially homogeneous liquid of this composition will, on cooling, first produce crystals of diopside and consequently move towards An down the Di + L liquidus curve till it reaches E. At E, isothermal crystallization of Di and An crystals will ensue until the liquid is gone and the Di:An ratio of the solids reaches 80:20.

Complete equilibrium melting in such a system as Di-An follows the reverse path of complete equilibrium crystallization. A mixture of crystals of Di and An, say in the ratio 40:60 will, when heated to T_4 (Fig. 5.4), begin to melt with the production of a eutectic liquid of composition E. Crystals will melt in the ratio Di:An 58:42 until all the Di is gone; the lever rule shows that 69% of the mass can be melted in this way without a change of either liquid composition or temperature. With continued input of heat, the liquid dissolves An crystals and hence becomes richer in An, moving along the liquidus curve as the temperature rises. At T_3, the percentage of liquid is

80%, as it was on crystallization, and at T_2, the last An crystal is taken into solution and the liquid may be heated without further chemical reaction.

Fractional Crystallization and Melting

The crystallization and melting processes described above are *equilibrium* processes, in which all of the crystals present remain in perfect equilibrium with the liquid. We may imagine another process, perfect *fractional* crystallization in which crystals are removed from equilibrium with the liquid as soon as they are formed. This process has a comparatively minor effect in a binary eutectic system, but is of much importance in many other types of systems. In the crystallization of Di-An, using the example of Fig. 5.4, removal of An crystals from the main body of liquid, perhaps by settling or floating, has no new effect on the composition of the liquid during its trip down the An + L liquidus curve. The only effect of perfect fractionation of An is to hide An crystals from view, so that if we could sample the liquid at any temperature we should see only a trace of An crystals, and we might think the liquid had always had the bulk composition of its liquidus composition at the moment of sampling. After the liquid has reached the eutectic E, removal of Di and An crystals will of course have no effect on the composition of the liquid, and both the liquid and the rock which it produces will have the eutectic composition. Probably the only geologically meaningful application of this fractional crystallization process is that liquids or rocks appearing to have eutectic compositions may have started out with bulk compositions far from the eutectic, only to have lost crystals continuously or intermittently during crystallization.

For the process of fractional crystallization in general, it is convenient to define the *instantaneous solid composition* (ISC) as the composition of the solid being crystallized from the liquid at any moment. In the example just given, the ISC consists simply of An crystals at first. When the liquid reaches point E, the ISC jumps discontinuously from An to E, the eutectic ratio of diopside and anorthite. We shall occasionally refer to such jumps in the ISC as "rock hops", and it should be emphasized that such discontinuities in the ISC path can occur *only* with fractional crystallization. In equilibrium crystallization, the ISC is identical to the TSC, and the TSC path must be continuous in *all* processes. In the cited example, there are two discrete ISCs and one continuous TSC, which starts at pure An and travels along the 1274°C isotherm towards E until it reaches the bulk composition.

Perfect fractional melting is a process in which liquid is continuously removed from the crystals as soon as it is formed. Using the example of $Di_{40}An_{60}$ as a solid starting mixture, perfect fractional melting will yield liquid of composition $Di_{58}An_{42}$, at isothermal conditions, as long as any diopside crystals remain. Imagine that all this liquid is withdrawn into a bucket, which we will label the *total liquid composition* (TLC). During this time, the TSC moves from $Di_{40}An_{60}$ to Di_0An_{100}, which is precisely the reverse of the crystallization process while the liquid is at E.

As soon as the last crystal of diopside is converted to liquid, however, the remaining solid mass consists only of An crystals. *No further melting* can take place until the temperature rises from 1274 to 1553°C, the melting point of pure An, at which point all the An melts isothermally; the liquid composition "hops" from the *E* to pure An. The reason for this different behaviour compared to equilibrium melting is that the eutectic liquid is not now available as a reservoir into which the An crystals may dissolve. The temperature gap in fractional melting is of interest in phase equilibria because it emphasizes the fact that, in equilibrium melting, An crystals dissolve in the liquid at temperatures far below their own melting point. When in fractional melting the pure An crystals do melt, we may imagine that their fractional liquid is withdrawn into the TLC bucket, and hence that the process will run to completion when the TLC equals the bulk composition.

In this latter part of the process, fractional melting is strikingly unlike equilibrium melting, since there is a large temperature interval over which no liquid is generated. Under equilibrium conditions, the entire original bulk composition would have been melted at about 1387°C, whereas fractional melting leaves a substantial crystalline residuum of pure An which is far more refractory than the initial bulk composition. This is a geologically important consequence, since it implies that fractional melting even under isothermal conditions in the mantle may leave behind crystalline residua which are unlikely to melt again unless a very large amount of heating occurs.

For fractional melting in general, it will be convenient to define the *instantaneous liquid composition* (ILC) as the composition of the liquid being formed, at any instant, from the crystals. The ILC is precisely analogous to the ISC; in the example at hand, it originates at *E* and, after a hiatus in melting, jumps (hops) discontinuously to An. The *total liquid composition* (TLC) is the sum over all ILCs produced during a melting process, and it must equal the bulk composition of the system at the end of melting. Thus in the present example, the TLC begins at *E* and, with the eventual addition of liquid An, moves towards An until it reaches the bulk composition. We never speak of the TLC in connection with any process except fractional melting, for in all other processes the TLC simply equals liquid, *L*. Similarly, the ILC is confined to fractional melting; in all equilibrium processes and in fractional crystallization, ILC = TLC = *L*.

To recapitulate the behaviour of the four kinds of paths in crystallization and melting, the TSC, TLC and *L* paths are always continuous; ISC and ILC paths may be discontinuous, providing us with rock hops and liquid hops.

Complications

It is now time to face the fact that Di-An is not a truly binary system. This is true largely because Di is not a pure compound of composition $CaMgSi_2O_6$ in the presence of certain impurities, in this instance Al_2O_3. Before enlarging on this statement, however, it is appropriate to make some general remarks on limited solid solubility.

Binary Solid Solutions in General

Many binary eutectic phase diagrams are known to have the form shown in Fig. 5.5, in which the compounds *A* and *B* exist in the pure state only at their respective melting points. Such a geometry has been likened to an upside-down fat man with a pair of pants. At all lower temperatures, the solid phase A_{SS} is a limited solid solution (subscript ss) towards *B*, and the solid phase B_{SS} is a limited solid solution towards *A*. The mutual solubilities of solid *A* and solid *B* increase with falling temperature from the melting points of pure *A* and pure *B*, reaching maxima at the eutectic temperature. The curves which give the composition of A_{SS} in equilibrium with liquid and the composition of B_{SS} in equilibrium with liquid are the *solidus* curves. Below the temperature of the eutectic, T_e, the mutual solubilities of *A* and *B* commonly *decrease*, reaching zero at 0 °K. The curves which give the composition, at any temperature, of A_{SS} and B_{SS} in equilibrium with each other are part of a curve termed a *solvus*. The region between these limbs of the solvus is often referred to as a *miscibility gap* (NOT an immiscibility gap, which is a contradiction in terms). The solvus curve in this case is intersected by a beginning-of-melting isotherm, but a

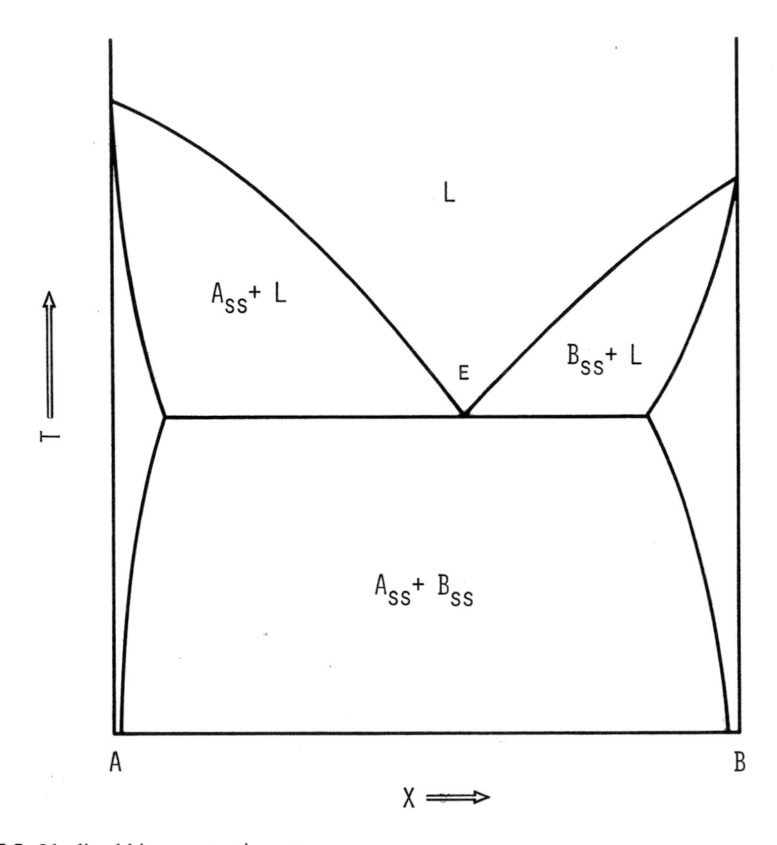

Fig. 5.5 Idealized binary eutectic system

solvus may be wholly within the solid field or, rarely in silicate systems, wholly within the liquid field.

We shall not dwell further on the properties of solvi at this juncture, except to point out that some degree of mutual solubility between solid end members of a binary system is inevitable at temperatures above 0 °K, and indeed must be present in Di-An. However, in this system, as in many geologically significant systems, mutual solubility is often not detected at ordinary levels of precision, and is therefore ignored.

Non-binary Solid Solution: The Role of Al in Pyroxene

There are several ways in which Al^{+++} may enter a pyroxene structure. That it does enter in nature is shown by the fact that most natural pyroxene analyses show Al_2O_3 as a major constituent. The mineral augite is characterized in Dana's *Textbook of Mineralogy* as "aluminous pyroxene" (Ford 1932). A number of hypothetical "pyroxene molecules" containing Al have been proposed as end members towards which diopside might show solid solution. Among these is CaTs, the *calcium Tschermak's molecule*, $CaAlAlSiO_6$. This compound is not found as a mineral, but it can be synthesized in the laboratory, and studies by the quenching method show that single crystals can be grown having compositions towards CaTs from Di. CaTs being, like An, a member of the system CaO-Al_2O_3-SiO_2, it is clear that An is capable of furnishing the constituents of CaTs for solid solution in Di. It is almost certain (Kushiro and Schairer 1970) that the diopside crystals observed in the system Di-An incorporate significant amounts of CaTs "stolen" from anorthite.

The Tschermak substitution, very significant in mineralogy, is the coupled substitution $[AlAl]^{6+} = [MgSi]^{6+}$, using the exchange components $AlMg_{-1}$ and $AlSi_{-1}$. Hidden in this code is the implicit fact that one half of the exchange is in the octahedral site, i.e. $AlMg_{-1}$, but we note that by itself this reaction carries a net charge of +1. The other half of the exchange is in the tetrahedral site, i.e. $AlSi_{-1}$, which carries a net charge of −1; clearly, the total exchange is electrically neutral, but only by virtue of the fact that the two different kinds of crystallographic sites can transfer their deficits or excesses of electrons between each other.

If the pyroxene structure can accommodate Al in this way from anorthite, can anorthite accommodate some of the extra MgSi kicked out of the diopside? Well, perhaps so. If we run the reaction in reverse, $[MgSi][AlAl]_{-1}$, we will remove all the Al from anorthite and end up with a hypothetical Al-free feldspar component, $CaMgSi_3O_8$. Evidence has been found to suggest small amounts of this substitution into feldspar in high-temperature experiments with diopside (Murphy 1977).

A Relevant Graphical Treatment

The effect of stealing CaTs from anorthite component in the liquid or crystals to make more pyroxene is to enrich liquids in silica. This fact is already evident from the fact that CaTs is low in silica by virtue of containing tetrahedral Al. The total exchange can be studied chemographically, by examining part of the system CaO-MgO-Al_2O_3-SiO_2 (CMAS). This quaternary system can be represented compositionally as a three-dimensional figure, a tetrahedron, shown in perspective in Fig. 5.6. This is an *X-X* plot, i.e. a strictly compositional plot to examine the purely geometrical relationships between components and phases. The front face of the tetrahedron is the ternary system CaO-Al_2O_3-SiO_2 (CAS). The floor is the ternary system CaO-Al_2O_3-MgO (CAM); the left rear face is the ternary system CaO-MgO-SiO_2 (CMS), and the right rear face is the ternary system Al_2O_3-MgO-SiO_2 (AMS). A line (altitude) bisecting the front face from SiO_2 to CA (i.e. CaO·Al_2O_3) contains the composition of both An and CaTs. The position of An can easily be plotted in the CAS triangle as follows. An is CaO·Al_2O_3·2SiO_2, hence the *C*:*A* ratio is 1:1, hence An must lie on the CA-*S* line. The CA:*S* ratio is 2:2, hence An plots halfway along the CA-*S* line. (Another trick is to plot, in turn, lines from each corner of the CAS triangle: from *C* to *A*:*S* = 1:2, and from *A* to *C*:*S* = 1:2, the line from *S* to *C*:*A* = 1:1 having already been plotted.) The position of CaTs is plotted in similar fashion: CaTs is CaO·Al_2O_3·SiO_2, with *C*:*A*:*S* 1:1:1, so the composition plots at the centre of gravity of the triangle, i.e. at the junction of lines from each apex to the 1:1 point on each opposite side.

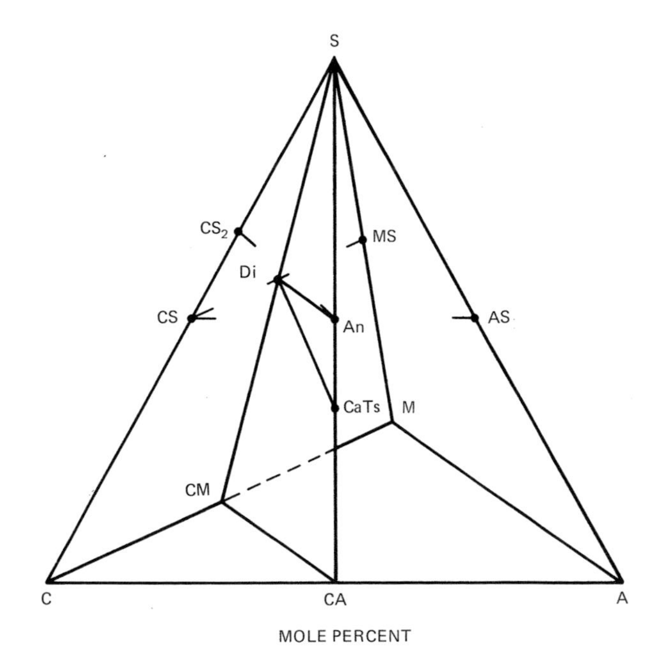

Fig. 5.6 The tetrahedron CMAS, the system CaO-Al_2O_3-MgO-SiO_2. The CM-CA-*S* plane is shown in Fig. 5.7

A further point of chemographic interest is that An lies midway on the line from CS ($CaSiO_3$; wollastonite, Wo) to AS (Al_2SiO_5; kyanite, sillimanite, andalusite).

The left rear face (CMS) of the tetrahedron contains a bisecting line *S*-CM on which lies the composition of Di. Di is $CaO{\cdot}MgO{\cdot}2SiO_2$, or CMS 1:1:2; like An, it plots halfway along the altitude of the triangle. Of collateral interest, it also plots halfway between CS (= Wo) and MS ($MgSiO_3$; enstatite), as every schoolchild knows.

It now appears that the system Di-An can be treated successfully in terms of the plane CM-CA-*S*, which contains all the solid phases in the tetrahedron of importance to our problem. This plane, a ternary system, is sketched in Fig. 5.7a. Figure 5.7b is a distorted blow-up of the central portion of CM-CA-*S*, showing qualitatively the expected solid solution from Di towards CaTs. The join[10] Di-An lies across the CM-

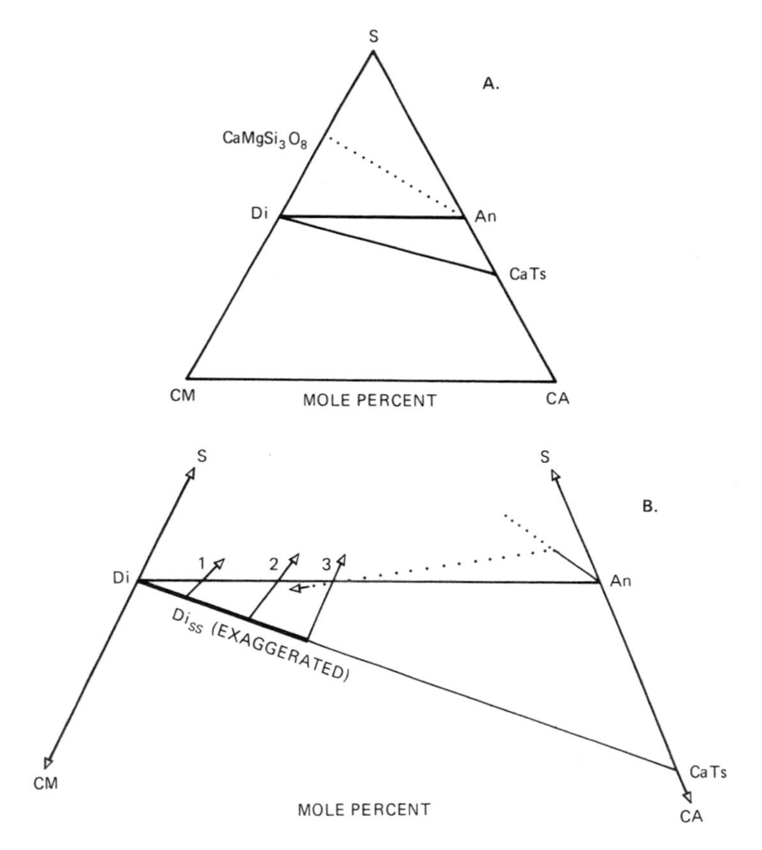

Fig. 5.7 Solid solution of diopside towards CaTs (calcium Tschermak's molecule). **A** shows the entire CM-CA-*S* plane, and **B** shows the central portion of that plane. Dotted lines show effects of solid solution towards $CaMgSi_3O_8$ from An

[10]When considered as part of a larger system (CMAS), a subsystem such as Di-An is commonly referred to as a join.

CA-*S* plane, and one may imagine the *T-X* plot of Fig. 5.1 arising out of the page from this line. Temperature information is thereby lost, the *T-X* plot being projected along its *T* axis onto the page. Compositional information is preserved. We see that bulk compositions lying truly in Di-An, such as 1, 2, and 3, will yield crystals, not of pure Di, but of solid solutions towards CaTs, or Di_{SS}. Since these crystals lie on the CM-CA side of the Di-An line, they must coexist with liquids, indicated by arrowheads, on the *S* side of the Di-An line. The tie line connecting two phases (Di_{SS} and *L*) in such a ternary plot must pass through the bulk composition if no other phase is present. Conversely, any second phase must lie on a straight line connecting the first phase and the bulk composition, by the same reasoning: everything present must add up to the bulk composition.

As discussed above, anorthite shows some solid solution as well, towards the join CMS and probably towards the component $CaMgSi_3O_8$ (Murphy 1977). Such a tendency is illustrated by the dotted lines in Fig. 5.7. If we assume that bulk composition (3) in Fig. 5.7b represents the cotectic composition, then the crystallization of anorthite will tend to counteract the effect of CaTs in the diopside, but as the arrows show, the anorthite effect is weak in its ability to counteract the silication of the cotectic liquid.

Petrologic Consequences

It is clear from this graphical analysis that entry of Al into Di causes enrichment of the liquids in SiO_2; upon cooling, they must become progressively more and more silica-oversaturated with respect to the join Di-An. This is true of all liquids from the eutectic *E* to, but not including, pure diopside. This is an interesting discovery, since we shall be much concerned with the over- or under-saturation of liquids with silica during fractional crystallization, and clearly the entry of CaTs into diopside must cause oversaturation in this case.

As to phase rule considerations, it is apparent that the system Di-An is at least ternary, perhaps rigorously confined to the plane $CaO{\cdot}Al_2O_3 + CaO{\cdot}MgO + SiO_2$ in the system CMAS. The phase rule must then be written $W_p = 4 - \phi$, so that all the equilibria we have discussed are of a higher variance by one than we had assumed. The assemblage $Di_{SS} + L$ is divariant ($W_p = 2$); and the assemblage Di + An + *L* is univariant ($W_p = 1$), and not a eutectic assemblage at all. A four-phase assemblage, probably Di + An + SiO_2 + *L*, is required for a ternary isobaric eutectic, and the three-phase assemblage Di + An + *L* merely lies on a line leading to the four-phase point. The non-binary nature of Di-An was noted and discussed by Osborn (1942), who found the solidus, shown as an isotherm in our Fig. 5.1, to curve downward slightly towards the Di side of the diagram.

Summary of Basic Principles

Diopside and anorthite, two important components of basaltic magma, form an approximately binary system of the eutectic type whose lowest melting point, 1274°C, is more than 100°C below that of Di alone and more than 250°C below that of An alone. This temperature, where Di, An, and *L* all coexist, and where a unique composition ($Di_{58}An_{42}$) is completely liquid, is not far above those measured in natural basaltic lavas (1200 ± 30°C for many basalts). Intermediate mixtures of Di and An appear to form a reasonable, though greatly oversimplified, model of basalt. From this model, one learns that fractional crystallization does nothing to the ultimate composition of the liquid, although it can produce liquids approaching the "eutectic" composition which may be separated in space from their fractionated crystals. Fractional melting of diopside anorthite mixtures is capable of producing substantial quantities of "eutectic" liquid while leaving behind a highly refractory anorthite residuum which will remain unmelted in a further temperature rise of well over 250°C.

Incorporation of Al into diopside, almost surely as $CaAlAlSiO_6$, may help to account for a silica-oversaturation trend in some basaltic liquids. It also lends itself to a straightforward example of chemographic analysis in a multicomponent system, a technique which is of great value in thinking about both rocks and the complex model systems which represent them.

G-X Diagrams

The *G* stands for the Gibbs function or the Gibbs free energy, a function which is minimized at equilibrium. The partial molar free energy or chemical potential μ can be used as well. The *X* stands for composition, as before. In *G-X* diagrams, *X* is best plotted in mole units rather than weight units. The purpose of discussing *G-X* diagrams is multiple. In the first place, they give us an intuitive understanding of what phase diagrams are all about, an understanding which in fact can prevent us from drawing incorrect phase diagrams. In the second place, they emphasize the fact that phase diagrams are mere symbols of more profound thermodynamic truths, and they remind us that if the behaviour of *G* with composition is known as a function of *T* and *P*, then phase diagrams can be calculated a *priori*. Unfortunately this is not generally the case (although it is sometimes well worth doing), because very often the melting experiments are inherently more sensitive generators of information than the calorimeters used to find the Gibbs energy. On the other hand, much progress has recently been made in harmonizing the results from experimental studies into a large set of self-consistent thermodynamic data in which the principle of *G-X* diagrams, the minimization of the free energy at equilibrium, allows calculation of the compositions of *all the coexisting phases* (for example, by Ghiorso and Sack 1995). Here

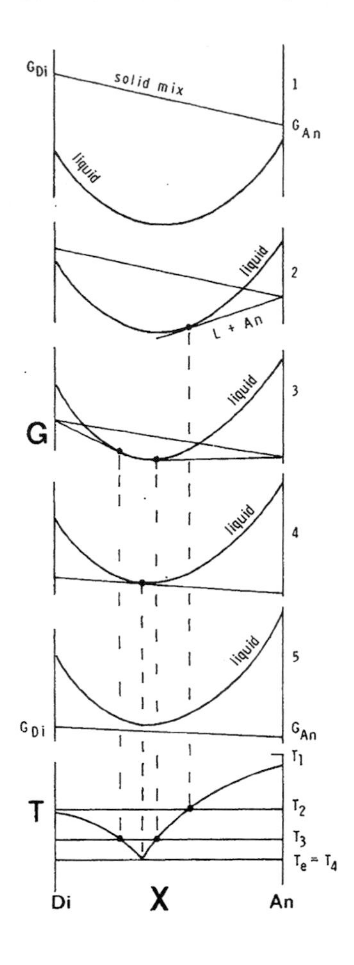

Fig. 5.8 Superimposed Gibbs energy (*G-X*) and *T-X* diagrams for the system diopside-anorthite, ignoring solid solution

we shall emphasize the use of *G-X* diagrams as intuitive guides to the nature of phase diagrams.

The bottom part of Fig. 5.8 is a schematic *T-X* diagram of Di-An. Above it are drawn five *isothermal* *G-X* diagrams for different temperatures marked on the *T-X* diagram (T_5 is not marked, but lies below T_4). In all five *G-X* diagrams, the Gibbs function for the liquid is described by a continuous, convex-downward curve. This curve changes position and shape with temperature, but is always convex down. It shows that the free energy of a melt of a pure compound is locally maximum, and is diminished by dilution with a second component, to a minimum at some special value of *X*. The free energy of a mechanical mixture of two pure solid compounds is directly additive, so it varies linearly as a function of molar *X* (for discussion see Zavaritskii and Sobolev 1964).

In part 1 at the top of Fig. 5.8, the temperature (T_1) is above the melting point of An and no crystals are present. The *G-X* diagram proclaims this because, at equilibrium, *G* is minimized, and for all *X* the melt has lower *G* than any combination of crystals. Moreover, note that *G* of An_L falls below *G* of An_S, wherein *L* and *S* signify liquid and solid. This means that anorthite melt is stable relative to anorthite crystals. At the melting point of An, the *G* values of *S* and *L* would coincide. If anyone wonders why the liquid and solid curves are placed as they are, the answer is, in Eddington's immortal words, because I put them there. They are drawn so as to be compatible with the known *T-X* diagram. *In principle* they could be drawn from experiments involving heats of solution and entropies, but in fact they are drawn to explain the observed geometry of the melting diagram, and they are schematic.

In part 2 of Fig. 5.8, representing temperature T_2, a tangent to the liquid curve can be drawn from the *G* of pure An crystals; for any point on this tangent, *G* for the system is minimized relative to *G* for either liquid or crystalline mixture alone. This signifies the stability of anorthite crystals with liquid of a certain composition given by the point of tangency. To the left of this point, liquid has everywhere a lower *G* than any combination of crystals, and the system there is evidently still liquid. The isotherm T_2 in the *T-X* diagram portrays the conventional statement of these facts. At T_3, two tangents to liquid occur, one each from diopside crystals and anorthite crystals. There are now two $S + L$ equilibria at this temperature, separated by a small region in *X* where liquid alone is stable. Finally, at T_4, the eutectic temperature, Di, An, and eutectic liquid all lie on a single line, signifying the invariant coexistence of these three phases. At T_5, somewhere not far below T_4, any mechanical mixture of solids is minimized relative to liquid, and no liquid occurs. Note that the free energy of the liquid is by no means undefined at T_5; it is just not minimized anywhere with respect to solids, so liquid is not stable.

Further examples of *G-X* diagrams are given in the succeeding chapters.

Chapter 6
Plagioclase: The System An-Ab

Introduction

The feldspar of common basalts is by no means pure anorthite, but instead an intermediate plagioclase near An_{50}. This means that we could much improve the Di-An model of basalts if we could study a model of Di-An_{50}, or in other words part of the ternary system Di-An-Ab. It would be folly to do this, however, without first examining the nature of all the binary combinations, An-Ab and Di-Ab as well as Di-An, already discussed.

The system An-Ab, or the plagioclase system, turns out to be a binary system with continuous and complete solid solution between the end members.[1] As such, it is a famous example of such a system, and an excellent one for illustrating the properties of such systems as well as for revealing some important geological principles, most notably that of fractional crystallization. The plagioclase system was first successfully studied by Bowen (1913), although it was the subject of the first investigation at the Geophysical Laboratory (Day and Allen 1905), a study remarkable for its pioneering efforts and frustrations before the advent of the quenching method.

Preparation and Melting of Albite

Anorthite has been discussed in Chap. 5; we need to discuss only albite before turning to the binary system. Albite is not a straightforward composition to synthesize, and its melting point determination encounters severe problems.

Albite, $NaAlSi_3O_8$, is related to anorthite, $CaAl_2Si_2O_8$, by the exchange reaction $[NaSi][CaAl]_{-1}$. It is thus more siliceous and less aluminous, and contains an alkali

[1] As far as we know this is strictly true at high temperatures and low to moderate pressures. It is not true at metamorphic temperatures, or at high pressures.

S. A. Morse, *Basalts and Phase Diagrams*,
https://doi.org/10.1007/978-3-030-97882-2_6

cation in place of an alkaline earth cation. Stated in terms of oxides, the composition is $Na_2O{\cdot}Al_2O_3{\cdot}6SiO_2$, or NAS 1:1:6. The synthesis is in principle like that of anorthite; the sources of silica and alumina are identical, and the source of Na_2O is Na_2CO_3 converted from the bicarbonate $NaHCO_3$. The first step is to make a sodium disilicate glass ($Na_2O{\cdot}2SiO_2$) from the carbonate and pure silica. This glass lies in the binary system Na_2O–SiO_2, and forms a compound whose melting point is agreeably low, 874 $\pm$ 3°C (Schairer and Yoder 1971). The mixture of carbonate and silica must be heated very slowly, over many days, to bring about the slow escape of CO_2; if the heating rate is too fast, an appreciable amount of the volatile Na will escape as well, and even with extremely slow heating, a small deficiency in final weight is usually noted, which is made up by an appropriate small addition of sodium carbonate. The final disilicate glass is crystallized, to avoid weighing errors due to the absorption of H_2O, since the glass is very hygroscopic.

Albite glass is made by weighing and mixing appropriate proportions of sodium disilicate, silica, and alumina ($Na_2O.2SiO_2 + 4SiO_2 + Al_2O_3$), and heating these slowly to the beginning of melting at 788$\pm$ 2°C, followed by heating to 1200C, with many successive fusions at 1500–1600C, separated by crushings, till all alumina is dissolved and the glass is optically homogeneous. The procedure is reported by Greig and Barth (1938) and with a discussion of alkali loss, by Schairer and Bowen (1956).

The melting point of albite is difficult to determine because of the reluctance of albite glass to nucleate crystals of albite, and the difficulty is encountered in melting albite crystals. The difficulties can be partly overcome by ordering the viscous melt for several months at successively lower temperatures above the melting point. Melts treated in this way will form sparse albite crystals when held for a few days at 1110°C, whereas freshly prepared melts are so disordered that no crystals will form even in 5 years at 1025°C (Schairer and Bowen 1956). The ordering process has been aptly described as "acclimation" by Schairer (1951), and appears to fall in the class of pre-freezing phenomena discussed by Ubbelohde (1965, Chapter 12), although some have ascribed the supposed ordering process to the introduction of heterogeneous nuclei during the crushing and reheating of the albite glass.

The high viscosity of albite melt not only slows down the ordering process in the melt, but also hinders melting, so that albite crystals can be superheated over short time scales. The melting point must therefore be determined by observation, in quenched charges, of whether the sparse albite crystals are growing (faceted) and therefore below the melting point at the temperature of the experiment, or dissolving (rounded) and therefore above the melting point during the experiment.

All of these troubles have persisted to the modern era owing to the difficulty of obtaining definitive thermodynamic data for the albite composition, including values for the compressibility of the liquid, with which to refine the difficult direct experimental results. The resolution of all these matters by Anovitz and Blencoe (1999) and discussed further by Lange (2003) results in a well-defined melting point at 1100°C at 1atm and 1245°C at 10 kbar (to be discussed much later!).

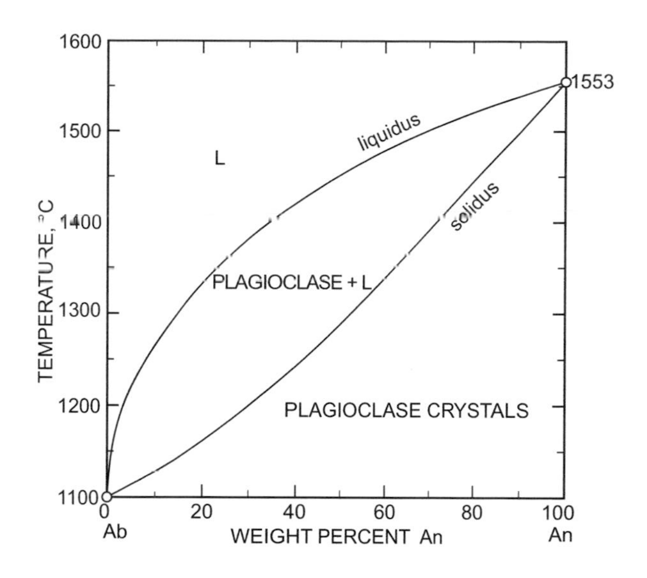

Fig. 6.1 Plagioclase melting relations: the system anorthite (An)-albite (Ab) (after Bowen 1913). The melting point of An is from Osborn (1942), and that of Ab from Lange (2003) as discussed in the text

Liquidus and Solidus

The phase diagram for melting in An-Ab, as determined by the quenching method, is shown as a *T-X* plot in Fig. 6.1. The upper curve is the liquidus curve, which connects the melting point of An at 1553°C to that of albite at 1100°C, and gives the composition of the liquid at any temperature. The lower curve is the solidus curve, likewise connecting the two melting points, which gives the composition of the crystalline solution of plagioclase feldspar at any temperature. Together, the two curves form what is often called a *melting loop*, or *S* + *L region*. The region above the liquidus is a one-phase region of liquid alone, and that below the solidus is a one-phase region of crystals alone. In the quenching method, the solidus is located by the beginning of melting, and the liquidus by the beginning of crystallization, so the only compositional control is that of each bulk composition. There was no direct measurement of crystal or liquid composition in Bowen's work.

Variance

The number of components is two (An, Ab), so at constant (e.g. atmospheric) pressure $W_p = 2 + 1 - \phi = 3 - \phi$. Above the liquidus or below the solidus, the system consists in one phase, *L* or *S*, and $W_p = 2$; in these regions, both temperature and composition must be specified to describe the system completely.

No single phase lies within the loop, but any point there represents a bulk composition at an isotherm, the intersection of the isotherm with the liquidus giving the liquid composition, and the intersection of the same isotherm with the solidus giving the composition of coexisting crystals. Both liquidus and solidus therefore denote two-phase equilibrium, and along these curves $W_p = 1$; these are univariant curves. Stipulation of a single variable, temperature, is sufficient to describe the system completely as consisting of a liquid, say An_{34}, coexisting with crystals, say An_{71}, at a temperature of 1400°C. The system is always univariant when both crystals and liquid are present; there is no invariant point between An and Ab. This means that fractional crystallization can exert a profound change on the composition of liquid, as we shall see.

Lever Rule

A distorted sketch of part of the plagioclase loop is shown in Fig. 6.2. The dashed vertical line labelled BC denotes the chosen bulk composition. At temperature T_1, the bulk composition lies at the liquidus, and the system is essentially all liquid, although saturated with a trace of plagioclase crystals. The length of the lever l represents the amount of liquid. At the lower temperature T_2, the system is about half crystallized, as the equality of levers s and l indicates; the fraction of liquid present is given by $l/(s + l) = 0.5$. At temperature T_3, the liquid is virtually gone, and the system is virtually solid, as indicated by the existence of lever s only.

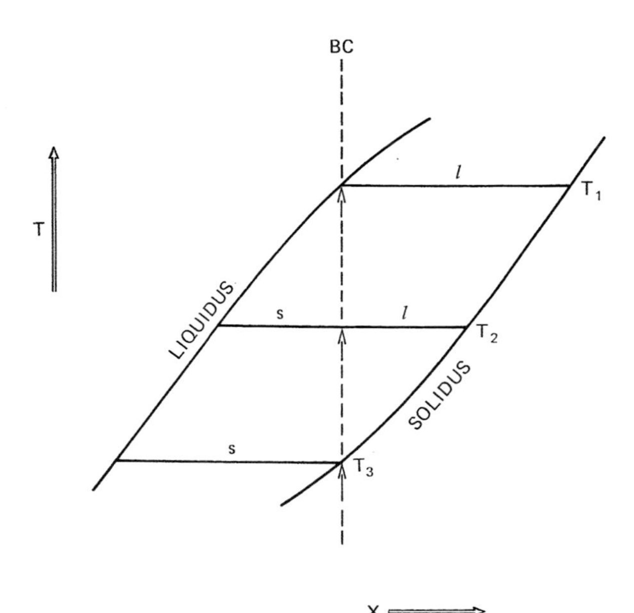

Fig. 6.2 Illustration of the lever rule in a binary loop

Equilibrium Crystallization

True reactive equilibrium crystallization requires continuous reaction between crystals and liquids. Slow diffusion in plagioclase crystals actually prohibits this reaction in the real world, so we must suspend our disbelief for a moment in order to discuss the idealized concept of equilibrium crystallization.

The An-Ab phase diagram is redrawn in Fig. 6.3 with a few isothermal tie lines to illustrate the course of crystallization of a bulk composition $An_{60}Ab_{40}$. Upon cooling from above the liquidus, a melt of this composition begins to crystallize at 1476°C, where it first attains the liquidus. The first crystals to form, at this temperature, have the composition An_{86}; clearly, they are very calcic, and far different from the liquid in composition. As heat is removed and the temperature lowered, the liquid composition changes towards Ab along the liquidus curve and the total crystal composition changes towards Ab along the solidus curve. The change in liquid composition results from the removal from the liquid of crystals more calcic than the bulk composition. At the same time, as the diagram shows, an increase of Ab in the liquid necessitates an increase of Ab in the crystals. There is a partition ratio $D = Ab_S/Ab_L$ which varies uniformly throughout the diagram, as plotted in the upper part of Fig. 6.3. When plotted against the solidus, this partition ratio forms an

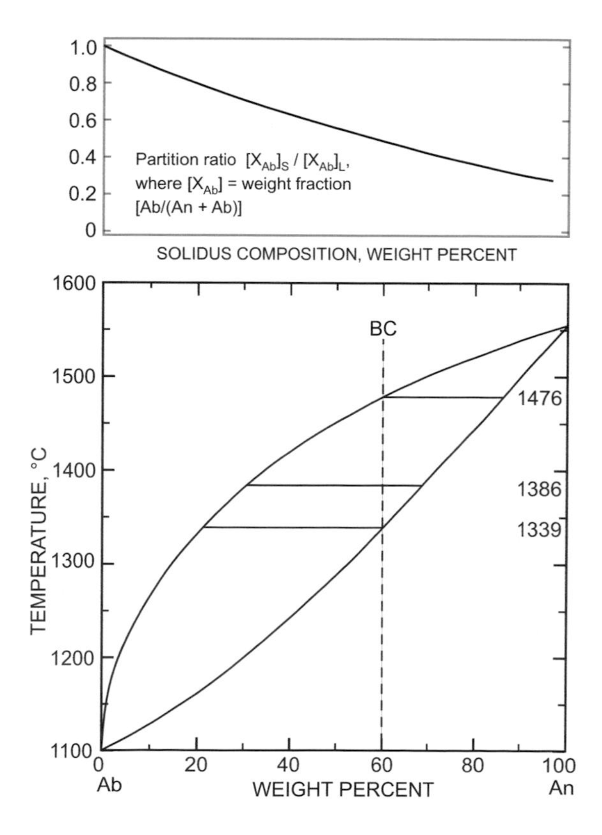

Fig. 6.3 Crystallization in the system An-Ab

almost straight line between the limits of a function called the *exchange coefficient* K_D at the An end to 1.0 at the Ab end. In this case K_D is defined as

$$K_D = \left[(\mathrm{Na/Ca})^S\right] / \left[(\mathrm{Na/Ca})^L\right] \tag{6.1}$$

and it has a value of about 0.3 at pure An. The relation between the *partition coefficient D* and K_D is given by

$$D^{S/L}_{X_{\mathrm{Ab}}} = \left[\left(K_D \cdot X^S_{\mathrm{An}}\right) + \left(X^S_{\mathrm{Ab}}\right)\right] \tag{6.2}$$

where X_{An} is the mole fraction An/(An + Ab). A linear relation for D against solidus composition (implying constant K_D) is found for many experimental binary solutions. This principle is called *linear partitioning in binary solutions* and there will be ample opportunity to explore it further as we go along.

The change of total crystal composition to more and more Ab-rich values, so as to maintain equilibrium with liquid, requires that early-formed calcic crystals *react* with liquid to change, in the solid state, to more sodic crystals. This reaction involves tetrahedral $\mathrm{SiAl_{-1}}$ exchange as well as octahedral $\mathrm{CaNa_{-1}}$ exchange in the crystal structure, and if equilibrium is to be maintained, the temperature of the system must fall slowly enough to permit thorough intracrystalline diffusion and reaction. As the size of the crystals increases, these processes will become more and more of a problem. In dry experimental and natural systems, it is easy to show that this reaction can never go to completion in plagioclase crystals of a few millimetres' size in all of geologic time. However, it can easily go to completion for $\mathrm{FeMg_{-1}}$ in such reactive crystals as olivine.

The idealized equilibrium course of crystallization is therefore a reaction course of crystallization, reaction of crystals with liquid proceeding in infinitesimally small steps as the temperature is lowered in infinitesimally small steps. When this happens, it is a nice example of what is called a thermodynamically reversible path.

Still pretending that the reaction can occur, the crystallization and reaction process continues, with falling temperature, until all the liquid is used up. As the final lever shows, this must happen the moment the total crystal composition has reached the initial bulk composition, $\mathrm{An_{60}}$ in our example, at a temperature of 1339°C and a liquid composition of $\mathrm{An_{20}}$. Any further lowering of temperature results only in the cooling of a mass of crystals having composition $\mathrm{An_{60}}$.

Fractional Crystallization

As may be surmised from the comment above concerning diffusion and reaction within feldspar crystals, it would not be totally surprising if the temperature were to fall too fast for reactions to proceed to completion. Such a process would result in zoning of the feldspar crystals, a feature so common in igneous rocks that its absence is remarkable. The failure of reaction leads, in one degree or another, to *fractional crystallization* of a profoundly different nature from that in a binary eutectic system such as Di-An. Fractional crystallization, by zoning or physical removal of crystals (perhaps by floating or sinking), is probably never a perfect process in nature, but it may approach perfection, and an idealized perfect model is amenable to theoretical treatment and hence useful as a limiting case.

Starting again with a melt of bulk composition An_{60}, crystallization again begins at 1476°C with the separation of crystals of An_{86}. With perfect fractional crystallization these crystals never react in the least with the liquid, but are isolated; let us say we put them in an imaginary bucket. With falling temperature the next batch of crystals is slightly more sodic than the first, but it also fails to react with the liquid. At any instant, the crystals just being formed have the equilibrium composition given by the solidus, so fractional crystallization is a process involving thermodynamic equilibrium on a microscopic (crystal) scale, although not on a macroscopic (bucket) scale. The fractionally removed crystals constitute the instantaneous solid composition (ISC), and the locus of the ISC with time and falling temperature is the solidus curve, which is an *ISC path*. The isolation of the ISC prevents the exchange reaction $[NaSi][CaAl]_{-1}$ from occurring. Failure of this reaction therefore conserves Na, Si (or normative Ab) in the liquid and causes the maximum possible depletion of Ca, Al (or normative An) in the liquid. The total solid composition (TSC) resulting from fractional crystallization is a numerical construct, whose value is given by the sum over all ISCs so far produced. The TSC may be determined algebraically, and the lever rule applied, as shown later in this chapter. The important thing to note for present purposes is that the TSC always lags behind the ISC. By the time the liquid has reached such a temperature and composition as to produce crystals of An_{60}, the total solid composition is still far more calcic than this. A system is never fully crystallized until the TSC equals the bulk composition. Since the total solid composition is more calcic than the bulk composition, it is clear that a good deal of liquid remains, having a composition of An_{20}, whereas in the equilibrium process the last of the liquid was just disappearing at this composition. Continued perfect fractional crystallization results in the production of more and more sodic crystals, in equilibrium with more and more sodic liquid, until both liquid and crystals reach the composition of pure albite, at which point the liquid is used up. The total solid composition produced by this process is An_{60}, the bulk composition, but the actual crystals produced range from An_{86} to An_0 in such a distribution as to yield a mean composition of An_{60}. The liquid composition during the process has ranged from An_{60} to An_0. The liquid has continually decreased in quantity, but the very Ab-rich fractions may amount to an appreciable part of the original. Quite logically, the amount of Ab-rich residual liquid varies directly with the amount of Ab in the starting bulk composition. If a sample of the fractionating liquid were to be drawn off and isolated late in the process, a significant amount of albite-rich rock might be produced.

Fractional crystallization is thus capable of producing a range of liquids all the way from the initial composition of the melt to pure albite, a result distinctly different from that of equilibrium crystallization. The presence of very albite-rich plagioclase in interstitial patches in basaltic rocks shows that the fractional crystallization process operates with considerable efficiency in nature, and leads to the proposition that at least some acidic rocks (those high in silica and alkalies) may have been produced by fractional crystallization.

Equilibrium Melting

Equilibrium melting follows the reverse of the path of equilibrium crystallization. For crystals of An_{60}, melting begins, on heating, at a temperature of 1339°C with the production of a liquid An_{20}. With further heating, both the melt and the crystals change composition towards An, along the liquidus and solidus curves respectively, and the process terminates at 1474°C, when the liquid attains the composition An_{60} and the last crystal to disappear has the composition An_{86}. If the system were examined at any instant of time during the process, it would be impossible to tell whether a melting or a crystallization process was taking place.

Fractional Melting

If, at the instant of melting, each fraction of liquid could be removed from the crystalline assemblage, a fractional melting process would result, in which the remaining crystals would be continuously changed towards anorthite, which they would finally reach. In our example, the mean composition of all the liquid generated would be An_{60}, the bulk composition, but the successive fractions would range from An_{20} at the start to An_{100} at the end. Continuous removal of liquid modifies the lever relationship. Each fraction of liquid is generated by enriching residual crystals in Ca, Al. The next fraction of liquid is generated at a higher temperature, and behaves as though the melting process were beginning all over again. The succession of crystal compositions, which are left as residua, changes along the solidus. There are no hiatuses in the melting process. Completion of melting is not achieved at 1476°C, as it was in equilibrium melting, but only at 1553°C, the melting point of anorthite.

The effect of fractional melting is to leave behind crystalline residua, and over a certain range (An_{60}-An_{86}) these would be indistinguishable from settled crystals produced on fractional crystallization. However, the residua produced on fractional melting attain more calcic compositions than could possibly be produced by crystallization of an An_{60} melt, although only at very high temperatures.

Assimilation

The plagioclase system furnishes a fine opportunity for discussing the possibilities of changing the composition of liquids by the assimilation of crystals. As an exercise, imagine a half-crystalline mixture, say a bulk composition of An_{50} at a temperature of 1386°C, into which an extraneous crystal of albite is now introduced. The albite crystal, whose melting point is 1100°C, will clearly melt, changing the liquid slightly towards albite in composition. The calories required for this process will, to a first approximation, be a large number corresponding to the heat of fusion of albite, and a smaller number corresponding to the specific heat of pure albite melt over the range—from 1100°C to nearly 1386°C. Presuming these calories to be furnished by the melt, the process will occur only by virtue of the crystallization of some calcic feldspar, nearly An_{69} in composition. The latent heat of crystallization of calcic plagioclase must furnish the latent heat of fusion of albite and an additional increment of heat for warming the newly formed albite liquid. The temperature at the end of the process will be slightly lower than it was before the albite is added, as is implied by the formation of some calcic plagioclase crystals. Clearly, contamination of an intermediate melt by albite can change the composition of the residual liquid, but it cannot increase the quantity of liquid. The process is by nature self-defeating, requiring the crystallization of a greater mass of liquid than can be gained by the melting of the albite. One may infer that contamination of magma by less refractory material is a likely process near intrusive contacts, but one which accelerates the solidification of the magma. In our example, we have assumed the contaminant to be at its melting temperature, whereas in nature the contact rock would be a good deal colder, and much heat would be lost by the magma in raising the albite to the melting temperature. The process of *assimilation* with accompanying *fractional crystallization* long ago recognized by Bowen has now become familiar under the name of an *AFC* process.

The assimilation of more calcic feldspar is a somewhat different proposition. Again assuming $T = 1386°C$, $BC = An_{50}$, $L = An_{31}$, $S = An_{69}$, suppose a crystal of exactly An_{69} is introduced, having previously been raised to 1386°C. Such a crystal is in equilibrium with the liquid and the other crystals at this temperature, so no reaction will take place. The sole result is that the bulk composition of the system is now changed slightly towards An, so the crystallization process will terminate at a slightly higher temperature than before, the liquid having to expend some excess energy on making over the new crystal of An_{69} as crystallization proceeds. Now suppose instead that a crystal of An_{80} is introduced. This crystal, if melted, would have been in equilibrium with a liquid of An_{50} at a temperature of 1450°C; it is clearly not in equilibrium with our liquid An_{31} at 1386°C. The crystal will react with the liquid, giving up Ca, Al in exchange for Na, Si. This will enrich the liquid in Ca, Al, and the phase diagram tells us that this will require a total crystal composition slightly more calcic than An_{69}, and a temperature slightly above 1386°C. Evidently, the reaction of *hot* An_{80} to a less calcic composition takes place with the evolution of heat; it is an exothermic reaction. In adding a calcic crystal, we have moved the bulk composition towards An, but in remaking it, we have also used up some liquid. The resulting ratio of crystals to liquid is now somewhat greater than it was before, and this assimilation process is, again, a self-defeating one in that it uses up liquid.

The effect of the above process on the crystal-liquid lever is not evident without a quantitative example. The following example is adapted, with slight numerical changes, from Bowen (1928, p. 186), which should be consulted for a more complete discussion. Fifty grams of melt having the composition An_{50} is cooled till the bulk composition just intersects the liquidus, at 1450°C, with formation of a few crystals of An_{81}. At this point, 50 g of crystals of An_{90}, also heated to 1450°C, are added. If equilibrium were reached isothermally, the result would be crystals of An_{81}, almost all of which were made over from An_{90}, and a liquid of An_{50} composition. The bulk composition, however, is now An_{70}, and the lever rule gives $F_L = 11/31 = 0.355$, $F_S = 20/31 = 0.645$. This loss of liquid ignores the heat effect. The present crystalline mass contains $64.5 \times 0.81 = 52.25$ g An, and $64.5 \times 0.19 = 12.25$ g Ab. The original crystals of An_{90} contained 45 g An and 5 g Ab. Taking Bowen's latent heats of fusion of An and Ab as 104.2 and 51.7 cal/g, respectively, we may calculate the heat effect of the isothermal process. We have gained from the liquid 14.5 g of solid, namely 7.25 g An and 7.25 g Ab, and we have therefore gained $7.25\ (104.2 + 51.7) = 1130$ cal, which would have to be removed to keep the system at 1450°. If instead this heat remains within the system, we may calculate the temperature rise, assuming a heat capacity for the system of 0.32 cal/g and using the latent heats of fusion stated above. This temperature rise comes out to about 10°C, hence the final temperature would be 1460°C. At 1460°C the crystals have a composition of An_{83} and they amount to 56 g, with liquid ($An_{53.5}$) amounting to 44 g. We have thus used 6 g liquid to convert 50 g An_{90} crystals to 56 g An_{83} crystals. Clearly, the reaction of calcic crystals with liquid is such as to affect the crystalline material more strongly than the liquid. The liquid becomes only slightly more calcic while decreasing appreciably in quantity. Assimilation of foreign material affords little prospect of accounting for a widespread variety of magmas, although it nicely accounts for many effects observed near contacts.

The Lever Rule with Fractional Crystallization and Melting

Use of the lever rule requires knowledge of the total solid composition (TSC), the bulk composition (BC), and the liquid composition (*L* in crystallization, TLC in fractional melting). In fractional crystallization, the TSC is not given directly by the phase diagram, and in fractional melting, the TLC is not given. The problem of using the lever rule with these fractional processes therefore reduces to the problem of finding the TSC or the TLC. This problem can be solved in various ways. The first method chosen here is to apply the Rayleigh fractionation equation with variable exponent and solve it iteratively. This historical formalism will serve as a preview to a more elaborate treatment, linear partitioning in binary solutions, to follow later.

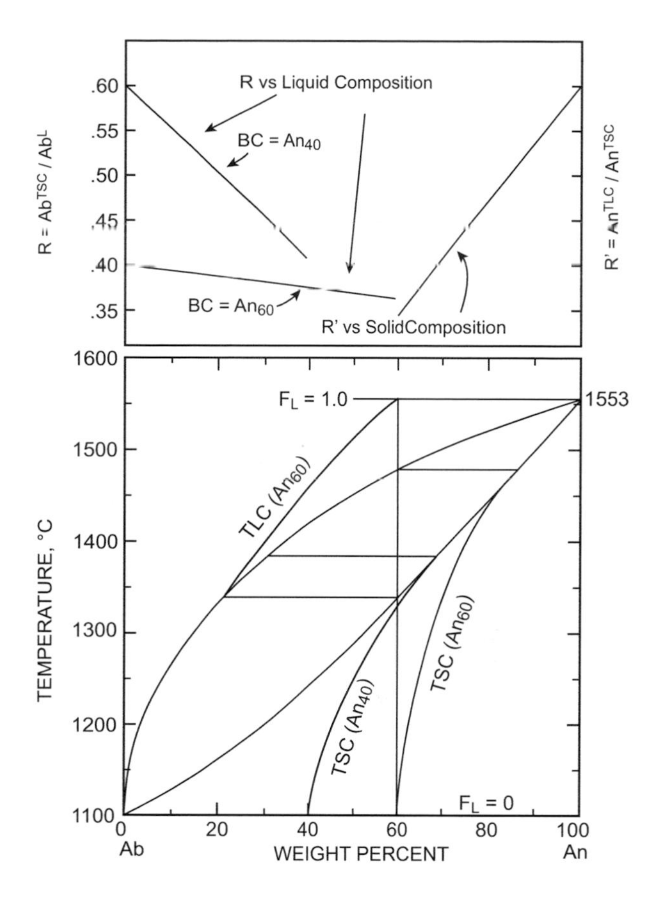

Fig. 6.4 Examples of fractional crystallization in the plagioclase system

The Rayleigh (1896) equation can be written for the purpose at hand

$$C = C_0 \cdot F^{(D-1)} \tag{6.3}$$

where C is the liquid composition, C_0 is the initial liquid composition, F_L is the fraction liquid, and D is the Nernst partition coefficient, C_S/C_L. For fractional crystallization, it is convenient to assign the mole fraction of Ab component as the composition variable. The value of D then ranges from 0.3 in the limit at pure An to 1.0 at pure Ab, as shown in the top part of Fig. 6.3. As numerical fractionation proceeds, the value of C_0 is continually reset from the inverse relation $C_0 = C/F_L{}^{(D-1)}$, using the current values of C and D. The TSC is found at each stage of the calculation from the relation

$$X_{Ab}^{TSC} = X_{Ab}^{L} - \left[\frac{\left(X_{Ab}^{L} - X_{Ab}^{BC}\right)}{(1 - F_L)}\right] \tag{6.4}$$

Figure 6.4a illustrates several examples of fractional crystallization. The initial and final points on the TSC path are known; the first is the solidus composition, in the case of the right-hand path $An_{85.5}$, and the last is the bulk composition, e.g. An_{60}, because the total of all crystals produced must equal the bulk composition. The final TSC is achieved only at 1100°C, the temperature of final crystallization. It is noteworthy that the TSC path drops steeply at first and then runs almost horizontally to the bulk composition as large amounts of liquid very near pure albite crystallize ISCs that are also very near pure albite. The paths illustrated in Fig. 6.4a are sufficient to estimate the behaviour of other TSC paths by interpolation between the examples. A lever is shown in the figure for bulk composition An_{50} at about 1370°C: it yields $F_L = 0.70$.

In the case of fractional melting, it is the TLC path that is desired, the TSC path being the solidus. Again, the initial and final points are known, the first being on the liquidus and the last being at the bulk composition at 1553°C, because the sum of all liquids produced must equal the bulk composition just as the last trace of liquid is formed at 1553°C. The TLC path between the initial and final points is found by a Rayleigh calculation analogous to that used for fractional crystallization, using X_{An} as the composition variable and using the ratio of the mole fraction of An in the liquid to that in the solid as the value of D. Examples for three bulk compositions are shown in Fig. 6.4b. Again, a substantial fraction of liquid near the composition of the conserved component is produced, although this effect is somewhat smaller than in the case of fractional crystallization because the plagioclase loop is narrower near the An end.

More detailed discussions of binary solutions and the lever rule with fractional crystallization and melting are given in Morse (2000, 2015).

G-X Diagrams

In Fig. 5.8 of the last chapter, we saw a conventional explanation of the *G-X* relations in a binary eutectic system. The development of such an analysis for binary solid solutions originated with H.W. Bakhuis Roozeboom around 1890, and was given its present graphical treatment by Van Rijn van Alkemade (1893). Among the earliest users of such diagrams in geology were Day and Allen, whose monograph with Iddings on "The isomorphism and thermal properties of the feldspars" (1905) was the first product of the young Geophysical Laboratory of the Carnegie Institution of Washington. This is a remarkable paper because although the authors determined only the liquidus of the plagioclase system, they were able to conclude through the use of Roozeboom's principles that it was indeed a complete binary solid solution series. In the course of doing so, they gave a very fine exposition of *G-X* diagrams. Let us now turn to the plagioclase system and Fig. 6.5. In this Roozeboom Type I system, we can no longer speak of a mechanical mixture of pure crystals, and so we have two solution curves in *G-X* space, one for liquid and one for solid. Crystallization begins at T_1 with the coincidence of S and L curves for pure An. For all other

Fig. 6.5 *G*- and *T*-*X* diagrams for the plagioclase system

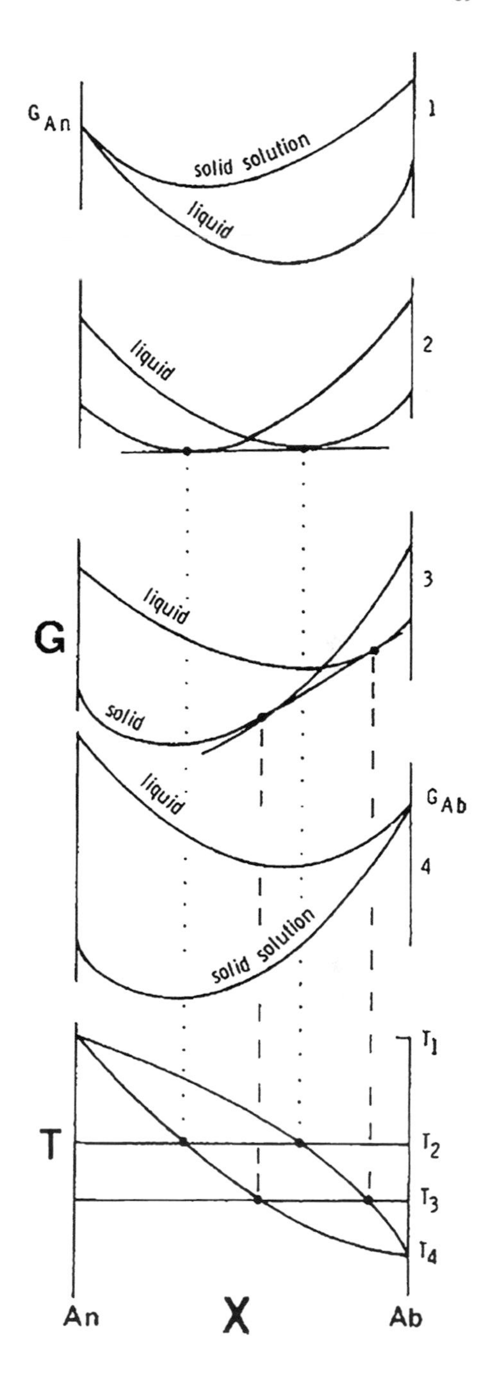

values of X, liquid is stable relative to solid; minimum G is achieved on the L curve relative to the S curve. At T_2, minimum G is obtained on a tangent to the two curves, and any point on this tangent signifies coexistence of crystalline solution and melt. The composition of the crystalline solution is given by the tangent on the solid curve; that of the melt by the tangent on the liquid curve. To the left of the crystal tangency, crystalline solution is everywhere minimized in G relative to liquid; this part of the system is solid. To the right of the liquid tangency, the liquid curve minimizes G; this part of the system is liquid.

At T_3, the picture is qualitatively the same, but the points of tangency have migrated further towards Ab. Finally, at T_4, the curves unite at the melting point of pure Ab and the rest of the system is solid. At lower temperatures, the curves become farther apart. It should be noted both for this system and for the previous one that the absolute values of G change directly and continuously with temperature, but at different rates for solids and liquids. This is what causes the solid curve to "overtake" the liquid curve with falling temperature. It is also worth noting here that despite the experimental difficulties of locating the solidus accurately, see what a problem it would be (at T_3 for example) to locate the tangent points accurately by correct placement of the G-X curves. This is why phase diagrams are often a powerful source of thermodynamic data.

Systematics of Binary Solutions

Here we extend the idea of making straight lines that stand as code for the display of a complete binary loop, using *linear partitioning in binary solutions*. The derivation of this idea is mentioned above at Eqs. (6.1) and (6.2); it is more fully described in a review paper by Morse (2015). The principle is found to apply to most binary solutions of petrologic interest, and even to binary solutions within multi-component systems. The only really interesting deviation from linearity among common petrologic binary solutions is that of the plagioclase feldspars! Nevertheless, it seems appropriate to explore the general case here.

For a solid phase S and liquid phase L we define the *partition coefficient* D set always to be ≤ 1.0 and hence

$$D_{X_1}^{S/L} = X_1^S / X_1^L \tag{6.5}$$

where the *mole fraction* X_1 is the normalized fraction of the low-melting component (1), i.e.

$$X_1 = n_1/(n_1 + n_2) \tag{6.6}$$

where n is the number of moles and component (2) is the high-melting component. The notation may apply in other cases with different names, to units of mass, gram-atoms, or oxygen-normalized cations. Now generalizing Eq. (6.1), the exchange coefficient K_D becomes

$$K^{S/L}_{D(n_1/n_2)} = \frac{(n_1/n_2)^S}{(n_1/n_2)^L} \tag{6.7}$$

and using this, we obtain

$$\begin{aligned} D^{S/L}_{X_1} &= K^{S/L}_{D_{1/2}} \cdot \left(1 - X^S_1\right) + X^S_1 \\ &= \left(K_D \cdot X^s_2\right) + X^S_1 \end{aligned} \tag{6.8}$$

which is the linear partitioning equation relating the partition coefficient to the exchange coefficient via the crystal compositions.

This relationship is conveniently used to plot D against X^S_2 and to test for linearity. When that occurs, the value of K_D is given at the intercept where $X^S_2 = 1.0$. For plagioclase as calculated by Bowen (1913), using his value of the latent heat of fusion for anorthite, ΔH_f(An) and for albite ΔH_f(Ab) from Robie et al. (1978), the result for plagioclase is as shown in the top part of Fig. 6.3. Instead of a single straight line, the data are best fit by two straight lines with a node at An_{60}. This detailed version along with further discussion leading to a single but curved line of partitioning is given by Morse (2015), along with other historical details. Most of the other binary solutions discussed in this book have straight-line partition slopes. The linear partitioning principle can be useful for calculating fractional crystallization histories.

Historical Note: A Bowen History Updated

In 1913 N. L. Bowen published "The melting of the plagioclase feldspars," a treatise that is the rock-bottom foundation of all modern igneous petrology. In this document, Bowen not only determined the liquidus and solidus temperatures of the plagioclase loop, but also used the dynamics of the day to analyse the result. In this exercise, he proceeded not merely to assume the provisional enthalpies of fusion of An and Ab, but ultimately to *extract* these values from the experimental array. He then turned around and conducted an error analysis (Bowen's Table V) by changing each value of the enthalpy of fusion by 10% and recalculating the implied temperature for each of the liquidus and solidus. In this exercise, he showed that the result of the 10% variation was "well beyond the limits of error of the temperature measurements".

In passing, it is worth our notice to cite two of Bowen's classic statements, as follows (both from p. 590):

> Van Laar derives a more rigid relation which contains factors for the heats of mixing in both phases, but in its application he finds it necessary to neglect these and the equations then reduce substantially to the form given above.

And again,

> If the calculated mean molal latent heat of melting of anorthite is divided by the formula weight, the result is $29000 \div 278 = 104.2$. Åkerman and Vogt have found by direct measurement that the latent heat of melting of anorthite is 105 cal. per gram, which agrees well with the calculated value. The extraordinary agreement is, of course, in part pure accident.

NB "in part"!

Bowen's Figure 3 shows the calculated liquidus and solidus curves for plagioclase along with the experimental data. All the points lie on the lines except three solidus points close to the Ab end member, which lie at lower temperatures than observed. A new melting point for Ab will bring the solidus about to the Bowen data.

Present Work

The current exercise to find a modern working diagram for plagioclase is basically that of Bowen (1913) with a major modification in the melting point of Ab from 1118°C of Schairer and Bowen (1956) to 1100°C from Anovitz and Blencoe (1999; see discussion by Lange 2003). The melting point of An (1553°C instead of Bowen's 1550°C) is taken from Osborn (1942). The liquidus and solidus curves were first recalculated from the latent heats of fusion using Bowen's value for ΔH_f (An) = 121.3 kJ and the value of ΔH_f (Ab) = 56.6 from Navrotsky et al. (1980). The *liquidus* curve so calculated fits Bowen's data almost exactly, especially at An_{11} with the new melting point for Ab at 1100°C. The *solidus* curve also benefits from the new end point and forms a continuous curve instead of the former straight line at high values of An followed by a curve to the Ab end point (e.g. Morse 1994, p. 63).

There is an objection to the use of modern data for the latent heat of fusion when they were determined on glass (quenched melt) which is frozen at the glass transition. In consequence, the heat of fusion calculated from solution calorimetry of glasses, as done in the recent literature including Robie et al. (1978), is not the ΔH of melting (e.g. Bottinga and Richet 1986). There is, however, a purely empirical approach in which we may test the results of new data inputs against the experimental T-X data of Bowen (1913) as modified by the new melting temperature at pure Ab. This procedure reflects the original use of the phase diagram from which Bowen extracted the heats of fusion, as discussed above.

Still using Bowen's ΔH_f (An) = 121.3 kJ, and the new end point at Ab, the linear partitioning plot is linear in two segments joined at An_{60} and a K_D value of 0.3. However, using instead an updated value of ΔH_f (An) = 134.6 kJ (Navrotsky et al. 1980), the partitioning line becomes a single curve with $K_D = 0.276$. The data near the An end of the array are typically scattered because an error of only 0.001 in X_{An} has a large effect on the result.

This marriage of old and new data continues the Bowen's principle of testing for a thermodynamic quantity by fitting to the experimental data. In this comparison, there is no *empirical* evidence for rejecting the 1980 result, despite its origin from glass. As a result, we have a model for plagioclase alone to compare with more complex bulk compositions. The value of $K_D = 0.276$ is not greatly different from the 0.26 of the diopside-saturated plagioclase system.

Chapter 7
Diopside-Albite: A Complex System

Introduction

It would be folly to suppose that diopside-albite is a binary system. We have already found, in Di-An, that diopside shows incorporation of Al when that element is present. In this case, it is to be expected that diopside will steal Al from albite. The bad news does not end with this, for the albite will also steal Ca from diopside to make plagioclase, and the resulting effects on residual liquids are both important and difficult to analyse.

The principle that pure albite does not crystallize from liquids containing Ca is one enunciated by Bowen (1945, p. 88) and called the "plagioclase effect". The plagioclase effect continues to operate after the original endowment of normative anorthite is used up, and when this happens (presumably by theft of Ca from normative augite) the inevitable result is to produce an excess of alkali silicate in the liquid, as we shall see.

Excess of alkali over the normative amount which is required to form feldspar is a situation dependent on the alkalies/alumina ratio, and when it occurs, liquids are said to be *peralkaline*. When an excess of silica also occurs, as in the case at hand, liquids are said to be *acid*[1] as well as peralkaline; rhyolitic rocks of this nature include pantellerite and comendite. Analysis of the system diopside-albite furnishes, at least in a general way, insight into how such peralkaline acid trends may develop.

And it may be said that when it comes to natural plagioclase fractionating with pyroxene, the degree of fractionation depends to some degree on the type and amount of the pyroxene being crystallized, and that, in turn, is a function of the silica activity. Said it may be; now to be ignored.

[1] The term "silicic acid" for silica is archaic and has nothing to do with hydrogen ion concentration. A better term would be silicic.

S. A. Morse, *Basalts and Phase Diagrams*,
https://doi.org/10.1007/978-3-030-97882-2_7

Pseudobinary *T-X* Diagram

The phase relations of diopside-albite have been studied by Schairer and Yoder (1960), who plotted the results in a *T-X* diagram (Fig. 7.1) which may be termed pseudobinary. Several curious features of this diagram immediately reveal its nonbinary nature. Although the liquidus curves intersect more or less sharply, their junction does not constitute an isobaric eutectic, for a substantial range of crystals + liquid remains below this point. In the albite-rich portion of the diagram, the plagioclase liquidus rises from that of pure albite (1118°C, following Schairer and Bowen 1956) to a maximum of 1148°C, and then falls to 1133°C at $Ab_{91}Di_9$ where it joins the diopside liquidus. The plagioclase liquidus then continues to fall in temperature, where covered by the diopside + liquid field. The phase diagram is to be read directly, without the usual implications of binary diagrams. At a composition of $Ab_{75}Di_{25}$, for example, the diagram simply announces the facts that, on cooling, Di crystals appear at 1226°C, that these are joined at 1122°C by plagioclase crystals, and that complete solidification does not occur until somewhere around 1050°C. To the left of $Ab_{91}Di_9$, taking as an example $Ab_{95}Di_5$, plagioclase crystals appear, on cooling, at 1148°C, and are joined at 1096°C by diopside crystals; complete solidification (beginning of melting) occurs only at 1028°C.

Application of the phase rule to this situation quickly reveals the nonbinary nature of the system. Supposing it were binary, $c = 2$, and $W_p = 3 - \phi$, so that a three-phase assemblage of diopside, plagioclase, and liquid should be invariant. Instead, the diagram shows a range as large as 90°C over which this three-phase assemblage occurs. Moreover, the assemblage shows a wide compositional range, so it must be at least divariant. The components therefore must number at least four. The reasoning below suggests in fact that five components are required for adequate representation of all phases in this system.

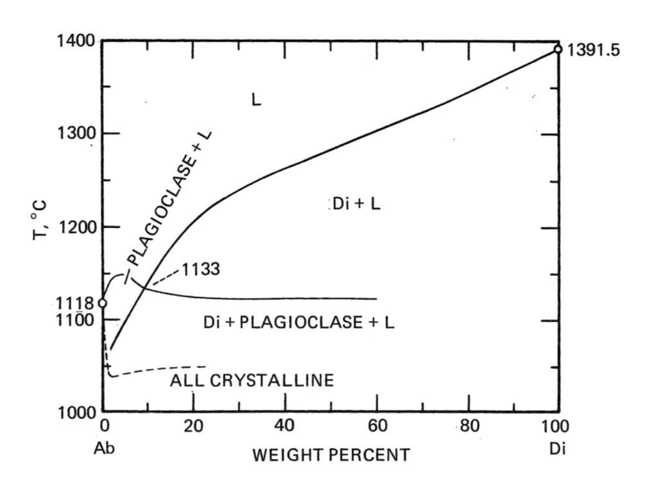

Fig. 7.1 The system diopside-albite, after Schairer and Yoder (1960). This is not a binary system, but merely a *T-X* "road map" of phase relations encountered along the Di-Ab composition line

Components and Liquid Trends

The system diopside-albite contains the elements Na, Ca, Mg, Al, Si, and O. We have no particular reason to believe that oxides would not serve as well, so we may choose as our first set of components the oxides Na_2O, CaO, Al_2O_3, MgO, and SiO_2. This is like our old friend CMAS (Chap. 5) with N (i.e. Na_2O) added. These oxides must surely suffice for describing the system, but they may not be the minimum valid set. To choose the most appropriate components, we should give some attention to the expected compounds and the effect on the liquid of removing them.

Let us assume that the feldspar produced in this system is rigorously binary, a member of the Ab-An series. Let us further assume, for the moment, that Al enters diopside in the form of CaTs, $CaAl_2SiO_6$. We shall then need Di, CaTs, Ab, and An to represent the compositions of the crystals. Removal of such crystals from bulk composition which lie rigorously in Di-Ab will result in liquids depleted in CaTs and An, and our object is to inquire how the compositions of such liquids may be represented. Without worrying about quantitative considerations, we want to know qualitatively in what sort of chemographic space these liquids will lie. To discover this, we can make a few arbitrary subtractions, just as done with the ternary analysis of Di-An.

Using the abbreviations *C*, *A*, *M*, *S*, *N* for the oxides, and choosing a bulk composition of 2 Di + 4 Ab, we get the initial oxide composition shown in the top row of Table 7.1. From this we subtract one mole of CaTs, $CaO{\cdot}Al_2O_3{\cdot}SiO_2$ (1:1:1), and then one mole of An, $CaO{\cdot}Al_2O_3{\cdot}2SiO_2$ (1:1:2). The result can then be expressed in terms of enstatite, sodium disilicate, and quartz (or 2 enstatite, 2 Na_2O and 11 SiO_2). Evidently, the entry of CaTs and An into crystals must lead to enrichment of the liquid in MgO, Na_2O, and SiO_2.

The liberation of MgO and SiO_2 suggests the further possibility, indeed likelihood, that the pyroxene is not merely a solid solution towards CaTs from Di, but also

Table 7.1 Effect of removal of Ca-Tschermak's molecule and anorthite from a bulk composition in Di-Ab

	C	*A*	*M*	*S*	*N*
Initial, 2 Di + 4 Ab	2	2	2	16	2
Less CaTs	−1	−1		−1	
	1	1	2	15	2
Less An	−1	−1		−2	
Result	0	0	2	13	2
Less 2 En			−2	−2	
	0	0	0	11	2
Less 2 disil.				−4	−2
				7	
Less 7 Q				−7	
				0	

CAMSN are the oxides of Ca, Al, Mg, Si, and Na. CaTs is $CaAl_2SiO_6$, An is $CaAl_2Si_2O_8$, En is $MgSiO_3$, Disil. is $Na_2Si_2O_5$, and Q is Quartz. For discussion, see text

Table 7.2 Effect of removal of Mg-Tschermak's molecule and anorthite from a bulk composition in Di-Ab

	C	*A*	*M*	*S*	*N*
Initial, Di + 4 Ab	1	2	1	14	2
Less MgTs		−1	−1	−1	
	1	1	0	13	2
Less An	−1	−1		−2	
Result	0	0	0	11	2
Less 2 disil.				−4	−2
				7	0
Less 7 Q				−7	
				0	

Abbreviations as in Table 7.1

a solid solution towards En, as we know natural augites to be. With a view to supplying Ca for plagioclase, we could also treat the Al in pyroxene as Mg-Tschermak's molecule, $MgAl_2SiO_6$, instead of CaTs. Table 7.2 shows the result of this assumption, starting with an initial bulk composition of 1Di + 4Ab. The result is similar to that of Table 7.1 except for the absence of MgO in the residual liquid.

Natural diopsidic augites commonly show CaTs when their analyses are recalculated, but Mg is usually recalculated as En. The model of Table 7.1 is probably somewhat more realistic than Table 7.2, if the "released" $MgO{\cdot}SiO_2$ is considered as residing in the pyroxene rather than the liquid. Both models leave little doubt that the liquid will be enriched in Na_2O and SiO_2, as a result of the incorporation of Al into pyroxene and Ca into plagioclase. A little consideration will show that this conclusion is valid for crystals composed dominantly of Di and Ab, instead of CaTs and An, since addition of diopside and albite to both sides of the equations represented by Tables 7.1 and 7.2 has no effect apart from dilution.

In order for a quantitative treatment to be given, the true compositions of experimental pyroxene and plagioclase crystals would have to be known. To date, these have not been explored in detail. Without such data, it is difficult to do much more with this system. Schairer and Yoder (1960) found no evidence for phases other than diopsidic pyroxene and plagioclase in runs thought to be completely crystallized. This could not be correct, according to our analysis, unless excess Na_2O and SiO_2 were taken up in one of the crystals, presumably plagioclase. Instead, it may be that small amounts of sodium silicate and quartz are present but undetected in the runs. Alternatively, undetected glass might still be present.

Petrologic Considerations

The system diopside-albite forms an important part of our model for basalts, the basalt tetrahedron, being a part of the critical plane of silica undersaturation, which is the ternary system Di-Fo-Ab. The critical plane is so called because it divides the two regions of undersaturated model rocks, and it is an important tenet of the classification of Yoder and Tilley (1962) that this plane has genetic as well as taxonomic significance. The critical plane (actually some nearby plane) may not be penetrated by liquids during fractionation at low pressures, and this notion is strongly implied by successions of natural lavas which tend to remain solely on one side or the other of a natural analogue of the critical plane.

To show experimentally that the critical plane has this genetic function, it is thought sufficient to show that the plane is a thermal barrier, which means that liquidus temperatures should fall, away from the plane, either towards silica or towards nepheline. If this were so, fractional crystallization could only drive liquids away from the plane, not through it.

Schairer and Yoder (1960) have concluded that the system is in fact close to a thermal barrier, because when small amounts of either nepheline or silica components are added, liquidus temperatures are lowered, and fractional crystallization must drive liquids away from the plane. Moreover, complete crystallization of such mixtures shows either nepheline or a silica mineral as crystalline phases, depending on whether the mixture was under- or over-saturated, respectively. Experimentally, therefore, Di-Ab is a thermal barrier, and Di-Ab-Fo may then also be one.

The experimental evidence is not completely in harmony with our theoretical analysis of the course of residual liquids in Di-Ab. The former suggests Di-Ab as a thermal barrier, but the latter suggests that bulk compositions in Di-Ab must yield liquids rich in soda and silica. If the theoretical analysis is correct, it implies that if a thermal barrier does exist, it must be one involving CaTs-bearing pyroxene and An-bearing plagioclase rather than pure Di and Ab. The appreciable three-phase melting intervals in Di-Ab show that the non-binary nature of the system is by no means trivial; there must be a significant liquid vector towards Na_2O-SiO_2-rich compositions, and the late stages of crystallization may be more important than liquidus relations in evaluating fractionation effects.

In summary, it has to be concluded that Di-Ab may to a first approximation be considered a thermal barrier, and that some natural analogue of this join may indeed be a thermal barrier. Di-Ab affords a reasonable basis for interpreting the origin of peralkaline silicic liquids by fractional crystallization from basaltic compositions.

Chapter 8
Diopside-Anorthite-Albite

Introduction

The system Di-An-Ab allows us to study a basaltic model consisting of diopside and an intermediate plagioclase, a model which is much closer to the real thing than we have studied before. The fact that the two joins Di-An and Di-Ab are not binary means of course that Di-An-Ab is not a three-component, or ternary, system. The reasoning of the last chapter suggests that it must be described as a quinary ($c = 5$) system, CMASN. The non-ternary nature of the system means that liquids at low pressure will always lie on the SiO_2-rich side of this plane, driven there by crystals somewhat richer in Al_2O_3 than the plane. When the plagioclase composition is near An, it is expected that the Di_{SS} will have a composition approximately towards CaTs from Di, and as the plagioclase approaches Ab, the Di_{SS} will become somewhat more magnesian. These effects mean that, as a part of the generalized critical plane of the basalt tetrahedron, Di-An-Ab has the same properties as Di-Ab, so we should expect that liquids might become oversaturated with silica by fractional crystallization in this system, but not undersaturated.

Having said this, it is now convenient to ignore for a while the non-ternary nature of the system and treat it as ternary for purposes of analysis. For intermediate compositions in the system, this is not a serious over-simplification as far as geometry is concerned, although it should always be remembered that crystals and liquids do not lie rigorously in the plane Di-An-Ab.

Properties of Triangles

Quantitative treatment of a ternary system requires that compositions be plotted in an *X-X* plot, conveniently a triangle. Temperature information will then have to be contoured on this triangle, so the result will have precisely the properties of a

S. A. Morse, *Basalts and Phase Diagrams*,
https://doi.org/10.1007/978-3-030-97882-2_8

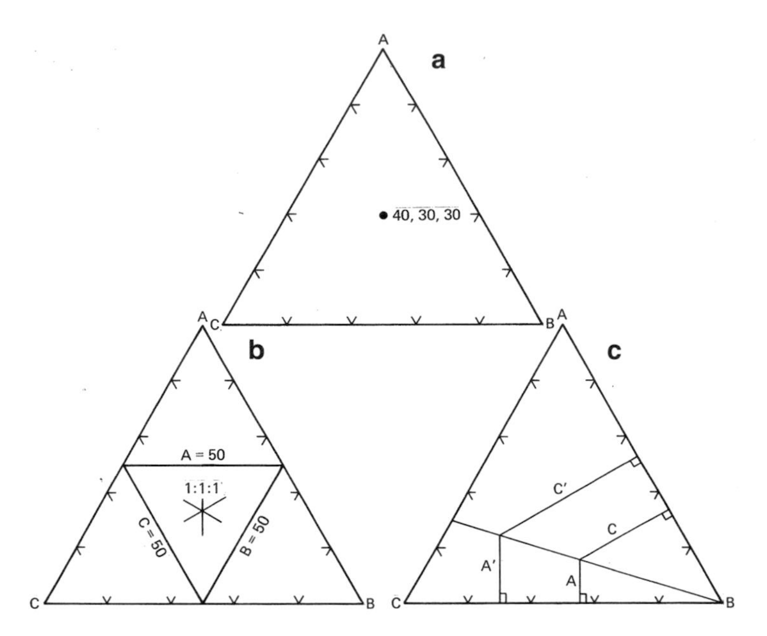

Fig. 8.1 Plotting of points and ratios in a triangle. (**a**) The plotted point has the composition 40% *A*, 30% *B*, 30% *C*. (**b**) Location of 50% lines and the center of gravity. (**c**) Proof that a line radial from *B* is a line of constant *A*:*C*

topographic map, with two dimensions serving to locate points geographically (compositionally) and a third, projected dimension serving to summarize height (temperature).

It is common to plot compositions in an equilateral triangle, although other triangles have the same properties and can be used equally well. An equilateral triangle representing components *A*, *B*, and *C* is shown in Fig. 8.1a. The corner labelled *A* signifies 100% *A*, and so on. The side opposite *A* is the locus of all points devoid of *A*, or in other words it is a binary system *B*-*C*. To plot a given amount of *A*, a distance from *BC* along the altitude towards apex *A* is measured; since the altitude represents 100% *A*, then half the altitude represents 50% *A*, and so on. A point *ABC* may then be plotted by finding the intersection of any two altitudes ("ladder rungs"), such as an amount of *A* and an amount of *B*. The point A_{40} B_{30} C_{30} is plotted, for example, in Fig. 8.1a. It is to be noted that ternary plots are concerned only with variables of constant sum (i.e. which can be recalculated to 1, or 100, or any convenient sum); a corner may not, for example, represent a ratio.

It is useful for "eyeball" plotting of points to note that the lines representing 50% *A*, 50% *B*, 50% *C* form an equilateral triangle whose apices lie at the midpoints of the legs of the parent triangle (Fig. 8.1b). A point representing the 1:1:1 composition falls at the centre of gravity for both triangles.

A very useful property of triangles, already mentioned in Chap. 5, is that a line from any apex such as *B* is a line of constant ratio of the other two components, such as *A*:*C*. The proof of this is shown in Fig. 8.1c; the ratio of altitudes *A*:*C* is clearly the same as the ratio of altitudes A':C', by similar triangles. This property of constant ratio for apical lines means that the removal of some component such as *B*, say by crystallization of a phase *B*, drives the remaining material (liquid) directly away from *B*.

Deduction of the Ternary *T-X* Projection

Having some knowledge of each of the three bounding binary systems, we are in a position to sketch the general relations in the ternary system, provided there are no interior compounds or other surprises. A convenient way to do this is to surround the ternary composition triangle with each of the three binary *T-X* plots folded down into the plane of the paper, so that their *T* axes remain normal to the sides of the triangle. This is done in Fig. 8.2. The reference temperature at the base of each *T-X* diagram is arbitrarily taken as 1100°C. Our goal is to map the liquidus surfaces in the ternary plot: we shall not be able to see through these to map the solidus surfaces, although a solidus map could be constructed instead of a liquidus map.

The obvious things to plot first are intersections of liquidus surfaces. The first of these is the eutectic in Di-An, at $An_{42}Di_{58}$. When projected along the *T* axis, this falls

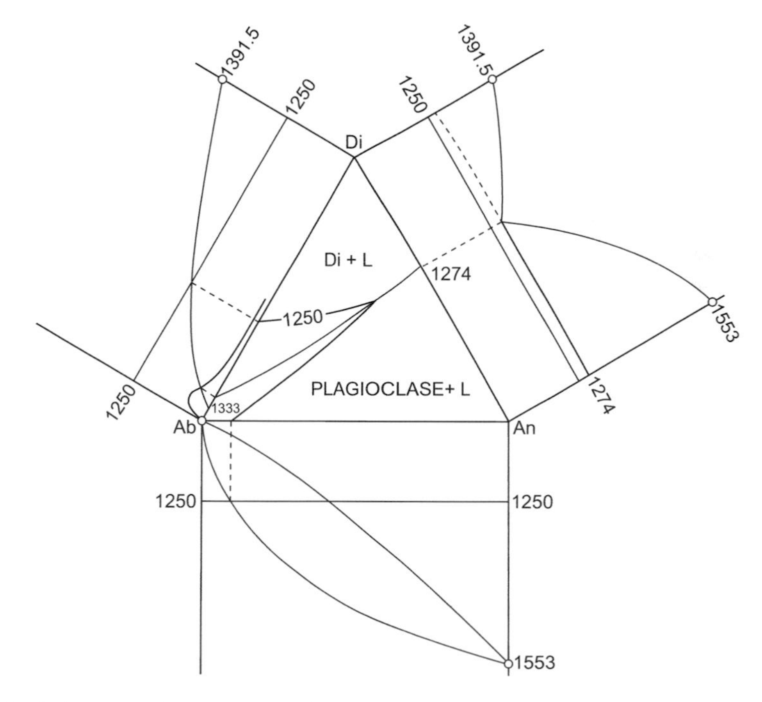

Fig. 8.2 Deduction of the ternary *X-X* plot of Di-An-Ab from the three bounding *T-X* binaries. The central triangle is the *X-X* plot, and the *T-X* plots, joined to the triangle at the 11000 °C isotherms, may be considered folded-down walls of a *TX* prism of which the triangle is the base

at the 42:58 point on the Di-An leg of the triangle. This point represents the 3-phase equilibrium Di+An+L. An analogous point occurs in Di-Ab, and although this point is by no means a eutectic, it does represent a 3-phase equilibrium, Di+Ab+L.[1] Both these points represent, in general, Di+Plag+L. In the binary systems (if they were binary), this assemblage would be invariant. In a ternary system, $W_p = 4 - \phi$ and the assemblage should be univariant. The two supposedly invariant points then must be connected by a univariant line running through the triangle. There is no reason for such a line to be straight, so we shall draw it curved; it is the heavy line in the triangular part of Fig. 8.2.

There is no liquidus intersection for An-Ab. The resulting ternary diagram is very simple. There appear to be two primary fields, one in which Di appears first on the liquidus, and another in which Plag appears first on the liquidus. These primary phase fields are exactly analogous to their binary counterparts in Di-An and Di-Ab. The line separating the two fields is a field boundary (or cotectic), supposedly also a univariant line. This line slopes in temperature from 1274°C at Di-An to 1133°C at Di-Ab. It is like a stream on a topographic map. To summarize thermal information about the liquidus surfaces on the ternary diagram, we shall have to draw contours of temperature. One of these is drawn in Fig. 8.2, for the temperature 1250°C. The contour can be accurately located at two of the three edges of the triangle. At the third, Di-An, it does not intersect a liquidus surface. The contour must, however, cut the field boundary somewhere between Di-An and Di-Ab. Recalling the "rule of *V*'s" from topographic maps, the contour must go through an inflection where it crosses the "stream" (field boundary), and the "*V*" must point "upstream" (towards higher temperature). A shape such as drawn in Fig. 8.2 is therefore consistent with the information available to us. Other contour lines could be sketched in after the same fashion.

The likening of the field boundary to a stream is quite apt, since the field boundary represents the locus of all liquids in equilibrium with both diopside and plagioclase, and these liquids must clearly move down-temperature along the line with crystallization. And just as the stream must touch both its banks, so must the liquid on the line be in contact with and saturated with both species of crystals.

To summarize the geometry of our deduced system Di-An-Ab, a perspective drawing (Fig. 8.3) is useful. The temperature axis is vertical, so the complete model is a triangular prism arising from the *X*-*X* base. The field boundary and the 1250°C isothermal plane are shown in the drawing, along with some other, schematic temperature contours. A single isothermal (1250°C) *tie line* between a liquid L and Di crystals is also shown.

In normal use, the liquidus surfaces and temperature information are projected parallel to the T axis onto the *X*-*X* plane, and such ternary phase diagrams are therefore *T*-*X* projections. They are capable of rigorous and quantitative interpretation, as we shall see.

---

[1] Actually $Di_{SS} + Ab_{SS} + L$, but we ignore for the time being the non-ternary nature of the system.

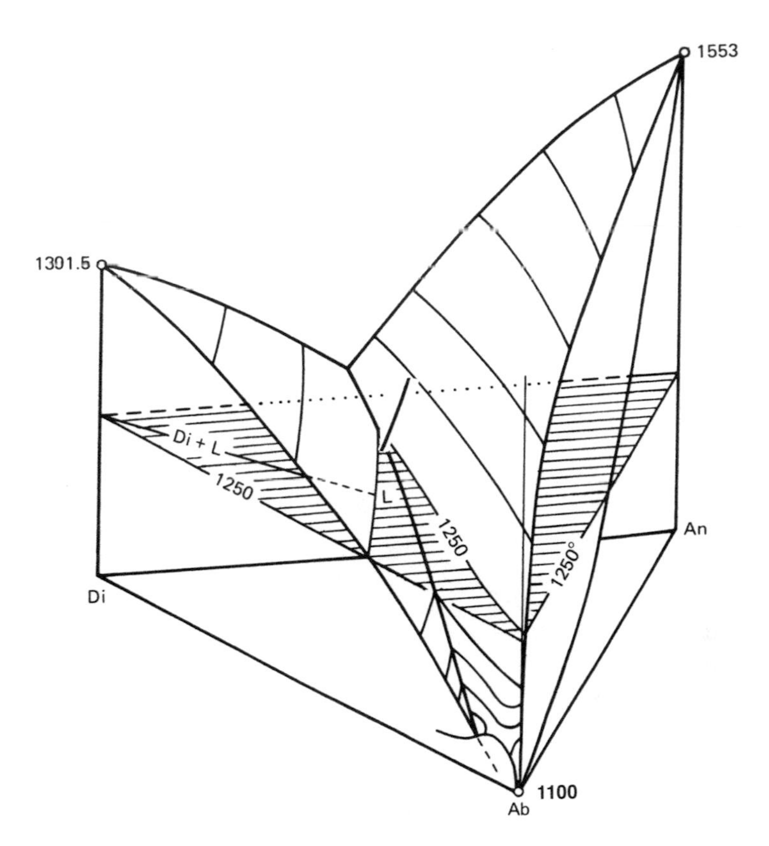

Fig. 8.3 Perspective drawing of the ternary *T-X* prism Di-An-Ab-*T*. The two intersecting liquidus surfaces are shown ruled with temperature contours. An isothermal plane (1250°C) is shaded

The Experimental Ternary Diagram

The diagram which we have deduced may be correct in principle, but it is not very useful, since it surely has no quantitative reliability. The position of the field boundary must be located by experiment, and the compositions of plagioclase solid solutions in equilibrium with a given liquid must be determined experimentally. The experimentally determined phase diagram, modified slightly from that of Bowen (1915), is shown in Fig. 8.4.

The maximum number of coexisting phases indicated by the phase diagram is 3, and the minimum variance at constant pressure is therefore 1. The univariant nature of the plagioclase diagram therefore exerts its influence in the ternary system, and since the addition of the component diopside is accompanied eventually by the phase diopside, the lower limit of variance is the same in the ternary system as in An-Ab. This implies that fractional crystallization, as in An-Ab, can have a profound effect on the course of the liquid.

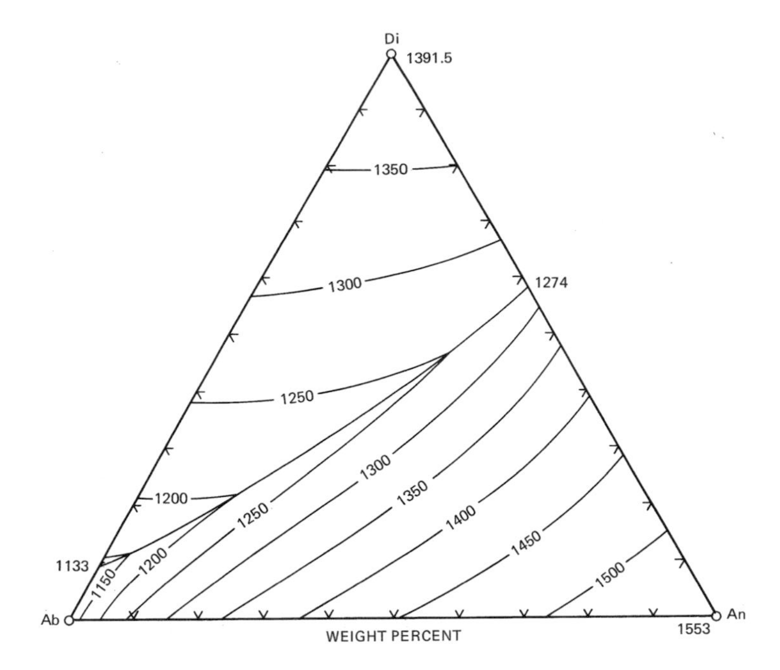

Fig. 8.4 Phase diagram, after Bowen, of the system Di-An-Ab. The Di-Ab sideline is revised after Schairer and Yoder (1960) and the Di-An sideline after Osborn and Tait (1952). See also Kushiro (1973)

Equilibrium Crystallization

The Di + L Field

Considering equilibrium crystallization, and assuming a ternary system, let us start with a bulk composition of $Di_{60}An_{20}Ab_{20}$, in the Di + *L* field. This composition is labelled *A* in Fig. 8.5a. When solid, it must consist of diopside and plagioclase of composition An_{50}. A melt of this composition is cooled, until it reaches the Di + *L* liquidus surface, at about 1300°C (point *A*, Fig. 8.5), at which point crystals of Di begin to form. Removal of Di from the liquid causes the latter to move directly away from Di towards $An_{50}Ab_{50}$. This process continues until the liquid reaches the field boundary, at about 1235°C (point B, Fig. 8.5), at which point a plagioclase also begins to crystallize.

Now it must be determined by experiment what composition this plagioclase has; the original work tells us that the feldspar in equilibrium with Di and liquid at this temperature is An_{80}. A new *tie line* is established between the liquid and plagioclase of this composition, and this tie line also generates a *three-phase triangle* Di-An-*L*, shown in Fig. 8.5b. The long leg of this triangle, Di-An_{80}, is the tie line connecting crystals, and the apex on the field boundary represents the composition of liquid. Further crystallization of both Di and calcic plagioclase must drive the liquid to the

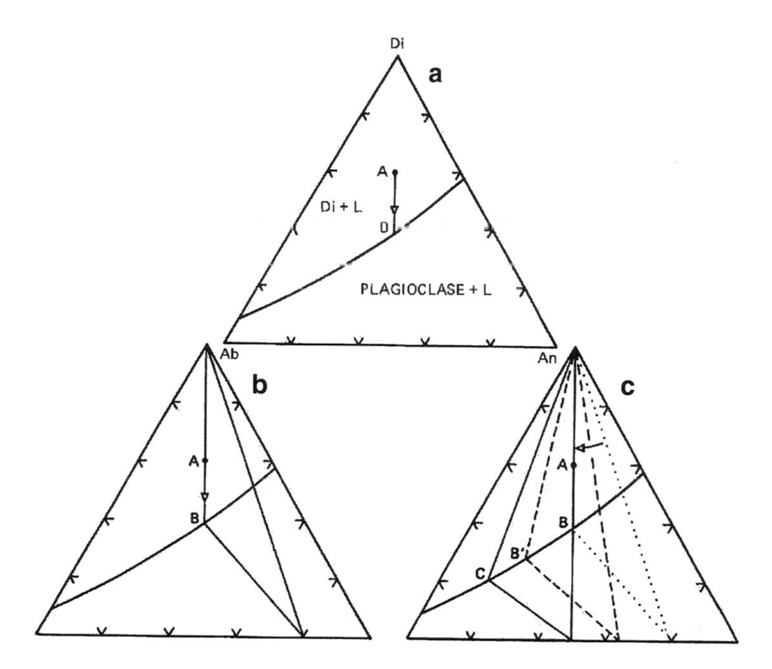

Fig. 8.5 Three stages in the equilibrium crystallization of a mixture *A* in Di-An-Ab. (**a**) Crystallization of Di has driven the liquid directly away from Di from *A* to *B*. (**b**) Same temperature as *A*, showing first three-phase triangle, Di-*B*-plagioclase. (**c**) Motion of the three-phase triangle from initial to final position. The motion is a rotation around Di as a pivot

left, but the coexistence of three phases means that the system is univariant, and therefore that the liquid must move in univariance, i.e. along the field boundary. As the liquid moves thus towards Di_9Ab_{91} in response to crystallization, it demands equilibrium with more and more sodic plagioclase, so (we shall pretend) the latter continually reacts with liquid, as in An-Ab. As both liquid and plagioclase become more sodic, Di remains fixed in composition, and the three-phase triangle *pivots* to the left around Di as a pivot point (Fig. 8.5c). The rate at which this motion occurs is a function of the rate of heat loss and the reaction time required for the plagioclase crystals. During the sweep of the three-phase triangle, both diopside and plagioclase increase in amount. Although we know nothing of the rate, we do know when the process terminates, for at the moment when the last iota of liquid disappears, the plagioclase must have the initial bulk composition of An_{50}, and experimental evidence shows that the last liquid has the composition shown by point *C*, Fig. 8.5c, which is about An_{19} in terms of its potential plagioclase composition. At this juncture, it is to be noted that the long leg (the solid-solid leg) of the three-phase triangle cuts the bulk composition (*A*), and the total solid composition coincides with the bulk composition. The tie line $C - An_{50}$ has a slightly flatter slope than the preceding plagioclase-liquid tie lines, and we learn from the experiments that this is a general rule in this system, except near the Ab corner. The temperature of final crystallization is given for liquid *C* from the contours of Fig. 8.4 as about 1200°C.

The Projected Plagioclase Loop

It is to be remarked that the coexistence of An_{50} crystals with liquid at 1200°C in this case is quite unlike the case of the binary system An-Ab; in the latter, a liquid at 1200°C would coexist with crystals of about An_{32}. The presence of diopside clearly has a drastic effect on the temperature-composition relations of plagioclase and liquid. This effect can be appreciated readily by reference to Fig. 8.6, in which the *projected* plagioclase loop for diopside-saturated liquids is compared with the binary An-Ab plagioclase loop. The Di-An-Ab loop is projected from Di onto the An-Ab-T face of the T-X prism (Fig. 8.3). The liquids lie on the field boundary in the ternary system, and the crystals lie in the plagioclase prism face. The projected loop is plotted by simply reading the An/(An+Ab) ratios of both ends of L-Plag tie lines (as in Fig. 8.5c) at various temperatures. The diopside-saturated loop is a good deal flatter, ranging from 1274°C, the Di-An eutectic temperature at the calcic end, to 1133°C, the Di + Ab + L liquidus at the sodic end. The loop is also somewhat narrower in the intermediate region, compared to the pure plagioclase loop. For example, a plagioclase composition of An_{70} coexists with a diopside-saturated liquid of An_{37} potential plagioclase composition at 1223°C, whereas in the pure system

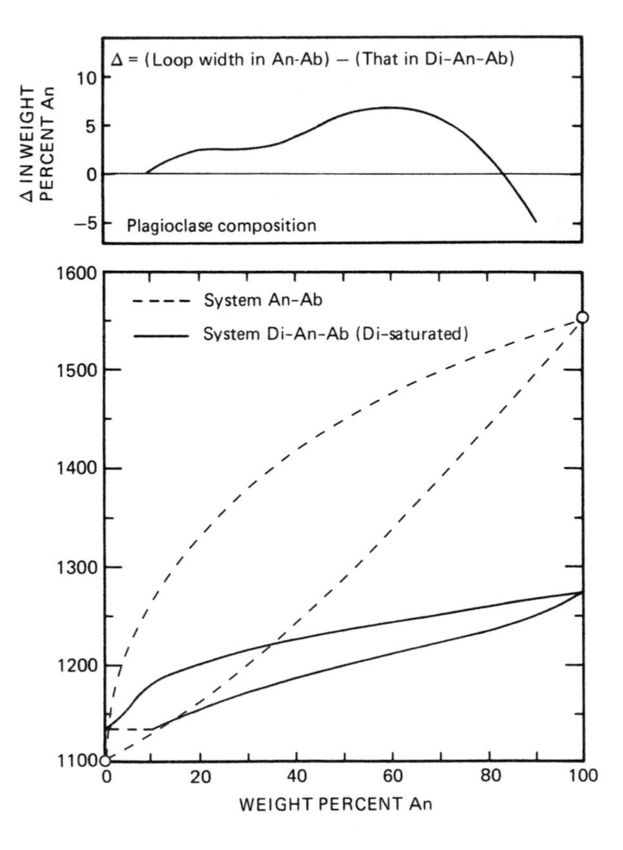

Fig. 8.6 Projected plagioclase loop in Di-An-Ab compared with pure plagioclase. The lower diagram shows the two loops, and the upper one monitors the change in loop width as a function of the solidus compositions

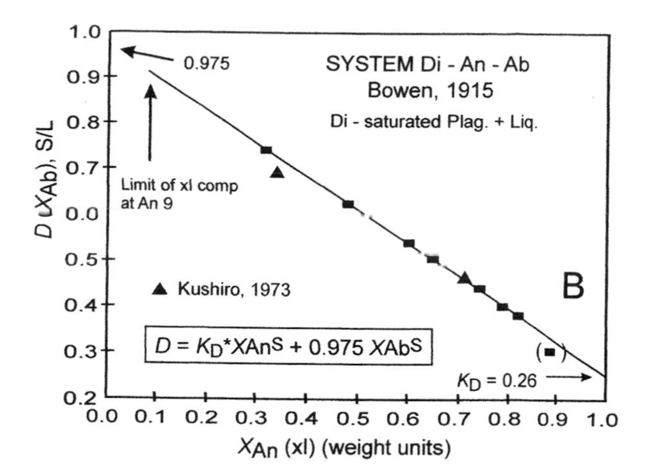

Fig. 8.7 Linearity of Bowen's (1915) diopside-saturated plagioclase-liquid partition coefficients with other data

such a plagioclase coexists with a liquid of An_{31} at about 1380°C. The loop is narrower by some 6–7% An in the plagioclase composition range of most interest to basalts and gabbros, as shown by the graph at the top of Fig. 8.6.

Both the flattening and the narrowing of the loop in the presence of diopside are instructive in terms of basalt and gabbro genesis. The flattening illustrates very clearly the marked temperature effect of adding only one component, diopside. The resulting crystallization temperatures are very close to those observed in natural basaltic lavas, such as 1150°C for olivine + augite + plagioclase + L (Peck et al., 1966). The flattening of the loop also has an important effect on the rate of change of plagioclase composition as a function of temperature (Willie, 1963), as compared to crystallization in the plagioclase field alone. Moreover, liquids in the plagioclase field move approximately along lines of equal Ca content, whereas liquids on the field boundary move almost directly across contours of equal Ca. It therefore follows that Ca is depleted much more rapidly when diopside and plagioclase crystallize together than when plagioclase crystallizes alone (Morse, 1979a).

The partitioning of An and Ab components between plagioclase and liquid on the field boundary constitutes an important limiting condition on the paths of liquids in the plagioclase field. Bowen (1915) recognized the importance of this partitioning and called the plagioclase-liquid tie lines "three-phase boundaries". He put considerable effort into their determination, with the remarkable result shown in Fig. 8.7. All but one of Bowen's points fall on a straight line with a correlation coefficient $r^2 =$ 0.993. Two additional points determined by Kushiro (1973) are also plotted on the diagram, but not included in the regression. This correlation has several important consequences. First, it misses the ideal value of 1.0 at pure Ab (Di-saturated). Instead, it requires that for pure Ab(Di), the plagioclase composition is $\sim An_9$, a result estimated only much later by Schairer and Yoder (1960) and Kushiro (1973). Second, the regression defines the limiting value of K_D (Eq. 6.1) at pure An(Di) as 0.26, very similar to that in the pure plagioclase system An-Ab (0.3; Chap. 6).

For an ideal binary solution with constant K_D, the partition coefficient D is related to K_D and composition by the simple relation given in Eq. (6.2). In this case, however, the lack of constraint to 1.0 at the An-free end requires a modification to Eq. (6.2):

$$D^{S/L}_{X_{\mathrm{Ab}}} = K_D\left(X^S_{\mathrm{An}}\right) + 0.975X^S_{\mathrm{Ab}} \tag{8.1}$$

as indicated in Fig. 8.7.

The theoretical linearity of the partition coefficient against solidus composition (Eq. 8.1) was not known to Bowen (Morse, 1996). If it had been, he would almost certainly have deduced that the system Di-An-Ab was not ternary, and that the liquids left the ternary plane in the direction of Na_2O and SiO_2, as later deduced (Schairer and Yoder, 1960; Kushiro, 1973).

The Plagioclase + L *Field*

Point *D*, Fig. 8.8a, represents a bulk composition of (An_{65}) 85%, Di 15% and the plagioclase must have a composition An_{65} at the end of equilibrium crystallization (still pretending that this might occur!). This initial melt lies in the plagioclase +

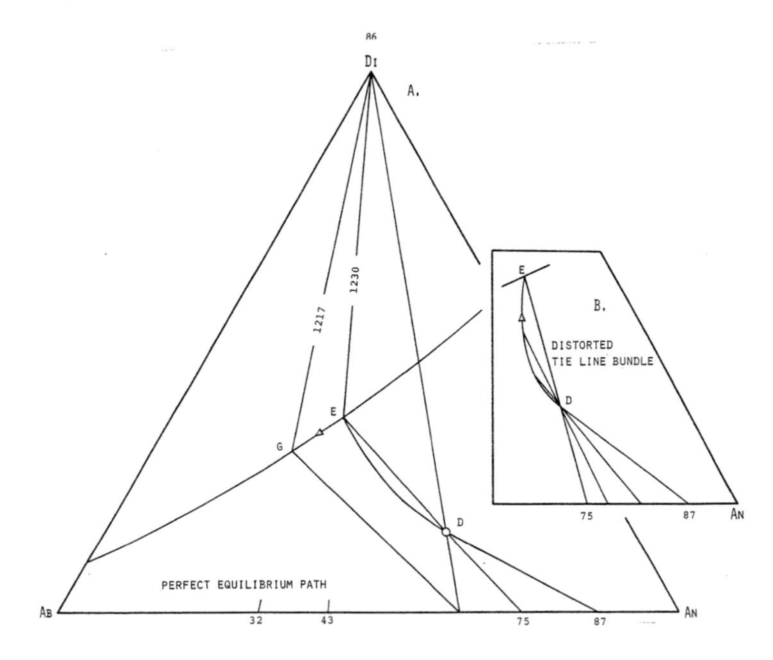

Fig. 8.8 (**a**) Perfect equilibrium crystallization of mixture *D* in Di-An-Ab. The liquid path is curved from *D* to *E*, and from *E* to *G*, where crystallization is complete. The inset (**b**) shows the constraint on the liquid path imposed by the plagioclase composition and the bulk composition *D*

L field, and the path of the liquid is not as simple as that in the Di + L field. Upon cooling, the melt intersects the liquidus surface at about 1420°C, with the production of some plagioclase crystals of about An_{87} as the initial tie line indicates (the recovery of correct tie lines within the plagioclase field will be treated later). Removal of this plagioclase composition from the liquid tends to drive the liquid directly away from An_{57}, but simultaneously, as this happens, crystals and liquid react to convert the plagioclase to a slightly more sodic composition. Therefore, the liquid may not move directly away from An_{57} for more than an instant of time, and the liquid must in fact follow a curved path. The restrictions on this path are shown in a distorted sketch, Fig. 8.8b. The tie line connecting crystals and liquid must at all times pass through the bulk composition D. As the crystals become more sodic, the tie line rotates clockwise about the bulk composition as a pivot. Some sample tie lines at different stages of the process are shown in Fig. 8.8b.[2] The curved path terminates at the field boundary at a point (E) shown by the experiment to have a potential plagioclase composition of An_{43}, at a temperature of about 1230°C, and the plagioclase coexisting with this liquid has the composition An_{75}. This plagioclase-liquid tie line is the last one to cut the bulk composition, for with the onset of diopside crystallization, a three-phase triangle is formed, and the rest of the crystallization process is identical in principle to the late stages of crystallization beginning in the Di + L field. The three-phase triangle pivots about Di, and the process is terminated when the solid-solid leg cuts the bulk composition, at which point the plagioclase crystals have just reached An_{65}, the last drop of liquid has composition G (about An_{32} in potential plagioclase) and the total solid composition is Di 15%, (An_{65}) 85%.

The curved path D-E has a further interesting property. The tangent to this path at any point, extended to the plagioclase join, gives the fictive composition of what is being removed from the liquid to react with the plagioclase crystals and convert them to more sodic compositions. It is important to distinguish this "fictive plagioclase", being removed from the liquid at any instant, from the actual plagioclase crystal composition, which is the integrated composition of all crystals produced and reacted during the course of crystallization up to the moment of consideration.

It is also noteworthy that the curved liquid path $D \rightarrow E$ is unique; no other bulk composition can yield the same liquid path, but each bulk composition follows its own unique course of crystallization. The path is dependent upon both the composition and the mass of crystals with which the liquid is in equilibrium.

A tangent to the liquid path E-G, when carried back to the solid-solid leg of a three-phase triangle, gives the fictive Di:Plag ratio of material being removed from the liquid at any instant. This ratio varies from about Di 50%, Plag 50% at E to about Di 45%, Plag 55% at G. At the moment when the liquid arrives at E, there is a discontinuous jump in this ratio from Plag 100% to Plag 50%, and the inheritance of this behaviour from that in the parent binary system Di-An (Chap. 5) is clear.

[2] For construction purposes, it is important to know that these tie lines are tangents to fractionation paths. The position of L along a tie line is given by the point of tangency.

Isothermal Sections

It is often useful to review the properties of ternary systems by means of complete isothermal sections which show the various two- and three-phase fields. Such sections become particularly important in the interpretation and application of experimental results. Isothermal sections can be constructed from any ternary liquidus diagram showing temperature contours, provided the slopes of tie lines between liquid and crystals are also known. Four isothermal sections for Di-An-Ab are shown in Fig. 8.9, and these illustrate in another way the equilibrium crystallization processes just discussed. The liquid region, at any temperature, is simply the area enclosed by the pertinent temperature contour on the liquidus diagram. For temperatures greater than 1274°C (the Di-An eutectic), the L region is a band across the diagram, separating the region of Di + L from the region of Plag + L (Fig. 8.9a). At temperatures less than 1274°C, the liquid region terminates at a point on the field boundary which is also an apex of the three-phase triangle Di Plag + L (Fig. 8.9b–d). Another apex of this three-phase triangle lies at the Di corner, and the third lies on the plagioclase join. The liquid region is a one-phase region, the three-phase triangle is a three-phase region, and there are three two-phase regions denoting either solid-liquid equilibria (Di + L, Plag + L) or solid-solid equilibria (Di + Plag). In two of the two-phase regions, tie lines radiate from the Di apex, and these can be drawn with no further information. In the Plag + L region, tie lines must be drawn according to the experimentally determined plagioclase and liquid compositions.

The three isothermal sections of Fig. 8.9b–d illustrate the sweep of the three-phase triangle across the diagram as it pivots about the Di apex with falling temperature. The L apex of the three-phase triangle follows the boundary curve shown on the liquidus diagram (Fig. 8.4).

The isothermal sections direct our attention away from the special bulk compositions discussed earlier and towards regions of coexisting phases instead. They show, with falling temperature, a progressively smaller field where bulk compositions are expressed as liquid only, progressively smaller Di + L and Plag + L fields, a progressively broader Di + Plag field. Within any one-, two-, or three-phase field, a variety of bulk compositions consist of the same phase or collection of phases at a given temperature. The experimental determination of isothermal sections aids in the construction of summary diagrams like the liquidus diagram.

Variance

The isothermal (isobaric) sections serve very well to illustrate the variance of the ternary system at various stages of crystallization and for various bulk compositions. Assuming a ternary system, $W_p = c + 1 - \phi = 4 - \phi$ and the 1-, 2-, and 3-phase regions are therefore isobarically 3-, 2-, and 1-variant respectively when temperature is also a variable, as in the liquidus diagram (Fig. 8.4). In the isothermal, isobaric

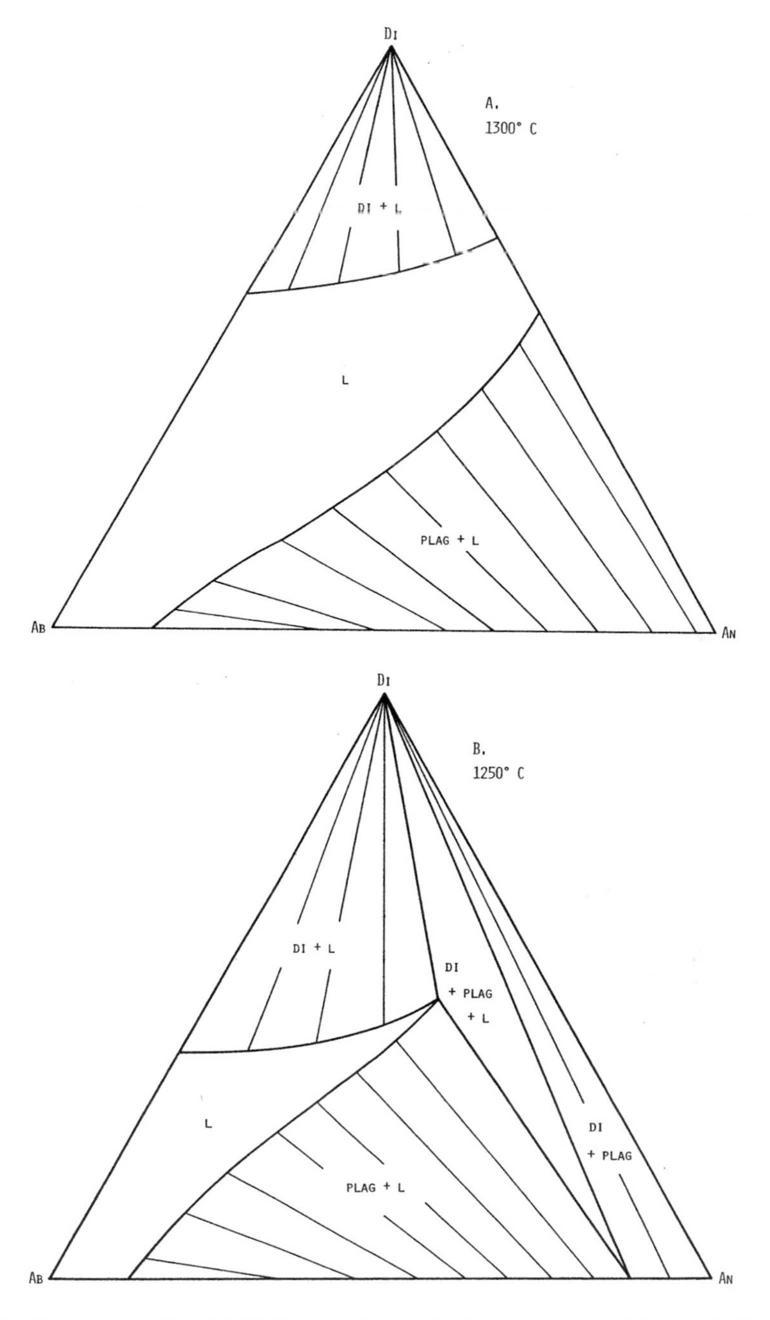

Fig. 8.9 Closure of the liquid field illustrated in four isothermal sections of Di-An-Ab. The three isothermal sections arbitrarily labeled *A*, *B*, *C*, and *D* illustrate the sweep of the three-phase triangle across the diagram with falling temperature

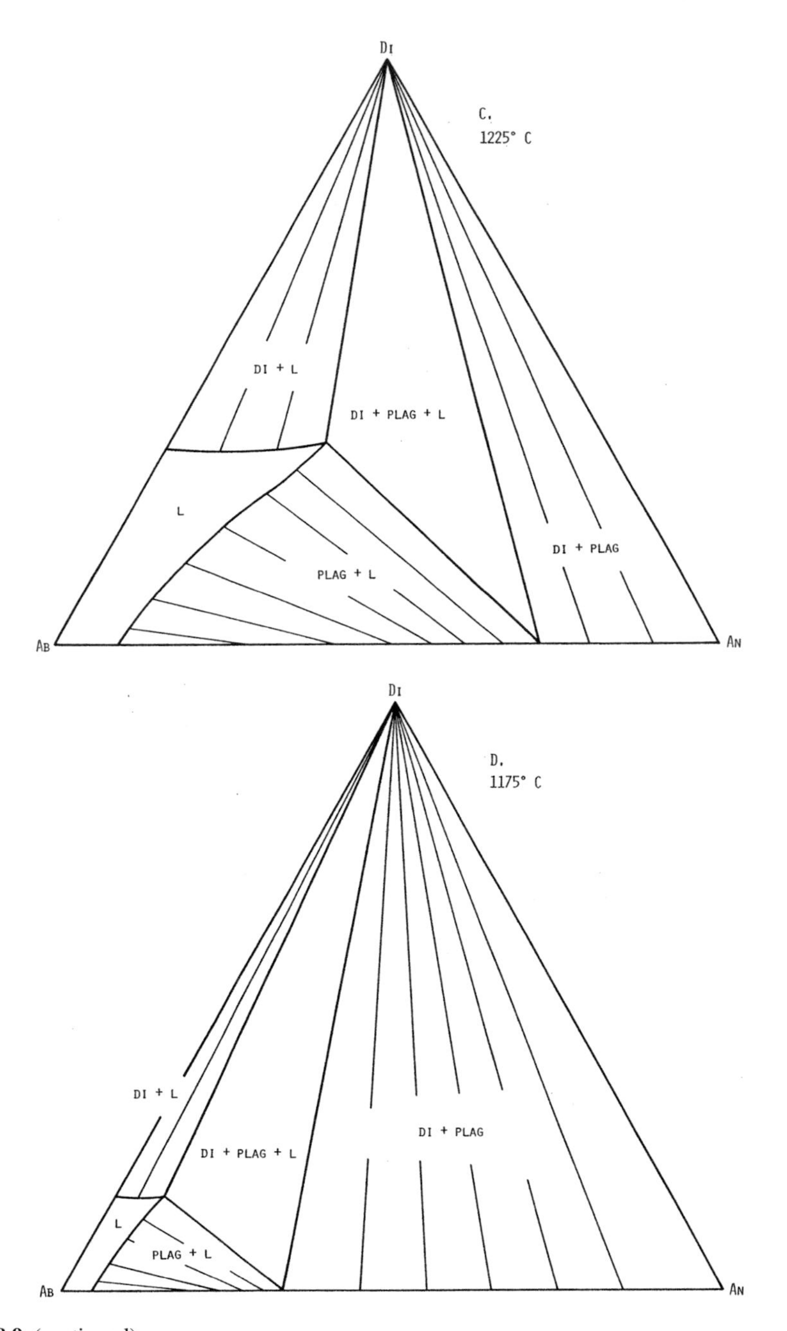

Fig. 8.9 (continued)

sections, however, we have $W_{p,t} = c - \phi$, or in other words a second arbitrary restriction on variance, and the 1-, 2-, and 3-phase regions are 2-, 1-, and 0-variant respectively. For example, in the 1-phase liquid region, which is isobarically, isothermally divariant, two composition variables must be specified to describe the system (i.e. to identify the liquid composition) or two such variables (such as the Di/An or Ab/An ratios) may change without changing the state of the system, namely liquid. In a 2-phase region, only one compositional variable need be specified to describe the system, either a plagioclase composition (for Plag + *L* and Plag + Di) or a liquid composition (for Di + *L*). And in the 3-phase triangle, the mere coexistence of 3 phases means that the compositions of all phases are completely defined; the phase compositions are invariant.

Completion of Equilibrium Crystallization

Any interior composition in Di-An-Ab can be expressed in terms of two solid phases, Di and a plagioclase feldspar of specific composition. When the system exists in these two specific phases, their tie line cuts the bulk composition and no liquid can be present. This is another way of saying that the last of the liquid (*G*, Fig. 8.8), on crystallization, will have been used up just as the last increment of rotation of the Di-Plag tie line takes place, to bring the tie line onto the bulk composition (*D*, Fig. 8.8). The last liquid present on crystallization will therefore, in general, lie in the supposedly ternary system on the univariant boundary curve, and the final stages of crystallization will be univariant, polythermal, as in the An-Ab system. The reaction $L =$ Di + Plag + cals will continue with falling temperature until the liquid is used up. The composition of the final liquid is dictated solely by the initial bulk composition, except that it must lie on the boundary.

Application of the Lever Rule

The lever rule may be simply applied in ternary systems to any assemblage of two phases. An isothermal tie line Di-*L* intersects the bulk composition *A* in Fig. 8.10a. The bulk composition divides the tie line into two segments, denoted *l* and *s* in the diagram, representing respectively the proportions of *liquid* and *solid*. The weight fraction of the system which has crystallized as Di by the time the liquid has advanced to point *L* is given by $s/(s + l)$. As with the binary lever rule (Chaps. 5 and 6), it is important to know which segment of the line represents liquid, and which represents solid. This is self-evident if it is recalled that the initial state, when the first crystal of diopside is just forming, is given by the tie line *A*-Di; the liquid lies at the bulk composition and the system consists almost solely of liquid. Therefore, the *l* or liquid segment is always that segment lying between the bulk composition and the crystalline phase. The *s* or solid segment grows with falling temperature while the *l* segment remains unchanged in this case.

In the case of the two phases plagioclase and liquid, the *s* and *l* segments both change length, as shown in Fig. 8.10b. In this case, two tie lines are shown for two temperatures, $T_1 > T_2$. P_1 and L_1 are the plagioclase and liquid, respectively, at

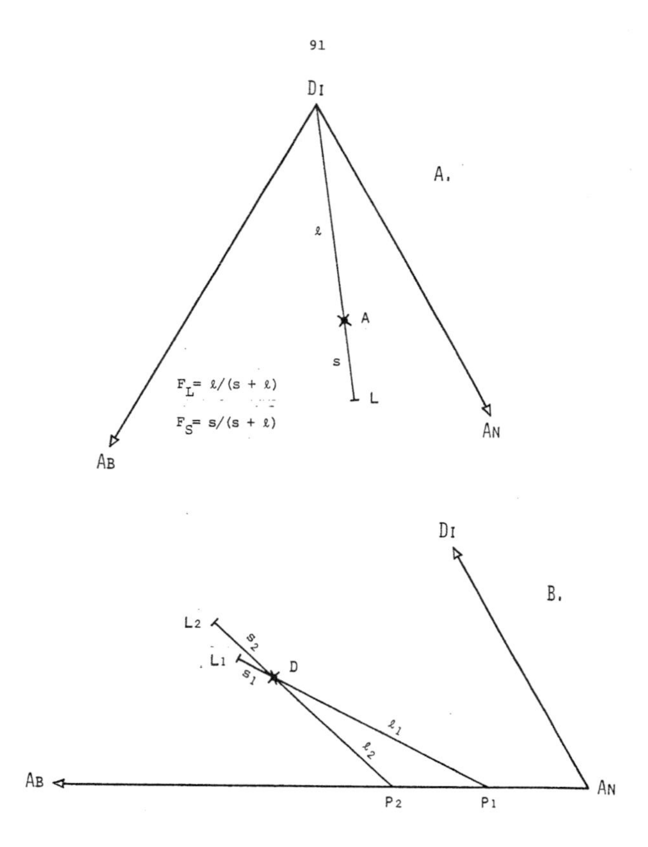

Fig. 8.10 Lever rule in a ternary system (Di-An-Ab). (**a**) Illustration from the diopside + liquid field. (**b**) Illustrations from the plagioclase + liquid field at two different temperatures

temperature T_1, and the fraction of the system represented by crystals is given by $s_1/(s_1 + l_1)$. A larger fraction of crystals is shown by the tie line P_2-L_2 for temperature T_2; not only has the length of the s segment increased over that at T_1, but the l segment has also diminished by virtue of the reaction of plagioclase from P_1 to P_2 with falling temperature. Therefore, the fraction of crystals, $s_2/(s_2 + l_2)$ is appreciably greater than at T_1.

When two phases are both solid, i.e. Di + Plag, the lever rule applies to give the fraction of each, in the same fashion as the *l*-*s* levers. The segment of the tie line between *A* and Di (Fig. 8.11) represents the fraction of plagioclase, and the segment between *A* and P_2 represents the fraction of diopside.

When three phases occur, all the ratios of interest can be determined by means of levers. The total solid composition (TSC) now lies within the ternary diagram on the Plag-Di leg of a three-phase triangle, and its position on this leg must be determined first, as follows. It is apparent that the tie line between the TSC and the liquid must always lie through the bulk composition, as in Fig. 8.11. Therefore, a line from L_1 through the bulk composition *A* locates the TSC on the leg Di-P_1. The segment *A*-TSC, by analogy with the segment *A*-Di in Fig. 8.10a, must represent the liquid fraction, while the segment *A*-L_1 represents the solid fraction. A convenient measure of the solid: liquid ratio is the fraction of liquid remaining, or F_L, which is given by *l*/

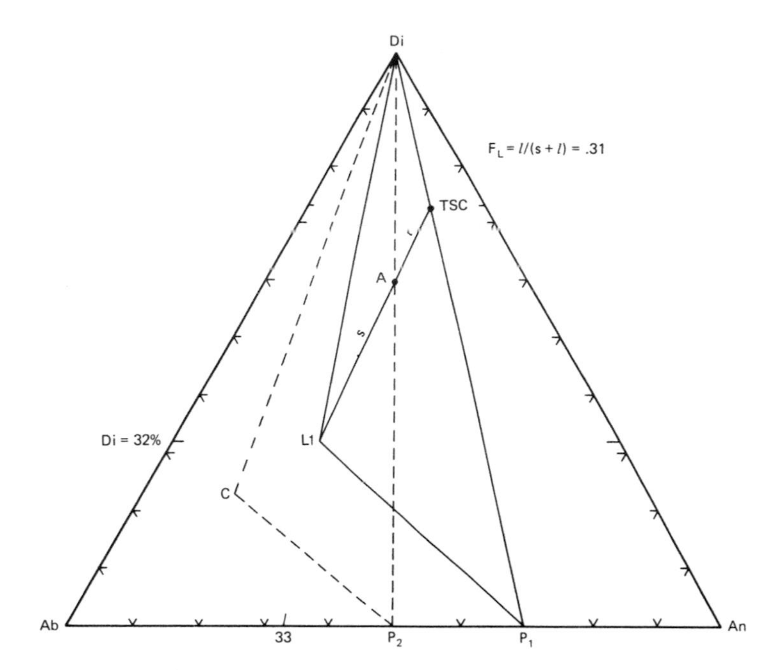

Fig. 8.11 Lever rule for a three-phase assemblage in Di-An-Ab. The position of the total solid composition (TSC) is found by drawing a line from the liquid (L_1) through the bulk composition (A) to the solid-solid leg (Di-P_1) of the three-phase triangle. The lever rule is then applied in the usual way, with the bulk composition as fulcrum. The *dashed* three-phase triangle refers to a lower temperature, at which crystallization is completed

($s + l$), and amounts to 0.31 in the illustration (Fig. 8.11). Among crystals, the fraction of plagioclase may be found by forming the fraction (Di-TSC)/(Di-P_1), but since in this case the solid-solid leg of the three-phase triangle is radial from one corner of the phase diagram, the ratio can be read directly from the coordinates of the diagram. The fraction of plagioclase is 0.27 in Fig. 8.11. In a similar manner, the exact composition of liquid L_1 can be determined from the coordinates of the diagram and the extension of the leg L_1-Di.

In summary, we may extract the following information from Fig. 8.11. A liquid, formed by equilibrium crystallization from bulk composition A, coexists with plagioclase and diopside, has composition L_1 (Di 32%, (An_{33}) 68%), and amounts to 31% by weight of the original mixture. The remainder of the original mixture has crystalline composition TSC, or Di 73%, (An_{70}) 27%.

The final three-phase triangle involving liquid C and plagioclase An_{50} is shown with dashed lines in Fig. 8.11. It is apparent, from the fact that one leg of this triangle cuts the bulk composition A, that crystallization has been completed, and $F_L = 0$.

A series of such analyses can be made for various stages of equilibrium crystallization, and the path of the total solid composition can then be drawn for the entire history of crystallization. This is done in Fig. 8.12, where the TSC path is shown as a

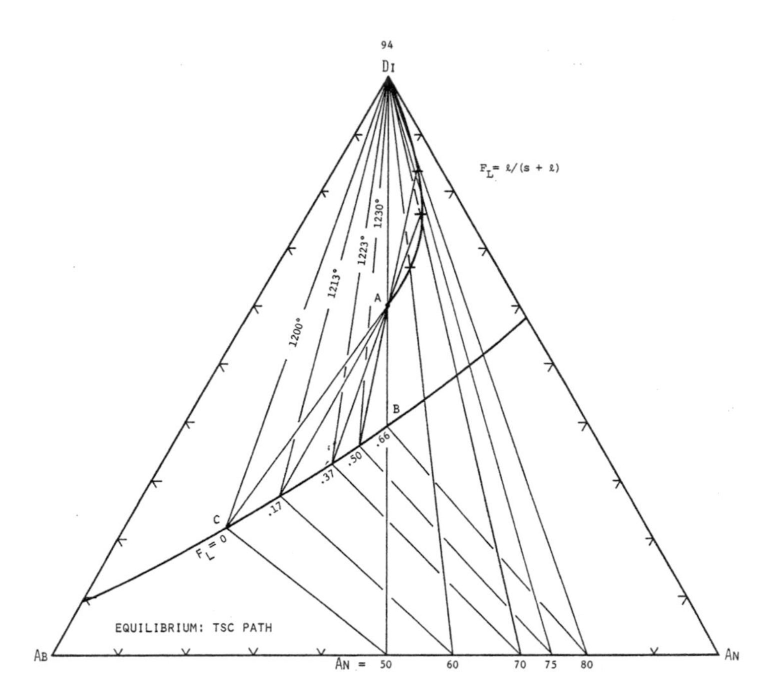

Fig. 8.12 Path of the total solid composition (TSC path): diopside field of Di-An-Ab. The TSC leaves Di when $F_L = 0.66$ and follows the curved path to *A*. The path is constructed from a series of TSC positions found as in Fig. 8.11.

heavy curved line. All the three-phase triangles from which points on this path were constructed are shown, and beside the liquid apex of each of these triangles is shown F_L, or the fraction of liquid remaining. It should be noted that the initial portion of the TSC path is represented by the point Di; only when the liquid has reached point *B* ($F_L = 0.66$) does the TSC move off the composition of pure diopside. In other words, 34% of the initial mass must have crystallized before the liquid reaches the field boundary. The remaining crystallization history can be read from the diagram in complete detail. For example, when $F_L = 0.37$, the liquid has composition Di 33%, (An_{37}) 67%, and the solids consist of Di 76%, (An_{70}) 24%, together constituting 63% of the initial mass. Such analyses may be useful in interpreting the crystallization history of both basaltic lavas consisting of crystals and glass, and gabbroic intrusive rocks whose textures suggest equilibrium crystallization.

Figure 8.13 summarizes the liquid and TSC paths for bulk composition *A*. It should be noted that the TSC path lying away from Di applies only to the crystallization interval *B-C* in terms of liquid composition.

A TSC path for the crystallization of the plagioclase-rich bulk composition *D* is derived in Fig. 8.14. The first portion of this path, corresponding to the motion of liquid from *D* to *E* on the field boundary, is confined to the plagioclase series between An_{87} and An_{75}. Thereafter, the TSC rises along a curved path to *D* as the liquid continues to produce diopside and plagioclase. This TSC path and the liquid

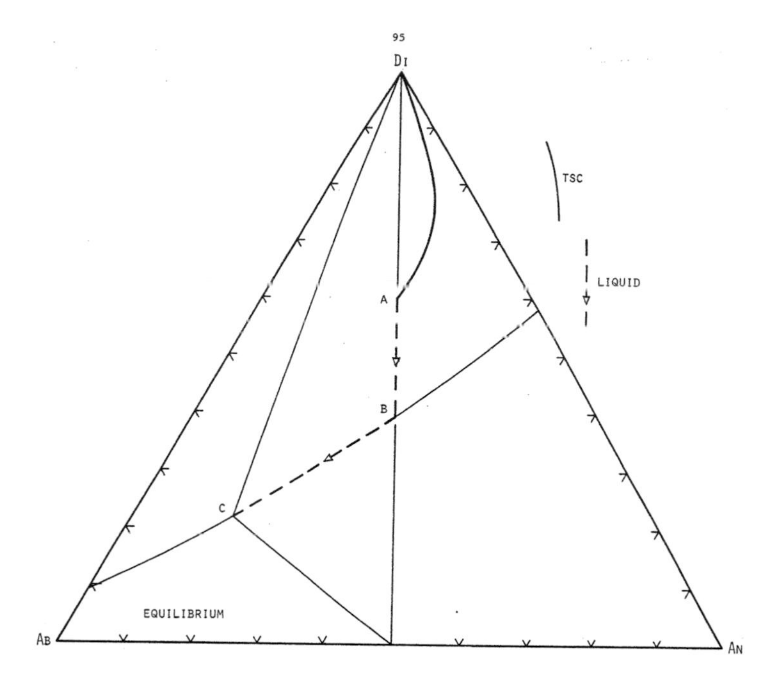

Fig. 8.13 Summary of liquid and TSC paths for bulk composition *A*. Note that the TSC remains at Di until the liquid reaches *B*

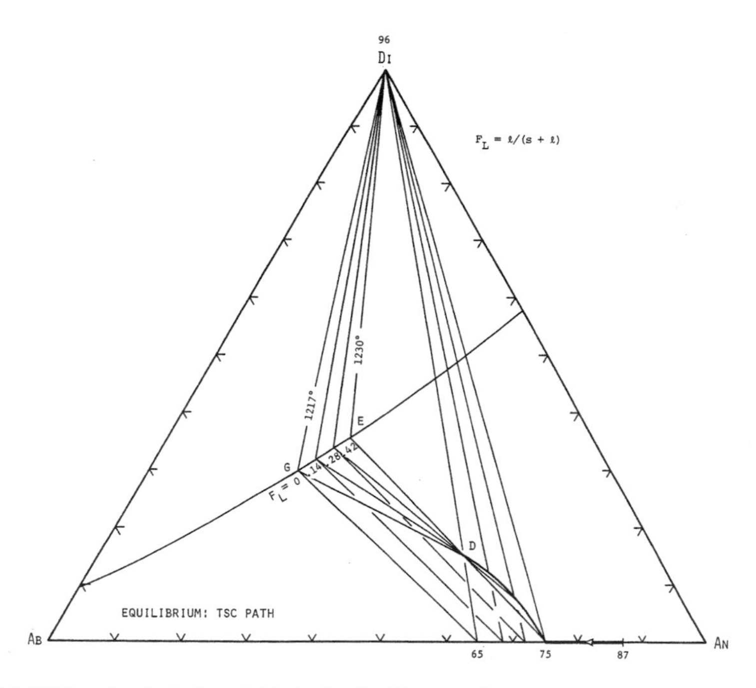

Fig. 8.14 TSC path: plagioclase field. As the liquid moves from *D* to *E* (path not shown) the TSC moves from An_{87} to An_{75} along the plagioclase sideline. While the liquid moves from *E* to *G*, the TSC moves from An_{75} to *D* along the curved path. The construction of the TSC path follows the same method as for composition *A* (Fig. 8.11)

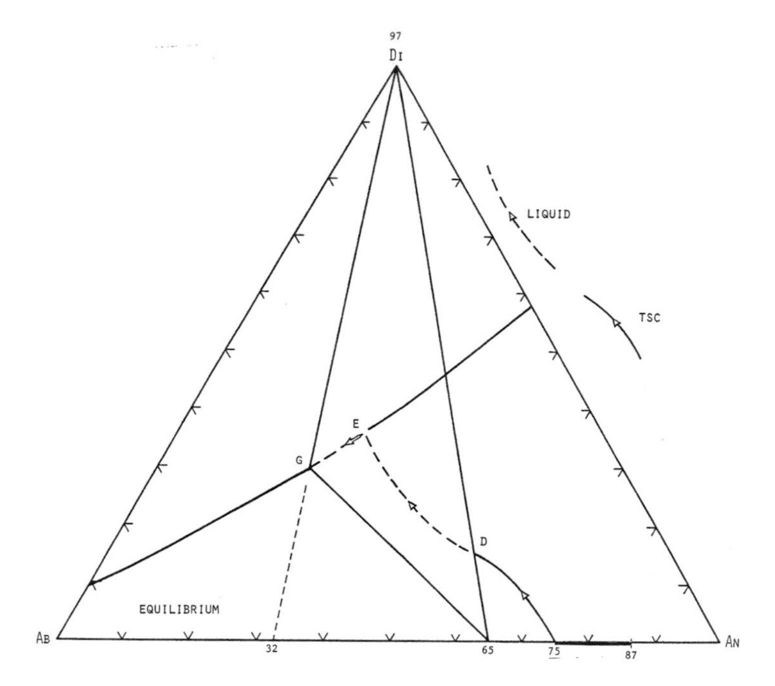

Fig. 8.15 Summary of liquid and TSC paths for bulk composition *D*. The TSC path along the plagioclase sideline is found from tie lines through *D* from the liquids along path *DE*

path which produced it (taken from Fig. 8.8) are compared in Fig. 8.15. The segment *E*-*G* of the liquid path corresponds to the segment 75-*D* of the TSC path.

Fractional Crystallization

The "ternary" system Di-An-Ab is strongly dominated by the plagioclase join in its crystallization behaviour, and as noted earlier, fractional crystallization has a profound effect on the late stages of crystallization, as in the plagioclase system itself. Moreover, we shall see that fractional crystallization of liquids within the plagioclase field has an immediate effect on the course of liquid composition. It is convenient to begin with the Di + *L* field, which is simpler.

Fractional crystallization in the Di + *L* field begins as it does in Di-An: crystals of Di are removed as soon as they form, and the liquid composition changes directly away from Di. At any instant, there is only an infinitesimal amount of crystalline Di in equilibrium with the liquid, and each new liquid composition therefore has all the properties of an initial liquid of new bulk composition. Accordingly, when the liquid reaches the field boundary, as at *B* in Fig. 8.5, it cannot be distinguished from an initial liquid formed in equilibrium with plagioclase and diopside. The textures of many basaltic and gabbroic rocks suggest simultaneous crystallization of pyroxene and plagioclase from an early stage of their history, and it is fair to suppose that these magmas may have been brought to the appropriate field boundary by fractional crystallization due to separation and perhaps sinking of pyroxene while the magma was being emplaced.

Continued fractional crystallization, now of plagioclase and diopside, drives the liquid to lower temperatures along the field boundary. Even though crystals may be removed as soon as they are formed, the liquid experiences equilibrium with both kinds of crystals, and cannot therefore leave the field boundary or its variance of one. The plagioclase crystals formed during this process have compositions given at any instant by the slope of experimentally determined Plag-*L* lines for the given isotherm. The array of plagioclase crystals formed over a period of time has a continuous range of compositions, from calcic to sodic, expressed either as successively accumulated homogeneous crystals of continuously differing composition, or as zoned crystals, depending on whether the fractionation mechanism involves physical separation of crystals or rapid cooling which prevents equilibration of crystalline cores with liquid. The history of the fractionating liquid is now exactly similar to that of fractionating liquids in An-Ab, the only differences being related to the presence of Di crystals and the lower temperatures due to the presence of the Di component. As in the pure plagioclase system, theoretically perfect fractional crystallization drives the liquid all the way to the An-free sideline Di-Ab; in truth, we are aware that the plagioclase effect will prevent crystallization of pure Ab in this Ca-hearing system, and cause Na-Si enrichment in the liquid, so the very late stages of fractional crystallization are scarcely worth discussing in terms of the ternary model. If the system were truly ternary, a eutectic in Di-Ab would be the ultimate destination of the liquid.

Fractional crystallization of liquids initially in the plagioclase field involves new principles of liquid paths. Recall that with equilibrium crystallization, liquids were at all times constrained to lie on a tie line (Fig. 8.8) which passed through the initial bulk composition to plagioclase crystals. In the case of perfect fractional crystallization, the liquid knows nothing of the initial bulk composition, and each new liquid composition behaves as a new initial liquid composition. The liquid need to yield only enough Ab component at a given instant to furnish the current plagioclase composition; it need not yield Ab component to make over all the previously-formed crystals, as it was required to do in equilibrium crystallization. The liquid therefore becomes Ab-rich more rapidly than in equilibrium crystallization, that is, it follows a more Ab-rich path, such as the one illustrated in Fig. 8.16. This path is curved, in as much as the composition of current plagioclase crystals changes continuously with falling temperature. However, the curvature is slight compared to that of the equilibrium path (Fig. 8.8). The exact path must be determined experimentally, by finding the slopes of plagioclase-liquid tie lines at different temperatures and compositions and constructing a series of isothermal sections such as those in Fig. 8.9. The fractionation path is a curve whose tangent at any temperature gives the composition of plagioclase crystals separating from the liquid; such a path can be constructed from the temperatures, liquid compositions, and tie-line slopes read from

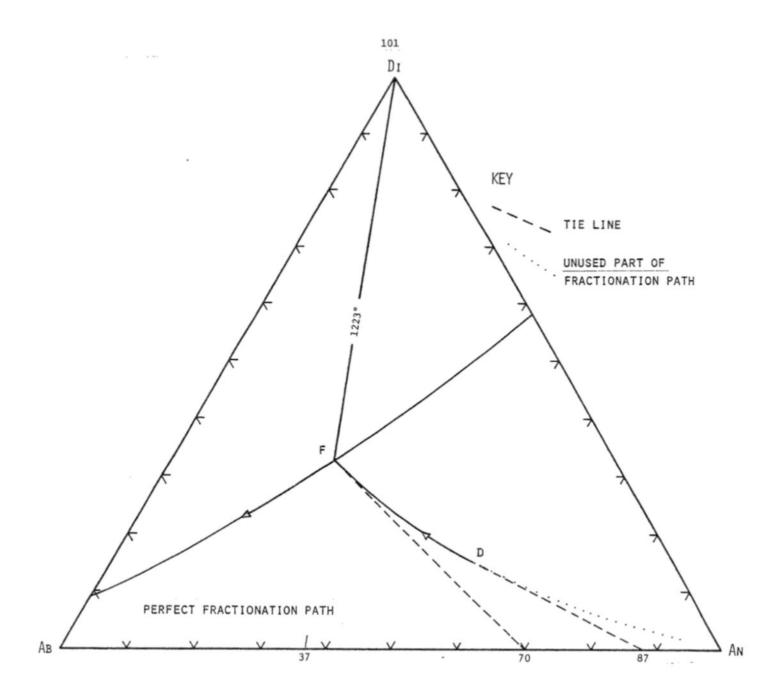

Fig. 8.16 Liquid path with fractional crystallization, bulk composition *D*. The plagioclase composition at any moment lies on a tangent to the fractionation path *D-F*. The liquid path extends from *D* through *F* to the Di-Ab sideline

the isothermal sections. In the plagioclase field, all fractionation paths emanate from the An corner.

In the example shown in Fig. 8.16, the initial plagioclase separating from liquid *D* is An_{87}, just as in the case of equilibrium crystallization. The fractionating liquid moves away from this plagioclase composition for only an instant of time, but then moves, at an infinitesimally lower temperature, away from an infinitesimally more sodic plagioclase composition. When the liquid reaches *F* on the boundary curve, it has a composition of Di 33%, (An_{37}) 67%, and is in equilibrium with traces of crystals of diopside, and plagioclase of composition An_{70}, given by the tangent to the fractionation path at *F*.

Inasmuch as the fractionation path is a function of *current* plagioclase composition only, in contrast to the *integrated* plagioclase composition controlling equilibrium crystallization, it is apparent that any liquid lying along the path *D-F* will follow the same path; the path does not define the initial liquid. This is unlike the case of equilibrium crystallization, in which every bulk composition yields a different path. In Fig. 8.16, *D* is a point lying on a fractionation path which extends backward (dotted) towards An. This whole path is shown as one of a family of fractionation paths in Fig. 8.17. These curves, *liquidus fractionation lines* (LFLs) of Presnall (1969), show the courses of liquids produced by fractional crystallization from any of a wide range of bulk compositions in the Plag + *L* field. Similar

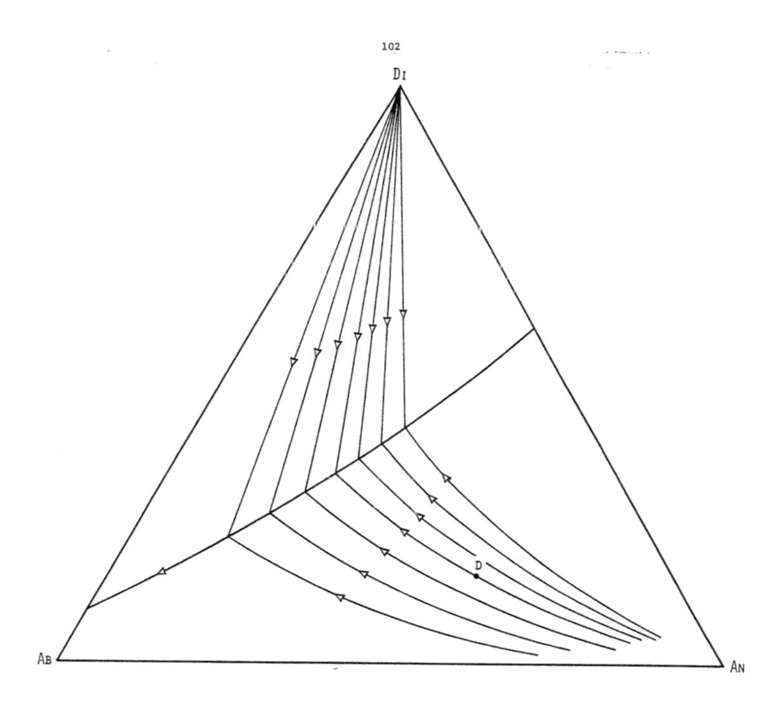

Fig. 8.17 Sample fractionation paths in Di-An-Ab. All paths in the plagioclase field emanate from An

fractionation paths (straight line LFLs) are shown in the Di + *L* field. Equilibrium paths may be found from a collection of fractionation paths, inasmuch as equilibrium liquids lie on a succession of fractionation paths, at points of tangency with tie lines through the bulk composition. An exceptionally clear discussion of these matters is given by Bowen (1941), as will be discussed later in the chapter.

The completion of fractional crystallization, once the liquid has reached the field boundary, is the same as with bulk composition *A*, in the Di + *L* field (Fig. 8.13); removal of plagioclase and diopside crystals from the liquid causes the latter to move down the field boundary, ideally until a eutectic in Di-Ab is reached, but in fact, eventually out of the Di-An-Ab ternary into chemographically more complex space. The liquid composition may thus cover an extreme range over a large range of temperature, compared to the equilibrium case.

Some aspects of equilibrium and fractional crystallization may be compared by inspection of Figs. 8.8 and 8.16, in which the initial bulk composition *D* is the same. The equilibrium liquid path reaches the field boundary at *E*, where plagioclase crystals have the composition An_{75} and the liquid has a potential plagioclase composition of An_{43}. The fractionation liquid path (Fig. 8.16) reaches the field boundary at *F*, where the analogous compositions are An_{70} and An_{37}, respectively. The two paths *D-E* and *D-F* are compared directly in Fig. 8.18. The most important difference between the two paths is that of the limiting final liquid composition,

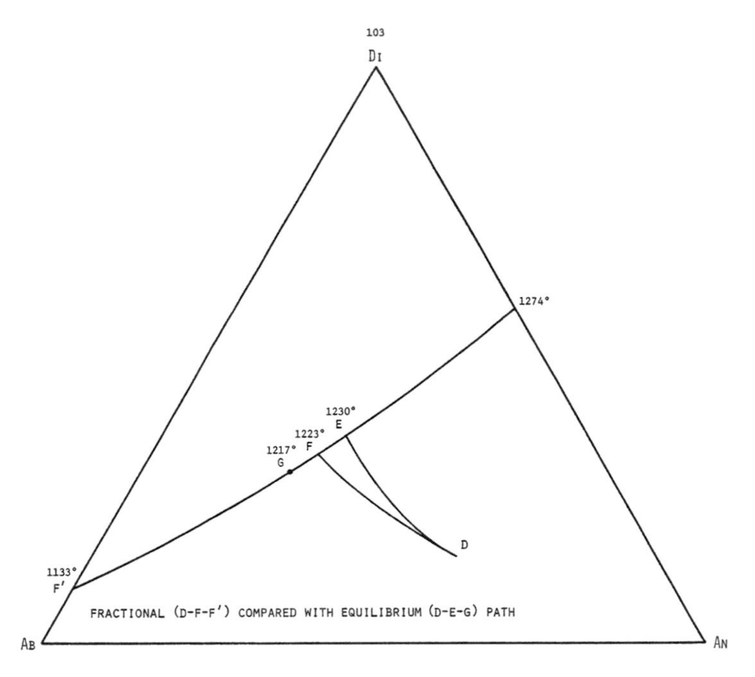

Fig. 8.18 Comparison of fractional and equilibrium crystallization paths for bulk composition *D*

point *G* in the equilibrium case and point *F*′ on the Di-Ab sideline in the fractionation case (Fig. 8.18).

Use of the lever rule requires the total solid composition to be correctly known. There are three stages in any fractional crystallization process when the TSC is accurately known: at the start, at any time until the liquid has just reached the field boundary, and at the end. The starting TSC is given by the solidus, i.e. a tie line from the liquid to diopside or a plagioclase. When the liquid has just reached the field boundary, or at any time before it does so, the TSC is given by a lever from the liquid through the bulk composition to the solidus. At the end of crystallization, the TSC is equal to the bulk composition. The only stage of crystallization for which the TSC is not immediately obtainable by construction is the stage during which the liquid lies on the cotectic *L*(Di, Plag).[3] In this book, we shall use a Rayleigh calculation for finding the TSC precisely for cotectic liquids, as described later in the chapter.

Fractional crystallization in the Plag + *L* field yields liquid *F* from bulk composition *D*, Fig. 8.19, by crystallization of plagioclase crystals ranging from An_{87} to An_{70} in composition. Now liquid *F* and the mean crystalline plagioclase composition must, together, add up to bulk composition *D*; therefore, a line (a lever, not a tie line) from *F* through *D* to the plagioclase sideline yields the mean plagioclase composition, in this case An_{80} (Fig. 8.19), slightly more calcic than the mid-range value

[3] *L*(Di, Plag) signifies “liquid in equilibrium with crystals of diopside and plagioclase”. It is a convenient shorthand for denoting an equilibrium without regard to the direction of reaction.

105

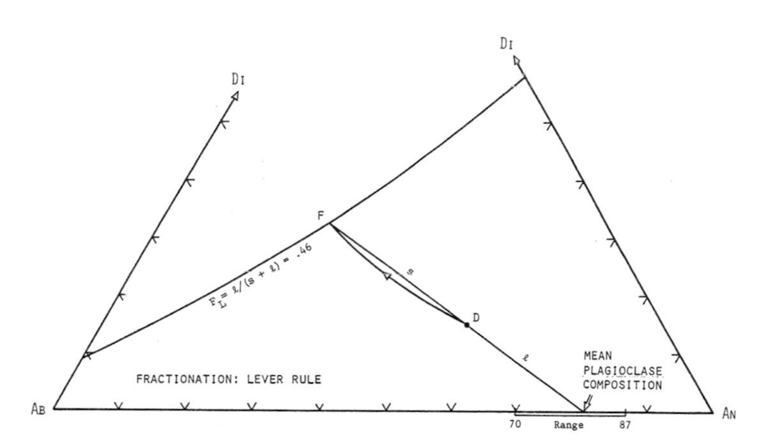

Fig. 8.19 Lever rule with fractional crystallization. A line from liquid *F* through bulk composition *D* defines the mean plagioclase composition, which is a TSC

between An_{70} and An_{87}. Measurement of the two segments of this line, with application of the lever rule, yields the value $F_L = 0.46$ (54% solidified), which is to be compared with $F_L = 0.42$ at *E* in the equilibrium case (Fig. 8.14). The fraction of liquid remaining can be similarly determined for any liquid along the fractionation path *D-F*.

As the liquid moves from *F* towards the sideline Di-Ab, we have no direct graphical way of measuring the integrated plagioclase composition. We can be sure, however, that the TSC path rises from the plagioclase sideline at first and curves over to meet the bulk composition at the end of crystallization. Furthermore, we may rigorously constrain the tangent to the TSC path at the moment that it leaves the plagioclase sideline, and at the moment that it reaches the bulk composition. These constraints are shown in Fig. 8.20. The governing rule, applicable to fractional crystallization in all ternary geometries, is that *the leading tangent to the TSC must run to the current ISC at all times*. The only remaining problem, then, is to define the ISC. This is easily done with precision for the two limiting cases.

When the liquid first reaches the field boundary at *F*, we note (Fig. 8.19) that the plagioclase ISC has just reached An_{70}, as determined from the tangent to the liquid path, the liquidus fractionation line or LFL. Now, however, the liquid will proceed down the cotectic curve, and the ISC must hop to the tangent to the new liquid path, which lies on the cotectic. We now appreciate yet another rule of fractional crystallization: *the ISC always lies on the tangent to the liquid path*. The combination of these two new rules can be expressed in the following code:

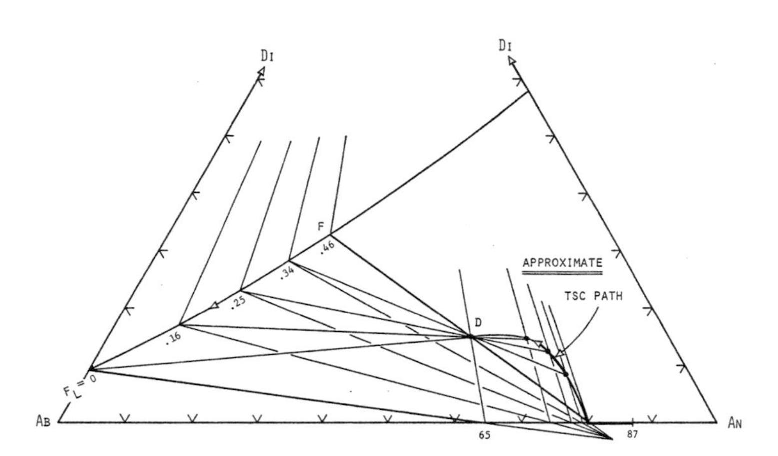

Fig. 8.20 Calculated TSC path for fractional crystallization of bulk composition *D*. The path along the plagioclase sideline is shown as the ISC and TSC. The curved path is generated from tangents to the varying ISC compositions and ends by moving towards the last ISC at the Di-Ab sideline. The fraction of liquid remaining at any stage of the process is related to the position of the liquid on the cotectic and the corresponding location of the TSC along its path

$$\begin{aligned} \text{ISC} &= \text{following tan (L)} \\ &= \text{leading tan (TSC)} \end{aligned} \tag{8.2}$$

or, the ISC follows the liquid and leads the TSC.

With this knowledge, the ISC at $F_L = 0.46$ can be constructed to lie on the tangent to the cotectic at *F* and simultaneously on the *S-S* leg Di-An_{70}. We need only construct these two lines and plot their intersection to locate the first "gabbroic" ISC, containing appreciable clinopyroxene in addition to plagioclase. Towards this ISC the TSC must now run.

Similarly, when the liquid reaches the Di-Ab sideline, the current ISC must lie on the tangent to the cotectic liquid path and on some Di-Plag tie line representing the plagioclase in equilibrium with the sideline liquid, An_9. The final ISC, therefore, lies just slightly below the cotectic curve and on a line from Di to An_9. We may extend the line from the final ISC to the bulk composition *BC* to constrain the approach of the TSC to the *BC*.

Using these two constraints, we can sketch a plausible TSC path for the gabbroic liquids on the cotectic. Our sketch will be more realistic if we recall from the system An-Ab (Fig. 6.4) that there is a significant component of motion in the TSC towards nearly pure Ab at the end of crystallization, depending on the Ab content of the bulk composition. We will therefore allow the path to remain close to the initial constraint at first, and curve towards the *BC* only near the end. Eventually, we shall show by a Rayleigh calculation that this configuration is correct.

Now having a plausible and rigorously constrained TSC path, we can run levers from any liquid composition through the *BC* to the TSC, and calculate F_L from any of these. Two levers are shown in Fig. 8.20 to illustrate this possibility. It will be clear that the results begin to lack precision as the TSC nears the *BC*, but that is what computation is for. Surely, the precision is all one could wish for in a graphical exercise.

Rock Paths (ISC Paths)

The TSC path in equilibrium crystallization is an example of a crystal path (Presnall, 1969) or, to bring home the relevance to nature, a *rock path*. A rock path describes the changing assemblage of crystals produced during crystallization or melting. Although in the case of equilibrium crystallization the rock path is identical with the TSC path, it is not so in fractional crystallization, where the TSC path is merely a geometrical convenience. We have already pointed out that the tangent to the liquid path at any point contains the composition of material being instantaneously removed from the liquid. In the case of equilibrium crystallization, this material is fictive, and has no physical expression save as an incremental part of the TSC. In the fractional case, however, the material being removed is physically isolated and preserved. This material constitutes the instantaneous solid composition (ISC), and the fractional rock path is the ISC path. The ISC lies at all times on the tangent to the liquid path and on the tangent to the TSC path. Figure 8.21 illustrates the fractional rock path for bulk composition *D*, and also illustrates the generality that fractional rock paths are discontinuous. The first part, labelled rock path 1 in the figure, results from the liquid path from *D* to *F* (Fig. 8.19) where the liquid produces plagioclase only from An_{87} to An_{70}. As soon as the liquid reaches *F*, diopside begins to crystallize, and the tangent to the liquid path is now the tangent to the field boundary. The material being instantaneously removed from the liquid is now gabbroic rather than anorthositic, and can be identified as the point where the tangent to the liquid path cuts the solid-solid leg connecting Di with An_{70}. A discontinuous "rock hop" thus occurs between the first and second parts of the rock path. The remainder of the rock path (Fig. 8.21) is smoothly continuous until the liquid is exhausted. The fractionation stages F_L are identified in the figure along the rock path as well as at the liquid apex of each 3-phase triangle. These numbers emphasize the familiar expectation that the plagioclase composition in rocks produced by fractional crystallization will be more calcic than are the liquids from which they crystallize.

The second part of the rock path in Fig. 8.21 lies very close to the field boundary, which may give the impression that rock paths in general tend to follow liquid paths. This is not necessarily the case, and is true here only because the field boundary is smooth and so nearly straight. A sharply curved field boundary would yield a quite divergent rock path. However, it is of geological interest to note that, because of the

109

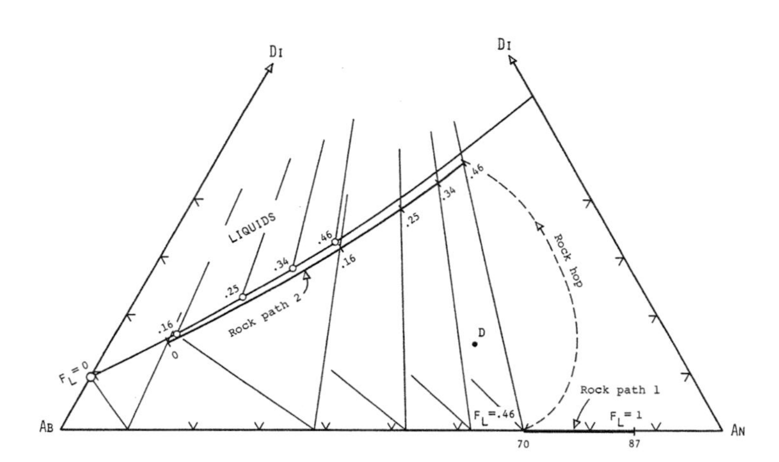

Fig. 8.21 Rock paths in fractional crystallization. The "rock hop" from ISC 1 to ISC 2 is a discontinuous change from pure plagioclase to gabbro. Rock path 2 terminates within the ternary diagram because of the "plagioclase effect"; note the tie line from An_9. In a truly ternary system it would terminate on the Di-Ab sideline

near-straightness of the field boundary, model basaltic magmas may be expected to yield fractional rock paths not unlike the liquid paths. In terminating the rock path in Fig. 8.21, the non-ternary nature of the system has been recognized, and the final plagioclase formed by the liquid lying in the Di-Ab sideline is assumed to have a composition near An_{10}. In a truly ternary system, the rock path would terminate where the liquid path terminates, as it would, for example, in the system An-Ab.

The rocks formed along the diopside-saturated rock path are very similar in plagioclase composition and mafic content to the natural rock series gabbro-diorite-monzonite-syenite.

Equilibrium Melting

We consider two cases, in which the bulk rock composition is alternatively diopside-rich and plagioclase-rich. In the first case, we may take composition *A*, Fig. 8.5 as the rock composition. Figure 8.5c shows that point *C* is the last liquid left on crystallization, so it must therefore represent the beginning of melting liquid composition for bulk composition *A*. Melting therefore begins, in this plagioclase bearing diopside rock, at a temperature of 1200°C, and as heating continues, the liquid changes towards *B* (Fig. 8.5) at the expense of plagioclase and diopside crystals. At *B*, plagioclase is used up, and the liquid changes composition towards *A* as heat is added. The path is therefore the reverse of equilibrium crystallization.

The second case is a duplicate of the first, in principle. The bulk composition *D* in Fig. 8.8 may be considered as the rock. Melting begins at 1217°C, with production of liquid *G*. With continued melting, liquid achieves composition *E*, and finally *D*, after Di crystals have disappeared.

In both cases, equilibrium melting is the exact reverse of equilibrium crystallization, levers and TSC paths may be used at any stage of the process to determine the quantitative yield of liquid.

It is of particular note that melting of any homogeneous *bulk* composition is initiated at the *equilibrium* termination of crystallization temperature, which is the *solidus* for the rock in question. The solidus is not the lowest temperature point on the liquidus surface, such as 1133°C on Di-Ab, unless the rock itself is a heterogeneous mixture containing domains of that special, low melting composition.

The lesson to be learned from equilibrium melting is that it would be impossible to distinguish a magma so formed from a magmatic residue of a crystallization process.

Fractional Melting

The initiation of fractional melting is the same as in the equilibrium cases cited above. However, continuous removal of liquid occurs, and the melting path is unlike the reverse of the equilibrium crystallization path. The first increment of melting of bulk composition *A* in Fig. 8.13 or Fig. 8.22 yields liquid *C*, and drives the TSC away from *AC*. The liquid-TSC tie lines are not constrained to pass through *A*, however, and each new TSC acts as a new solid composition about to melt for the first time. The resulting *reverse* TSC path is unlike the crystallization TSC path, being a curve whose tangent at any point runs to the composition of liquid being fractionally removed. Plagioclase, as in the system An-Ab (Chap. 6), is not completely dissolved until it reaches the composition of pure anorthite, because fractional removal of liquid changes the crystalline composition continuously towards An. Fractional melting therefore yields a series of liquids all the way along the field boundary to Di-An, as shown in Fig. 8.22, and the path of solid residua leads sharply towards Di-An before curving towards tangency with it. The TSC lies at all times on a Di-Plag leg of a 3-phase triangle: a number of these are suggested at the bottom of Fig. 8.22. When all of the Ab components are finally exhausted by fractional melting, the TSC lies at Di[4] and the liquid lies at the Di-An

[4]The TSC path terminates at pure Di. This may be proved by noting that the *L*-TSC tie line is tangent to the TSC path: consider several three-phase triangles successively closer to the Di-An sideline. For any infinitesimally small component of Ab in the liquid, the TSC must have a component of motion towards the Di-An sideline, and therefore cannot lie on that sideline at all.

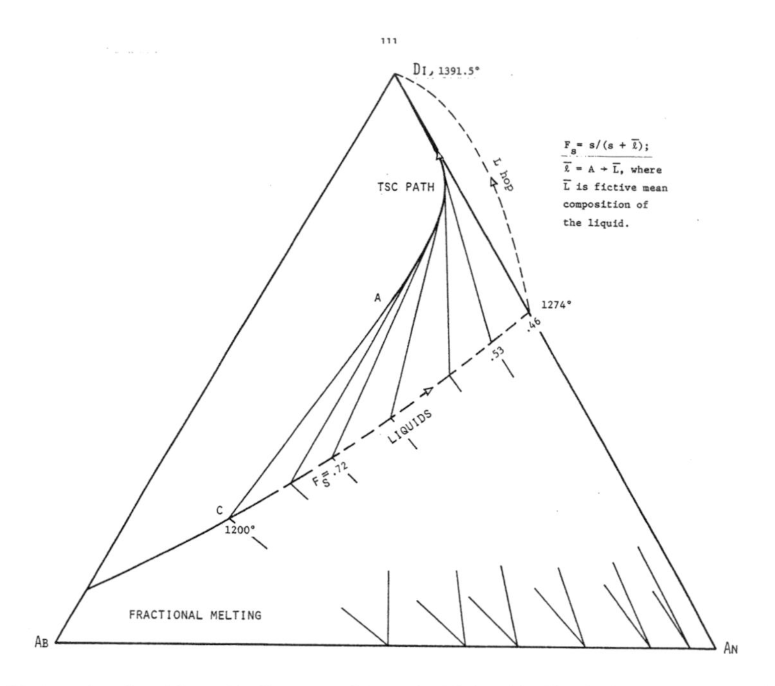

Fig. 8.22 Fractional melting of bulk composition *A* in Di-An-Ab. The fictive mean composition of the liquid removed $(\bar{L})$ always lies on the extension of a line from the TSC through *A*. This mean composition is not shown in the diagram, to avoid confusion, but its locus runs from *C* to a point directly beneath *A*, and then approaches *A* while Di is being melted

eutectic, at 1274°C. No further melting takes place until the melting point of Di itself, 1391.5°C, is reached. The entire melting path is then characterized by a series of univariant liquid fractions on the field boundary from 1200°C to 1274°C, followed after a "liquid hop" by a pure diopside liquid at 1391.5°C. No liquid of composition *A* is generated, nor is any liquid lying off the field boundary, except the diopside liquid. Fractional melting therefore produces a result very different from equilibrium melting, including a large hiatus in melt production, which is probably a natural barrier to further melting.

Fractional melting of plagioclase-rich compositions may produce an even larger melting hiatus. This is illustrated for bulk composition *D* (Di 15%, (An_{65}) 85%) in Fig. 8.23. Initial liquids lie on the field boundary from *G* to *H*, and their fractional removal drives the TSC backwards to An_{78} on the plagioclase join. In this initial melting interval, covering a small temperature span from 1217°C to 1235°C, diopside crystals are completely used up. Note that the last liquid of this series (*H*, Fig. 8.23) has the model composition of basalt. The solid residuum now consists solely of plagioclase, An_{78}, and no further melting can take place until the temperature of 1430°C is reached, after which the remainder of fractional melting involves only the plagioclase series (Chap. 6). It is doubtful if such temperatures as 1430°C are achieved in the crust and upper mantle, and this large melting hiatus is likely to bring the melting process to an end. Such a production of gabbroic (strictly,

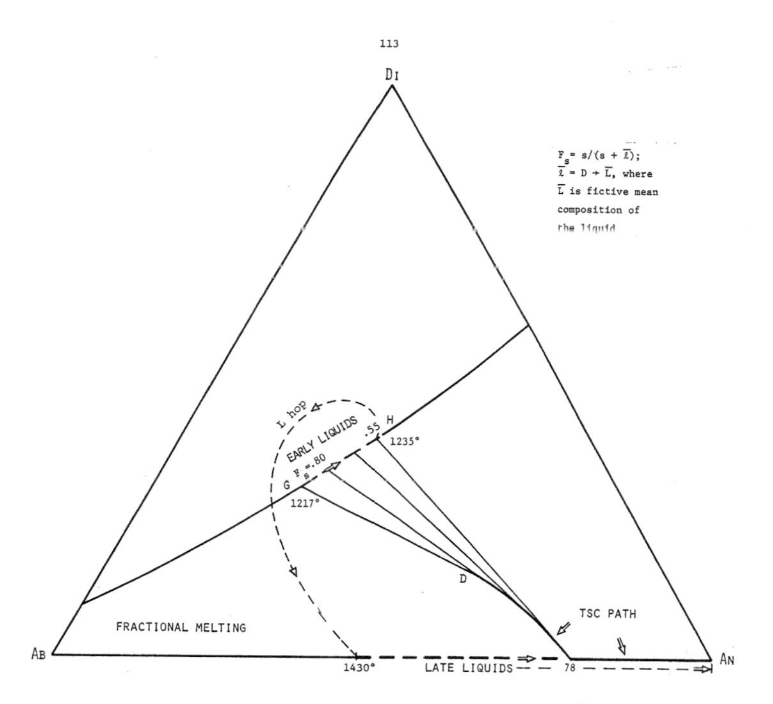

Fig. 8.23 Fractional melting of bulk composition *D* in Di-An-Ab. See Fig. 8.25 for further details

leuconoritic) melt from gabbroic anorthosite, leaving an anorthosite residuum, has been invoked to explain certain field relations around the anorthosite massifs of Egersund, Norway (Michot, 1960). The interpretation is open to debate, but the phase diagram does indicate that the process is at least possible, and likely to be terminated upon the incorporation of all Di into the liquid (Fig. 8.23).

It should be noted that equilibrium melting yields no hiatuses, because liquid is continuously available to act as a solvent for crystals. Perfect fractional melting requires continuous removal of liquid, hence a continuous change in the composition of any given reaction volume. The nearest natural approach to fractional melting may be a succession of small-batch processes, and the results will be in some way intermediate between those of equilibrium and fractional melting. The principle, embodied in fractional melting, of arresting the melting process by consumption of the whole amount of one phase, may very well operate effectively in such batch processes.

Complications Due to Non-ternary Nature

We have already seen in Chaps. 5 and 7 that the systems Di-An and Di-Ab are not strictly binary, but must be analysed in terms of four or five components for rigorous interpretation. It is therefore clear that the system Di-An-Ab, which incorporates and joins the two other systems, cannot be strictly ternary. A rigorous analysis of the system must at the very least incorporate the components needed for Di-An and Di-Ab individually, namely CaO, Al_2O_3, MgO, SiO_2, and Na_2O. Kushiro (1973) concludes that the residual liquid must include components of sodium silicate, silica, and enstatite, suggesting the reaction

$$6NaAlSi_3O_8 + 3CaMgSi_2O_6 = 3CaAl_2Si_2O_8 + 3Na_2O \cdot 8SiO_2 + 3MgSiO_3 + 7SiO_2 \quad (8.3)$$

or

$$Ab + Di = CaTs + Na\ silicate + En + silica. \quad (8.4)$$

Enrichment of residual liquids in SiO_2 holds for both Di-An alone and Di-Ab alone; enrichment in Na_2O holds for Di-Ab alone. Crystallization of Di in Di-An-Ab will therefore result in silica enrichment, and crystallization of plagioclase will tend to cause enrichment in Na_2O over and above that required by available Al_2O_3 to make the Ab molecule in the liquid. This Na_2O enrichment results from the "plagioclase effect" of Bowen (1945) already mentioned as the principle that pure Ab cannot crystallize from a melt which contains Ca. This effect is probably minor while intermediate plagioclase is crystallizing, but significant for Ab-rich liquid compositions.

Summary of Basic Principles

The model system Di-An-Ab illustrates many of the most important principles of liquidus phase diagrams, and the appreciation of these principles is demanding enough upon the reader that we ought to pause and take stock of where we stand, both with regard to phase diagrams and with regard to rock genesis.

Any ternary diagram has certain geometrical properties which allow simple, quantitative interpretation of liquid paths and TSC paths. Perhaps the most powerful of these properties is that a straight line radiating from any apex is a line of constant ratio of the other two apices. Thus we observe that in equilibrium or fractional crystallization in the Di + L field, the liquid path is a straight line away from Di.

The continuous variability of plagioclase composition with temperature causes all liquid paths in the Plag + L field to be curved; more strongly so in the case of equilibrium crystallization, since the liquid must make over earlier-formed plagioclase, and less strongly in the case of fractional crystallization. No two equilibrium paths coincide; each different bulk composition has its own equilibrium path. Fractionation paths, on the other hand, are the same for a whole variety of bulk compositions which lie on the path.

When a liquid has reached the field boundary from either primary field, the equilibrium is univariant (at constant pressure), and the liquid is forever after constrained to the field boundary in both equilibrium and fractional crystallization. In equilibrium crystallization, the end of crystallization is reached when the solid-solid leg of a three-phase triangle cuts the bulk composition. A similar termination exists for fractional crystallization, but the mean rather than the momentary plagioclase composition must be used as one end of the solid-solid leg.

The lever rule may be used to determine the solid/liquid ratio at any stage of equilibrium crystallization, and also the ratio of plagioclase to diopside among the solids, always by using the bulk composition as the fulcrum, and the solid composition on an apex (Di), sideline (Plag) or solid-solid leg of a three-phase triangle (Di + Plag) as one terminus of the lever, the liquid being the other terminus. Tangent constraints allow the lever rule to be used to approximate the progress of fractional crystallization.

In fractional crystallization, rock paths are discontinuous, and their later stages tend to mimic the continuous liquid paths if the field boundary is nearly straight.

TSC paths within the ternary diagram are always curved, the curvature being more pronounced for the fractional case than for the equilibrium case. In equilibrium melting, the TSC path and the liquid path are both simply the reverse of the equilibrium crystallization paths. In fractional melting, the TSC path is unlike the reverse of either equilibrium or fractional crystallization, and the liquid path is discontinuous, from which we may infer that when one solid phase is used up, the melting process is effectively halted, at least for a long time.

The ternary model for Di-An-Ab is imperfect because the Di actually contains CaTs and no plagioclase of composition $NaAlSi_3O_8$ may occur in the Ca-bearing system. The model can be made qualitatively adequate by stating that residual liquids tend to become enriched in Na_2O and SiO_2 as a result of these complexities of crystal composition.

Model "basalts" composed of equal amounts of plagioclase and diopside fall almost on the boundary curve in Di-An-Ab, and a similar conjunction in nature would explain why many basaltic rocks carry phenocrysts of both a pyroxene and a plagioclase, or have textures which otherwise suggest simultaneous growth of both pyroxene and plagioclase. Fractionation of such model basalts yields solid rocks of gabbroic composition grading towards feldspar-rich, pyroxene-poor rocks analogous to some syenites, or else crystalline networks of zoned minerals among which may be found "trachytic" (or at least Ab-rich) patches representing highly fractionated residual liquid. Fractionation leads towards late-stage liquids or rocks which in two important ways resemble those often formed in association with basalt: they are both feldspar-rich and alkali-rich. The ternary system therefore serves as a better model for the fractionation of basaltic magma than any of our previous models. Diagrams with plagioclase do not make very good models for genesis of basalt magma in the mantle, because plagioclase is unstable in most regions of interest. However, the melting paths may serve as good models for certain deep crustal melting processes, provided these occur and can be identified in the field.

Lever Rule with Fractional Crystallization and Melting

The system Di-An-Ab is sketched in Fig. 8.24, with a bulk composition *BC* composed of (An_{60}) 85%, Di 15% in the plagioclase + liquid field. It is assumed for convenience that the system is ternary. A segment of a liquidus fractionation line *BC-F* is shown dotted. While the liquid moves from the bulk composition *BC* to *F*, the instantaneous solid composition spans the range An_{84} to An_{65}. As long as the total solid composition remains on the plagioclase sideline, there is no problem in rigorously determining F_L; for example, when the liquid has just reached *F* in the figure, a line from the liquid through the bulk composition to the plagioclase sideline defines the total solid composition, and the lever rule yields $F_L = 0.49$.

Now the trouble starts. As diopside (Di) crystallizes, the total solid composition must rise into the triangle along some such path as shown in Fig. 8.20, constrained to follow the ISC. As in the plagioclase system, we have the initial and final states

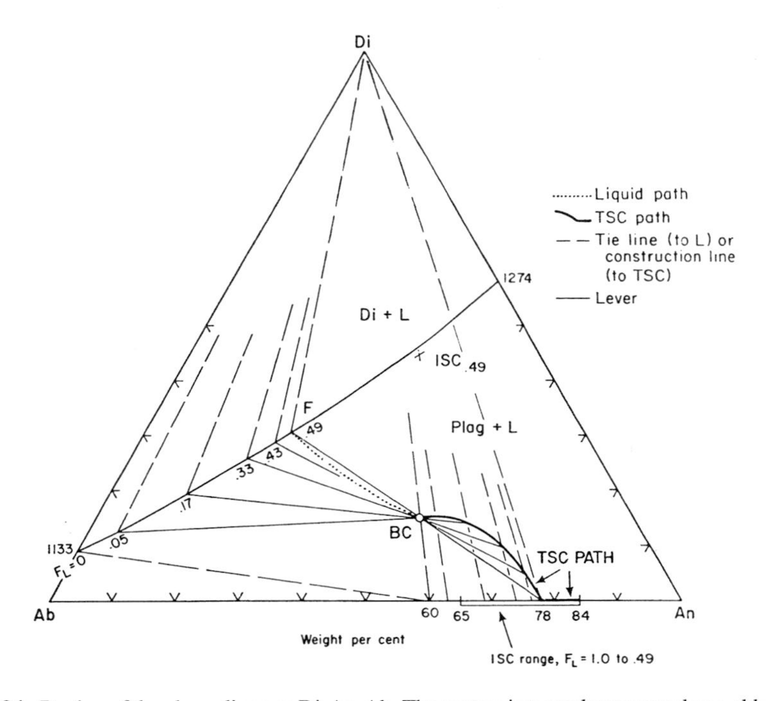

Fig. 8.24 Portion of the phase diagram Di-An-Ab. The system is treated as ternary here, although it is not. The bulk composition *BC* is (An=) 85%, Di 15%. Truncated lines radial from the Di apex are like the legs of three-phase triangles, but the Di-Plag legs meet the total solid composition of the plagioclase, not the solidus composition. The *L* (Plag) fractionation path is shown dotted. The remaining *L* (Plag, Di) path follows the field boundary. The Ab^{TSC} values for the Di-Plag legs are found from a graph (not shown) of *R* versus liquid composition. The positions of TSCs on these legs are found from levers passing through *BC* from the liquids. These levers yield the fractions of liquid (F_L) indicated. The TSC path connects the derived TSC points. The point labelled ISC is the instantaneous solid composition for the liquid at $F_L = 0.49$; it is located by the tangent to the liquid path and the (Di-An_{65}) leg of a three-phase triangle. Reproduced by permission from Morse (1976)

corresponding to F_L, = 0.49 and $F_L = 0$ (we ignore all the nonternary effects). For the plagioclase component of the ISC, the partitioning is obtained as in Eq. (8.1) above, and noted in Fig. 8.7. For the Di content of the ISC, the algorithm used for the variation of the weight fraction of plagioclase in the ISC was $F_{Pl} = 1.091X_1 - 0.181$, where X_1 is the weight fraction Ab/(An+Ab) in the liquid.

In order to apply the Rayleigh equation as in Chap. 6, we must first allow for the fact that, to a first approximation, the crystallization of Di does not affect the ratio An/(An+Ab) in the plagioclase or the liquid. In this sense, therefore, Di acts as an inert component. Its crystallization reduces the amount of liquid left, but has no effect on the plagioclase composition. The value of the *wholerock* partition coefficient in such a case is given by

$$D_{X_{Ab}}^{WR/L} = F_{Pl}{}^{*}D_{X_{Ab}}^{PLAG/L} + (1 - F_{Pl}) \tag{8.5}$$

where F_{Pl} is given in the algorithm above. The input value of $D^{PLAG/L}$ *is continuously varied* according to Eq. (8.1), and the value of wholerock *D continuously reset* from Eq. (8.6), as the Rayleigh fractionation exercise proceeds. The calculation is done by decreasing F_L in small steps (such as 0.001) and recalculating the ISC and TSC at each step. The TSC is given by

$$X_1^{TSC} = X_1^L - \left[\left(X_1^L - X_1^{BC}\right)/(1 - F_L)\right] \tag{8.6}$$

where X_1 is, as before, the fraction of the low-melting component and *BC* is the bulk composition. The result of the fractionation exercise is shown in Fig. 8.24, where it is seen that about 9% liquid remains when the liquid on the cotectic reaches the sideline. A significant amount of liquid therefore remains having a composition near pure Ab + Di.

Fractional melting in Di-An-Ab yields at first a series of liquids (ILCs) lying on the cotectic, and a TLC path which is, strictly speaking, a set of mixtures lying along a curve whose tangent always leads to the current ILC. In practice, in this system, the cotectic curve is so nearly straight that the TLC path nearly coincides with the cotectic and is indistinguishable from it at normal scales of illustration (Fig. 8.25). The TLC path is rigorously defined by levers through the bulk composition from the solidus fractionation line (TSC path) to the TLC as long as two phases remain in the TSC. If the bulk composition contains diopside in excess, the final leg of the TLC path must run straight towards the Di corner to the bulk composition (Fig. 8.22).

Only in cases where the TSC leads to the plagioclase sideline must the final stages of the TLC path be constructed by reference to fractional fusion in the plagioclase system. The TLC in this case approaches the bulk composition along a line segment of some length leading towards nearly pure An, as shown in Fig. 8.25. When the TSC has reached a composition of $An_{99.9}$, $F_L = 0.937$, so there is still 6.3% of the original solid left to melt. The calculation of this TLC path is done by running the numerical calculation for pure plagioclase, and multiplying the calculated values of F_S by the fraction solid when Di is exhausted, in this case 0.53. The calculation

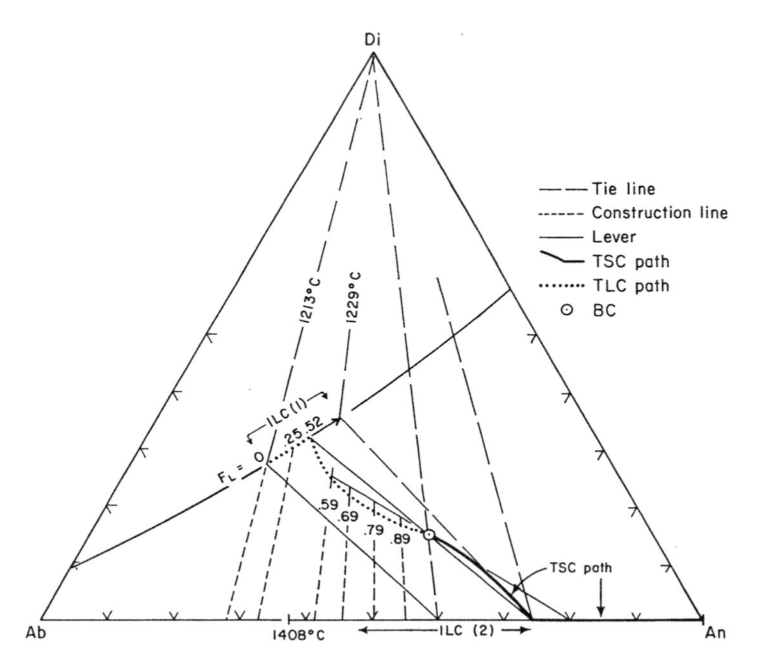

Fig. 8.25 Fractional fusion in Di-An-Ab, using the same bulk composition as in Fig. 8.24. The dotted line is the total liquid composition (TLC) path. Two instantaneous liquid composition (ILC) paths are shown, one on the field boundary and one on the plagioclase sideline. The TSC is driven away from *BC* by tangents to ILs. The TLC path is constructed from the ratio *R*′, which yields An^{TLC}, and from intersections with levers through the bulk composition from the total solid composition. Sample levers are shown at $F_L = 0.52$ and 0.59. Reproduced by permission from Morse (1976)

furnishes the solidus plagioclase crystal composition, so the TLC may be found by running a lever from the crystal composition to the bulk composition and solving for the total lever length $(s + \ell) = \ell/F_{L.}$ The TLC then lies at the end of that lever. The plagioclase composition of the TLC is then defined by the end of the lever, as shown by the short-dashed lines in Fig. 8.25.

About half the mass of this bulk composition can be extracted, in the limit, as haplodioritic liquid before diopside is used up. The nearly 200°C increase needed to produce further melting of plagioclase is probably adequate insurance against such an event in nature. Rare blocks of pure plagioclase found in some bodies of anorthosite may represent such solid residua from fractional melting.

Derivation of Fractionation and Equilibrium Paths

Fractionation paths (lines) are simply found by tangents to moving liquid or solid compositions, depending on whether the process is crystallization or melting. Equilibrium paths are, in turn, derived from fractionation paths. The plagioclase field of

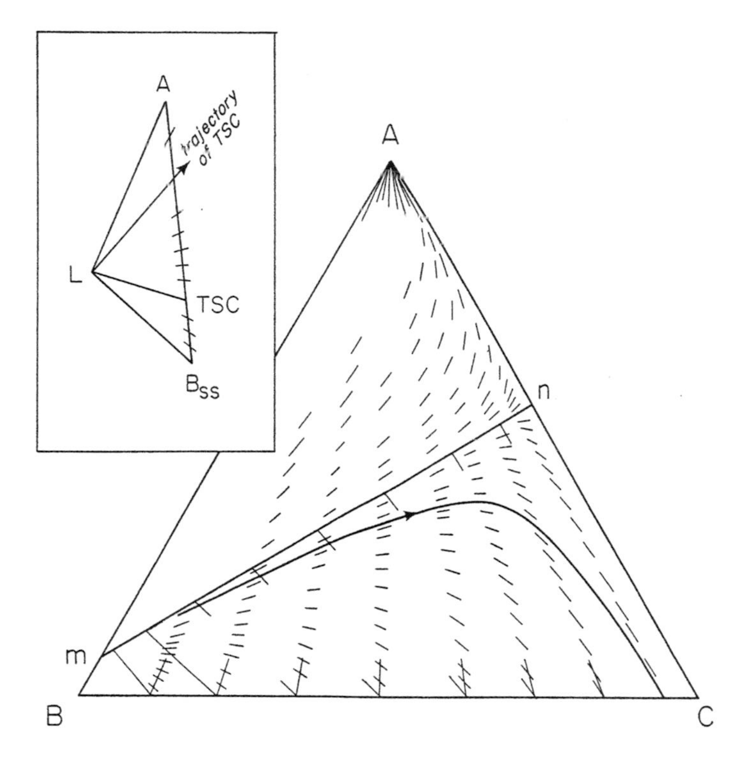

Fig. 8.26 Derivation of solidus fractionation lines (Presnall, 1969). Inset shows a three-phase triangle whose liquid lies on the cotectic *mn*. Any point on the solid-solid leg is a possible TSC in equilibrium with that liquid, and with fractional melting all TSCs would move directly away from *L* at first. Short lines radial from *L* along the solid-solid leg are therefore tangents to continuous TSC paths. An array of such tangents and a sample solidus fractionation line are shown in the main figure

the system Di-An-Ab serves very well as a general case with which to illustrate the principles. It is convenient to follow here the terminology of Presnall (1969), whose work may be consulted for further examples.

We shall use an idealized version of the system Di-An-Ab as an example, and it is convenient to begin with the derivation of solidus fractionation lines in fractional fusion. Consider the system *ABC*, Fig. 8.26, which mimics the system Di-Ab-An. The inset shows an isothermal three-phase triangle ALB_{SS}, where *L* is the liquid apex and AB_{SS} is the solid-solid leg. This leg contains a continuum of possible TSCs that are in equilibrium with liquid *L*. Fractional fusion must drive each of these TSCs, initially, directly away from *L*. Accordingly, all lines radial from *L* in the vicinity of the solid-solid leg are trajectories of TSCs and tangents to TSC paths in fractional fusion.

A map of such tangents for the ternary system may be constructed by drawing a number of three-phase triangles (which must be known from experiment) and drawing, radial from the liquid apex, an array of short tangents to TSC paths on each solid-solid leg. Such an array is shown in the main portion of Fig. 8.26. Tick marks on the cotectic *mn* indicate the location of liquid apices and L-B_{SS} legs of the three-phase triangles, and the collections of short tangents from each liquid are arranged along solid-solid legs whose ends are shown in the diagram. A solidus fractionation line is drawn through the array of short tangents so as to fall smoothly among them. Any number of other such lines (along which TSC paths must lie) could be drawn in a similar manner. Note that above the cotectic *mn*, all solidus fractionation lines lead directly to *A*, whereas below *mn* they all lead to the sideline *BC*. Line *mn* is itself a solidus fractionation line in this idealized case, but the curved cotectic in the system Di-An-Ab is not.

The case of fractional crystallization is not confined to the univariant equilibrium $L(AB_{SS})$, but involves divariant equilibria such as $L(B_{SS})$. Accordingly, much more experimental information on tie line slopes is needed for a satisfactory derivation of liquidus fractionation lines. Figure 8.27 shows an isotherm which is the locus of all

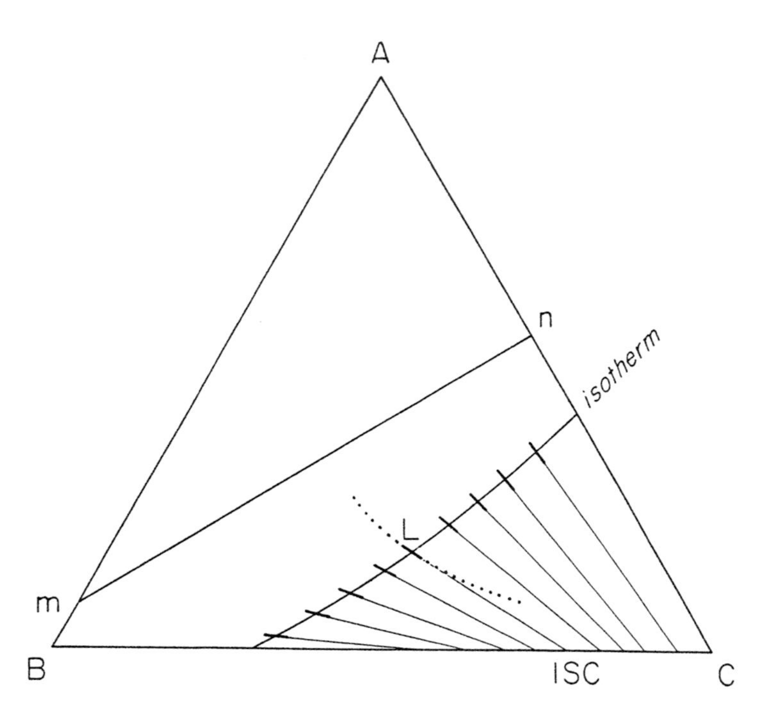

Fig. 8.27 Derivation of liquidus fractionation lines (Presnall, 1969). Crystal liquid tie lines are known from experiment to lie as shown from liquids along the isotherm. Each tie line, where it passes through the isotherm, is tangent to a liquid path in fractional crystallization (a segment of one being shown as a dotted line). An array of such tangents for different isotherms is shown in the next figure

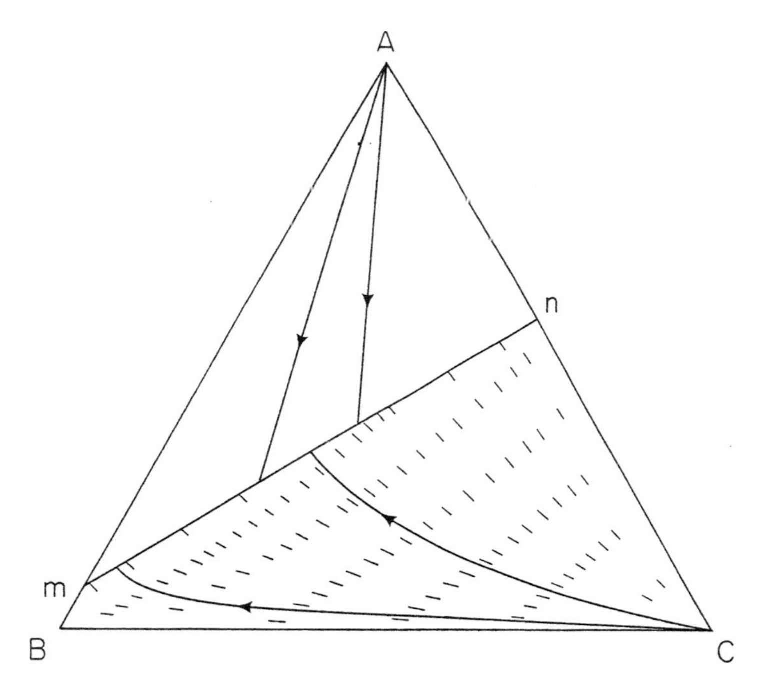

Fig. 8.28 Array of short tangents on isotherms, which collectively define the trajectories of liquidus fractionation lines. Two such lines are shown in full

liquids saturated with B_{SS} at that particular temperature. It is determined experimentally that the crystal-liquid tie lines have the slopes shown. A sample liquid L is in equilibrium with crystal ISC. Upon fractional crystallization, L must move, initially, directly away from ISC. An infinitesimally short line through L towards ISC is therefore part of a liquidus fractionation line, and the line L-ISC therefore gives the tangent to a fractionation path at L. Such a fractionation path is shown as a dotted line. The template of tie lines allows one to draw an array of such short tangents lying on the isotherm.

To complete the template of short tangents for purposes of drawing liquidus fractionation lines, the above exercise must be repeated for a variety of different isotherms. A map representing such a construction is shown in Fig. 8.28, where two sample fractionation paths are drawn in. Fractionation paths above mn are, of course, simply straight lines radial from A.

Further details on the construction of liquidus fractionation lines, and the derivation from them of equilibrium liquid paths, are to be found in Bowen (1941) and Roeder (1974).

Equilibrium Paths

It is simple to derive an equilibrium path from an array of fractionation paths, and in practice, that is how it is done. Consider the point *X* in Fig. 8.29a; it is a bulk composition, and the exercise involves crystallization. The curve lying near *X* is a fractionation path (liquidus fractionation line). The straight line passing through *X* is a tie line tangent to the fractionation line at *L*, which is the liquid end of the tie line. The other end is the crystal composition, TSC. Since the bulk composition *X* may be expressed in terms of *L* and TSC, the tie line is a lever. The point *L* lies simultaneously on the fractionation line and (because the tie line passes through *X*) on an equilibrium path. Evidently, the equilibrium path is the locus of all points such as *L* which lie on fractionation lines at points where tie lines through *X* are tangent to

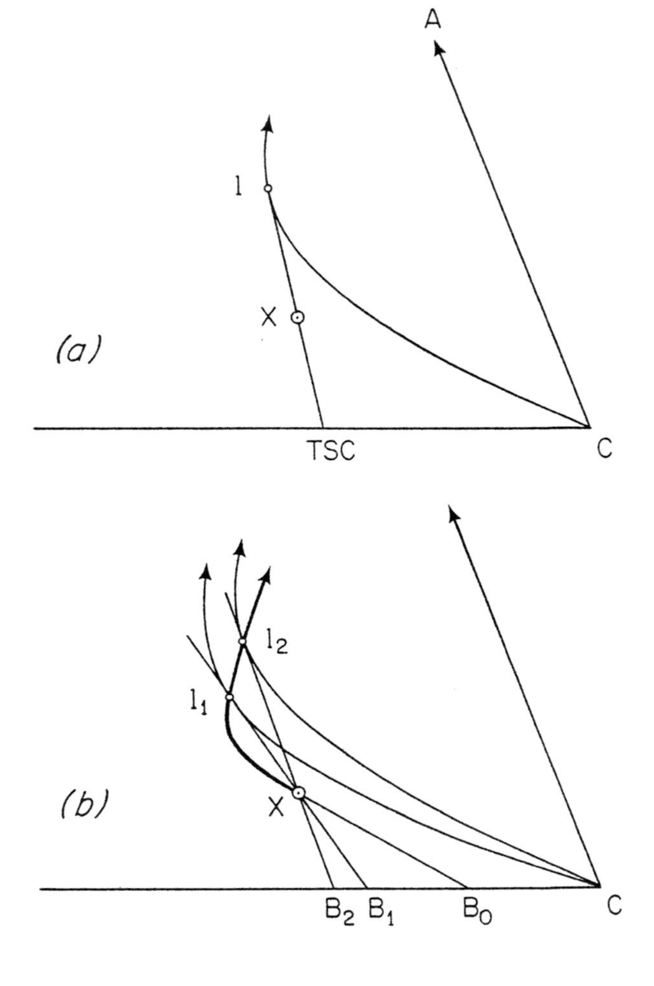

Fig. 8.29 Derivation of equilibrium liquid paths from liquidus fractionation lines (curvatures exaggerated for clarity). (**a**) Tie line through bulk composition *X* is tangent to fractionation path at ℓ, which is a point on the equilibrium path. (**b**) An array of points ℓ derived as in (**a**) define the equilibrium liquid path

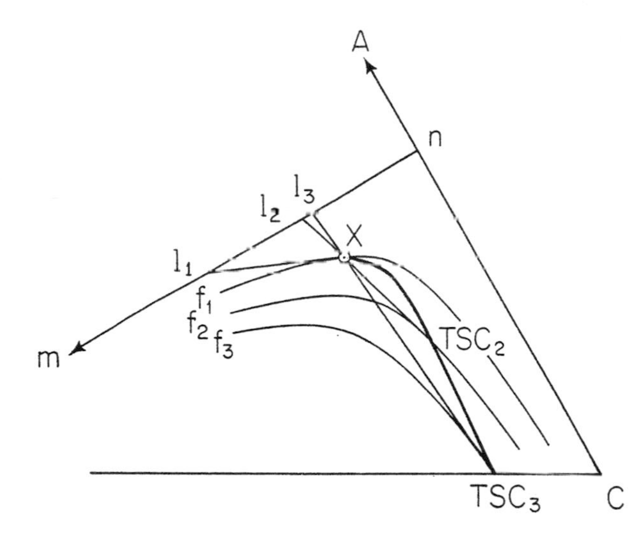

Fig. 8.30 Derivation of equilibrium TSC path from solidus fractionation lines (after Presnall, 1969). *X* is a bulk composition, the first TSC on melting and the last TSC on crystallization. Tie lines from liquids are tangent to solidus fractionation lines at *X*, TSC 2, and TSC 3

those fractionation lines. To illustrate this, two such points are located in Fig. 8.29b. The tie lines L_1B_1 and L_2B_2 are tangent to fractionation lines at L_1 and L_2. The equilibrium path is then drawn so as to run from *X* through L_1 and L_2. B_0 is the initial TSC, and B_1 and B_2 are successive TSCs at lower temperatures. For further details, see Bowen (1941).

An assembly of solidus fractionation lines may be used to derive the equilibrium TSC path followed during either melting or crystallization. In Fig. 8.30, curves f_1, f_2, and f_3 are solidus fractionation lines similar to that on Fig. 8.26. *X* is the bulk composition. Tangents to the fractionation lines which pass through *X* lead backward to liquids L_1, L_2, and L_3 on the cotectic *mn*. The points of tangency are labelled *X* (which is TSC 1), TSC 2 and TSC 3. Since these points of tangency lie simultaneously on tie lines through *X* and on the fractionation lines, they lie on the equilibrium TSC path (heavy line). Note that this is the same whether the process is melting ($X \rightarrow TSC_3$) or crystallization ($TSC_3 \rightarrow X$), involving liquids L_1, L_2, and L_3. It may be noted further that in Fig. 8.14 we found the crystallization TSC path directly by running tie lines from cotectic liquids through the bulk composition to the solid-solid legs of three-phase triangles. If you go back to the start of this section you will see that the construction of solidus fractionation lines also involves three-phase triangles, and hence the present construction (Fig. 8.30) is directly related to the earlier one.

Chapter 9
Forsterite-Diopside-Anorthite: A Basaltic Analog

Introduction

Our models for basalt have so far dealt only with plagioclase and pyroxene, the two dominant minerals in most basalts. We have ignored olivine and calcium-poor pyroxene, but these are fundamentally important minerals in any classification scheme of basalt, and are furthermore centrally involved in our study model, the basalt tetrahedron (Chap. 2). The present chapter explores a phase diagram containing iron-free olivine, namely forsterite, thereby illuminating some important principles of rock genesis as well as some simple but useful new principles of phase diagrams.

It may be noted that we are avoiding iron-bearing systems in our initial models: there are two reasons for this. One is that most basaltic rocks contain more moles of Mg than of Fe, and the Mg end-members of ferromagnesian solid solution series are therefore more suitable as first-approximation models. The second reason is that Fe, having at least two important oxidation states, requires special experimental and theoretical treatment which we shall consider in due time, but can cheerfully avoid for the moment. The most important result of adding Fe to our model systems will be to lower temperatures of crystallization towards and very close to those actually observed in natural lavas.

In now considering magnesian olivine, we are taking a big step away from fantasy and towards reality for the following reason. The source region of basalt magma in the mantle consists of an olivine-rich rock resembling peridotite. We know this in part because when very high degrees of melting occur at high temperatures, olivine-rich liquids such as picrite and komatiite are produced. We know it also because in some circumstances, nodules of source rock such as garnet peridotite are brought to the surface by vigorously rising magmas. Because the mantle source rock is rich in olivine, every basaltic melt has at some stage been in contact with olivine, and must owe many of its characteristics to this contact.

S. A. Morse, *Basalts and Phase Diagrams*,
https://doi.org/10.1007/978-3-030-97882-2_9

Preparation and Melting of Forsterite

It should be a simple matter to prepare forsterite (Fo, Mg_2SiO_4) by combining appropriate weights of pure MgO and pure SiO_2 to yield the molar ratio 2MgO: SiO_2. Like most everything apparently simple, however, this little scheme has a catch: the melting point of forsterite turns out to be more than one hundred degrees higher than that of the platinum crucibles, furnace windings, and thermocouples usually employed in the laboratory. A glass, therefore, is not readily made. Bowen and Andersen (1914), and later Grieg (1927), used furnaces and crucibles made of iridium, and optical pyrometers to overcome these difficulties. Crystalline forsterite can be made reasonably well by sintering the $2MgO{\cdot}SiO_2$ mixture at high temperatures for a long time. The product is likely to contain some unreacted MgO and SiO_2 even after several months, but with patience it can be made homogeneous enough to use as a component of other mixtures. Liquid can be made in an iridium apparatus, but it is of such low viscosity and easy nucleation that it cannot be quenched to a glass; quench-crystals form instead.

The melting point of forsterite, as determined by the above authors, is accepted as 1890°C, but an uncertainty of perhaps $\pm 25°$ is normally assigned to this value because of the experimental difficulties.[1]

Two Sidelines: Forsterite-Diopside and Forsterite-Anorthite

The interior of the ternary system Fo-Di-An can be predicted from the nature of the three sidelines, as was done in Chap. 8 for Di-An-Ab, but the prediction is somewhat crude owing to complexities in the sidelines. We shall skip the predictive exercise this time in order to get on with the matter at hand, but a brief presentation of the sidelines is in order.

Forsterite-diopside was found by Bowen in 1914 to be a binary eutectic system with a eutectic at $Fo_{12}Di_{88}$, 1387°C, just slightly below the melting point of Di itself. As shown in Fig. 9.1, the system is actually non-binary with a peritectic at $Fo_{10}Di_{90}$, 1388.5°C (Kushiro and Schairer 1963).

Forsterite-anorthite is non-binary over some of the composition range, because spinel ($MgAl_2O_4$) appears at the liquidus. The equilibrium diagram of Fig. 9.2, modified from Andersen (1915) by Osborn and Tait (1952), shows the phase relations encountered along the Fo-An join. Point *C* represents equilibria among Fo, An, Sp, and *L*; it looks like a eutectic, but application of the phase rule leads to the following result:

[1]The value of 1890°C is compatible with more recent high-pressure experiments by Davis and England (1964).

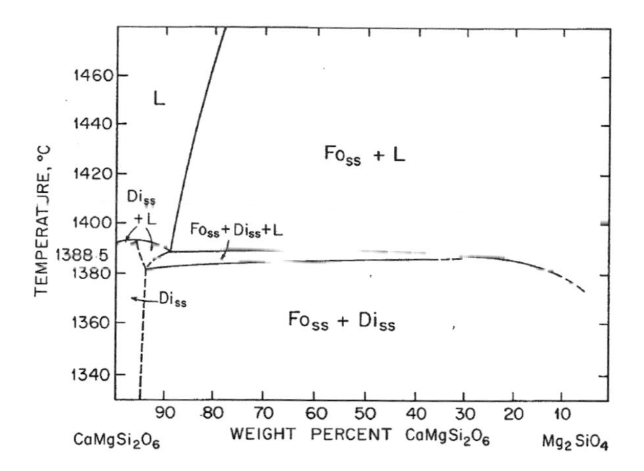

Fig. 9.1 Revised equilibrium diagram of the system diopside ($CaMgSi_2O_6$)-forsterite (Mg_2SiO_4), from Kushiro and Schairer (1963). The system is not binary because of solid solution in both forsterite (Ca as $CaMgSiO_4$) and diopside (Mg as $MgSiO_3$ and perhaps Mg_2SiO_4). A further description of solid and liquid composition along this join may be found in Kushiro (1972), discussed in Chap. 12 of this text. For the purposes of this chapter, the join will be treated as binary, with a eutectic (which is really a peritectic) at 1388°C

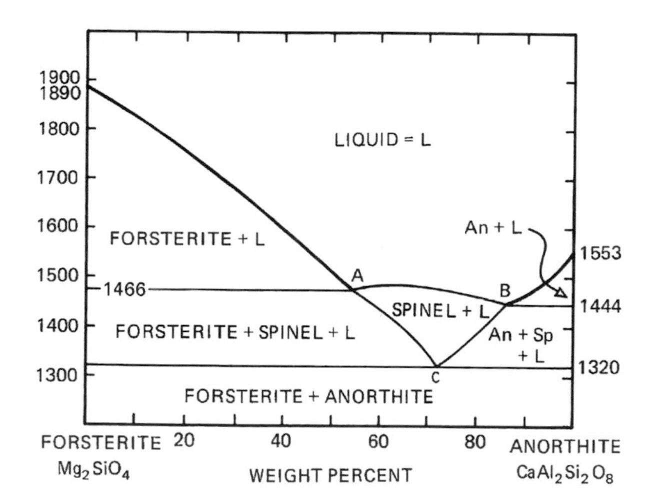

Fig. 9.2 Equilibrium diagram of the system forsterite-anorthite. The system is non-binary between and below points *A* and *B*, and in this region should be read like a road map rather than a conventional *T–X* plot (see text)

$$W_p = c + 1 - \phi$$
$$= 3 - \phi, \phi = 4, W_p = -1 \text{ (imaginary)} \quad (9.1)$$

and in order for the variance to be zero or greater, *c* must be increased to at least 3. Letting spinel be an added component, $W_p = 4 - \phi = 0$, and the point *C* is an isobaric invariant point. It is not a eutectic, because the composition of the liquid cannot be represented as a positive combination of the solid phases, but instead lies outside the diagram, towards SiO_2 (because spinel is a non-silicate). Point *C*, for this reason, is called a peritectic point. The points *A* and *B* in Fig. 9.2, and the liquidus between them, also represent equilibria with spinel, and here, too, the liquid does not lie in the composition plane Fo-An.

To see where the liquid does lie in the Sp + *L* field, we would need to draw the quaternary CMAS system and some of its interior surfaces. However, we can easily find the sense of motion of this liquid by means of a subtraction diagram (Table 9.1). The table shows that removal of spinel from a bulk composition consisting of Fo + An yields a liquid enriched in Di and SiO_2. The reaction can be written formally as

$$\underset{(An)}{CaAl_2Si_2O_8} + \underset{(Fo)}{Mg_2SiO_4} - \underset{(Sp)}{MgAl_2O_4} = \underset{(Di)}{CaMgSi_2O_6} + \underset{(silica)}{(SiO)_2} \quad (9.2)$$

Now it is a matter of experience that spinel is unstable in the presence of much excess silica, and therefore it can be expected that when SiO_2 is sufficiently concentrated in the liquid by the removal of spinel, a reversal of the above reaction will ensue, converting spinel back into other crystalline phases. This is precisely the secret of point *C*, below which Fo + An coexist without any spinel. The history of crystallization in the spinel + *L* region of Fig. 9.2 can therefore be characterized as follows: crystallization of spinel causes *L* to leave the diagram into a more silica-rich space. With cooling towards 1320°C, however, spinel begins to react with liquid, drawing the latter back towards the plane of the diagram, which it finally re-enters at point *C*, becoming exhausted just as the last bit of spinel is used up. Analogous events occur in parts of Fo-Di-An. In equilibrium crystallization, spinel is only a transient phase, eventually becoming used up by reaction with liquid. Removal of spinel by fractional crystallization can, however, exert a permanent influence on the course of liquids. We shall postpone further discussion of spinel equilibria to the end of the chapter.

Table 9.1 Effect of removing spinel from a 1:1 bulk composition of anorthite and forsterite

	C	*A*	*M*	*S*	
	1	1		2	=An
+			2	1	=Fo
	1	1	2	3	=BC
−		1	1		−Sp
	1	0	1	3	
−	1		1	2	−Di
				1	=SiO_2

Reaction: An + Fo = Sp + Di + SiO_2
Or: An + Fo − Sp = Di + SiO_2

Treatment of Fo-Di-An as a Ternary System

The diagram of Osborn and Tait (1952) is shown in Fig. 9.3. It is not difficult to recognize the relations near the sidelines, which themselves consist of the three systems Di-An (Chap. 5), Fo-Di, and Fo-An, just discussed. The first two are eutectic systems, and their invariant points must therefore be the termini of univariant lines in the ternary system. These two univariant lines, from the eutectics in Di-An and Fo-Di, meet at point E (1270°C) in the ternary diagram, where they are joined by a third univariant line, thus forming what appears to be a *ternary eutectic point* (point *E*). It is a property of isobaric ternary eutectic points that they are isobarically invariant, consist of four phases, and mark the junction of three isobarically univariant curves. In this case, the phases are Fo, Di, An, and *L*, and by the phase rule

$$C = 3, W_{\mathrm{p}} = c + 1 - \phi = 4 - \phi, \phi = 4 \tag{9.3}$$

$$W_{\mathrm{p}} = 0 \tag{9.4}$$

The three univariant curves each represent equilibria among three phases: Fo + Di + *L*, Di + An + *L*, and Fo + An + *L*. The latter is a curve which we have not seen

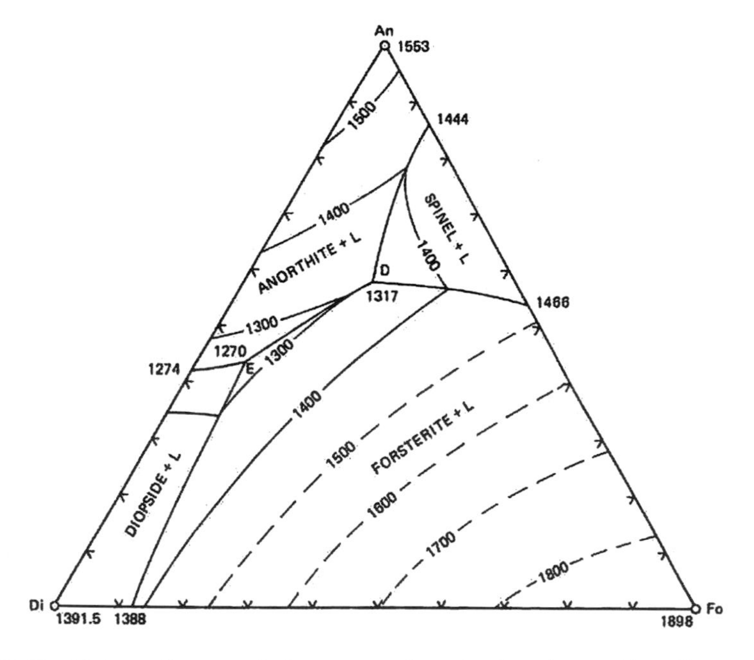

Fig. 9.3 Equilibrium diagram of the system forsterite-diopside-anorthite. The system is non-ternary for reasons discussed in the text. Point *E* resembles (and very nearly is) a ternary eutectic

before, but which we might have deduced from the existence of point *C* in Fig. 9.2, which represents coexisting Fo + An + *L* + Sp. This prediction would have been justified because, as we have already seen, the removal of spinel causes the liquid to move towards Di and SiO_2, so as it moves towards Di in the ternary plot of Fig. 9.3, it must eventually achieve a silica richness sufficient to destroy spinel by reaction, hence yielding the Fo = An + *L* equilibrium.

The ternary point *E* fails of being a true eutectic because of the CaTs content of the diopside, which we have previously encountered. The composition of the Di cannot be exactly represented in the Fo-Di-An plane, and therefore the liquid at *E* cannot lie exactly in that plane. Point *E* is called a piercing point rather than a eutectic, because the liquid lies outside the plane of the diagram. The non-ternary effect is, as before, an enrichment of the liquid in SiO_2. To a first approximation, however, the system behaves as a ternary eutectic system in this region, and we shall treat it as one for the present.

A ternary eutectic is thermally like a catch-basin into which streams (univariant lines) run. All liquids must move down-temperature in primary phase fields and along univariant lines to the eutectic, where they complete their crystallization isothermally. Conversely, all solid assemblages begin to melt with production of liquid of eutectic composition, recalling to us the original definition of a eutectic (Guthrie): the lowest melting point of a system.

The thermal properties of a ternary eutectic, ternary univariant lines, and a ternary divariant surface may be appreciated by examination of Fig. 9.4, which shows the temperature at *E*, Fig. 9.3, compared to temperatures along the field boundaries Fo +

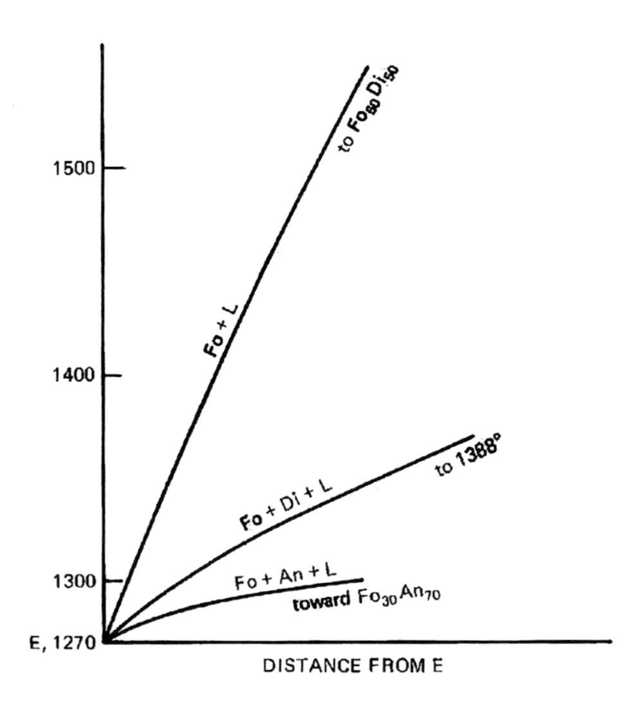

Fig. 9.4 Temperatures along various paths approaching point *E*, Fig. 9.3, to illustrate the effects of the number and kinds of solid phases on melting or crystallization temperatures

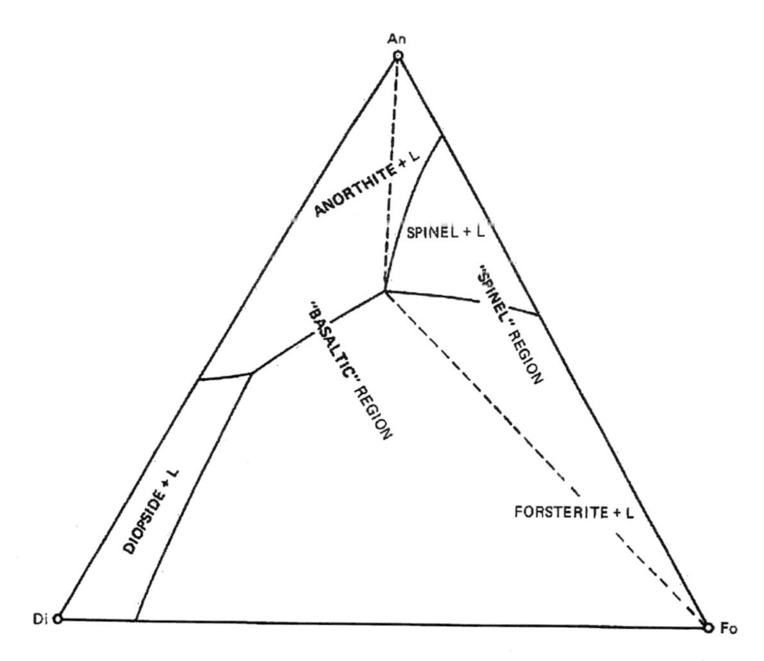

Fig. 9.5 "Basaltic region" and "spinel region" of the system Fo-Di-An. Bulk compositions to the right of the dashed line will encounter spinel equilibria on crystallization: other bulk compositions will not

An + *L*, and to temperatures along the field boundaries Fo + An + *L* and Fo + Di + *L*, and to temperatures along the surface Fo + *L* towards the $Fo_{50}Di_{50}$ composition. By analogy to topography, the "valley walls" (Fo + *L*) are very steep compared to the "stream gradients" represented by the univariant field boundaries. Accordingly, unless we are to expect enormous rates of heat production in the earth, we will not expect liquids to lie very far from a field boundary—we will not expect them in general, to climb far up "valley walls".

There are four primary fields in Fig. 9.3: Fo + *L*, Di + *L*, An + *L*, and Sp + *L*. We reserve the Sp + *L* field for later discussion, along with liquids which encounter spinel equilibria, as shown in Fig. 9.5. This figure shows that the phase diagram may be divided into spinel and non-spinel parts: any initial liquid in the area enclosed by the dashed lines An-D-Fo will encounter spinel; any initial liquid outside that area will not. We will call the non-spinel part the "basaltic region".

Equilibrium Crystallization

Crystallization in any of the three primary fields in the basaltic region is a simple affair, obeying the same rules as that in the Di + *L* region of Di-An-Ab (Chap. 8). We may illustrate the process by considering a bulk composition in the Fo + *L* field, $Fo_{30}Di_{30}An_{40}$, point *G*, Fig. 9.6.

Crystallization of Fo begins, on cooling, at a temperature of 1450°C, and drives the liquid directly away from the Fo corner, thus preserving a constant ratio of Di to An. The liquid eventually reaches the field boundary with An at 1290°C where the liquid has a composition of $Fo_{13}Di_{37}An_{50}$. At this point An begins to crystallize, generating the three-phase triangle An-Fo-*L*. Two legs of this triangle are shown within the diagram of Fig. 9.6: the other leg is the sideline An-Fo. The liquid now moves in univariance ($W_p = 4 - \phi$, $\phi = 3$, $W_p = 1$) along the field boundary towards *E*, driven by the crystallization of An + Fo. As this happens, the three-phase triangle cannot pivot, because of the fixed composition of Fo and An, so instead the angles change as *L* moves towards *E*. The liquid finally reaches *E*, 1270°C and at this moment Di begins to crystallize.

Now the phase rule tells us that the 4-phase equilibrium Di + Fo = An + *L* is isobarically invariant ($W_p = 0$), and this means that neither the temperature nor composition of the liquid at *E* may change as further heat is removed. Evidently, then, crystals must form in the exact ratio of Fo:Di:An of the liquid composition *E*, until the completion of crystallization. This is exactly the same behaviour as we saw in Di-An, treated as a binary eutectic system. The completion of crystallization is an isothermal process.

The TSC path for this crystallization process is also illustrated in Fig. 9.6. The TSC is stationary at Fo as the liquid moves from *G* to the field boundary, and application of the lever rule places achievement of the field boundary at $F_L = 0.80$. In other words, 20% of the system has crystallized as forsterite up to this point. When

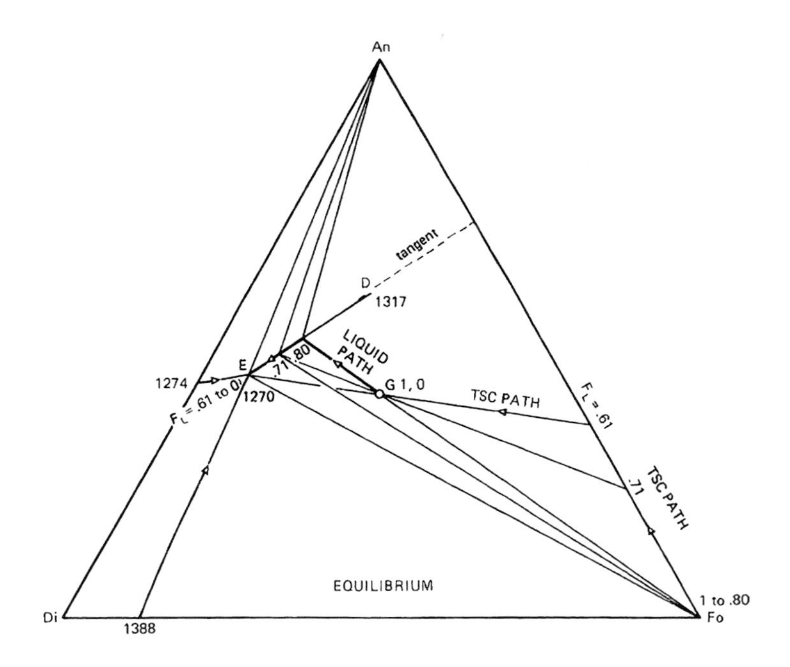

Fig. 9.6 Equilibrium crystallization of bulk composition *G* (FDA = 30,30,40). $F_L = 1$ when the liquid composition is at *G*, and $F_L = 0$ when the TSC reaches *G*. The heavy line is the liquid path

the liquid first lies on the field boundary, the *fictive* material being removed from the liquid has the ratio $Fo_{29}An_{71}$, as given by the tangent to the liquid path (field boundary). Anorthite crystals now begin to accumulate in the TSC, which then moves towards An along the sideline. This process continues until the liquid reaches E, at which point the TSC, found as always by a line passed from the liquid through the bulk composition G, lies at $Fo_{65}An_{35}$; the lever rule gives $F_L = 0.61$ at this point. The system now consists of liquid, 61%, Fo crystals 25%, and An crystals 14% by weight; the rock is a mafic allivalite (basic olivine-plagioclase rock), and the liquid has an invariant, model-basaltic composition. On further cooling, the TSC now moves directly towards E, as Fo, Di, and An crystallize in the ratio of E, until the TSC coincides with G at the completion of crystallization.

Identical principles govern the equilibrium crystallization of a more mafic bulk composition H, Fig. 9.7, $Fo_{50}An_{15}Di_{35}$. In this case, removal of Fo from the liquid drives the latter to the Fo + Di + L field boundary, and An is the last crystalline phase to appear, when the liquid reaches E. The liquid and TSC paths are shown in the figure, and should be compared with those in Fig. 9.6.

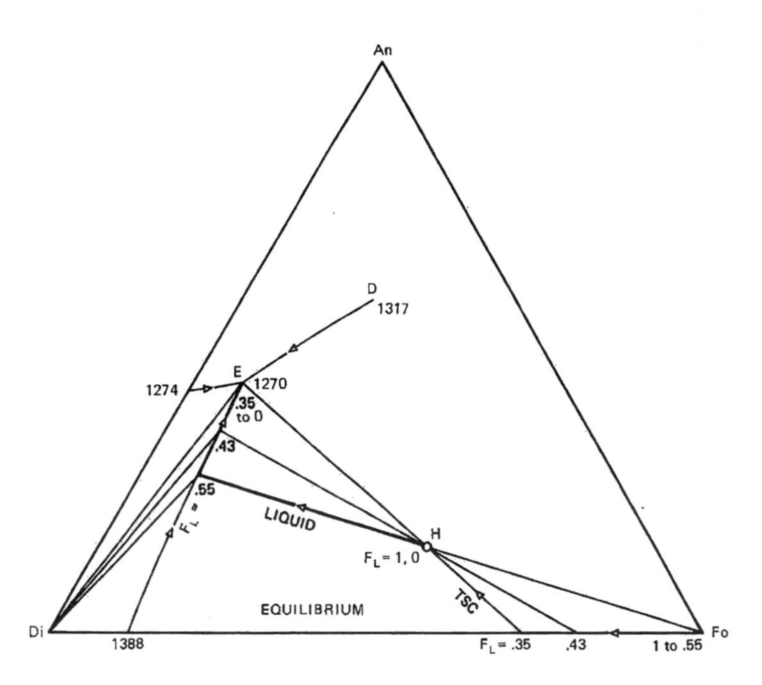

Fig. 9.7 Equilibrium crystallization of bulk composition H (FDA = 50,15,35). The liquid lies at H when $F_L = 1$, and the TSC lies at H when $F_L = 0$. The heavy line is the liquid path

Fractional Crystallization

Once the equilibrium crystallization paths and processes are understood, then fractional analogues are easily deduced. Figure 9.8 illustrates the fractional process for a liquid of initial bulk composition *G* (see Fig. 9.6 for comparison). The liquid path is identical with the equilibrium case, in as much as no solid solution occurs. The rock path is discontinuous, and is given at all times by the tangent to the liquid path.

The TSC path is the same as in the equilibrium case, and yields the same fractions of liquid remaining, F_L. Thus from $F_L = 1$ to 0.80, the rock composition is solely Fo; a dunite is produced. From $F_L = 0.80$ to 0.61, the rock composition is that of an allivalite (point 2, Fig. 9.8), having around 70% An and 30% Fo; the change from dunite to allivalite at $F_L = 0.80$ is an instantaneous "rock hop". At $F_L = 0.61$, another rock hop occurs, as the liquid reaches *E*. The rock composition now jumps to *E*, and olivine gabbro is produced. The first and third stages of the rock path are indistinguishable from the equilibrium case, but the second stage of plagioclase-rich allivalite represents a product unlike that found at any stage of equilibrium crystallization.

Natural rocks like those at (2) in Fig. 9.8 occur; more commonly they carry intermediate (Na-bearing or Fe^{+2}-bearing) plagioclase and olivine and are called *troctolites*. The Kiglapait layered intrusion in Labrador consists of early troctolites

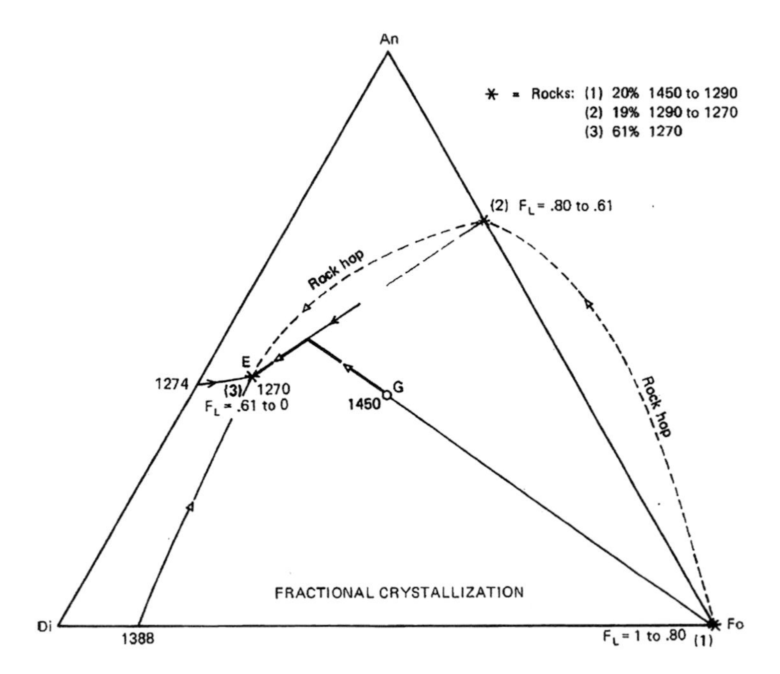

Fig. 9.8 Fractional crystallization of bulk composition *G*. Note that the liquid path (heavy line) is identical with that in the equilibrium case (Fig. 9.6), whereas the ISC follows no continuous path, but resides at three points in sequence, the points being separated by instantaneous, discontinuous "rock hops"

overlain by olivine gabbros (which then grade into ferrosyenites); the transition from troctolite to olivine gabbro is sudden, and it is a discontinuous rock hop precisely like that between (2) and (3) in Fig. 9.8.

Fractional crystallization of a liquid such as *H*, Fig. 9.7, embodies the same principles. The reader may wish to verify by her own analysis that the rock compositions produced are dunite (Fo), olivine pyroxenite (Di + Fo), and gabbro of composition *E*.

Equilibrium Melting

Still ignoring the Sp + *L* field, we may say that the equilibrium melting process for a crystalline assemblage in Fo-Di-An is just the reverse of the equilibrium crystallization process (Figs. 9.6 and 9.7). For any such composition, melting begins at 1270°C, with production of liquid having the composition of *E*. At the end of this isothermal stage, one of the crystal species Fo, Di, or An is exhausted in the mixture, and the liquid moves off *E* along one of the three univariant field boundaries. Upon exhaustion of a second crystal species, the liquid moves in divariance across a primary phase field, dissolving the last crystal species until the bulk composition is reached. The liquid path and the TSC path are both continuous. It is good practice to analyse for oneself the equilibrium melting process for bulk compositions such as *G* and *H* (Figs. 9.6 and 9.7).

Fractional Melting

With continuous removal of liquid, the melting behaviour is significantly different in only one important respect: at the termination of isothermal production of liquid having composition *E*, there ensues a hiatus wherein no further melting takes place with rising temperature until the solidus of one of the sidelines (Fo-An, Fo-Di, or Di-An) is reached. Examples are shown for bulk crystal compositions *G* and *H* in Figs. 9.9 and 9.10. In both of these, it may be observed that the TSC path on fractional melting is the same as that on equilibrium melting, or the reverse of that for crystallization. Again, it should be stressed that this holds true only when there is no solid solution, for we recall that in Di-An-Ab the TSC paths are somewhat different for fractional and equilibrium melting.

In Fig. 9.9, liquid *E* is produced isothermally until the TSC reaches the Fo-An sideline. The amount of liquid *E* produced on melting is 61% of the mass of the system (bulk composition *G*), and this liquid is indistinguishable in composition and amount from that produced on fractional crystallization. Any further melting involves only the sideline Fo-An. At this point it is well to recall the existence of a spinel field, Fig. 9.2. Melting in this pseudobinary sideline begins at 1320°C (after a 50°C rise in temperature from *E* in the “ternary” system), with production of

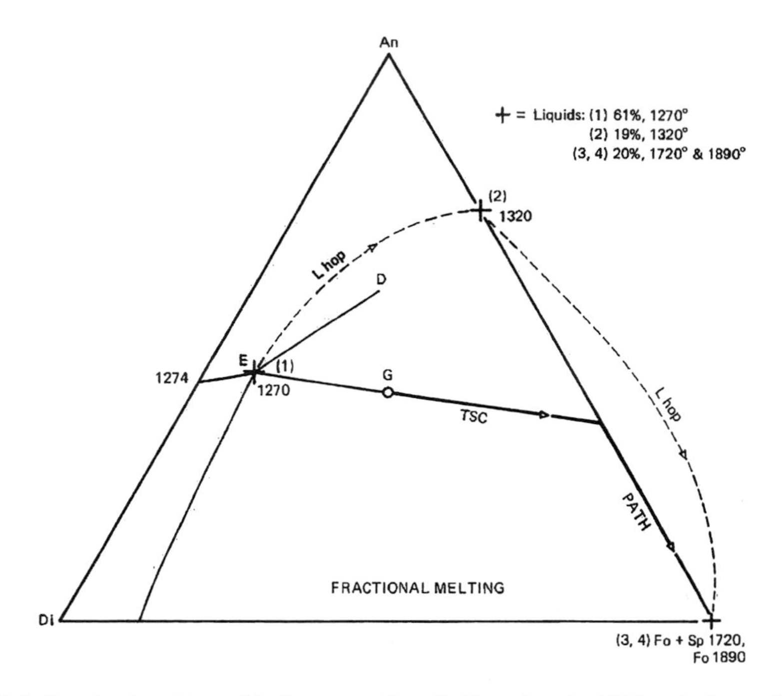

Fig. 9.9 Fractional melting of bulk composition *G*. Note that the TSC path (heavy line) is continuous and is the reverse of the equilibrium crystallization TSC path (Fig. 9.6), whereas the liquid resides at three points in sequence, separated by instantaneous "liquid hops"

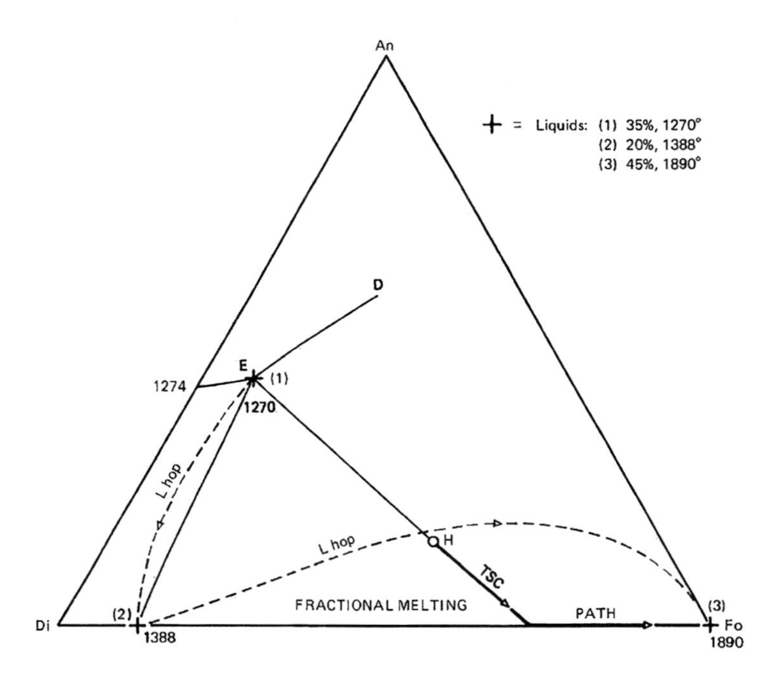

Fig. 9.10 Fractional melting of bulk composition *H*. The principles of Fig. 9.9 are duplicated here

liquid C, at $Fo_{28}An_{72}$. The reaction is Fo + An = Sp + L. Now the composition of the liquid is fixed at C, and the TSC changes towards Fo, away from An and C, with a small component of motion towards Sp. In other words, some spinel is being built up in the crystalline residue. The reaction continues until An is exhausted, when the TSC consists of Fo and Sp crystals. No further melting takes place until the Fo-Sp eutectic is reached (at about 1720°C). Here a new invariant liquid is produced until spinel crystals have disappeared, and at this point no further melting takes place until Fo itself melts at 1890°C. These facts are summarized by two "liquid hops" in Fig. 9.9, where liquid No. 2 corresponds to point C in Fig. 9.2. It is a *coincidence* that liquid No. 2 lies very close to the fractionally produced rock No. 2 in crystallization (Fig. 9.8), but it is instructive to note that a temperature of 1320°C is required to make liquid of composition 2, Fig. 9.9, whereas the *rock* of composition 2, Fig. 9.8, is produced at temperatures from 1290°C to 1270°C. The moral of this story is that simplified rock compositions can be attained at lower temperatures by fractional crystallization than by fractional melting followed by equilibrium crystallization.

Fractional melting of composition H is illustrated in Fig. 9.10, which the reader should attempt to analyse for himself. Notice again the coincidence between a fractional liquid composition at 1388°C, Di_{88} Fo_{12}, and the rock composition $Di_{89}Fo_{11}$ produced at lower temperatures (1320°C to 1270°C) by fractional crystallization. It is now apparent that this coincidence is not entirely fortuitous, but is caused by the near-straightness of the field boundary Fo + Di + L, so that the tangent to the liquid path (field boundary) on crystallization nearly coincides with the origin of the field boundary at the sideline eutectic.

The lever-rule liquid mass fractions on fractional melting are nearly identical to the solid mass fractions on crystallization, because of the near-straightness of the field boundaries. It must not be assumed that they are identical. They may be rigorously determined as shown by the following example, using Fig. 9.10. When the TSC has reached Fo-Di, the lever rule shows that 35% of the initial mass H has been liquified. Sixty-five percent is therefore still solid, and composed of Di + Fo crystals. Application of the lever rule from point 2, 1388°C, along the Di-Fo sideline, gives about 31.5% liquid of composition (2), 1388°C, when the TSC has reached Fo. This is 31.5% of the Di-Fo mass, but 31.5 $\times$ 0.65 of the *initial* mass H, or about 20% of the initial mass obtained as a second liquid at 1388°C.

Discussion of Melting

One very interesting feature of ternary eutectic melting, whether of the fractional or equilibrium type, is the large volume of eutectic liquid produced isothermally, as at E in Figs. 9.9 and 9.10. We have seen in Fig. 9.10 that a bulk composition having only 15% of its initial mass as An yields 35% of its mass as liquid E, which is basaltic in nature. It can easily be shown by means of similar triangles that the liquid yield is 35% whenever the initial An content is 15%, and whatever the initial Fo:Di ratio. The basaltic region of the Fo-Di-An ternary diagram may, then, be contoured

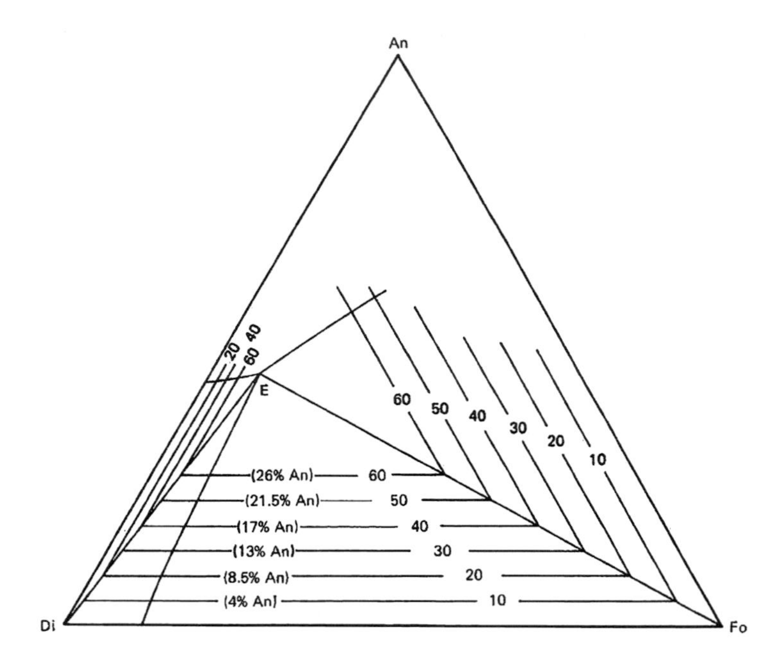

Fig. 9.11 Percentage of eutectic liquid produced on melting in Fo-Di-An. The contours show that the percentage of liquid *E* is sensitive only to the content of the component beyond *E* in the bulk composition. The contours also show the amount of liquid left (F_L) when the liquid first arrives at *E* on crystallization. Non-ternary effects are ignored

as to the mass yield of liquid *E* for varying bulk compositions, as in Fig. 9.11. This figure suggests that an ultramafic rock composed of olivine and diopside with only a trace—4%—of plagioclase would yield 10% of its mass as isothermal basaltic liquid, which is probably sufficient to become mobilized and emplaced elsewhere as magma or lava. It is, then, an important general property of eutectic systems that a meaningful amount of eutectic liquid is likely to result from melting (or crystallization, for that matter) even for bulk compositions very near one of the sidelines.

The geological importance of this result deserves brief examination. It might be argued that the production of liquid isothermally is of little importance, since the latent heat of fusion is bound to require many more calories than the mere heating of the rock, whose heat capacity is far lower than the latent heat of fusion. On the other hand, it must be said that the latent heat represented by the fusion of only 10% of the mass is likely to be less than or equivalent to that required to raise the temperature of the whole system by 3°C. Therefore melting may well be terminated in nature at or near the point where isothermal production of liquid ceases, because one solid phase is used up.

On the other hand, it is very unlikely that a truly eutectic relation will be encountered in dry natural rocks, whose minerals are, without any important exception, solid solutions. Recalling the Plagioclase diagram (Chap. 6) and Di-An-Ab

(Chap. 8), we have learned that the eutectics are not expected to occur in the presence of solid solution. There will be, therefore, a restricted likelihood of truly isothermal liquid production in nature. We cannot say there is *no* likelihood of isothermal melting in nature, because the presence of H_2O in the system will generate another type of eutectic behaviour, as we shall ultimately see. Even in the absence of a truly isothermal melting process, however, it is correct to say that when a four-phase point such as *E* becomes a piercing point because of solid solution, it retains certain eutectic-like properties. Melting in the presence of three solid phases may involve the input of fewer calories than continued melting after one of these solid phases has disappeared.

The principles of eutectic melts are, then, important to bear in mind for concepts of magma generation, but difficult to evaluate in completely general terms.

Effects of Non-ternary Nature

We have learned that Di-An is not strictly binary, due to incorporation of Al in Di (as CaTs), and therefore it is clear that point *E* in Fig. 9.3 cannot truly be a ternary isobaric eutectic, because the compositions of Di and *L* do not lie strictly in the Fo-Di-An plane. Note, however, that liquids approaching *E* along Fo + An + *L* are truly ternary until Di actually begins to crystallize at *E*. Note further that crystallization of diopside will not materially change the Fo-Di-An ratio of the liquid from that of *E*, a fact which is borne out by the experimental difficulty of showing *any* range of temperature at *E*. In truth, we are aware that the liquid must move off the Fo-Di-An plane towards SiO_2,[2] but evidently it does not move very far. Accordingly, we can expect some tendency towards silica-saturation in natural analogues of the Fo-Di-An model, although this tendency may be cancelled by other effects. The non-ternary nature of the system when the liquid projects to *E* is far less important than the effect of adding Ab to the plagioclase, for example.

A more drastically nonternary behaviour is seen when the Sp + *L* field is encountered. In terms of basalt genesis, this field is a somewhat specialized problem, first because of the relative scarcity of (olivine + plagioclase)-rich liquid, and second, because addition of Fe^{+2} to the olivine or SiO_2 to the system tends to eliminate the Sp + *L* field. For these reasons, a detailed treatment of the Sp + *L* field is somewhat of a digression from our central purpose, and so it is included as an addendum to this chapter for future reference, or for those who are curious about it.

[2] Away from SiO_2 according to Presnall et al. (1978), as discussed in Chap. 18. If their interpretation is correct, the effect of Mg in anorthite must outweigh the effect of CaTs in diopside, contrary to expectations.

Summary

The system Fo-Di-An is an interesting model of basalt because it contains a four-phase point E (Fo + Di + An + L) which comes very close to the olivine: augite: plagioclase ratio of many natural olivine basalts. By inference, the natural examples may achieve this ratio by a modest amount of fractional crystallization if their composition lies anywhere near the model four-phase point. Moreover, while plagioclase is not an expected phase in the upper mantle site of basalt magma generation, its presence in small quantity in a high-level ultramafic rock could cause the generation of appreciable quantities of basaltic liquid if melting were to occur in the deep crust or uppermost mantle.

The system treated here is of further interest as an example of a nearly ternary, nearly eutectic system. It serves as a model for discussion of crystallization and melting paths in ternary eutectic systems without solid solution, and it can be seen that these paths are simpler than those in systems where solid solutions play a role.

Addition of the single component Ab to the system generated the quaternary system Fo-Di-An-Ab, which is the critical plane of the basalt tetrahedron expanded to include An. This quaternary system is discussed by Yoder and Tilley (1962, p. 395).

Addendum: The Spinel + Liquid Field

The spinel + liquid field in Fo-Di-An requires special geometrical treatment, since the phases spinel and liquid do not lie in the Fo-Di-An plane.

An elementary comprehension of what goes on in the Sp + L field can be gained by means of three-dimensional sketches of a quaternary system which contains all the phase compositions of interest. What system should this be? Choosing all the oxide components of Fo, Di, and An, we find CaO, MgO, Al_2O_3, and SiO_2 (CMAS), which will surely suffice, as spinel ($MgAl_2O_4$, or $MgO{\cdot}Al_2O_3$) can be expressed as a member of the group. Figure 9.12 is a perspective view of the tetrahedral volume CMAS, within which is drawn the system Fo-Di-An. This system is part of the plane $CaO{\cdot}2SiO_2 - 2MgO{\cdot}SiO_2 - 2Al_2O_3{\cdot}SiO_2$, ($CS_2$ – Fo – A_2S). The idealized spinel composition lies on the MA edge, well out of the plane of the Fo-Di-An system. *Removal* of Sp from a liquid *in* that plane will, therefore, drive the liquid away from the Fo-Di-An plane and into the volume Fo-Di-An-SiO_2, which is shown as a distorted tetrahedron in the figure. A line in the figure from Sp through the Sp + L field of Fo-Di-An illustrates the fact that the liquid must lie in this distorted tetrahedral volume, and not in any other volume of the CMAS parent tetrahedron.

To examine the geometrical relations further, we may remove the tetrahedral volume Fo-Di-An-SiO_2 and redraw the part near the Fo-Di-An base, as in Fig. 9.13, in which the SiO_2 apex lies upward, out of the picture. Figure 9.13 is a perspective view, and Sp is plotted in an arbitrary position below the Fo-Di-An plane and

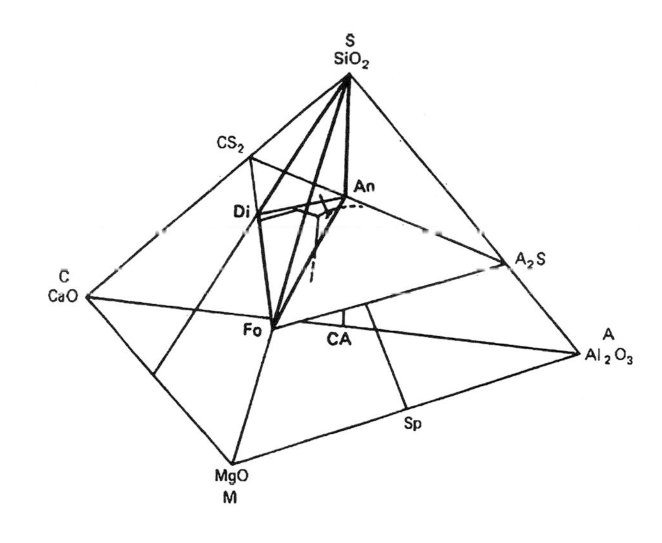

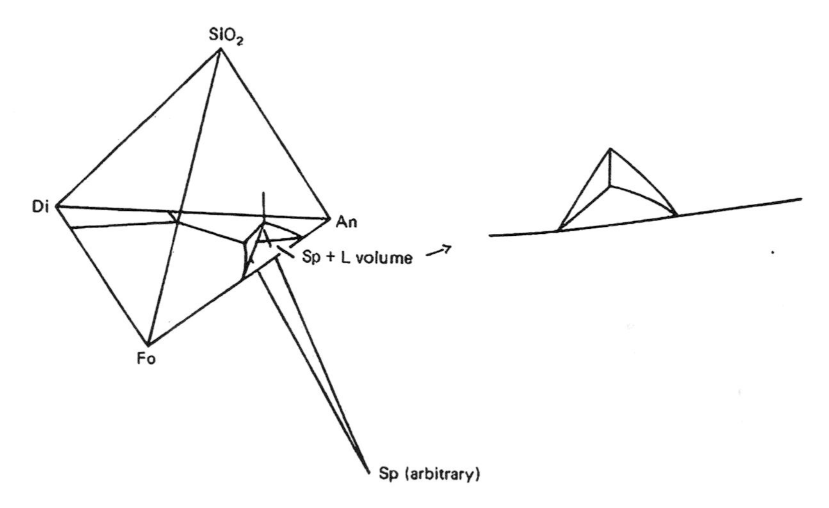

Fig. 9.12 Position of the Fo-Di-An plane in the C-M-A-S tetrahedron. The lower sketches show progressively closer views of the spinel + L volume shown in Fig. 9.13

towards the observer. Now it can be seen that the lines which bound the Sp + L region in the Fo-Di-An plane are simply the traces of intersection of that plane with the surfaces An + Sp + L and Fo + Sp + L. These two surfaces, in turn, intersect in the line Fo + An + Sp = L, and it is this line which generates the piercing point D in the Fo-Di-An plane. The surfaces do not terminate in Fo-Di-An, but extend below it, and in fact are part of a multi-faceted boundary surface of the Sp = L field which occupies a large volume around Sp in the CMAS tetrahedron.

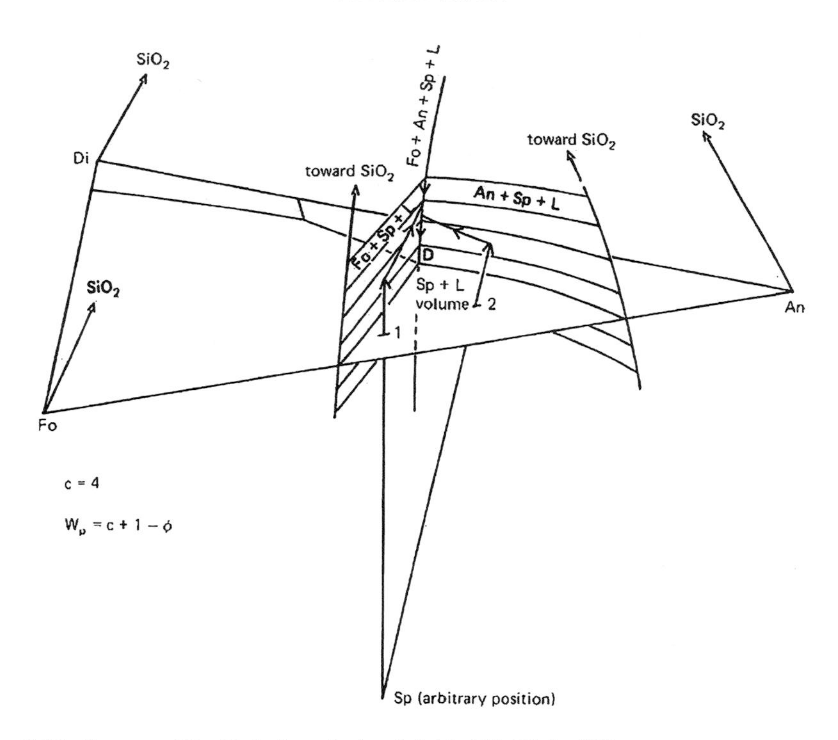

Fig. 9.13 Courses of liquids in the spinel = L field of Fo-Di-An-SiO_2

Moving along, two arbitrary bulk compositions in Fo-Di-An are labelled 1 and 2 in Fig. 9.13. Both lie in the Sp + L field, which means that when a liquid of composition 1 or 2 begins to crystallize, Sp is the first crystalline phase to form. Removal of Sp from the liquid drives the latter out of the plane Fo-Di-An, towards a more SiO_2-rich composition, until it hits one of the surfaces Fo + Sp + L or An + Sp + L. The liquid then continues along this surface in divariant equilibrium as it crystallizes either Fo or An in conjunction with spinel, as shown by arrows in the figure. Crystallization continues until the liquid reaches the four-phase, isobarically univariant ($W_p = 1$) curve Fo + An + Sp + L. Now at this point, the liquid finds itself in an awkward position. It is told that it must maintain equilibrium with Fo + An + Sp, yet no tangent to the univariant curve cuts the Fo – An – Sp plane in the CMAS tetrahedron. This univariant equilibrium cannot, therefore, represent mere crystallization by the reaction L = Fo + An + Sp + cals. There is only one way in which the liquid can remain on the univariant curve, and that is by eating Sp, not producing it, and the reaction then becomes L + Sp = Fo + An + cals. (Another way of saying this is to say that the tangent to the liquid path does cut Fo-An-(minus Sp), in which (minus Sp) is an imaginary point produced by an inversion of the Sp position through the L position.)

The result of the reaction $L = \mathrm{Sp} = \mathrm{Fo} = \mathrm{An} + \mathrm{cals}$ is that L now moves towards Sp rather than away from it, by the simple expedient of *resorbing* or dissolving the previously formed spinel crystals. In other words, the liquid is too silica-rich to permit spinel to persist, and instead, spinel reacts with the liquid to permit formation of more Fo +An. The spinel reaction terminates just as the liquid reaches D, for at this point all of the previously formed spinel is used up. and the system now consists only in Fo + An + L where L lies at D. Since Fo, An, and D are all coplanar in Fo-Di-An, it is evident that the system is now truly ternary, and crystallization proceeds as previously described.

In the equilibrium crystallization process just described, it is seen that an exact balance exists between the amount of Sp formed and the amount dissolved, so that the liquid returns precisely to the Fo-Di-An plane. Any fractional crystallization of Sp would, then, destroy this balance by locking up Sp crystals in the rocks and leaving excess SiO_2 in the liquid, which would ultimately be expressed as the mineral enstatite (= Fo + SiO_2) or the minerals enstatite and quartz. Removal of Sp would thus cause silica saturation or oversaturation of the liquid.

Equilibrium melting of compositions rich in Fo and An begins with production of eutectic liquid at E at 1270°C (see Fig. 9.6), until Di crystals are used up. The liquid then dissolves Fo and An while moving towards the peritectic D, which it attains at 1317°C. Spinel crystals now form on melting, by the reaction Fo + An = Sp + L, and the liquid leaves the Fo-Di-An plane towards SiO_2 (Fig. 9.13). When either Fo or An crystals are completely dissolved, the liquid leaves the univariant line in Fig. 9.13 and continues along one of the divariant surfaces An + Sp + L or Fo + Sp + L. When the last of Fo or An crystals is dissolved, the liquid dissolves spinel until it returns to the Fo-Di-An plane, at which point all spinel is used up and the system consists of liquid only. This is exactly the reverse of equilibrium crystallization.

Fractional melting is quite different. As long as any Di crystals remain, liquid E is produced at 1270°C, and the TSC moves directly away from E to the Fo-An sideline. No liquid of peritectic composition D is produced. The next liquid is liquid C of Fig. 9.2, at 1320°C. Continuous extraction of this liquid takes place, so that crystals of Fo An continue to generate liquid C on melting. However, small amounts of crystalline spinel are generated continuously by the reaction Fo + An = Sp + L. Therefore when, for example, An is used up, the next liquid is that of the Fo Sp eutectic at about 1720°C. If the final crystalline residue is Fo, the final liquid is formed at 1890°C; if the final crystalline residue is Sp, the final liquid is formed at 2135°C, the melting point of pure spinel. Of course these are temperatures vastly higher than expected in the upper mantle.

Fractional melting therefore ignores the Sp + L field in Fo-Di-An and produces liquids E and C, of which E is the only one of high petrological interest. A crystalline residue of Fo + An would probably be immune to further melting in the crust and upper mantle.

The above principles of equilibrium and fractional crystallization and melting apply equally well to compositions initially outside the Sp + L field in the plane Fo-Di-An, i.e., any composition in the triangular area An-Di-Fo of Fig. 9.5. Liquids in the Fo + L field, for example, crystallize Fo on cooling until they hit the Fo + Sp + L surface where it intersects the Fo-Di-An plane. Further cooling yields crystals of Sp as well as Fo, driving the liquid towards SiO_2-rich compositions along the Fo + Sp + L surface, till they reach the Fo + Sp + An + L univariant line, where the spinel reaction begins as before, bringing the liquid back into the Fo-Di-An plane at D.

Chapter 10
Incongruent Melting: The System Forsterite-Silica

Introduction

Up to now, we have concentrated on developing the phase relations in and around the critical plane in the basalt tetrahedron, described in Chap. 2. This critical plane is the plane which separates nepheline-normative bulk compositions from hypersthene-normative ones. Hypersthene is modelled, in the fundamental basalt tetrahedron, by its magnesian end member, enstatite, $MgSiO_3$ (or $MgO \cdot SiO_2$). Enstatite appears at the apex of another triangular plane in the tetrahedron, the plane Di-Ab-En, called the plane of silica saturation by Yoder and Tilley (1962). This plane marks the transition from (enstatite + forsterite)-normative bulk compositions to (enstatite + quartz)-normative bulk compositions. It also truly marks the division between solid, equilibrium assemblages of enstatite + forsterite, to the left, and enstatite + quartz, to the right. However, it does not precisely divide the crystal + liquid equilibria En + Fo + L and En + Q + L. Instead, an important phenomenon exists near the plane, wherein the assemblage forsterite + L extends to more silica-rich compositions than the plane itself, with or without enstatite, thus providing a possibility for liquids to become silica-oversaturated by fractional crystallization of forsterite. To see why this is so, we must now examine the so-called incongruent melting behaviour of enstatite in the system Fo-SiO_2.

Incongruent melting occurs when a compound, in this case enstatite, does not melt directly to a liquid of its own composition, but instead undergoes a dissociation reaction to another solid phase plus liquid.

The nomenclature of materials having the composition $MgSiO_3$ is complicated by the existence of several crystalline polymorphs of different symmetry. We shall call the composition itself enstatite (En), as a molecular or normative name. The actual crystalline phase which appears in experiments at high temperatures is an orthorhombic polymorph called protoenstatite (Pr): at lower temperatures, this inverts either to a second orthorhombic polymorph, enstatite, or a monoclinic polymorph

S. A. Morse, *Basalts and Phase Diagrams*,
https://doi.org/10.1007/978-3-030-97882-2_10

called clinoenstatite. Protoenstatite is not found in nature, indicating that one or both of the polymorphic transitions is rapid and proceeds with ease.

The enstatite composition may be made by sintering MgO and SiO_2, as with forsterite, or by thoroughly melting such a mixture and crystallizing it below 1500°C.

Incongruent Melting of Enstatite

When crystalline enstatite is heated, it melts at 1557°C, but as the experiments of Bowen and Andersen (1914) clearly show, the result is not simply liquid, but liquid plus crystals of forsterite. Specifically, Bowen and Andersen observed at 1556°C only crystalline $MgSiO_3$, and at 1558°C only glass and forsterite in their experimental charges. The melting reaction therefore cannot be

$$MgSiO_{3x1} + cals = MgSiO_{3liq}$$

but is instead of the type

$$Pr_{xl} + cals = Fo_{xl} + liquid,$$

in which the liquid must be more SiO_2-rich than $MgSiO_3$, because Fo is SiO_2-poor relative to $MgSiO_3$. This type of melting is termed incongruent, because the liquid is not coincident in composition with the phase which undergoes melting.

Having observed this incongruent melting behaviour, Bowen and Andersen were able to locate the liquid composition in the usual manner, by the quenching method. They located the completion-of-melting reactions, Fo + $L = L$, for several compositions near $MgSiO_3$, and traced this liquidus curve downward to its intersection with the 1557°C isotherm which represents the melting reaction of $MgSiO_3$. The result is shown in the binary diagram of Fig. 10.1, which is a portion of the full diagram MgO-SiO_2. The liquid generated at 1557°C contains about 61% SiO_2 by weight, compared to slightly less than 60% by weight of SiO_2 in $MgSiO_3$ (the molecular weights of MgO and SiO_2 are very near 40 and 60, respectively). The lever rule, applied to Fig. 10.1, shows that only slightly more than 5% Fo is generated by the melting reaction, along with 95% liquid.

The relations near the enstatite composition are shown more clearly in the slightly distorted sketch of Fig. 10.2. The composition of the liquid in equilibrium with Pr and Fo at 1557°C lies at *R*, a 3-phase, isobarically invariant reaction point. Point *R* is a special type of reaction point called a *peritectic point.* A peritectic point identifies the composition of a liquid which cannot be expressed as a positive combination of the solid phases with which it is in equilibrium. Instead, in this case, the liquid composition can only be expressed as Pr *minus* Fo. Graphically, this means that the peritectic point lies on the extension of a line connecting the two solid phases, unlike

Fig. 10.1 Phase diagram of the system forsterite-silica, after Bowen and Andersen. Note that the abscissa is scaled in terms of the components MgO, SiO_2

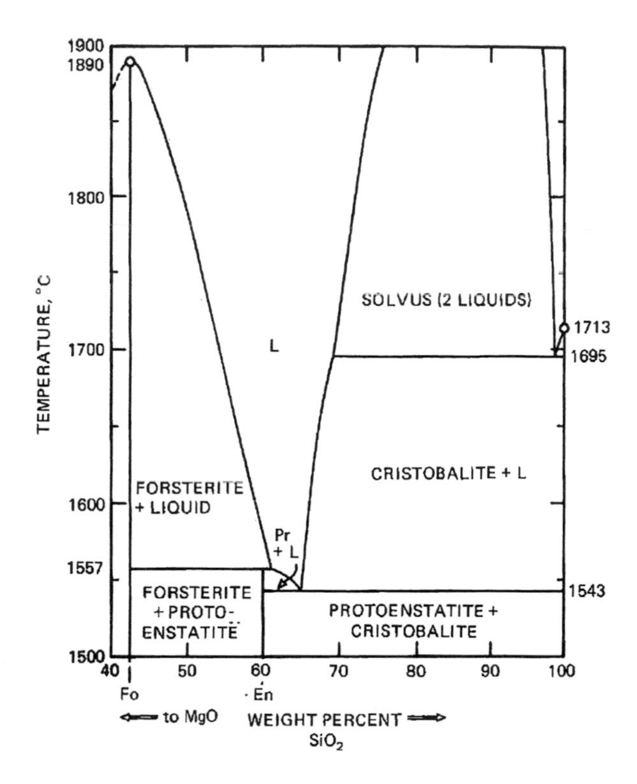

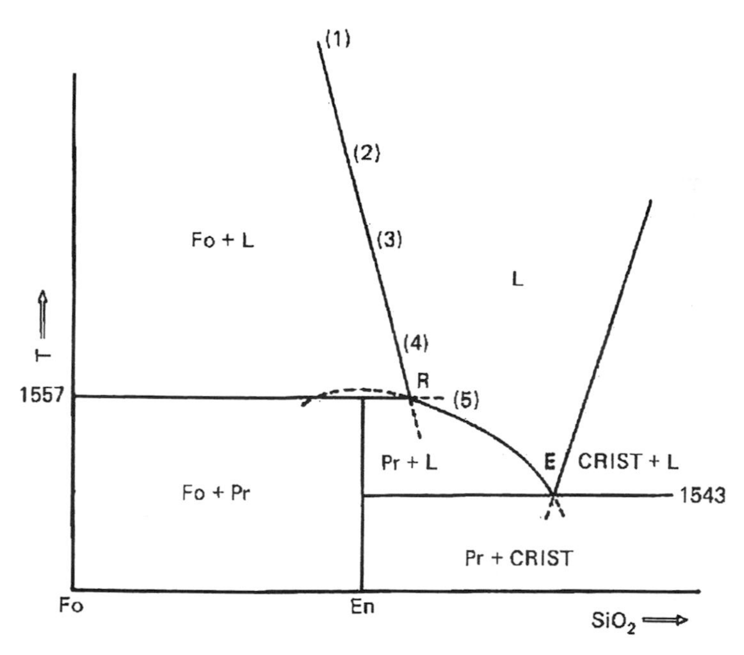

Fig. 10.2 Enlarged sketch of the protoenstatite region in Fo-SiO_2

a eutectic point, which lies between the two solid phases. Reactions which take place at a peritectic point are often termed *odd*, which can be remembered by associating it with the positive-negative combination of solid phases needed to express the liquid composition. Eutectic reactions, conversely, are said to be *even*.

To the right of point *R*, the phase diagram is of the simple binary eutectic type (see Di-An, Chap. 5), involving the phases protoenstatite, cristobalite and liquid. The liquidus curve $Pr + L = L$ passes through point *R*, and its metastable extension (dashed) shows schematically that the unrealizable, metastable, congruent melting point of $MgSiO_3$ lies on a maximum at a temperature slightly above 1557°C. By an accident of nature, this congruent melting point happens to be overlapped by the Fo $+ L$ field at 1 atm pressure, although at high pressures, as we shall eventually see, the Fo $+ L$ field recedes "behind" the En composition and the melting of $MgSiO_3$ becomes congruent.

Equilibrium Crystallization Paths

We begin with a liquid of bulk composition 1, Fig. 10.2, perceiving that this liquid lies to the Fo side of the En composition. Such a liquid, on cooling, first crystallizes Fo; the phases are now two, Fo + *L*, and the variance W_p is 1, constraining the further course of the liquid to the isobarically univariant Fo liquidus curve. Crystallization of Fo continues until the liquid reaches *R*, when the temperature is 1557°C. At this moment, a reaction sets in, as the liquid has become too rich in silica for Fo to be unambiguously stable. The reaction converts crystals of Fo to Pr, and may be expressed as

$$\underset{\text{(Fo)}}{Mg_2SiO_4} + \underset{\text{(silica in liquid)}}{SiO_2} = \underset{\text{(Pr)}}{Mg_2Si_2O_6} + \text{cals}$$

At this point the phases are three, and $W_p = 0$; the temperature and compositions of the three phases are invariant. The total solid composition (TSC), which was at Fo, now moves along the 1557°C isotherm towards the composition 1, by addition of Pr crystals. When the TSC reaches composition 1, the original bulk composition is now expressed as crystalline Fo + Pr, and no liquid therefore remains. Crystallization is therefore completed through reaction of the liquid at point *R* until it is exhausted.

Beginning at point 2, of composition $MgSiO_3$, a liquid cools until Fo begins to crystallize, and the events are exactly the same as in case 1, the liquid remaining finally at *R* as the TSC moves from Fo towards En along the 1557°C isotherm. In this case, however, the destruction of Fo by reaction with liquid is complete; the last drop of liquid converts the last crystal of Fo into Pr, and a monomineralic mass of pyroxene results.

A liquid of composition 3, containing more silica than the composition $MgSiO_3$, has a somewhat longer biography. This liquid also produces Fo crystals as it starts to crystallize, but when these have fully reacted with the invariant liquid at *R*, the TSC lies at the composition En, still short of composition 3. The liquid now coexists with only one crystalline phase, Pr, and is therefore free to move in univariance down the Pr + *L* curve towards *E*. The remaining crystallization history is just that of any binary eutectic diagram: the TSC remains at En as crystals of Pr are produced, until the liquid reaches the eutectic *E*, 1543°C. At this point, crystals of cristobalite begin to form, and the TSC moves towards *E* until it achieves the composition 3, when the liquid is finally exhausted. It is most important to note that the final result of equilibrium crystallization of composition 3 is an assemblage of protoenstatite and cristobalite. Of forsterite, which was the first crystal to form, there is no trace. All the forsterite has disappeared during the reaction when the liquid was at *R*. Under equilibrium conditions, no rock would preserve magnesian olivine crystals in the presence of a later silica mineral, because of the reaction relation between olivine and liquid. The completion of reaction insures that the disappearance of olivine will be followed by the appearance of a silica mineral.

A liquid of composition 4, coinciding with *R*, would produce, on cooling, only an infinitesimal trace of Fo at *R*, which would immediately react to Pr. Thereafter the course of the liquid would be the same as in the case of composition 3, the system behaving as a binary eutectic system.

Finally, a liquid of Composition 5 behaves solely as a liquid in a binary eutectic system, crystallizing only Pr along the initial liquidus, then Pr and Cr at *E*.

The silica-rich portion of the system beyond E is a binary eutectic region offering no complications. Bulk compositions in this region are probably rare, and irrelevant to basalts in any event.

In summary, equilibrium crystallization of bulk compositions in the neighbourhood of En will be likely to produce Fo as an initial crystalline phase (unless the liquid lies to the right of *R*), and some or all of the Fo crystals will inevitably be destroyed by reaction with liquid *R*. Such a reaction is seen in many rocks as a reaction rim of orthopyroxene around grains of olivine, testifying to the silica-saturated nature of the residual liquid, analogous to liquid *R*. This textural evidence of the olivine reaction relation in basalts is fundamental evidence for silica saturation, and forms a sufficient basis for applying the epithet *tholeiite* or *subalkaline basalt*. Persistence of olivine, in spite of reaction, can be caused by several circumstances. Under model conditions of equilibrium it can only mean that the bulk composition lies to the left of En in Fig. 10.2. Partial rims of orthopyroxene on olivine can be accepted as evidence for this condition. Complete rims, however, may indicate failure of complete equilibrium, wherein olivine persists only because it was shielded by the orthopyroxene rim from further reaction with liquid *R*. In such a case, the bulk composition may have been oversaturated, i.e. lying to the right of En, and the texture would represent a crystallization history intermediate between equilibrium and fractional crystallization.

Fractional Crystallization

Examination of the foregoing section and Fig. 10.2 will suggest that fractional crystallization may occur either through removal of olivine by sinking or flowage transport, or through the isolation of olivine grains from the liquid by shells of orthopyroxene. For all bulk compositions discussed above, perfect fractional crystallization will always result in the attainment of the eutectic *E* by the liquid. The residence of the liquid at *R* will be instantaneous, as the ISC undergoes a discontinuous hop from Fo to En. A final ISC hop to *E* occurs when the liquid reaches *E*. Fractional crystallization can only lead to silica-oversaturation of compositions like liquid 1, originally silica saturated. The incongruent melting of enstatite guarantees this result.

Melting

Melting of any mixture of protoenstatite and cristobalite crystals begins when the temperature reaches 1543°C, with the production of eutectic liquid *E*. When equilibrium is maintained, the liquid moves off along the curve *ER* (assuming the bulk composition lies to the left of *E*) as soon as the cristobalite is used up. Liquids along this liquidus curve continue to dissolve Pr until it is used up, if the initial bulk composition lies between *R* and *E*. If the initial bulk composition lies to the left of *R*, then the liquid must halt at *R*, while the reverse of the crystallization reaction occurs, i.e. Pr + cals = Fo + *L*; this is the odd, or incongruent melting reaction. This reaction ceases when Pr is used up, and then the liquid must follow the forsterite liquidus until Fo is used up, at which point the system consists only of liquid; melting is completed. We see, in this process, the manufacture of crystals of forsterite, a silica-undersaturated mineral, during the melting of a silica-oversaturated mixture. If equilibrium melting is arrested during this process residual crystals of Fo will remain set in the glass or matrix which represents liquid.

Melting of any mixture of forsterite and protoenstatite begins at 1557°C with the incongruent melting reaction, producing liquid *R* and more Fo at the expense of Pr. This reaction continues as the TSC moves towards Fo, and ceases only when Pr is used up and the TSC equals Fo. After this, melting continues, as before, as liquid along the forsterite liquidus dissolves Fo crystals until they are gone.

It is interesting to note that the melting of the silica-saturated assemblage Fo + Pr yields, by odd reaction, a silica-oversaturated initial liquid, which continues to form, isothermally, as long as any Pr remains. Thus in an equal mixture of Fo and Pr, somewhat more than half of the mass would melt isothermally to a silica-oversaturated liquid which, if removed and then observed, would show no clue of having come from a rock saturated with olivine. When you think of it, this is really a rather startling result, a bold stroke calculated to keep the student of nature on her toes.

The possibilities of fractional melting should be considered in light of the principles just discussed. Fractional melting of Pr + Cr mixtures yields liquid E which, fractionally removed, causes the TSC to move towards Pr at 1543°C. Melting ceases when Cr is used up, resuming again only at 1557°C, with a discontinuous liquid hop from *E* to *R*. Fractional removal of liquid *R* occurs during the melting of Pr, during which time Fo must accrue in the TSC, due to the incongruent melting of Pr. About 95% of the mass of Pr will have been removed as liquid when all the Pr is used up; the remaining 5% is now crystalline forsterite, which will not melt until a temperature of 1890°C is reached, an event not likely to occur in the earth's upper mantle. Fractional melting produces liquid hops and continuous rock paths, as we have seen before. However, in this particular system, the results of fractional melting are qualitatively and compositionally similar to the results of equilibrium melting, the chief difference being the absence of any liquids of composition other than *E* or *R*.

Summary

The melting of enstatite must be treated in the system Fo-SiO_2 because of the phenomenon of incongruent melting, which yields liquid more silica-rich than En, and crystals of forsterite. Fractional crystallization of liquids initially containing normative olivine as well as orthopyroxene will yield liquids containing normative quartz: hence the incongruent melting behaviour of enstatite furnishes an easy transit *towards* silica across the plane of silica saturation for liquids in the basalt tetrahedron. Perfect fractional removal of olivine must, indeed, produce silica-oversaturated residual liquids, and even imperfect fractional removal will be likely to do so. The rimming or corrosion of olivine crystals by orthopyroxene in many basaltic rocks testifies to the common occurrence of such fractionation, as does the frequent appearance of quartz in the mesostasis of such rocks. The melting of olivine-orthopyroxene rocks likewise yields a silica-oversaturated liquid which, if crystallized, will produce a silica mineral. The crystalline residuum of such a melting process will be enriched in olivine.

G-X Diagrams

Having got this far, it would be best to attempt a *G-X* diagram for Fo-SiO_2 before looking too hard at Fig. 10.3. Such an exercise teaches some profound truths about incongruent melting, and these are best learned by struggling along on one's own.

For those who have done this or are too impatient, Fig. 10.3 tells the story. The right half of the diagram simply describes a eutectic type subsystem and needs no further comment: see Fig. 5.8 for comparison. The T_1 section shows the equilibrium Fo + L as being minimized in G relative to protoenstatite (Pr, circle in *G-X* space)

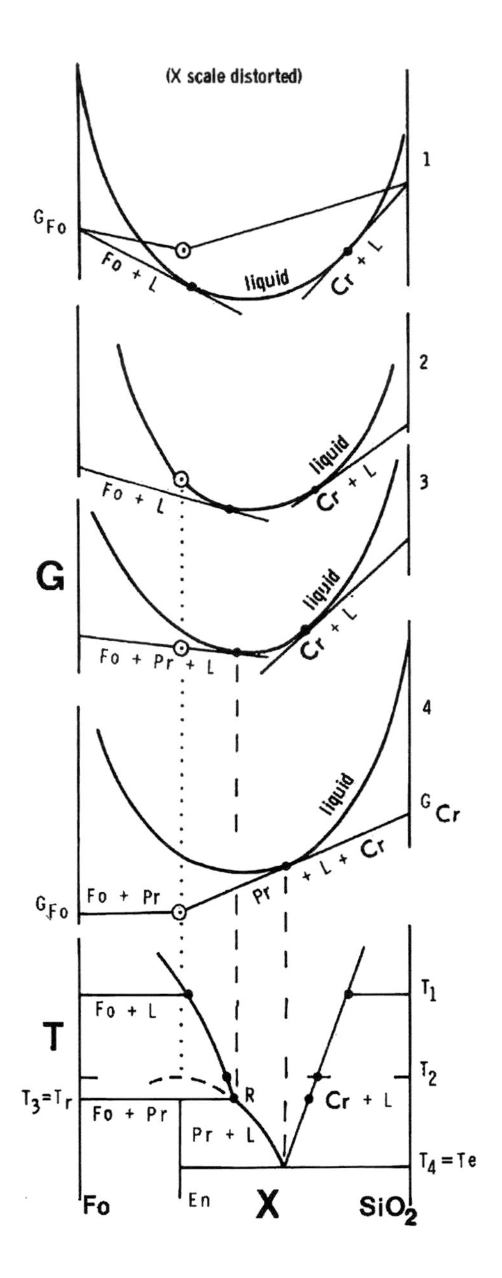

Fig. 10.3 *G*- and *T-X* diagrams for the system forsterite-silica

alone or combined with either Fo or SiO_2. T_2 is the metastable congruent melting point of Pr, which now lies on the liquid *G-X* curve but is still undercut by the stable pair Fo + *L*. Observe that nothing special is required in the shape of the liquid curve to account for incongruent melting. It is only the relative Gibbs energy of the participating solids which determines the incongruency, and by applying as little as 2 kbar pressure, we can bring the melting of enstatite clear of the Fo +

L equilibrium by sufficiently raising G_{Fo} relative to G_{En} to bring the tangent point on the liquid curve around to the left of the En composition. Section T_3, at the temperature of the reaction point R, shows that Fo, Pr, and liquid are collinear in G-X space. This condition implies their stable coexistence. A sketch between T_3 and T_4 would show a new Pr + L tangent isolating Fo from L. Section T_4 shows the eutectic condition.

Chapter 11
Forsterite-Anorthite-Silica: Incongruent Melting in a Ternary System

Introduction

In the ternary system Di-An-Ab, we saw that the complete solid solution of plagioclase had a dominant influence on the course of liquids in the ternary system. We shall now see that the incongruent melting of orthopyroxene in the system Fo-SiO_2 is carried over into a ternary system when a component such as anorthite is added. This ternary system is one of the most useful in illustrating the importance of the reaction relationship of olivine in igneous petrology. The system is the An-analogue of the right half floor of the basalt tetrahedron (Fig. 2.1), and as such is fundamental in establishing the role of the plane of silica saturation, modelled in this system by the line An-En.[1] This role is like that of a check valve: liquids which pierce the plane towards SiO_2 cannot return, at atmospheric pressure, because of the incongruent melting of En to Fo + *L*.

The system Fo-An-SiO_2 has often been used as a basis for discussing the origin of tholeiitic layered basic igneous complexes (e.g. Irvine 1970; Jackson 1970). It has a useful role to play in such discussions, and no doubt its higher pressure and Fe, Na-bearing equivalents will have an even more important role to play when they have been studied.

Binary Joins

The joins Fo-An and Fo-SiO_2 have already been discussed in Chaps. 9 and 10. An-SiO_2 was initially studied by Rankin and Wright (1915), and by Andersen (1915); it was re-determined by Schairer and Bowen (1947) to be a binary eutectic system with a eutectic at SiO_2 49.5% by weight, at 1368°C.

[1] This line was called a conjugation line in the original work by Andersen (1915), and by many others since. It is also an Alkemade line or a join. The term conjugation line has, over a period of time, become confused in the literature, and it is perhaps best to drop the name.

S. A. Morse, *Basalts and Phase Diagrams*,
https://doi.org/10.1007/978-3-030-97882-2_11

Primary Phase Regions

The ternary system determined by Andersen (1915) and shown in Fig. 11.1 could be rather accurately predicted from the bounding binary joins. It contains features reminiscent of Di-Fo-An and Fo-SiO_2, particularly the non-ternary Sp + *L* field, the two-liquid field between En and SiO_2, and the cristobalite-tridymite inversion. It is no surprise to find large fields of Fo, An, and SiO_2 polymorphs are adjacent to their respective corners. Two pairs of these fields meet in the cotectic field boundary curves *D-R* and *E*-1368°C; using conventional notation these curves could be designated *L* (Fo, An) and *L* (An, Tr), respectively. The third pair, Fo + SiO_2, is prevented from occurring by the intervention of the intermediate compound En, probably crystallizing as protoenstatite (Pr) everywhere in its primary phase field. This phase occurs together with An on the cotectic *L* (An, Pr) and with SiO_2 on the cotectic *L* (Pr, SiO, polymorph). All the curves so far mentioned are cotectics, in

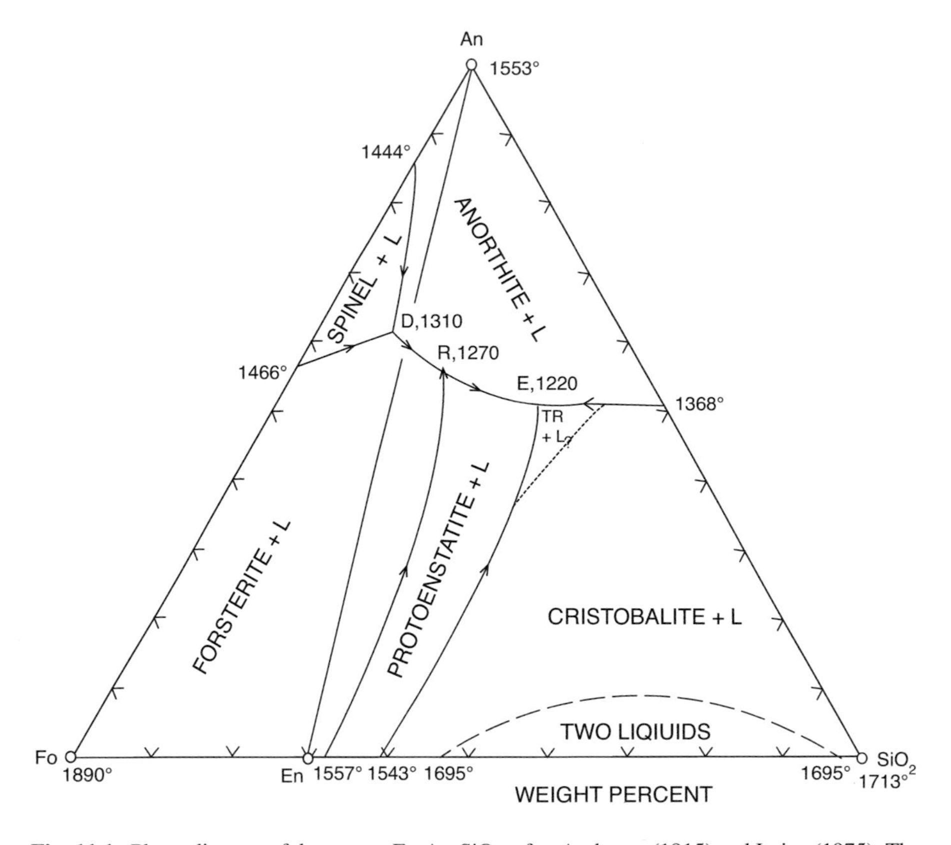

Fig. 11.1 Phase diagram of the system Fo-An-SiO_2, after Andersen (1915) and Irvine (1975). The tridymite-cristobalite boundary is highly speculative, based on Andersen's judgement that tridymite occurs below 1300°C. The inversion temperature is below the solidus on the An-SiO sideline (Longhi and Hays 1979)

which the reaction is *even*, as discussed in Chap. 10; for example, $L = \text{An} + \text{Pr}$, where all quantities are positive. The curve L (Fo, Pr), on the other hand, is odd or peritectic, the reaction being $L = \text{Pr-Fo}$, just as it is in the bounding system Fo-SiO_2. Geometrical reasons for knowing that this reaction remains peritectic all along the curve will be given shortly.

The spinel + L field plays the same role in this system as in Fo-Di-An. Spinel occurs far outside the composition plane of the system, so liquids in equilibrium with Sp lie off the plane in the opposite direction. Point D is a reaction point at which, if continuous equilibrium is maintained, the liquid returns to the plane Fo-An-SiO_2 just as all the Sp is consumed in reaction. By examination of Fig. 9.12, one may be able to discern that upon crystallization of spinel, the liquid moves into the volume Fo-An-CS_2-SiO_2 in the system CMAS, and returns to the plane Fo-An-SiO, when the solids consist only of Fo + An. Further discussion of the Sp + L field, although of some importance to certain picritic rocks, need not be attempted here. Our discussion will be confined to the "basaltic region" of the diagram in which spinel is not involved.

Cotectic Crystallization

Point E, Fig. 11.1, is an isobaric ternary eutectic which is the goal of all liquids in the system. Its temperature, 1220°C, is appreciably lower than that of the piercing point in Fo-Di-An, 1270°C, and comes within the range of natural lava temperatures, which are usually lower by virtue of having other components such as Fe and alkalies. The variance of point E is zero under the isobaric restriction, that is, $c = 3$ (F,A,S), $9 = 4$ (L, An, Tr, Pr), and $W_p - c + 1 - \phi = 4 - 4 = 0$. Converging at this point are three isobaric univariant curves, all cotectic, and crystallization of any bulk composition in the right half of the diagram (strictly anywhere to the right of a line from An to En) will drive liquids to one of these curves and thence to E, as in any ternary eutectic system. Application of the principles set forth in Chap. 9 will lead to a correct analysis of any such crystallization path, and both the continuous equilibrium and fractional cases of crystallization and melting are straightforward. Except for two examples to follow later, it is left to the reader to work out such examples for herself, preferably on paper, to keep rigorous track of TSC and other paths and to apply the lever rule. The cristobalite-tridymite transition is an internal phase transformation of silica, and proceeds with a relatively minor heat effect. When divariant liquids cross this isothermal boundary, they are briefly arrested at isobaric univariance while cristobalite inverts to tridymite (or vice versa on heating).

Equilibrium Crystallization

We shall consider all cases of crystallization and fusion involving the eight bulk compositions indicated in Fig. 11.2. All but two of these (*V* and *W*) involve peritectic crystallization. It is appropriate to follow Andersen, Bowen and many others in looking first at the sequence of compositions *P*, *Q*, *S*, *T* which span the range from silica undersaturation (*P*) through saturation (*Q*) to oversaturation (*S*, *T*).

In all exercises such as these, it is important to pause and consider, right at the start, how the bulk composition will be realized in the solid state after crystallization is complete. This requires an ability to look through the liquidus map, ignoring it completely, and identify the solid phases surrounding any bulk composition. Ignoring spinel, there are only four such solid phases to consider: Fo, An, Pr, and Tr (Cr never occurs in the final assemblage, because it inverts to Tr). Assuming equilibrium, bulk composition *P* will end up as a mixture of Fo, An, and Pr; *Q* as An + Pr, and both *S* and *T* as An + Pr + Tr. Now, looking at the invariant points *R* and *E*, it is evident from the field labels that liquid *R* coexists with Fo, An, Pr, while liquid *E* coexists with An, Pr, and Tr. Accordingly, we may correctly surmise that when olivine remains in the TSC, the final liquid will be *R*, and when Tr occurs

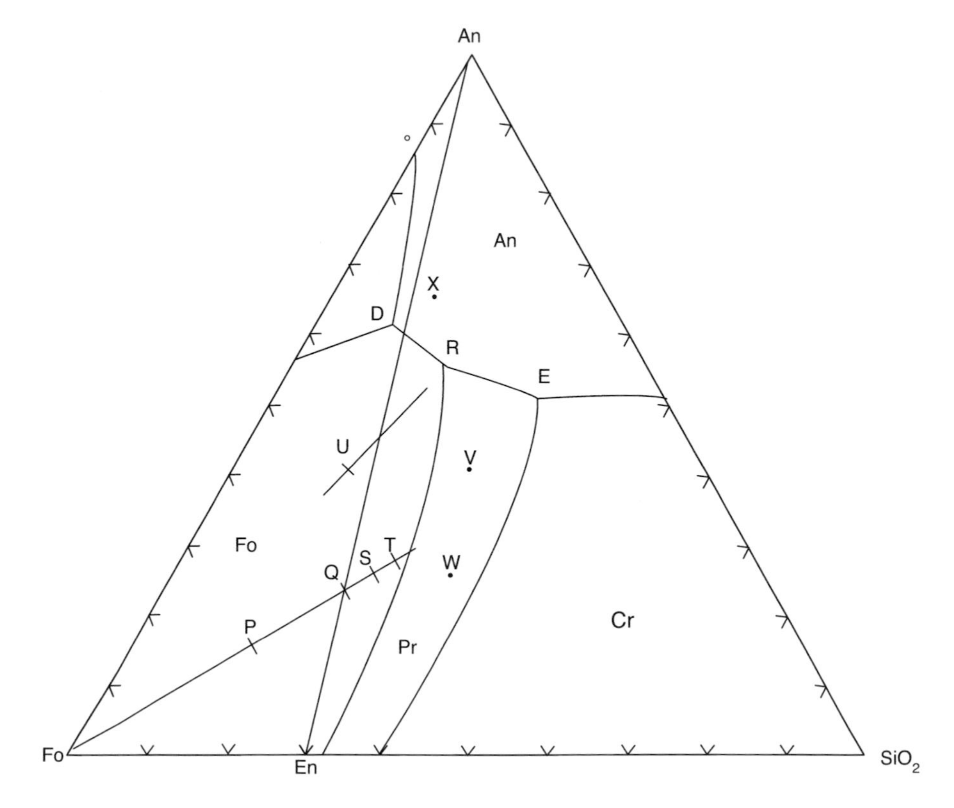

Fig. 11.2 Location of bulk compositions discussed in the text

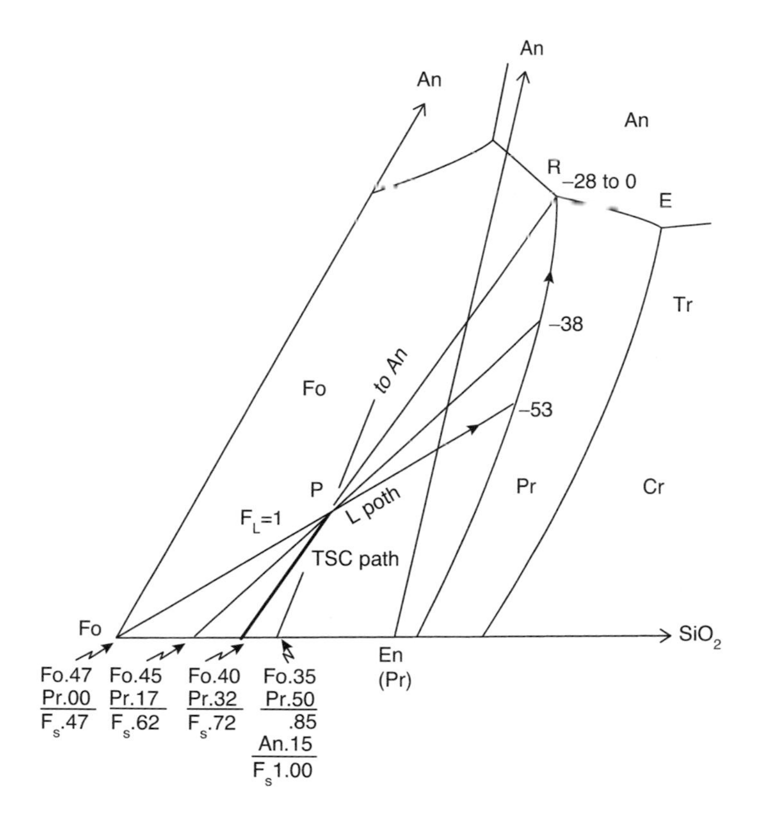

Fig. 11.3 Equilibrium crystallization of bulk composition *P*. The figures at the ends of levers show the compositions of the solids at various stages of crystallization

in the TSC, the final liquid will be *E*. Such a preview is always a helpful reminder of how things will proceed in a crystallization exercise.

The equilibrium crystallization of bulk composition *P* is illustrated in Fig. 11.3. Since *P* lies in the primary phase field labelled Fo, the liquid will at first move directly away from Fo, until it reaches the peritectic field boundary *L* (Fo, Pr). At this moment, $F_L = 0.53$, $F_S = 0.47$, the TSC being expressed as crystals of Fo. Pr now crystallizes, according to the field label, and the liquid is constrained by univariance to the field boundary. This field boundary is an exact analogue of point *R* in Fig. 10.2, Fo-SiO_2, where we saw that, because En lies *between* *L* and Fo, the reaction was odd, i.e. $L = \text{Pr} - \text{Fo}$, or $\text{Fo} + L = \text{Pr}$. This is an incongruent reaction. In the ternary case, the geometry appears at first confusing, since En (Pr) lies off the line *L*-Fo and cannot immediately be seen to lie "between" those phases. This confusion is easily rectified by looking at the *tangent* to the field boundary *L* (Pr, Fo), shown in Fig. 11.4b. As mentioned on an earlier page, the fictive material being removed from the liquid, here called the *fictive extract*, FE, must at all times lie on the tangent to the liquid path (here the field boundary). This tangent cuts the line

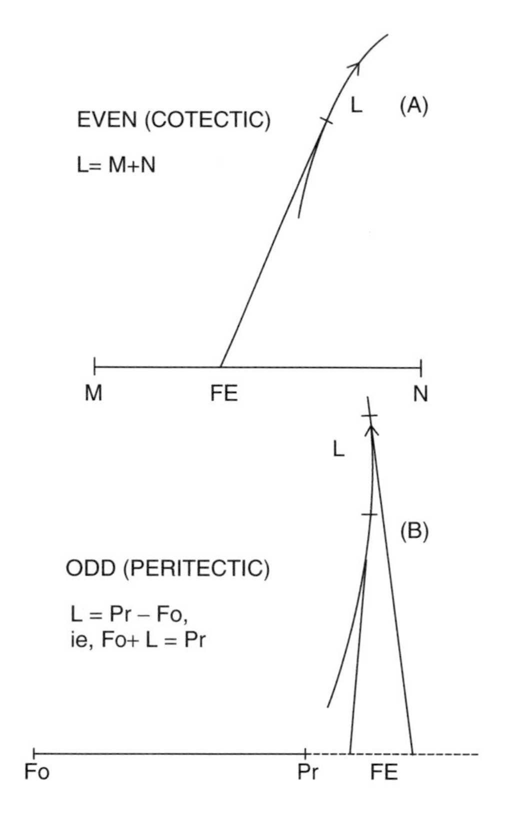

Fig. 11.4 Criteria for odd and even reactions. (**a**) Even reaction: the tangent to the liquid path falls between the two crystalline phases *M* and *N*. (**b**) Odd reaction: the tangent to the liquid path falls outside the line segment connecting the two crystalline phases

Fo-Pr beyond Pr, and Pr can now be seen to lie between Fo and the FE; hence the reaction is odd, as in the binary system Fo-SiO_2. It will be observed that *all* tangents to the peritectic curve *L* (Pr, Fo) lie outside the line segment Fo-Pr, and therefore that the curve is indeed peritectic or odd throughout its length. The ternary geometry for an even cotectic reaction is shown for comparison in Fig. 11.4a. In this case, the tangent to the liquid path cuts the line segment joining the phases *A* and *B*, and the FE is a positive linear combination of the two phases.

Back to Fig. 11.3. As the liquid moves towards *R*, crystals of Pr form at the expense of Fo + *L*, and the TSC moves towards Pr from Fo. Levers to the TSC path are shown at $F_L = 0.38$ and 0.28, and the ratio of the two solid phases is given at the TSC end of these levers. This ratio is determined by applying a lever between Fo and Pr with a fulcrum at the TSC. At $F_L = 0.28$, the liquid has reached *R*, an isobaric invariant point. As long as the system is invariant, the liquid cannot move; the TSC now approaches *R* and *P* simultaneously, as An and Pr crystallize and Fo is consumed. Crystallization ceases, as always, when the TSC equals the bulk composition. The composition of point *P* can be expressed in terms of the three solid phases Fo, Pr, and An, whose ratio is formed by a lever from An through *P*; the Fo:Pr ratio is taken from a lever on Fo-Pr, and the fraction of An is determined by the height along the lever from the base of the triangle to *P*.

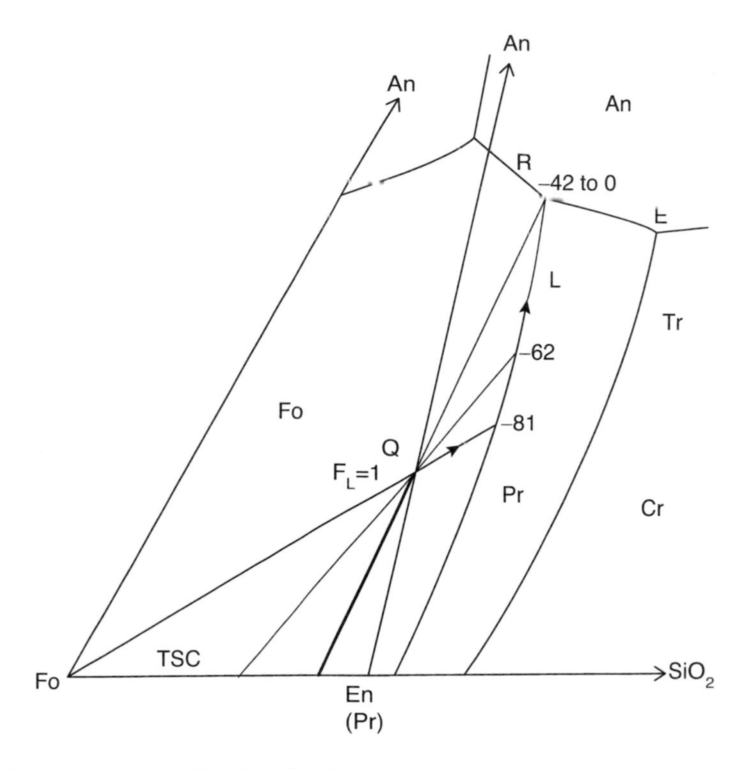

Fig. 11.5 Equilibrium crystallization of hulk composition *Q*

As a confirmation of the continually peritectic nature of the reaction involving olivine and pyroxene, note that the weight fraction of Fo decreases continually from 0.47 to 0.35, as shown in the lever calculations on Fig. 11.3.

The crystallization of bulk composition *Q* under continuous equilibrium conditions is illustrated in Fig. 11.5. The principles are identical with those applying to point *P*. Here, however, there is no excess of Fo, and the bulk composition can be expressed in terms of En and An alone. Liquid *R* and the last trace of Fo crystals are used up simultaneously, as the TSC reaches *Q*.

Bulk composition *S*, Fig. 11.6, must produce Tr to be completely solid, and its crystallization history is slightly more complex. When the TSC reaches the line An-En, the lever rule shows that 43% of the system is still liquid. Fo is now used up (TSC = An + Pr), so the system is univariant and the liquid must move from *R* along the cotectic *L* (An, Pr) to *E*. The FE of this reaction (not shown) lies where the tangent to the liquid path cuts An-En. The TSC moves continually along An-En towards the FE, until the liquid reaches *E*. Now Tr begins to crystallize, and continues to do so while the last 15% of the system is converted from liquid to solids. The FE now lies at *E*, in order to maintain invariance, and the TSC moves directly towards *E* until it reaches the bulk composition *S*.

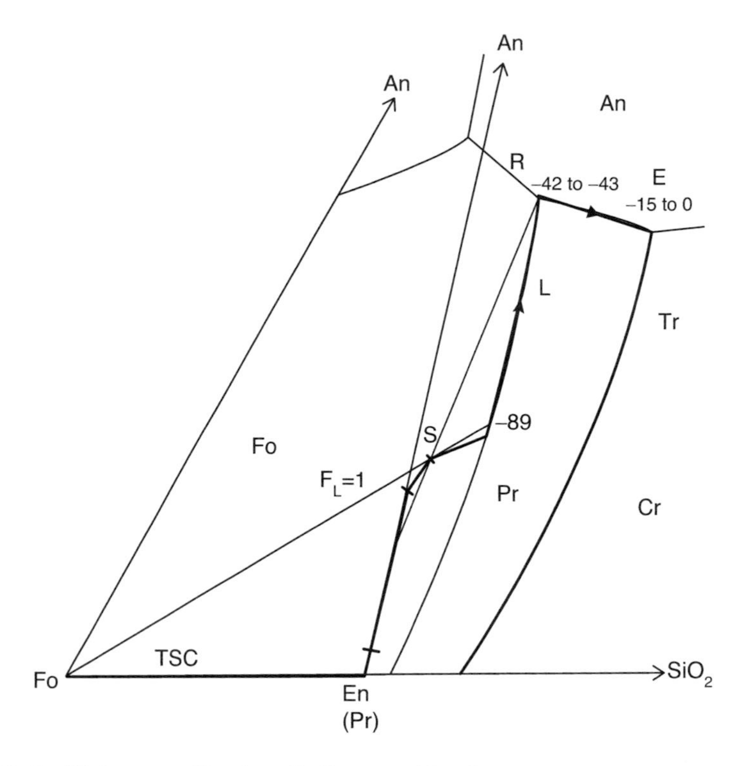

Fig. 11.6 Equilibrium crystallization of bulk composition *S*

Bulk composition *T*, Fig. 11.7, runs out of Fo before the liquid can get to it, i.e. at $F_L = 0.68$ when the TSC equals Pr alone. No longer constrained to the peritectic curve by Fo, the liquid now moves directly away from Pr across the divariant field Pr + *L*, and the TSC remains at Pr from $F_L = 0.68$ to 0.51. Now An joins the assemblage and the TSC moves along En-An just as it did with bulk composition *S*. The liquid reaches *E* at $F_L, = 0.28$, and remains there until TSC = *T*.

Note that failure of the olivine reaction at any stage of the process starting at *P* would result in fractional crystallization, giving rise to liquids *Q*. *S*. *T* in turn. This case is treated in detail somewhat further on.

Figure 11.8 calls attention to the limiting fields of bulk compositions whose daughter liquids do or do not reach *R*. All those lying to the left of the line En-*R* must produce liquid *R*: those to the right of this line (such as *T*) never do. Moreover, bulk compositions to the left of the line En-*E* must yield liquids which approach *E* via *L* (An, Pr), while those to the right of that line must approach *E* via *L* (Pr, Tr). The angles marked in the figure are the minimum angles at which isotherms in the field Pr + *L* may intersect the peritectic curve. These angles are defined, at any point, by the peritectic curve and the *L* path directly away from En, and if an isotherm were to intersect the field boundary at any smaller angle than α, the temperature would have to rise with the crystallization of Pr. This is impossible, for although there are

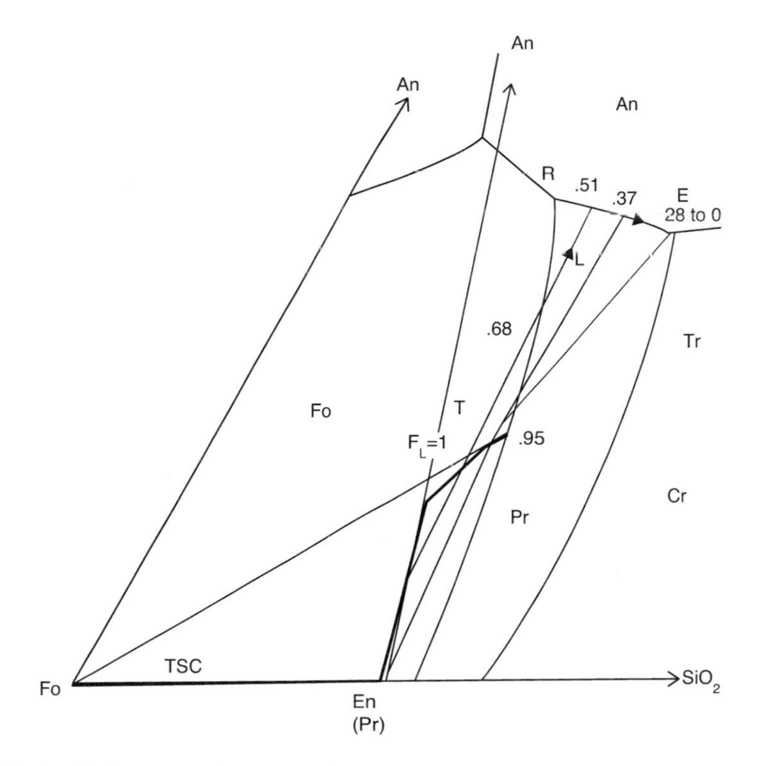

Fig. 11.7 Equilibrium crystallization of bulk composition *T*

examples of so-called "retrograde melting" in the laboratory, they all involve the input of energy in some other form than thermal energy, and we have here no such case. Liquid paths such as these, constrained away from a field boundary, are occasionally useful as guides in constructing isotherms on liquidus surfaces.

For those bulk compositions whose liquids never reach *R* in Fig. 11.8, it is interesting to note the unusual fact that these liquids eventually leave the *L* (Pr, Fo) field boundary and cross the divariant field. This is, of course, due to the peritectic nature of the curve and the loss of Fo (by reaction) from the equilibrium. Only peritectic curves can allow a liquid to leave in this manner; cotectic curves have no such property, and no liquid paths can cross the Pr field except directly away from Pr itself.

Several other bulk compositions yield crystallization paths which are straightforward yet instructive. Two are shown in Fig. 11.9. Composition *U* is so located that the liquid goes directly to *R*, whereupon reaction occurs and the TSC moves from Fo to *U* by addition of An and Pr simultaneously in the proper ratio. To verify that the reaction is odd when the liquid is at *R*, the reader can calculate for himself that Fo is consumed during the reaction, knowing that $F_{Fo} = 0.26$ when the liquid just reaches *R*, and calculating the final proportions of Fo, An, and Pr as done in Fig. 11.3. (Note from Fig. 11.2 that the An content of *U* is 40%.)

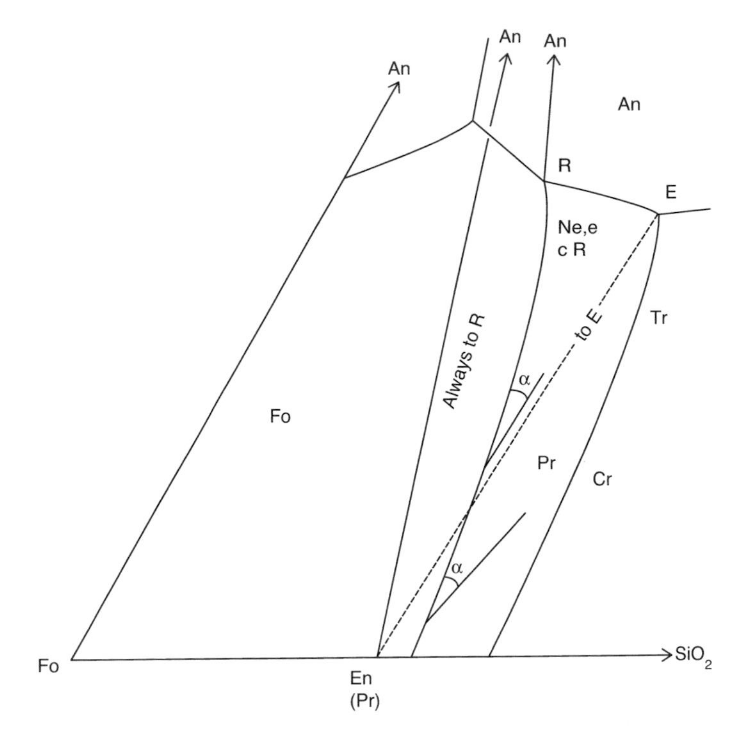

Fig. 11.8 Delineation of regions of bulk composition whose equilibrium liquids always reach *R* or never reach *R*. The angle α is the minimum angle at which an isotherm may intersect the peritectic curve

All liquids having bulk compositions to the left of UR will crystallize An before reaching the reaction point *R*. Those to the right of UR will behave like the *P*, *Q*, *S*, *T* group previously discussed.

Liquid *W*, Fig. 11.9, crystallizes Pr only until $F_L = 0.68$, when Tr joins the solids and the liquid is constrained to the cotectic. The FE, always tangent to the liquid path, forms its own path on En-SiO_2 as the tangent sweeps along that line. This is an even reaction, as the FE lies between the solid phases Pr and SiO_2. The TSC moves towards the FE, at all times, and its path towards Tr from Pr is shown. At $F_L = 0.50$, the liquid has reached *E* and is henceforth constrained there; the FE now also lies at *E*, and the TSC moves towards *E* to *W*.

Bulk composition *X* in Fig. 11.10 illustrates the case of a liquid which reaches the *L* (An, Fo) cotectic before it reaches point *R*. At $F_L = 0.82$, Fo begins to crystallize along with An, and the TSC begins to move along the line An-Fo. The liquid reaches *R* at $F_L = 0.65$ and remains there until $F_L = 0.39$, while the TSC moves directly towards *R*. Now all Fo is exhausted by reaction, and the liquid moves towards *E* with production of An + Pr. The liquid reaches *E* at $F_L = 0.12$, and completes its crystallization there. The rock produced is a slightly oversaturated anorthositic

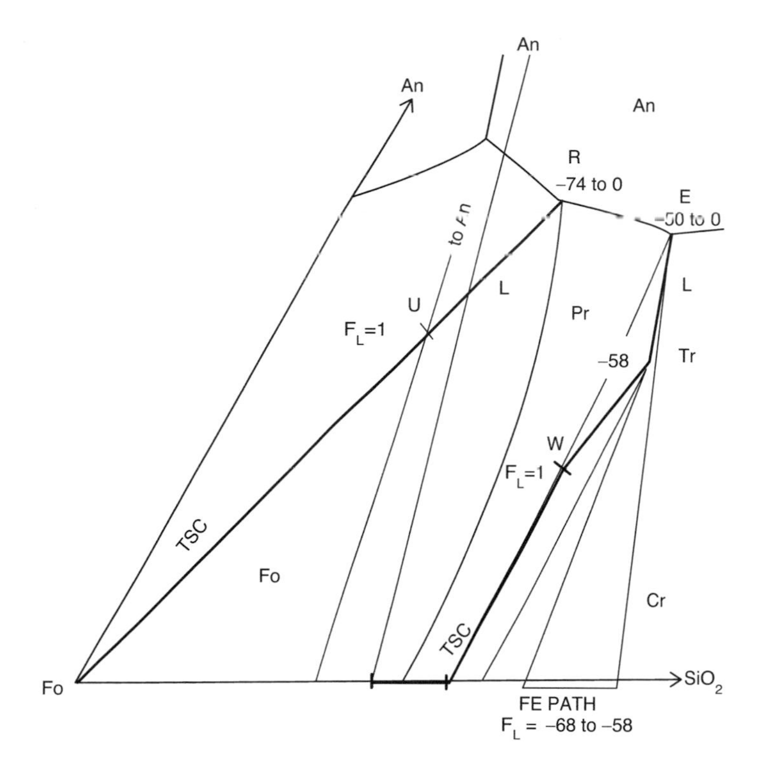

Fig. 11.9 Equilibrium crystallization of bulk compositions *U* and *W*

norite, all vestiges of olivine having disappeared by reaction. A very different series of rocks results from fractional crystallization, to be described shortly.

The equilibrium crystallization of liquid *V* is also illustrated in Fig. 11.10, and the reader may verify for herself the liquid and TSC paths for this composition.

Fractional Crystallization

There are important differences, both qualitative and quantitative, between fractional and equilibrium crystallization of any bulk composition whose daughter liquids involve a reaction relation. In the fractional process, no reaction is possible; all the olivine produced remains in the accumulated solids, and the liquid always reaches *E*. We may first examine in detail the fractional crystallization history of several familiar bulk compositions, and then comment on the differences between the cases of fractional and continuous equilibrium crystallization.

For bulk composition *P* (Fig. 11.11), the liquid follows the dotted path with fractional crystallization. At $F_L = 0.53$, *L* reaches the field boundary with Pr, but is not constrained to the boundary because Fo is isolated from reaction. The liquid

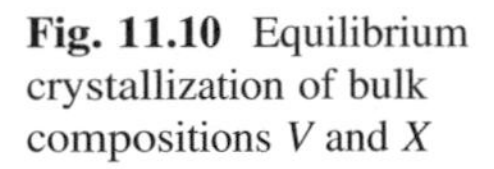

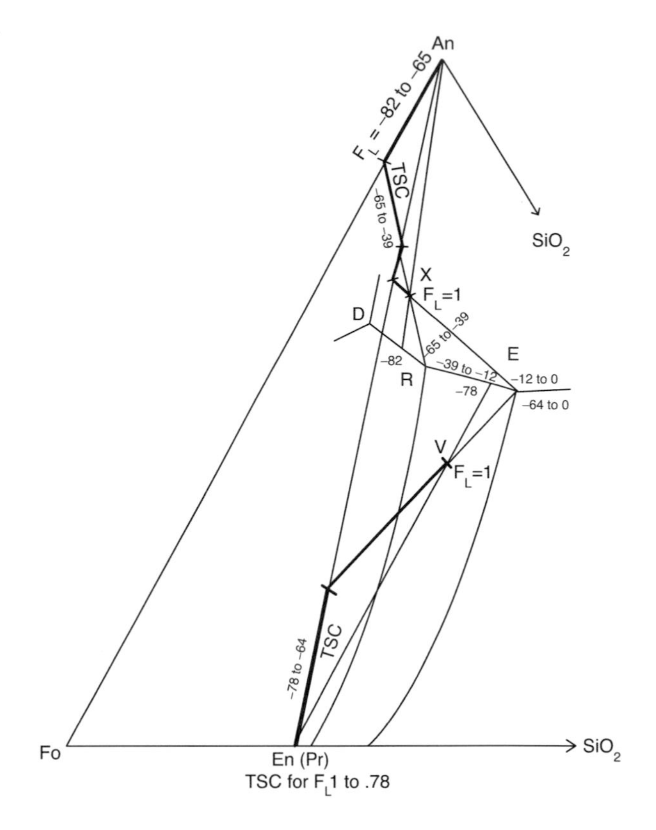

Fig. 11.10 Equilibrium crystallization of bulk compositions *V* and *X*

therefore moves directly away from Pr until it reaches the cotectic *L* (Pr, An), which it follows to *E*, which it reaches at $F_L = 0.20$. The four ISCs (i.e. model rocks) produced in this process are Fo, Pr, a mixture of Pr and An near the ratio 43:57, and a mixture of Pr, An, and Tr in the ratio of *E*. Analogous natural rocks would be dunite, bronzitite, norite, and, assuming some Na and K, quartz monzonite. The fraction produced of each of these model rocks may be calculated with the aid of the TSC path. This always runs towards the ISC, in this case from Fo towards Pr, then towards Pr and An, and then towards *E*. Levers applied to the TSC through *P* from any liquid along the fractionation path yield F_L or F_S, and the fraction represented by any given ISC is found from differences in successive values of F_L or F_S. Thus the initial ISC, Fo, amounts to 47% of the total mass, this being found directly. The next ISC, Pr, amounts to 24%, being 100 times the increment in F_S as the TSC moves directly towards Pr. The third ISC, Pr + An, amounts to 9%, again found from the difference in F_S during the time in which the TSC moves towards this ISC. The fourth ISC, *E*, amounts to 20%. The fraction of each ISC formed may not be found by applying a lever from the ISC, but must be calculated from the differences in F_L or *F*. Values of the fraction of each ISC for each of the bulk compositions *P*, *Q*, *S*, and *T* are posted in Fig. 11.11. The same liquid path is followed for all these bulk compositions, each of which in turn quite naturally yields

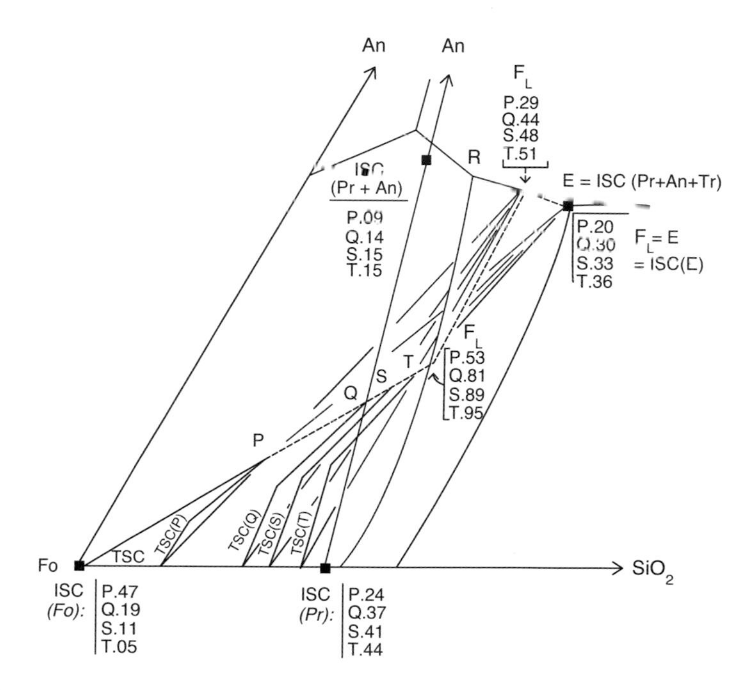

Fig. 11.11 Fractional crystallization of bulk compositions *P*, *Q*, *S*, and *T*. The figures posted near the ISCs are the fractions of each ISC produced for each of the bulk composition. The liquid path is dotted

less Fo and more Pr, Pr + An, and *E* compositions. By comparison with Figs. 11.3, 11.4, 11.5, 11.6, and 11.7, it can be seen that the amount of residual liquid *E* is much greater in the case of fractional crystallization, the increase being proportional to the amount used in the peritectic reaction under equilibrium conditions. It may also be seen that the model rocks (ISCs) produced in the later stages of fractional crystallization are very different from any rocks (TSCs) produced in the equilibrium process. Most noteworthy among these are the ISCs which correspond to norite and quartz monzonite in nature.

The fractional crystallization of bulk compositions *U*, *T*, *W*, and *X* is illustrated in Fig. 11.12. Again, the weight fraction of each ISC produced is posted for each bulk composition. For bulk composition *U*, the ISCs are Fo, An + Pr in the maximum range of ratios allowed by tangents to the curved path RE, and *E*. For *V*, the ISCs are Pr, Pr + An, and *E*. For *W* the ISCs are Pr, Pr + Tr (note the relatively large range of Pr:Tr ratios resulting from the sharper curvature of the *L* (Pr, Tr) cotectic), and *E*. For *X*, the ISCs are An, An + Fo (corresponding in nature to allivalite or troctolite), An + Pr, and *E*. All TSC paths converge towards *E* in their final stages, but it should be recalled that the TSC is composed of an ensemble of ISCs; no small homogeneous volume of rock has the composition of the total solids. The same TSC paths will be found in fractional fusion, discussed below, but in that case they do represent locally homogeneous compositions. Fractional crystallization of compositions *U* and

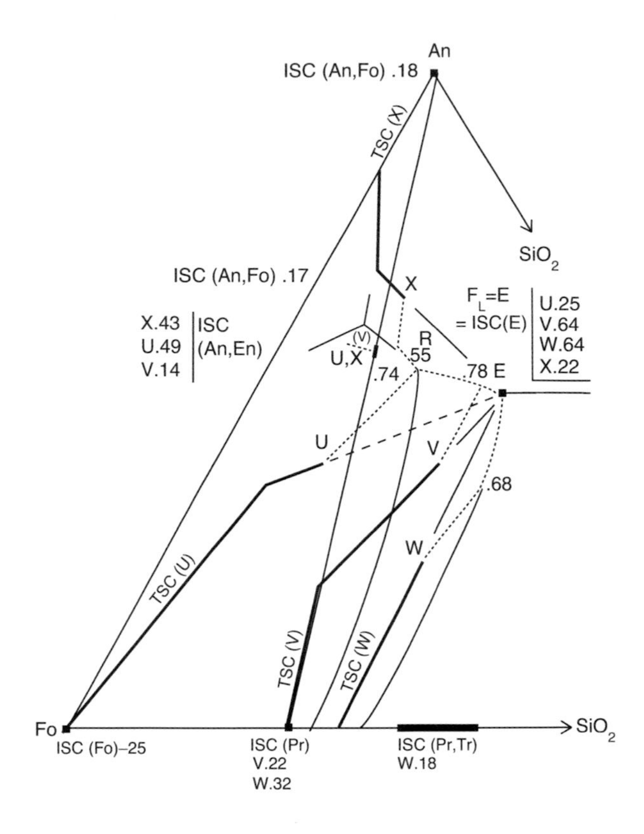

Fig. 11.12 Fractional crystallization of bulk compositions *U*, *V*, *W*, and *X*. The liquid path is dotted

X produces enhanced volumes of liquid *E* compared to the equilibrium process, as may be verified by comparison of the values of F_L posted in Fig. 11.12 with those in Figs. 11.9 and 11.10. On the other hand, no increase in the amount of liquid *E* results from the fractional crystallization of bulk compositions *V* and *W*, because no olivine is involved in the crystallization of these silica oversaturated bulk compositions. This emphasizes once again that it is the reaction relation of olivine which so profoundly affects the fate of liquids with fractional crystallization in this system with an incongruent-melting compound. This relation pertains only to those liquids which would reach *R* on equilibrium crystallization, as classified in Fig. 11.8.

The further details of fractional crystallization of bulk compositions *U*, *V*, *W*, and *X* are straightforward, and should need no elaboration beyond that to be found in Fig. 11.12.

Equilibrium Melting

No special diagrams are needed for analysis of equilibrium melting, which proceeds as always in precise reversal of the equilibrium crystallization process. Several of the previously discussed bulk compositions should be examined to appreciate that this is so. Both the crystalline and liquid products of equilibrium fusion are identical in composition and amount to those of equilibrium crystallization, and in deep-seated environments the two processes might well yield indistinguishable products.

Fractional Melting

When liquid is continuously removed as it is generated, the TSC always moves directly away from the liquid (ILC), either in a straight line or in a curved path, depending on whether the ILC is fixed (as it is in Fo-An-SiO_2) or continuously changing (as in systems with crystalline solutions). The paths that aid in analysing fractional melting are those of the TSC and total liquid composition (TLC), both of which are continuous, just as they both are in fractional crystallization. The ILC, like the ISC in fractional crystallization, jumps discontinuously from one discrete composition to another (in "*L* hops"). Unlike the crystallization ISC, however, the ILC has no continuous motion, even locally, like that produced in the ISC of crystallization by a sweeping tangent to a curved liquid path. In the basaltic part of the system Fo-An-SiO_2, there are only two TLCs, and these are the invariant points *R* and *E*. A single detailed example of fractional melting should suffice to establish the general principles for the entire system.

Bulk composition *T*, Fig. 11.13, lies within the solid-phase triangle En-An-SiO_2. It is composed, in the subsolidus condition, of En + An + Tr. Upon heating to 1220°C, liquid *E* forms at the triple junctions where all three solid phases coexist. With the addition of more heat, this liquid continues to form isothermally, driving the TSC directly away from E until it reaches the An-En tie line, or in other words, until tridymite is exhausted. This occurs at $F_L = 0.28$, and we may note that so far the process is the exact reverse of equilibrium crystallization (Fig. 11.7) or for that matter, fractional crystallization Fig. 11.11). This fact underscores a geologically important point, namely that so long as melting produces an invariant liquid the batch removal of a large volume of liquid is indistinguishable from the fractional or ideally continuous removal of that liquid. That is important because very small quantities of liquid, present as thin films or isolated pockets, are not likely to collect together and move out of the rock; continuous removal is unlikely in nature, but batch removal of as much as 28% melt is highly probable.

The TSC in our example, now being devoid of Tr and isolated from the solvent melt, will not melt further until the solidus temperature of An + En is reached, when liquid *R* (the second ILC) will form at 1270°C. The melting reaction is now An + Pr Fo + *L*, and Fo now accrues in the TSC as melt *R* is formed. The reaction is completely analogous to the reverse of crystallization of bulk composition *Q*, Fig. 11.5, and it will continue, whether or not continuous removal is effected, until An is exhausted in the solids. By construction of a TLC path and a lever to the TSC, this event is found to occur at $F_L = 0.52$. *Note especially* (see inset to Fig. 11.13) that the TLC path is a straight mixing line between *E* and *R*, and has nothing to do with the cotectic *L* (Pr, An). The amount of liquid *R* formed is evidently $0.52 - 0.28 =$

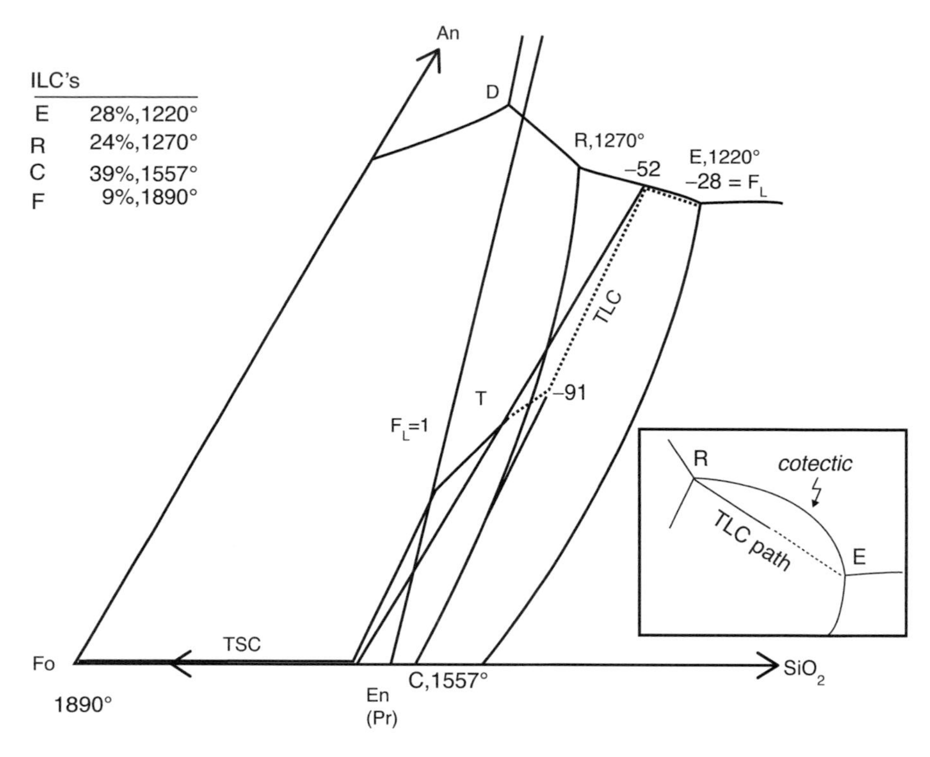

Fig. 11.13 Fractional melting of bulk composition *T*. The TLC path is dotted. *Inset*: The TLC path lies on the mixing line RE, not the cotectic curve from *R* to *E*

0.24. Again, it is to be observed that batch removal of this liquid at 1270°C would be equivalent to its continuous removal. Upon its removal, no further melting can take place until the solidus of En-Fo is reached at 1557°C, point *C* in Fig. 11.13. This is the peritectic point in the system Fo-SiO_2, and further fractional melting is confined to this binary system. The entire family of ILCs and their amounts is summarized in Fig. 11.13; the TLC moves next on the mixing line towards point *C* and finally towards pure Fo, and it is used solely for applying the lever rule.

Generalizing from the analysis of bulk composition *T*, it is possible to construct the summary diagram of Fig. 11.14, which shows the TSC paths of representative bulk compositions throughout the basaltic region of the phase diagram. All bulk compositions to the right of the lines An-*E* and *E*-En yield liquid *E* alone as the ternary liquid, while all others yield some quantity of liquid *R*, either in addition to liquid *E* or alone. TSC paths and tabular comparisons of the liquid fractions *E* and *R* of fractional crystallization and fractional melting are given for selected bulk compositions in the figure. Of particular note among the tabulated values of F_L is that certain bulk compositions produce none of a given liquid by one fractional process but an appreciable quantity by the other process. Specifically, silica undersaturated solid mixtures such as *P* and *U* yield no liquid *E* upon melting. but liquids

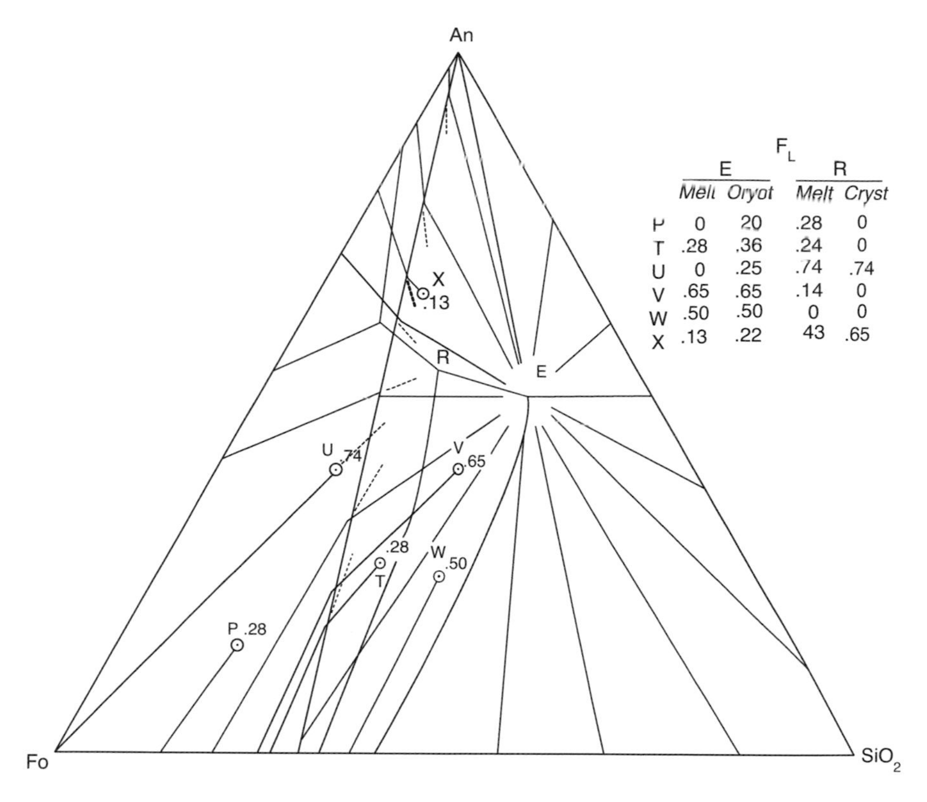

Fig. 11.14 Fractional melting summary diagram, showing TSC paths for the entire system Fo-An-SiO_2. All TSC paths are radial from *E* or *R*. The figures posted by each lettered bulk composition are the fractions of initial liquid, either *E* or *R*. The table compares fractions of liquids produced on fractional melting and fractional crystallization

of such bulk compositions yield plenty of liquid *E* upon fractional crystallization. No liquids below a line Fo-*R* yield any of liquid *R* on fractional crystallization, but the corresponding solid mixtures (down to the limiting line En-*E*) do produce liquid *R* upon melting.

Application to Basalts and Related Rocks

The system Fo-An-SiO contains representatives of all three of the major minerals of basalt: plagioclase, pyroxene, and olivine. It also contains a silica mineral which would be found as quartz in nature, and demonstrates the incompatibility of magnesian olivine and quartz in the presence of liquid. The Fo-An sideline is an analogue of the critical plane of undersaturation in the basalt tetrahedron (Fig. 2.1), and the En-An tie line is an analogue of the plane of silica saturation. By no process can liquids represented by the one atmosphere phase diagram fractionate *towards* the critical plane when Fo or Pr are crystallizing; in contrast, all liquids producing olivine will fractionate through the plane of silica saturation and reach *E*, a silica-rich eutectic. The principles exemplified by the phase diagram are directly pertinent to the crystallization of olivine tholeiites, norites (especially lunar norites which consist of nearly end-member An and En), hypersthene basalts, tholeiites, and quartz basalts or quartz gabbros. In particular, the reaction relation of olivine with liquids on *L* (Pr, Fo) or at *R* is comparable to that observed in natural olivine tholeiites. which frequently show early olivine crystals with orthopyroxene reaction rims wherever olivine occurred in close proximity to a sufficiently silica-oversaturated liquid. Preservation of olivine in such rocks, by armouring with plagioclase or reaction rims of pyroxene, is generally accompanied by a mesostasis or intersertal network of late granophyre or micropegmatite containing quartz. Thus in a single thin section, one may commonly observe the results of imperfect fractional crystallization beginning with the early removal of olivine from reaction and ending with a silica-rich residual liquid closely modelled by the point *E*. On the other hand, slowly cooled rocks of suitable chemistry may be found in which all traces of early olivine have been removed by reaction under conditions approaching continuous equilibrium crystallization.

The sequence of ISCs produced on fractional crystallization is closely modelled by the successive layered cumulate zones of many strongly differentiated layered intrusions. The experimental sequence, for tholeiitic liquids plotting anywhere in the primary phase field of olivine, is Fo, Pr, Pr + An, and *E*, corresponding to the following zones in an idealized natural layered complex: dunite, bronzitite, norite, and granophyre. Among famous examples of tholeiitic intrusions which exhibit all or some of these zones are the Stillwater complex of Montana, the vast Bushveld igneous complex of South Africa, the Muskox intrusion of the Northwest Territories, Canada, the Great Dyke of Zimbabwe, the enormous Dufek Complex of Antarctica, and the Windimurra Complex of Western Australia. There are also numerous more weakly differentiated sills such as the Triassic Palisades and Dillsburg sills of New Jersey and Pennsylvania. All of these can be found described in convenient volumes on layered intrusive igneous rocks by Wager and Brown (1968), by Cawthorn (1996), and by Charlier et al. (2015).

The system Fo-An-SiO_2 has been used extensively by Jackson (1970) and Irvine (1970), among others, as a model for the detailed crystallization history of such complexes. (It should be noted that a drafting error apparently originating with Morey (1964) has persisted in some of these later works, with the result that an arrow incorrectly points towards *R* on the cotectic RE.) The system Fo-An-SiO_2 is particularly appropriate to the study of anorthosite and related plutonic igneous rocks, which are composed principally of plagioclase and hypersthene, locally with lesser amounts of olivine or quartz. In the Nain complex of Labrador (Ryan 1990), for example, are several bodies showing upward gradation from olivine bearing rocks such as troctolite or leucotroctolite, through norite or leuconorite, to noritic rocks with interstitial quartz or micropegmatite. Apart from the plagioclase richness of these bodies, which may perhaps be accounted for by high pressure and

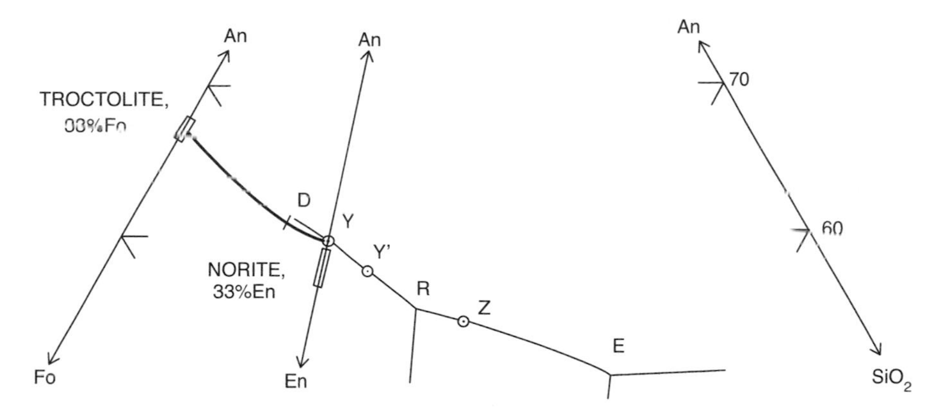

Fig. 11.15 Portion of the system Fo-An-SiO_2: location of several bulk compositions relevant to the anorthosite problem. The heavy TSC path is shown for the fractional crystallization of bulk composition *Y*

the mechanical concentration of plagioclase, the major rock sequences may be readily explained in terms of the model system Fo-An-SiO_2.

Point *Y*, Fig. 11.15, represents a liquid of noritic composition consisting of about 40% En, 60% An. The fractional crystallization of such a liquid yields ISCs analogous to the following rocks: troctolite 38%, norite 50%, and quartz monzonite 12%. If a slightly quartz-normative initial composition is chosen instead, such as *Y*′, the respective ISC percentages are 20% troctolite, 57% norite, and 23% quartz-bearing rock. Liquid *R* itself yields about 74% norite, 26% quartz-bearing rock, and liquid *Z* yields 50% each of norite and quartz-bearing rock. Large quantities of granitic rocks ranging from diorite through granodiorite to quartz monzonite (adamellite) occur in close association with anorthosites, in the Nain complex as elsewhere in the world. If we assume that liquid *E* is an approximate model for quartz monzonite or granite (it contains 26% quartz, 24% orthopyroxene, and 50% plagioclase), we can see from the above exercise that only one of the bulk compositions, *Z*, yields a subequal volume of norite and model quartz monzonite. Therefore, if the volumes of anorthositic (noritic) and granitic rocks are subequal, as they appear to be in terms of outcrop area at Nain, and if the two rock types are to be considered cognate, a parent liquid similar to *Z* is required, and this is quartz dioritic in nature, with 13% normative quartz. On the other hand, liquid *Z* never produces any olivine, so it cannot be the parent of anorthositic rocks that are underlain by troctolites. The two bulk compositions *Y* and *Y*′, saturated and slightly oversaturated respectively, both produce troctolites and smaller amounts of liquid *E* (12% and 23%). Each of these therefore makes a more appropriate parent magma for anorthosite, and each allows a small or moderate amount of granite rock to be cognate, but not a subequal amount. From such a simple model one is likely to prove little or nothing about the parent magma of anorthosite, but one can gain much insight into the constraints which more elaborate experimental models might provide, and the implications of even the simple model are probably worth taking seriously.

As a final application of the system, consider the proposal of Michot (1960) that a crustal melting event produced a noritic liquid and an anorthositic residuum. The phase diagram tells us that for any quartz-normative bulk composition, the first liquid on melting will be *E*, with 26% quartz, hence granitic. The only other liquid produced on fractional melting will be *R*, perhaps a quartz norite, and the residuum will become olivine bearing as orthopyroxene is destroyed. It therefore appears that the presence of quartz in the proposed product of the partial melt, combined with the presence of olivine in the crystalline residuum, would strengthen the argument for the proposed melting event, whereas the absence of either key mineral would weaken the argument.

It is especially noteworthy that no liquid of strictly noritic (An + En only) or troctolitic (An + Fo) composition is generated by fractional melting in this system, despite the fact that both rock types are abundantly produced by fractional crystallization of a considerate range of bulk compositions. Equilibrium melting can produce such liquids only at appreciably higher temperatures, unlikely to be reached in nature. It therefore follows that if liquids having the composition of norite occur in nature, they are most likely the products of melting in more complex systems or at higher pressures, or both. The abundance of noritic rocks both on the Earth and on the highlands of the Moon suggests that they are not so likely to be the products of a unique noritic liquid as they are to be the fractional crystallization products of a wider range of liquids having noritic affinity.

Chapter 12
Forsterite-Diopside-Silica: Pyroxenes and Their Reactions

Introduction

The system forsterite-diopside-SiO_2 (Fo-Di-SiO_2 or "FDS") is one of the most illuminating systems in basalt petrology; it is also one of the most complex, and indeed, it is only because of the very diligent reinvestigation of the system by Kushiro (especially 1972) and his colleagues that we can now begin to appreciate the subtleties of crystallization, melting, and reaction in this system. Like Fo-An-SiO_2, this system illustrates the reaction relationships of olivine and pyroxene, but in this case there are three varieties of pyroxene, all involved in reactions with olivine and to some extent with each other. These synthetic pyroxenes represent all the varieties of pyroxene occurring in silica-saturated basaltic rocks (augite, pigeonite, and hypersthene), and the system now under discussion forms a foundation for understanding many of the complex relations among these pyroxenes in nature. The system at atmospheric pressure can be applied to a large variety of natural basaltic rocks, including layered intrusions as well as volcanic rocks, and at high pressures, it furnishes several interesting and relatively simple models for generating the classical spectrum of basalt magma types in the mantle.

Aside from its central importance as a model for basaltic rocks and liquids, the system Fo-Di-SiO_2 furnishes a very large variety of principles relating to the deduction of liquid and crystal paths and reactions in ternary phase diagrams. It will be appropriate to pursue these principles in some detail, especially towards the end of the chapter, even though some of the deductions must be tentative because of a few lingering experimental uncertainties.

S. A. Morse, *Basalts and Phase Diagrams*,
https://doi.org/10.1007/978-3-030-97882-2_12

Bounding Joins

Fo-Di and Fo-SiO_2 were discussed in Chaps. 9 and 10, respectively. The join Di-SiO_2 was determined by Bowen (1914) to be binary, with a eutectic between diopside and tridymite near 15 wt.% SiO_2. The system was re-studied by Schairer and Kushiro (1964), who found that the equilibrium among diopside, tridymite, and liquid at this point is not eutectic but univariant, because diopside crystals do not have the composition of pure $CaMgSi_2O_6$, instead containing small amounts of both Mg_2SiO_4 and $MgSiO_3$ in solid solution. The liquid therefore lies in the general direction of $CaSiO_3$, and the univariant curve terminates in the system CaO-MgO-SiO_2 at 1320°C at a eutectic involving the added phase wollastonite ($CaSiO_3$). The same variability of diopside (i.e. Di_{ss} as we shall henceforth call it) causes the equilibrium among liquid, Di, and Fo to be univariant; the liquid is enriched in CaO, and terminates in a ternary eutectic with the phase åkermanite ($Ca_2MgSi_2O_7$) at 1357°C in the system CaO-MgO-SiO_2. Neither of these excursions of liquids from the system Di-Fo-SiO_2 affects the relations within the system, and they can be ignored for present purposes.

Ternary Compounds: Olivine and Pyroxenes

The only invariant, stoichiometric compounds in the system are the polymorphs of silica: tridymite and cristobalite.

Forsterite shows a small range of solid solution towards Ca_2SiO_4 (larnite), amounting to 1.6 wt.% of that component at its maximum, when Fo_{ss} coexists with Di_{ss}.

Diopside shows some solid solution towards Fo_{ss}, as well as a large range of solid solution towards $MgSiO_3$, analogous in nature to the augite composition range. The composition of Di_{ss} is variable with temperature and with the nature of the coexisting phases. The composition range of Di_{ss} found by Kushiro (1972) at 1390 + 2°C is shown in Fig. 12.1. This temperature is very near the liquidus point of pure diopside. The Di_{ss} phase occupies a lens-shaped field on the Fo side of the En-Di join, and at lower temperatures this lensoid field is enlarged lengthwise towards $MgSiO_3$ and sideways so as to include the En-Di join. Although the details of Di_{ss} composition variation are important to the interpretation of crystallization in the Di-rich part of the system, the variation towards En is the main feature to be noted for most of the discussions of this chapter.

Pyroxenes encountered along and near the join Di-En are of three types: diopside solid solution (Di), crystallizing in the monoclinic space group C2/*c* "iron-free pigeonite" (hereafter Pig), crystallizing in the monoclinic space group $P2_1/c$ and the orthorhombic protoenstatite solid solution (Pr_{ss}), which inverts on cooling to enstatite with space group *Pbca*. The *T-X* phase relations along the pyroxene join are shown in Fig. 12.2. This is a *pseudobinary* section, because while most of the

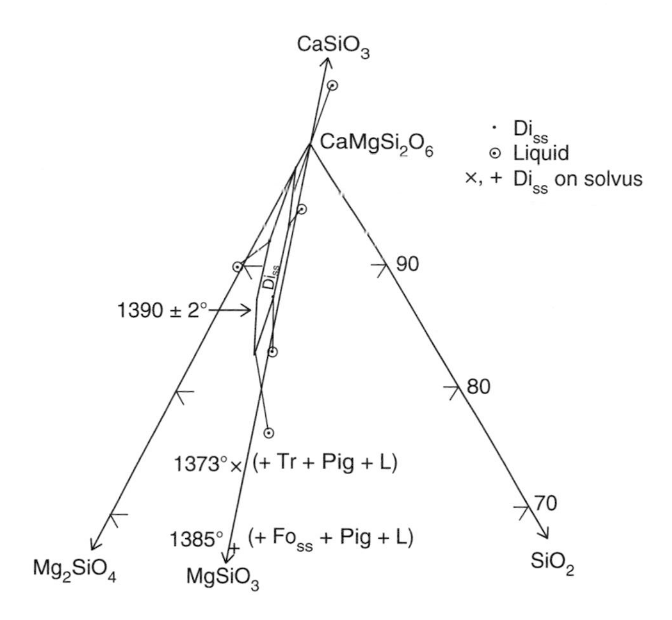

Fig. 12.1 Composition range of diopside solid solution (Di_{ss}) at 1390 ± 2°C (after Kushiro 1972). Fo_{ss}, forsterite solid solution; *L*, liquid; Pig, pigeonite; Tr, tridymite

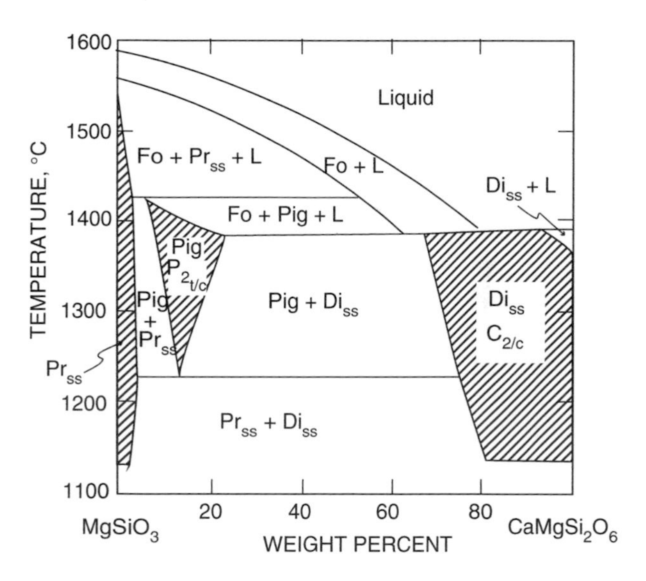

Fig. 12.2 *T-X* diagram of the join enstatite-diopside. Shaded areas are fields of homogeneous single pyroxene; they are separated by three solvi. Liquid curves are sections through surfaces, and in general they do not represent liquids in equilibrium with these pyroxenes. The solidus and liquidus of Di_{ss} are locally merged in the diagram because of the narrow temperature interval between them. Abbreviations as in Fig. 12.1. (After Kushiro 1972)

pyroxenes lie in or near the section (on or near the join), the liquids in equilibrium with pyroxenes do not. The liquidus relations may be ignored for present purposes; they will become apparent when the full ternary system is discussed, and suffice to say for now that the liquid curves in Fig. 12.2 are merely sections through the liquidus and peritectic surfaces in the ternary system. The main point of Fig. 12.2 is to show the relations among the pyroxenes. The three shaded regions show the compositional range of homogeneous pyroxene in each of the three structural groups. The limits of mutual solubility are solid-saturation curves, or *solvi*, which were briefly mentioned in Chap. 5. A narrow solvus bounds the region of coexisting Pr_{ss} + Pig, and a much wider solvus bounds the region of coexisting Pig + Di_{ss}. For any temperature, and any bulk composition within the solvus, the compositions of the two coexisting phases are invariant. Each single-phase (shaded) region is bounded upward by a *solidus* of the usual type; that for Di_{ss} cannot be shown at this scale, so it is merged with the liquidus in the drawing. The solidus for Pr is very steep, and indicates the very small range of Di substitution allowed by this phase. The solidus for pigeonite, combined with the liquid curve *L* (Fo, Pig), amounts to a truncated melting loop analogous to a portion of the plagioclase loop. The solidus for Di_{ss} will be discussed towards the end of the chapter. Note from the diagram, Fig. 12.2, that only two pairs of pyroxenes may coexist with liquid: Pr_{ss} + Pig, or Pig + Di. The pair Pr_{ss} + Di, found in nature as enstatite (or bronzite or hypersthene) plus augite, does not occur with liquid and here the phase diagram appears to fail us. However, it is to be noted that the pigeonite field terminates downward in temperature, and it can readily be imagined that the additional components of natural magmas could depress the temperatures of liquid-crystal equilibria below the stability of pigeonite, thus permitting equilibria among Pr_{ss}, Di_{ss}, and *L*.

The Ternary System

The system Fo-Di-SiO_2 was studied by Bowen (1914), and discussed extensively in his book (1928). The two-liquid field in the silica-rich part of the system was studied by Greig (1927). Bowen correctly determined the field boundary between pyroxene and silica minerals, shown in Fig. 12.3, but did not have access to the X-ray and other methods needed to distinguish the various species of pyroxene or their solvi. He assumed from the evidence at hand a continuous series between enstatite and diopside. The initial modern reinvestigation of the system extended over a period of more than 10 years, beginning with the recognition of a Ca-poor pyroxene-Di_{ss} boundary by Schairer and Yoder (1962), continuing through a revision of that boundary by Kushiro and Schairer (1963), the discovery of iron-free pigeonite by Kushiro and Yoder (1970), and the later revision by Kushiro (1972). The latter study represented a very great advance in technique, as it was based on electron probe analysis of glass and crystal compositions, using relatively few starting mixtures and runs. This technique allowed the very precise location of univariant curves and invariant points by direct analysis of the glass ("liquid") coexisting with the solid phases observed in the run products. It also permitted precise delineation of solid-phase composition limits, as in Fig. 12.1, which is based on this method.

The phase diagram shown in Fig. 12.3 is adapted closely from that of Kushiro (1972). The dominant features of this simple-looking diagram are the large field of

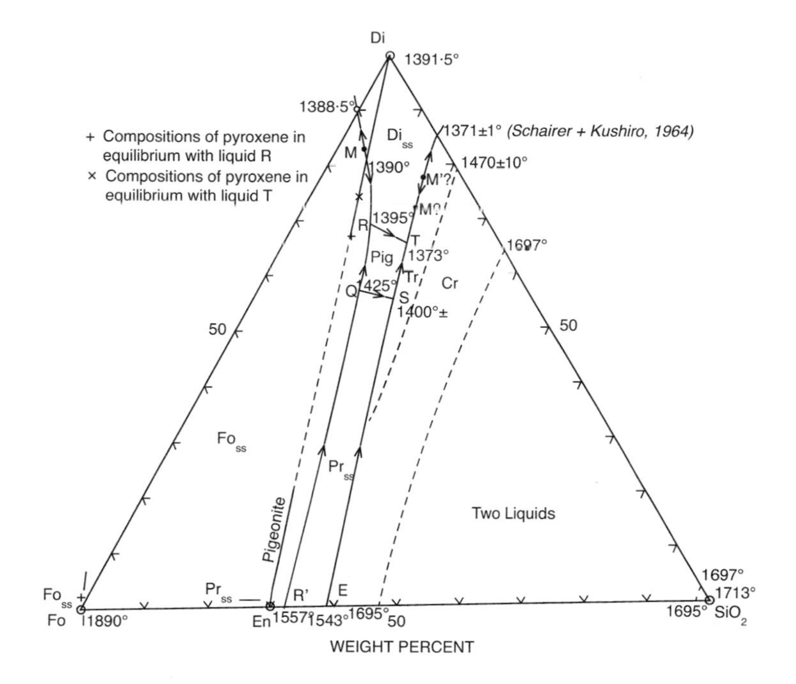

Fig. 12.3 Phase equilibrium diagram of the system Fo-Di-SiO_2 (revised by Kushiro 1972 after Bowen 1914, field of two liquids from Greig 1927). Points *m* and *M′* are added for reasons discussed towards the end of the chapter

Fo_{ss}, overlapping most of the length of the En-Di join, the univariant curves[1] *R′QRM* of liquids in equilibrium with olivine and a pyroxene, and the univariant curves *ESTmM′* of liquids in equilibrium with pyroxene and a silica mineral. Ignoring the geologically uninteresting fields of two liquids and cristobalite, there are five primary phase fields: Di_{ss}, Fo_{ss}, Tr, Pig, and Pr_{ss}. The pigeonite field is small, but all liquids must reach it unless the bulk composition is so restricted as to yield only Pr_{ss} or Di_{ss} with olivine or silica. All other liquids reach and then depart the Pig field on fractional crystallization.

The curve *QS* is a peritectic reaction curve, as we shall see, and points *Q*, *R*, *S*, and *T* are invariant reaction points, as we shall also see.

The Di-En join, constituting one edge of the plane of silica saturation in the basalt tetrahedron, separates the solid assemblages into two main groups, those with olivine and those with tridymite. A bulk composition may be expressed in the solid state in terms of one or the other of these antithetic minerals, plus one of the following five pyroxenes or pyroxene pairs: Di_{ss} alone, Di_{ss} + Pig, Pig alone, Pig + Pr_{ss}, or Pr_{95}

[1]The plural "curves" is correct, despite the fact that the boundaries of olivine, pyroxenes, and silica mineral look continuous at the scale of the diagram. Each of these boundaries is actually composed of three segments, one for each pyroxene species, and the intersections at *Q*, *R*, *S*, and *T* must all be junctions of non-collinear curves.

alone. There are thus ten possible multiphase solid assemblages, plus three coincidental single phase assemblages, Di_{ss} or Pig or Pr_{ss}.

The points labelled *M* and *M′* in the diagram are liquidus *maxima*, and the point labelled *m* is a liquidus *minimum*.

Equilibrium Crystallization

We shall consider only those bulk compositions in the primary phase field of Fo_{ss}. The analysis will in every case require knowledge of the tie lines between pyroxene solid solutions and liquids, just as in the case of plagioclase plus liquid in Di-An-Ab. In the present case, the solid solutions fall within the ternary system, rather than along one edge, but the principles are the same. The solid solutions are of limited range, however, and certain special compositions are known to occur with certain invariant liquids, so it is a simple matter to find a set of pyroxene-liquid tie lines which must be sufficiently close to the actual ones, even though these have not been determined in detail experimentally.

It is convenient to work with a distorted cartoon of the actual system, so that graphical relations can be displayed and analysed with clarity. Such a "toy" system is shown in Fig. 12.4; although the distortion is gross, the essential relationships of the

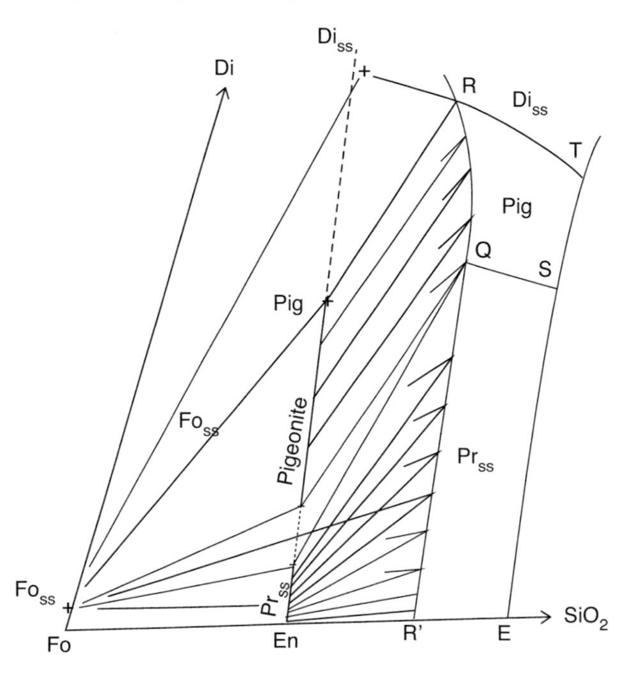

Fig. 12.4 Tie line template; slopes of pyroxene-liquid tie lines in a distorted (toy) sketch of the FDS system. Heavy tie lines are experimentally established, lighter ones are interpolated

actual system have been preserved. The labels of primary phase fields tell us that liquid *Q* coexists with crystals of Fo_{ss}, Pr_{ss}, and Pig. At the same time, the dotted gap signifies the miscibility gap (see Fig. 12.2) of pigeonite and protoenstatite solid solution, and we know that when these two crystalline phases coexist, they are mutually saturated. Liquid *Q* therefore lies at the corner of a three-phase triangle *Q*-Pig-Pr_{ss}, and two legs of this triangle establish the tie line slopes *Q*-Pig and *Q*-Pr_{ss}. At the other limit of Pr_{ss} compositions, the tie line must be En-*R*′, on the binary join Fo-SiO_2. It is sufficient for any purpose to assume that all the other Pr_{ss}-liquid tie lines are evenly distributed between the limiting ones En-*R*′ and Pr_{ss} (Pig)-*Q*. (As usual, the phase in parentheses denotes saturation of the preceding phase with the component bracketed.)

By a similar analysis, the tie lines between pigeonite solid solutions and *L* (Pig, Fo_{ss}) may be established within reasonable limits. Kushiro (1972) located the compositions of Pig and Di_{ss} in equilibrium with liquid *R* and with each other, by direct electron probe analysis. These solvus compositions are indicated by + signs on Fig. 12.4. We therefore have two known pigeonite-liquid tie lines, i.e. *R*-Pig (Di_{ss}) and *Q*-Pig (Pr_{ss}). These happen to be nearly parallel in the actual system (Fig. 12.3), and are drawn so in Fig. 12.4. The intermediate *L*-Pig tie lines are interpolated.

The entire suite of tie lines shown in Fig. 12.4 is valid only for liquids saturated with Fo_{ss}. For liquids saturated with SiO_2, a similar analysis can be performed using, instead, points *S* and *T*, and the curve *EST*. Figure 12.5 shows the results of such an analysis, assuming little change in the width of the Pr_{ss}-Pig solvus but a known and significant widening of the Pig-Di_{ss} solvus with falling temperature; again, Fig. 12.2 will aid in visualizing the solvus relationships. All liquids along *QS* lie at the apex of three-phase triangles *L*-Pr_{ss}-Pig, so the liquid-pyroxene tie lines must flatten as liquids move from saturation with Fo_{ss} to saturation with SiO_2. By use of

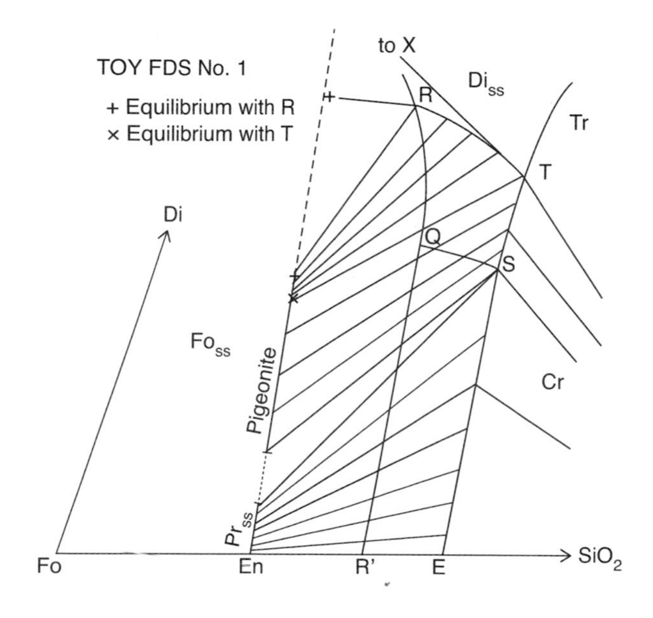

Fig. 12.5 Another tie line template; slopes of tie lines from pyroxene to silica saturated liquids. Conventions as in Fig. 12.4

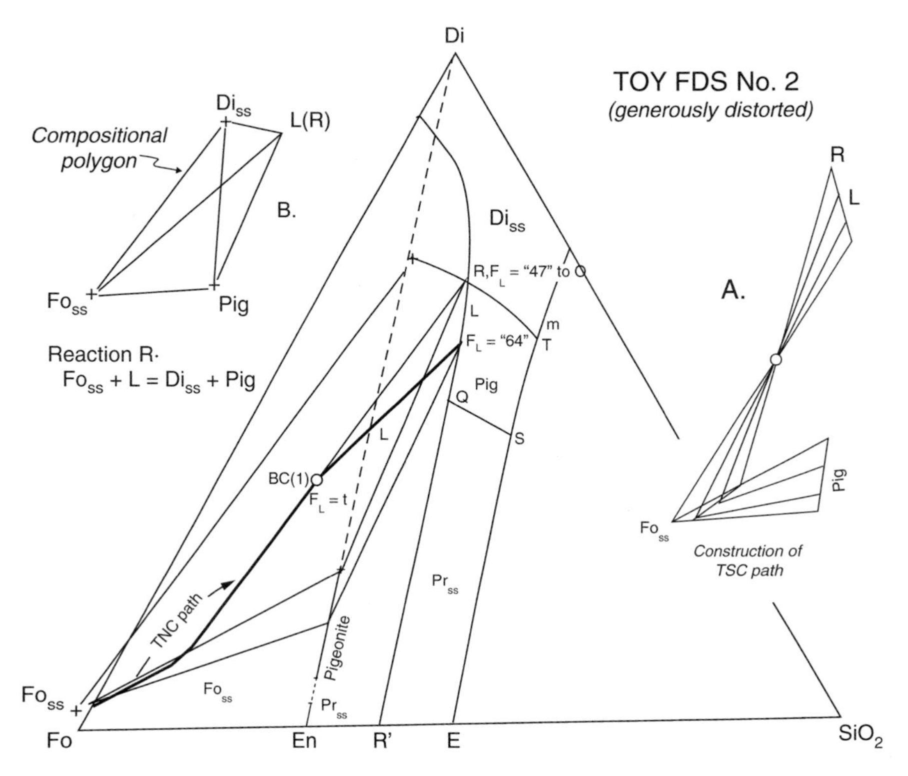

Fig. 12.6 Toy FDS diagram No. 2, with equilibrium liquid and TSC paths for bulk composition 1. Insets: (a) construction of TSC path (dashed) from Fo-Pig tie lines and solid-liquid levers; (b) analysis of reaction of liquid *R*. Quotation marks appear around values of F_L because the diagram is distorted

Figs. 12.4 and 12.5, any desired tie line between pyroxenes and liquid can be deduced, and we shall have frequent recourse to these figures as templates for tie line slopes.

We may begin the analysis of crystallization with bulk composition 1, Fig. 12.6, plotted on a second toy system, somewhat less distorted than the first. The liquid occurs in the primary phase field of Fo_{ss}, and initially moves directly away from Fo_{ss} while olivine crystallizes.[2] The liquid reaches the Pig field boundary at $F_L = $ "0.64" (quotation marks because the diagram is distorted), and now Pig, of composition to be determined by reference to a tie-line template like that of Fig. 12.4, begins to crystallize. The TSC now moves from Fo_{ss} towards Pig, as shown in inset A, as the

[2] In fact, the composition of Fo_{ss} plotted is that in equilibrium with liquid *R*. Slightly "purer" (less-Ca-rich) compositions occur with liquids unsaturated with Di_{ss}, and the crystal composition reaches pure Fo for liquid *R′*. For graphical convenience, we shall assume the plotted composition of Fo_{ss} to be invariant, although in fact there must be a small composition change with falling temperature, and the actual liquid path across the Fo_{ss} field would be concave downward.

liquid moves towards R. The reaction is evidently odd, $L = \mathrm{Pig} - \mathrm{Fo_{ss}}$, because the tangent to the liquid path lies on the extension of the tie line $\mathrm{Fo_{ss}}$-Pig; the liquid path is therefore peritectic. The isobaric variance, W_p, is 1. When L reaches R, the phases are four ($\mathrm{Di_{ss}}$, $\mathrm{Fo_{ss}}$, Pig, L), and $W_p = 0$; the fictive extract therefore lies at R, and the TSC moves directly towards R until it reaches the bulk composition. We are left with crystals of olivine (partly resorbed and reacted to pigeonite), pigeonite, and diopside solid solution, the latter tending to fill spaces between aggregates of olivine and pigeonite. The reaction at R must be odd, as shown in inset B. This "chemographic" diagram shows the relative positions of four phases participating in an equilibrium reaction. There is no way for the liquid to produce positive quantities of all three solid phases simultaneously, because it lies outside the composition triangle formed by those three phases, and cannot move away from them. The liquid must therefore produce a negative quantity of one phase, i.e. it must react with one of them, and the phase with which it reacts must lie across the tie line joining the other two phases, i.e. it must be $\mathrm{Fo_{ss}}$. The analysis boils down to finding which two tie lines cross, and writing the reaction in terms of the crossing tie lines: $\mathrm{Fo_{ss}} + L = \mathrm{Di_{ss}} + \mathrm{Pig}$. If a bulk composition lay exactly on the $\mathrm{Di_{ss}}$ + Pig tie line, the end result would be $\mathrm{Di_{ss}}$ + Pig, with no excess of either L or $\mathrm{Fo_{ss}}$. If the bulk composition lies to either side of that tie line, there will be a final excess of the phase towards which it lies, in this case $\mathrm{Fo_{ss}}$. This type of chemographic analysis of reactions finds extensive use in the development of phase diagrams with variable pressure, where the location and identity of reactions in P-T-X space can be deduced purely from a knowledge of the compositions of all the phases. We shall not dwell upon the matter here, but it is appropriate to introduce such chemographic principles at this point, with a view to their further use later on.

For further analysis, it is convenient to choose a group of bulk compositions, labelled 2 through 6 in Fig. 12.7. These are plotted on toy FDS No. 1, which will be used, along with the tie line templates (Figs. 12.4 and 12.5) in the succeeding analyses. The solid phase assemblage of each bulk composition is listed in Fig. 12.7. It will be seen that only one, No. 2, contains olivine, and that the rest contain tridymite, which means that liquids must reach the SiO_2 field boundary EST. The final TSCs for these latter compositions must lie on pyroxene-silica tie lines, as shown.

The crystallization of BC 2 is shown in Fig. 12.8. It is similar to the case of BC 1, except that $\mathrm{Pr_{ss}}$ is produced before Pig. The liquid moves away from $\mathrm{Fo_{ss}}$, and when it reaches the field boundary with $\mathrm{Pr_{ss}}$, the TSC begins to move towards $\mathrm{Pr_{ss}}$ from $\mathrm{Fo_{ss}}$. The composition of $\mathrm{Pr_{ss}}$ is taken from the tie line template of Fig. 12.4. The tangent to the L path lies on the extension of the tie lines $\mathrm{Fo_{ss}}$-$\mathrm{Pr_{ss}}$, so the reaction is odd, i.e. $L = \mathrm{Pr_{ss}} - \mathrm{Fo_{ss}}$, as in the binary system Fo-SiO_2. Evidently, recalling BC 1, the entire boundary $R'QR$ is peritectic. When the liquid reaches Q, $\mathrm{Pr_{ss}}$ contains a maximum of Di component (it lies on the solvus with Pig), and no further enrichment of L in Di can take place as long as $\mathrm{Pr_{ss}}$ remains in the TSC. The liquid therefore pauses at Q ($W_p = 0$), while a reaction takes place among the four phases $\mathrm{Fo_{ss}}$, $\mathrm{Pr_{ss}}$, Pig, and L. The chemographic relations of this reaction are shown in inset A. Pigeonite is an interior phase within the compositional triangle L-$\mathrm{Fo_{ss}}$-

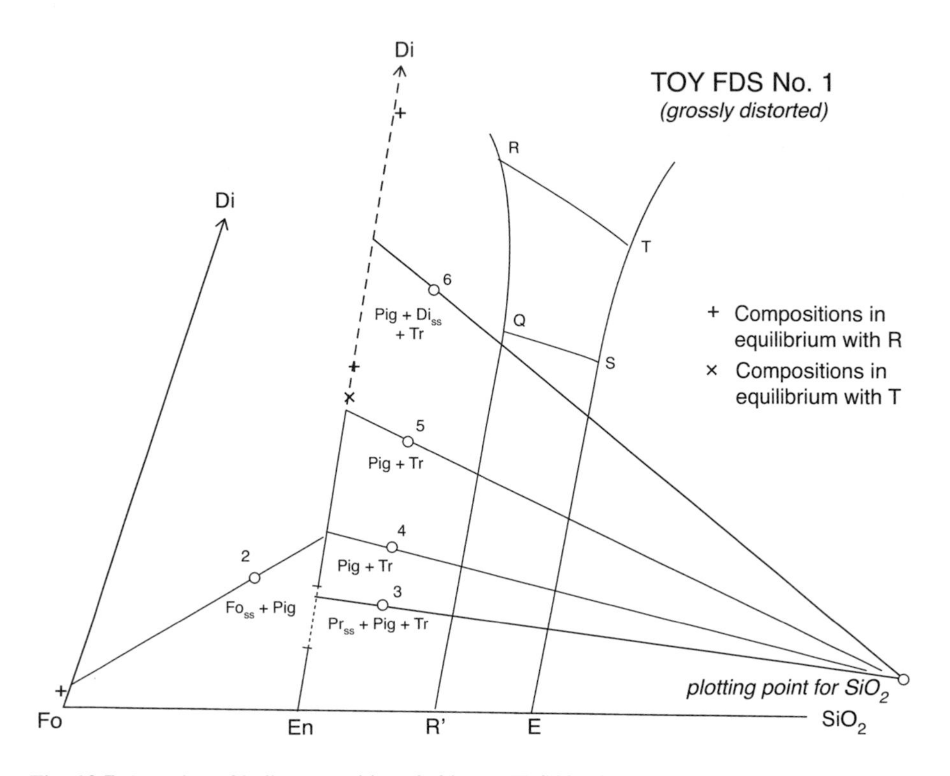

Fig. 12.7 Location of bulk compositions 2-6 in toy FDS No. 1

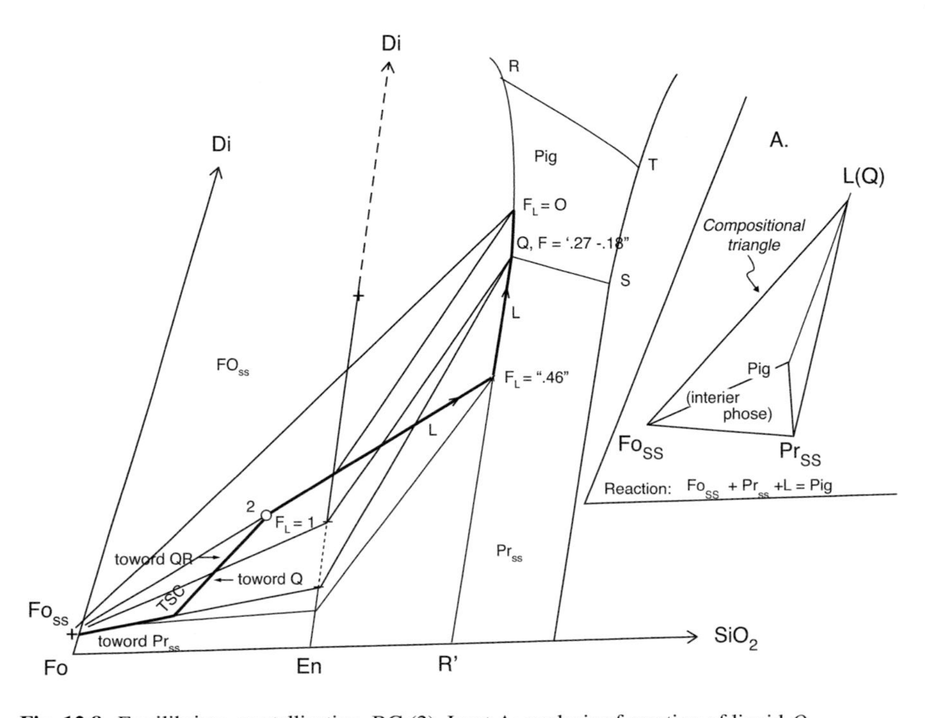

Fig. 12.8 Equilibrium crystallization, BC (2). Inset A: analysis of reaction of liquid *Q*

Pr_{ss}. There are no crossing tie lines in this case. The only chemical reaction that can be written is one in which Pig is created or destroyed from the three corner phases, in this case created with falling temperature. The chemographic relations indicate that *L*, Fo_{ss}, and Pr_{ss} are *all three* consumed in the production of Pig. The reaction is odd, and may be written $L = \text{Pig} - Fo_{ss} - Pr_{ss}$, or rearranged as in inset A of the figure. During the time of this reaction, the FE lies with the liquid at *Q*, and the TSC must move directly towards *Q*, by accruing Pig at the expense of Pr_{ss}. When all Pr is used up, the system returns to an isobaric variance of one, and the liquid now continues along the boundary *QR*, producing pigeonite at the expense of $Fo_{ss} + L$. During this time, the tie lines Pig-*L* may be inferred from Fig. 12.4, and the TSC path is slightly curved (concave upward) as the composition of Pig changes slightly. The position of the final liquid along *QR* is found from the tie line template, knowing that the final composition of Pig must lie on a line from Fo_{ss} through the BC. This course of crystallization illustrates that both Pr_{ss} and Pig have a reaction relation to olivine, and that an invariant condition occurs temporarily at the reaction point *Q*, while Pr_{ss} is in effect converted to Pig.

Bulk composition 3, Fig. 12.9, is the first of our silica-oversaturated examples. In the solid state, it consists of Pr_{ss} + Pig + Tr, and these products must be kept in mind during the analysis. We may note right away that only the invariant liquid *S* is in equilibrium with all three of these solid phases, so evidently the final liquid will be of composition *S*. The final pyroxenes will consist largely of Pig, with a small amount

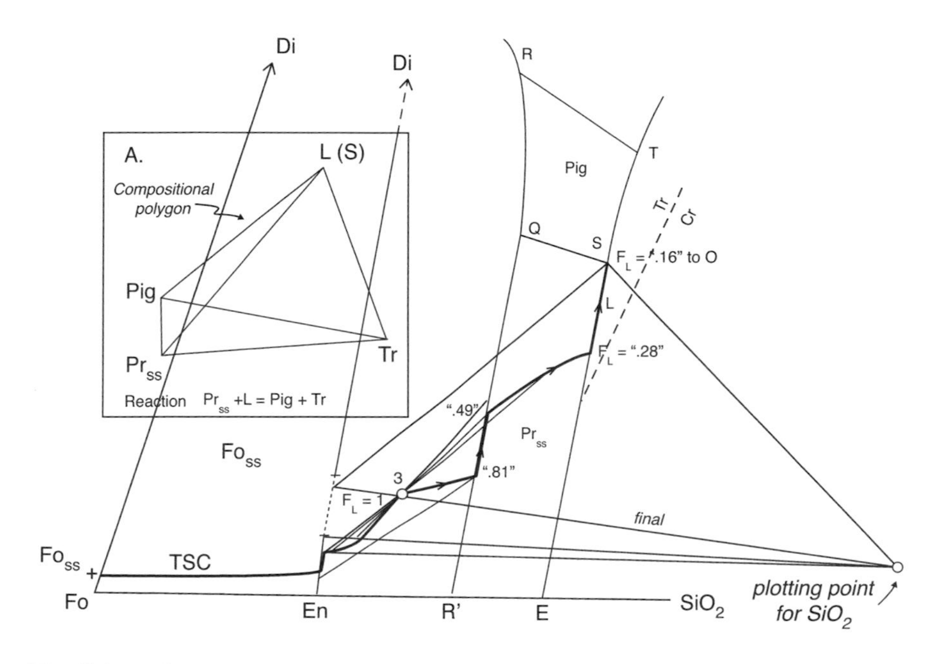

Fig. 12.9 Equilibrium crystallization, BC (3). Inset A: analysis of reaction of liquid *S*

of Pr_{ss}, as indicated by the pyroxene end of the solid-solid leg of the final three-phase triangle.

The liquid first crystallizes Fo_{ss}, and then reaches the Pr_{ss} boundary. The tie line template, Fig. 12.4, is used to determine the initial composition of Pr_{ss}. The liquid moves in univariance along the boundary towards *Q*, but does not reach *Q* because the TSC reaches Pr_{ss} at F_L = "0.49," when a tie line to Pr_{ss} from *L* passes through the BC. At this moment, all Fo_{ss} has been used up by reaction. The liquid now moves in divariance across the Pr_{ss} field, continually tied to varying Pr_{ss} through the BC. The path of the liquid across the field is determined by first finding its goal on *ES*. This is found by seeking a tie line on the template, Fig. 12.5, which passes through BC (3) to Pr_{ss} from liquids on the silica-saturated boundary *ES*. This is the initial tie line of liquid in equilibrium with a silica mineral, *L* (Pr_{ss}, Tr), and the final one of the equilibrium *L* (Pr_{ss}). The liquid compositions at F_L = "0.49" and F_L = "0.28" may thus be found rigorously, and a slightly curved path connecting them represents the equilibrium liquid path. (Rigorous construction of this path would first require construction of the fractionation paths from Pr_{ss}, but both these and the equilibrium paths are so nearly straight that a rigorous treatment would be a waste of effort.) As the liquid follows this path, the TSC moves along the Pr_{ss} composition line. At F_L = "0.28", Tr begins to crystallize, and the liquid is confined in univariance to the boundary *L* (Pr_{ss}, Tr). The TSC moves in a curved path along the solid-solid legs of a series of three-phase triangles, the initial and final legs of which are shown in the figure. The tangent to the liquid path cuts the Pr_{ss}-Tr tie lines, and the reaction is even, i.e. $L = Pr_{ss} + Tr$. At *S*, the liquid begins to produce Pig, the composition of Pr_{ss} having reached saturation in diopside component, and the liquid remains at *S* and at invariance until crystallization is completed. The final TSC path approaches *S* directly until it reaches the bulk composition. The reaction of liquid *S* is odd, as may be determined in inset A: *S* lies exterior to the triangle Pr_{ss}-Pig-Tr; the tie lines *L*-Pr_{ss} and Pig-Tr cross, and the reaction requires subtraction of Pr_{ss} to yield Pig + Tr. From the analysis, it is clear that the boundary curve *ES* is cotectic, but that the point *S* is peritectic.

Bulk composition 4, Fig. 12.10, consists of Pig + Tr in the solid state. The initial stages of crystallization are like those of BC 3, just discussed; the TSC moves from Fo_{ss} to Pr_{ss} while the liquid follows the peritectic boundary from F_L = "0.81" to "0.48". At F_L = "0.48", Fo_{ss} is exhausted, and the liquid moves across the Pr_{ss} field to a point on *QS* to be determined with the aid of Figs. 12.4 and 12.5 together. This point is the apex of a three-phase triangle Pr_{ss}-Pig-*L*. The liquid now follows *QS* in univariance, while the TSC moves across the gap towards Pig by virtue of accruing Pig crystals at the expense of Pr crystals. Inset A shows that the reaction is odd along this curve, as the tangent to *QS* cuts the extension of Pr-Pig. The ratio of Pig to Pr_{ss} increases as the liquid approaches and reaches *S*; the liquid now sits at *S* while the remaining crystals of Pr_{ss} are consumed and the TSC moves straight towards *S*. The reaction is the same as illustrated in inset A, Fig. 12.9, namely Pro + *L* = Pig + Tr. When the TSC reaches the limiting, En-saturated Pig-Tr tie line, all Pr_{ss} has been exhausted by reaction, and the liquid is free to move in univariance along *ST*, producing only Pig + Tr in an even reaction. The TSC moves on a succession of Pig-Tr legs, always tied to *L* through BC, along a slightly curved path until it reaches

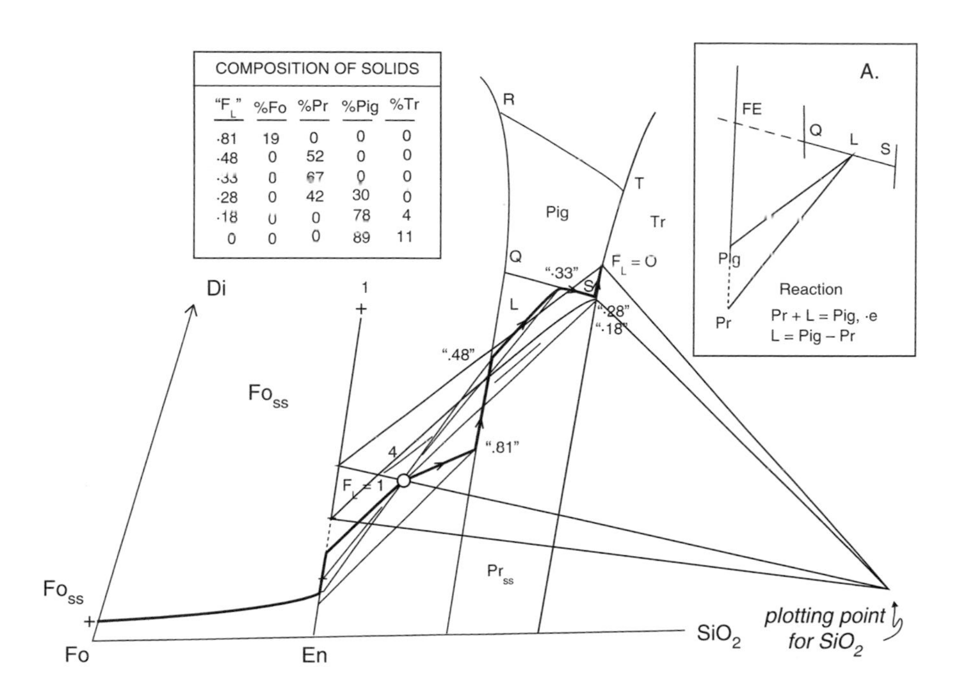

Fig. 12.10 Equilibrium crystallization, BC (4). Inset A: analysis of reaction of liquid on *QS*

BC. The final *L*-Pig tie line is found from the template, Fig. 12.5, using the final Tr-Pig leg to find the intercept on Pig. The table posted in Fig. 12.10 shows the evolution of the TSC from pure Fo_{ss} to Pig + Tr. Of special note is the massive destruction of Pr_{ss} while the liquid sits at S from 42% Pr at F_L = "0.28" to Pr at F_L = "0.18". During this process, Pig rises from 30% to 78%, as the fraction of solids rises from 0.72 to 0.82. Very little Tr (4%) is formed during this reaction.

Bulk composition No. 5, Fig. 12.11, also consists of Pig + Tr in the solid state, but the liquid path avoids *QS* altogether. (After some consideration of Figs. 12.4 and 12.11, it may be seen that this is so because BC 5 lies above the tie line *Q*-Pig (Pr), whereas BC 4 lies below that tie line. Liquids in the small triangle *Q*-Pr_{ss}-Pig (Fig. 12.4) always reach *Q*; those below *Q*-Pr (Pig) reach *QS* between *Q* and *S*, and those above *Q*-Pig (Pr) reach some point along *QR*, avoiding *QS*). The liquid produces Fo_{ss} and then Pr_{ss} until it reaches *Q*. The TSC now moves directly towards *Q*, until all Pr_{ss} is exhausted by reaction. Fo_{ss} still remains, as known by the lever *Q*-BC, so the liquid now moves along *QR*, in the reaction $L + Fo_{ss}$ = Pig. At F_L = "0.46", Fo_{ss} is exhausted and the *L*-Pig tie line cuts BC. The liquid now moves in isobaric divariance across the Pig field, to a point on *ST* determined by the use of Fig. 12.5. Now Tr joins Pig in the solids, and the TSC moves off the Pig composition line, in a curved path along successive Pig-Tr legs to the BC. The final *L*-Pig tie line is found from Fig. 12.5. The table posted in Fig. 12.11 summarizes the history of the TSC.

Bulk composition 6, Fig. 12.12, lies in the solid field of Di_{ss} + Pig + Tr. The liquid crystallizes Fo_{ss} until F_L = "0.84", at which time pigeonite begins to form. The TSC

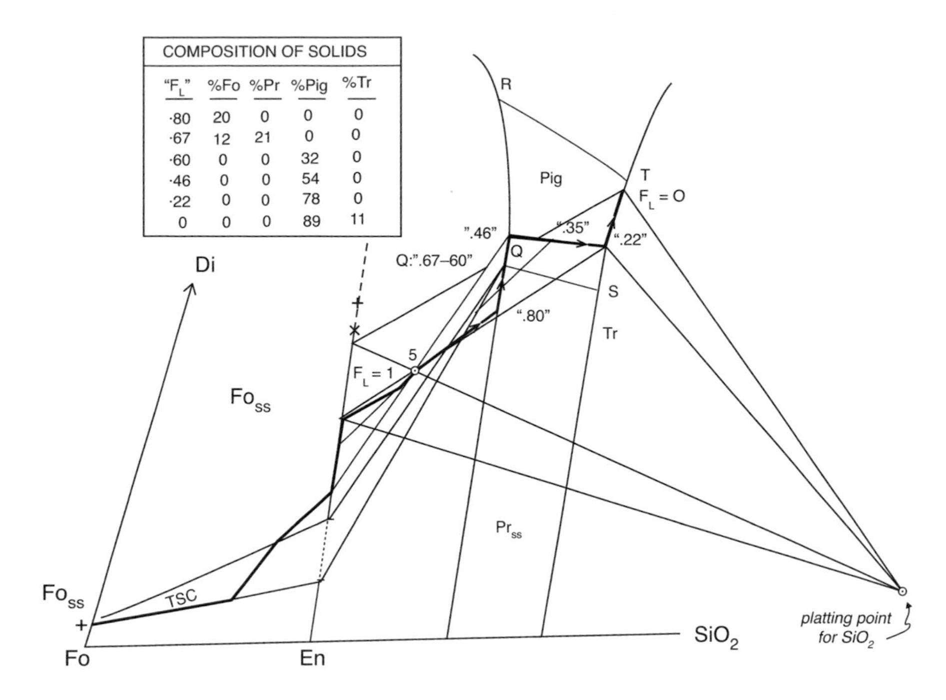

Fig. 12.11 Equilibrium crystallization, BC (5)

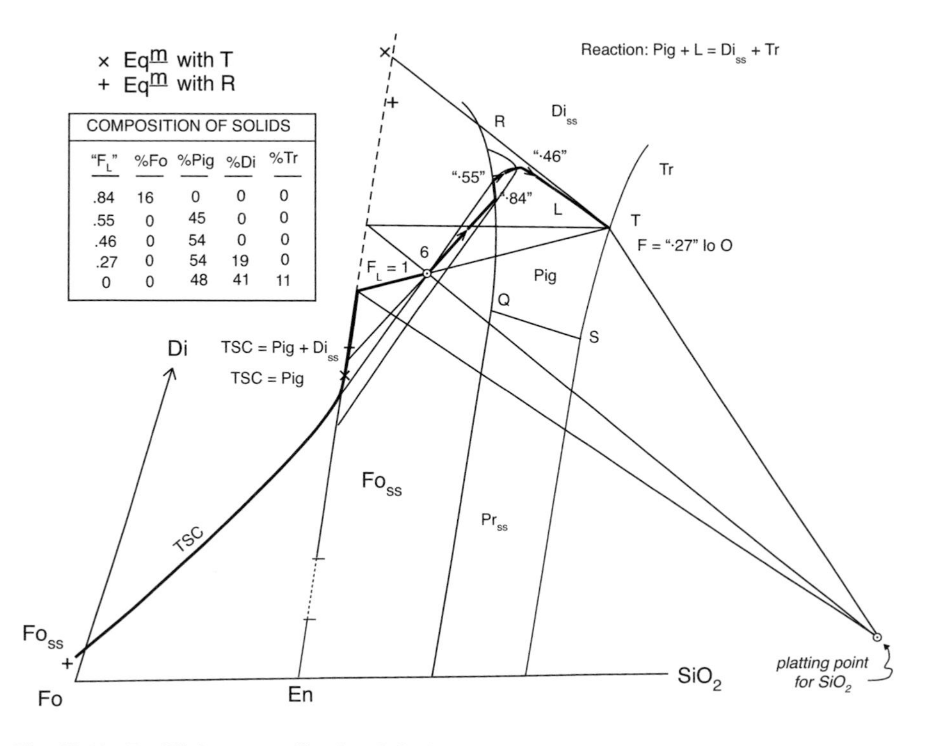

Fig. 12.12 Equilibrium crystallization, BC (6)

now moves towards Pig, which it reaches at F_L = "55" upon the destruction of all Fo_{ss}. The liquid now leaves the peritectic L (Fo_{ss}, Pig) and moves across the Pig field. The composition of Pig is still variable, and it reaches a point between the + and the X on the diagram just as the liquid reaches the Di_{ss} boundary. The two special pigeonite compositions indicated by these marks are the Di-saturated compositions on the solvus for liquids at R and T, respectively. Liquids along RT are in equilibrium with Pig and Di compositions between these pairs of marks in their respective composition regions. The liquid moves along RT, crystallizing mutually saturated Pig and Pi, while the TSC moves towards Di across the solvus. Note that the TSC is a two-phase mixture, and that no single pyroxene phase occurs within the solvus. The reaction along RT is even, but we reserve the proof of this until later, when it can be shown more clearly with another drawing. The liquid reaches T at F_L = "0.27"; here, the reaction is clearly odd, Pig + L = Di_{ss} + Tr, because T lies exterior to the triangle Di_{ss}-Tr-Pig. This invariant, four-phase equilibrium continues, as the TSC approaches T directly, until the latter reaches the BC. The table of TSCs shows complete destruction of Fo_{ss} and partial destruction of Pig by peritectic reactions. (Note that the gain of Pig between F_L = "0.46" and "0.27" is almost zero.) The final crystalline products will be larger crystals of Pig partly rimmed by Pi, and somewhat finer-grained interstitial masses of Di_{ss} + Tr (quartz at room temperature).

Summary of Equilibrium Crystallization

The liquid paths of BCs 1–6 have from two to five legs, and it is evident that equilibrium crystallization may be a highly complex process in this system. Olivine is partly or wholly destroyed by reaction along the peritectic curve $R'QR$, and Pr_{ss} may also be destroyed by reaction along QS. The determination of whether reactions are odd or even may be made with the aid of compositional polygons or triangles—so-called chemographic diagrams which simplify the writing of chemical reactions. The TSC paths in equilibrium crystallization are straight towards any invariant liquid, curved when they represent a mixture of two or more solid phases, and straight along the pyroxene join when the TSC consists of pyroxenes.

Fractional Crystallization

A few detailed examples will suffice to illustrate the principles of fractional crystallization in Fo-Di-SiO_2. Bulk composition 1 is illustrated in Fig. 12.13. The liquid moves away from Fo_{ss}, then away from Pig across the Pig field along a very slightly curved path until it reaches the Di_{ss} boundary RT. The liquid is constrained to this cotectic curve by crystallization of Di_{ss} + Pig, the latter in insignificant amounts. When the liquid reaches T, reaction is prevented by fractionation, and the liquid continues along the cotectic Tm, on which, with the crystallization of Di_{ss} and Tr, it

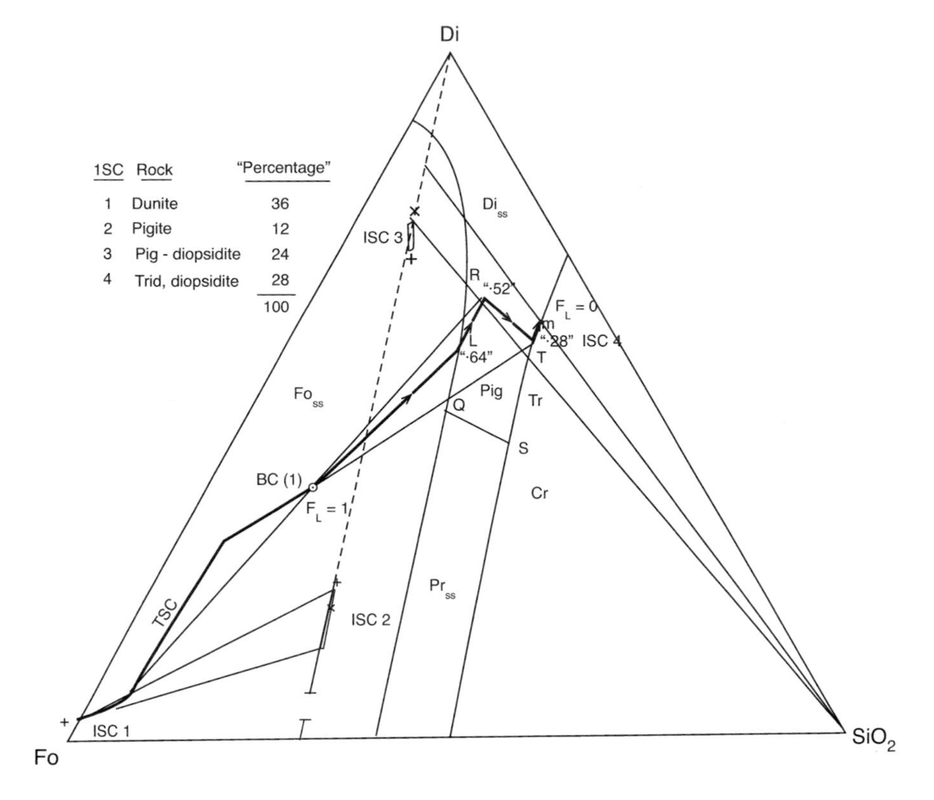

Fig. 12.13 Fractional crystallization, BC (1). *ISC*, instantaneous solid composition

moves to the *minimum m*, where crystallization is completed. More will be said later about the crystallization of Di_{ss} and the properties of the postulated minimum *m*, but suffice it to say for now that if such a minimum does indeed exist, it must be the goal of all liquids in the basaltic part of the system with fractional crystallization. Note that the liquid, in this example, is constrained to only two field boundaries (Di_{ss}-Pig and Di_{ss}-Tr), and passes quickly through invariant point *T*, finally reaching *m*, which is said to be invariant by restriction because it lies on a line between a special composition of Di_{ss} and SiO_2. For this equilibrium, the system can be described in terms of these two components alone, and $W_p = 0$ when $\phi = 2 + 1 = 3$. The ISCs produced during the fractional crystallization process are of special interest, as these are models of the rocks which would be produced by such a process in nature. The ISCs and their approximate amounts are posted in Fig. 12.13 (the distortion of the diagram accounts for the approximate nature of the figures). ISC 1 is Fo_{ss}: ISC 2 is Pig, of a composition range which can be determined from the tie line boundaries, Figs. 12.4 and 12.5; ISC 3 is Di_{ss} with a minor amount of Pig, and ISC 4 is a mixture of Di_{ss} and Tr which lies on the tangent to the liquid path *Tm*. Of course all other ISCs also lie on tangents to their respective *L* paths. The amount of each ISC produced is found by applying the lever rule to the TSC at each major stage of crystallization. The TSC path, in turn, follows successive tangents to ISCs, and is terminated at any stage by a lever through BC from *L*. Such a termination is shown for F_L = "0.52", where the TSC has just stopped moving towards Pig from Fo_{ss}. Another such termination is shown at F_L = "0.28", where the TSC has just stopped moving towards Di_{ss} and has begun to move towards ISC 4. The model rocks produced during fractional crystallization are tabulated in the figure; the natural version of "pigite" would probably be orthopyroxenite of the so-called "Stillwater type", i.e. inverted pigeonite with fat exsolution lamellae of augite in the former basal planes of pigeonite. The natural version of diopsidite might be described as clinopyroxenite. One would not expect to find a quartz-augite rock corresponding to "tridymite diopsidite", for by this stage of a fractional crystallization process, other components of natural magmas would become major, and one would expect to find abundant feldspars, and iron-enriched mafic minerals. Such a final-stage rock might well be a hedenbergite (or ferroaugite) quartz monzonite, and we can assume that rocks and liquids along *Tm* are oversimplified models of rocks having the general character of granitic rocks.

A very different suite of ISCs results from the fractional crystallization of BC 2, Fig. 12.14, a reasonably close model of an ultramafic magma rich in olivine and undersaturated with respect to silica. The liquid moves, as always, away from the ISCs, which are, in turn, Fo_{ss}, Pr_{ss}, Pig, Pig + Tr, and Di_{ss} + Tr. Natural varieties of the first three ISCs would be the monomineralic rocks dunite, orthopyroxenite of the Bushveld type (with thin (100) exsolution lamellae of augite), and orthopyroxenite

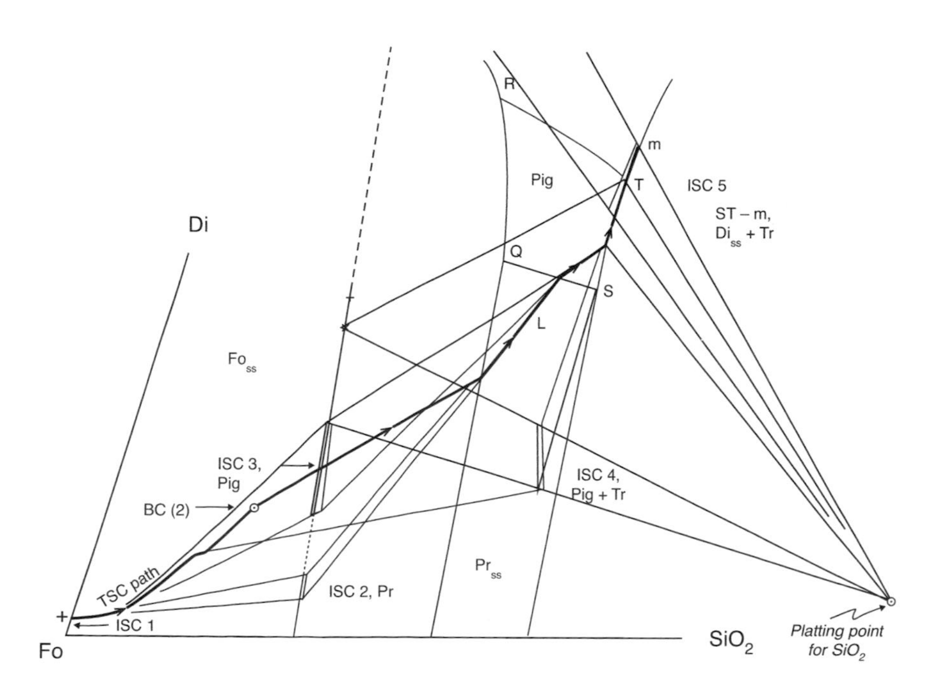

Fig. 12.14 Fractional crystallization, BC (2)

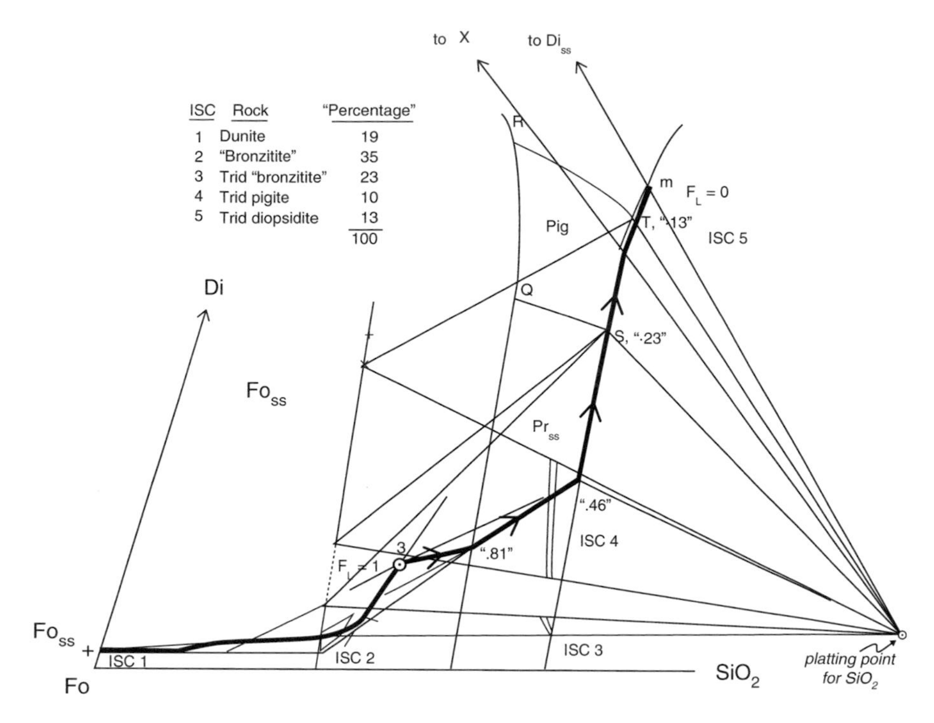

Fig. 12.15 Fractional crystallization, BC (3)

of the Stillwater type (with thick exsolution lamellae of augite in the former (001) planes of pigeonite, now "inverted" to hypersthene). ISCs 4 and 5 are more remote from typical natural rocks, but might be realized in nature as quartz diorite and quartz monzonite.

BC 3, Fig. 12.15, produces larger amounts of silica-saturated ISCs, consistent with its silica oversaturated location in the diagram. The ISCs and their amounts are tabulated in the figure, and it is left to readers to verify these for themselves, recalling that the TSC always moves towards the ISC, and is "stopped" at any stage by a lever through BC from *L*.

Fractional crystallization paths are summarized for five other bulk compositions in Fig. 12.16, drawn approximately to correct scale. The bulk compositions are denoted by their Fo-Di-SiO_2 ratios. ISC 1 is common to all bulk compositions, and in this case the initial Ca content of Fo_{ss} is correctly shown as varying with the Ca content of the BC. ISCs 2, 3, and 4 are shown only for BC 65-15-20; ISC 5 is common to all bulk compositions. Values of F_L are shown only for BC 65-15-20. TSC paths are shown for three of the five bulk compositions. The two silica-saturated bulk compositions are collinear from Fo_{ss} with their undersaturated counterparts; they share the same liquid paths. Liquid and TSC paths should be verified by inspection, or worked out where incompletely shown.

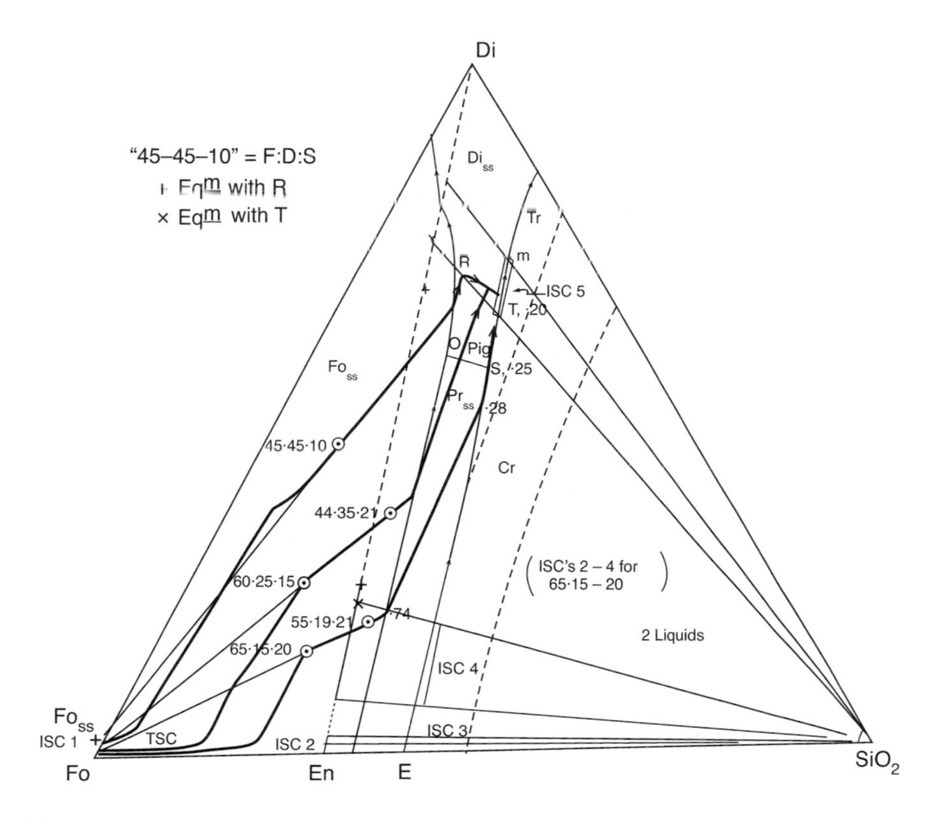

Fig. 12.16 Fractional crystallization of five bulk compositions, represented on a nearly scale-true diagram. F_L is shown only for the bulk composition Fo-Di-SiO_2 65-15-20. TSC paths are omitted for the two silica-oversaturated bulk compositions

Summary of Fractional Crystallization

Liquids produced in fractional crystallization from the Fo_{ss} field all arrive at *m*. The only special path common to all bulk compositions occurs along the cotectic curve *ESTm*. Each initial Di:SiO_2 ratio yields a different liquid path to *m*. Di-poor liquids reach *m* via the cotectic and *T*, while more Di-rich liquids reach *Tm* via the pigeonite field and cotectic *RT*. Failure of olivine and pyroxenes to react with liquid permits preservation of these phases as cumulate layers or isolated segregations, and increases the final amounts of SiO_2-producing liquids. Common natural rocks for which fractional crystallization in this system serves as a model are dunite, bronzitite (or orthopyroxenite of the Bushveld type), orthopyroxenite of the Stillwater type, and websterite. All residual liquids and ISCs are oversaturated in SiO_2, and serve as crude models for granitic liquids and rocks produced by fractional crystallization of ultramafic, basaltic, and intermediate liquids. It is interesting that even so mafic a liquid as 65-15-20 yields as much as 20% granitic residual liquid lying on the curve *Tm*.

Fractional Melting

Both equilibrium and fractional melting require knowledge of Pig-L and Pr_{ss}-L tie lines (Figs. 12.4 and 12.5) if the initial TSC contains one of these pyroxenes as the single pyroxene phase. The liquid composition at the beginning of melting is obtained from these tie lines. Equilibrium melting is the exact reverse of equilibrium crystallization, and need not concern us further. Fractional melting requires construction of the TSC path, and can easily be summarized after running through one example. This is shown in Fig. 12.17 for a bulk composition in the solid field of Pig + Tr. The figure is distorted, and the values of F_L are pertinent only to the distorted drawing, not to reality.

The tie line from SiO_2 through BC gives the initial composition of pigeonite in the TSC. By use of the tie line template, Fig. 12.5, the composition of liquid on ST in equilibrium with this pigeonite and Tr is found to be that labelled $F_L = 0$ in Fig. 12.17. Removal of this liquid causes the TSC to change directly away from it, but the successive liquids also move continually towards S, so that the initial TSC path away from BC is slightly curved. Eventually the TSC reaches the limiting Pig-Tr tie line, and further melting involves the production of Pr_{ss} in the TSC, by reaction. The liquid is now at S, in isothermal equilibrium with three solid phases,

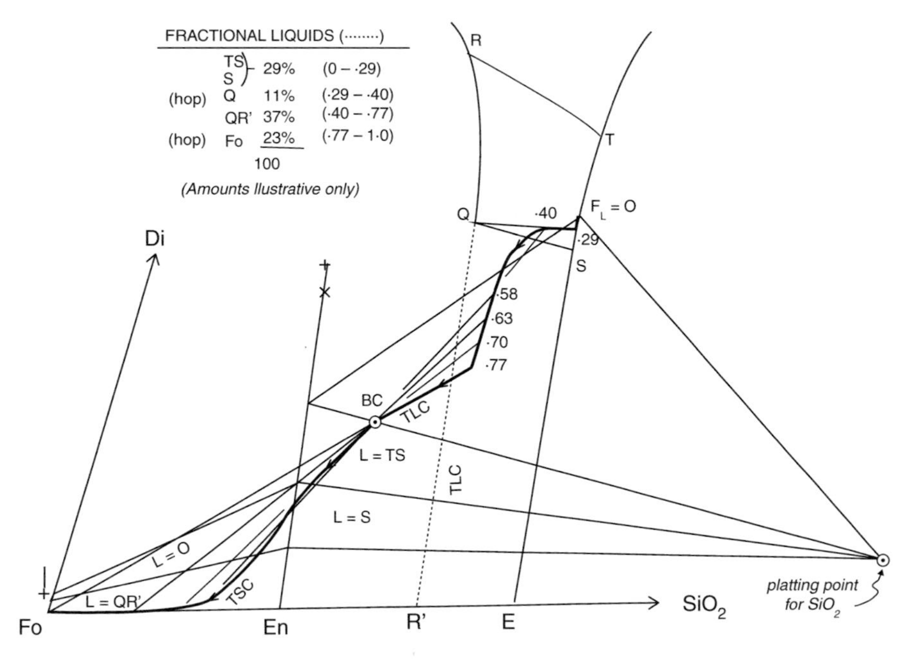

Fig. 12.17 Fractional melting, using toy FDS No. 1. The ILC path is dotted, and the TLC path is heavy. The identity of liquid being removed is labelled for each solid composition region. The TLC path is straight towards S while the solids lie in Pig + Pr_{ss} + SiO_2, and is straight towards Q between F_L 0.29 and 0.40

and the TSC moves directly away from *S*. After a short time, Tr is exhausted in the solids, and at F_L 0.29 the melting now involves Pr_{ss} + Pig, with production of Fo_{ss} in the solids. The liquid now lies at *Q*; the TSC moves directly away from *Q*, and the TLC moves directly towards *Q*. This melting process ceases when Pig is exhausted in the solids, at $F_L = 0.40$, and the liquid now moves along *QR′* with increasing production of Fo_{ss}. The TSC path towards Fo is now essentially like the TSC path in the Di field of Di An-Ab, Fig. 8.21, converging to pure Fo as the liquid reaches *R′* and as the composition of Pr_{ss} reaches pure En. The final instantaneous liquid has the composition of Fo, at 1890°C. The complete list of liquids (ILCs) is: *TS* from $F_L = 0$ to *S*, *S* alone, *Q* alone, *QR′*, and Fo. The model quantities of these liquids are posted in Fig. 12.17; they are obtained with the aid of the TLC path, which always runs towards the ILC, and which is terminated at any stage by a lever through BC from the TSC. Sample levers are shown at five different values of F_L TLC. The TLC path has straight segments towards *S*, towards *Q*, and towards Fo, and is elsewhere curved. The TLC may or may not be realized in nature as a homogeneous entity; in principle, it is possible to draw off the liquid continuously and mix it in an "inert" reservoir (such as a dunite). Alternatively, successive liquid fractions could be drawn off into separate conduits and preserved separately. The TLC is a graphical convenience, but at times may also represent a valid model for earth processes. Certain batches of liquid, especially the invariant liquids at *S* and *Q*, need not be drawn off continuously to achieve fractional melting. Batch removal of these liquids is equivalent to fractional removal. It is of interest to note that fractional *crystallization* tends to produce liquids which avoid the invariant points *Q*, *R*, and *S* (Fig. 12.16), while fractional melting may yield fair amounts of such liquids. Accordingly, a natural liquid which could be determined to be just saturated with olivine, orthopyroxene, and pigeonite could reasonably be a product of equilibrium crystallization or fractional melting, but less reasonably a product of fractional crystallization.

TSC paths in fractional melting are summarized in Fig. 12.18, where Di-rich compositions are ignored. Following the principles illustrated in Fig. 12.17, it can be seen that the major portion of the silica-oversaturated region is dominated by a bundle of TSC paths emanating from invariant point *T*. The silica undersaturated region is dominated by a bundle emanating from *R*. The liquid or range of liquids from which each TSC flees is labelled in each of the solid-phase regions. These TSC paths are the solidus fractionation lines of Presnall (1969).

Diopside-Rich Compositions

We have heretofore avoided detailed discussion of Di-rich compositions, because most of the geologically interesting parts of the system Fo-Di-SiO_2 can be provisionally treated without getting involved in the complex nature of Di_{ss}. It is important, however, to come to grips with these complexities eventually, both for geologic reasons and for experience in the principles of interpreting phase diagrams. Not all of the measurements necessary for a completely confident interpretation have yet been

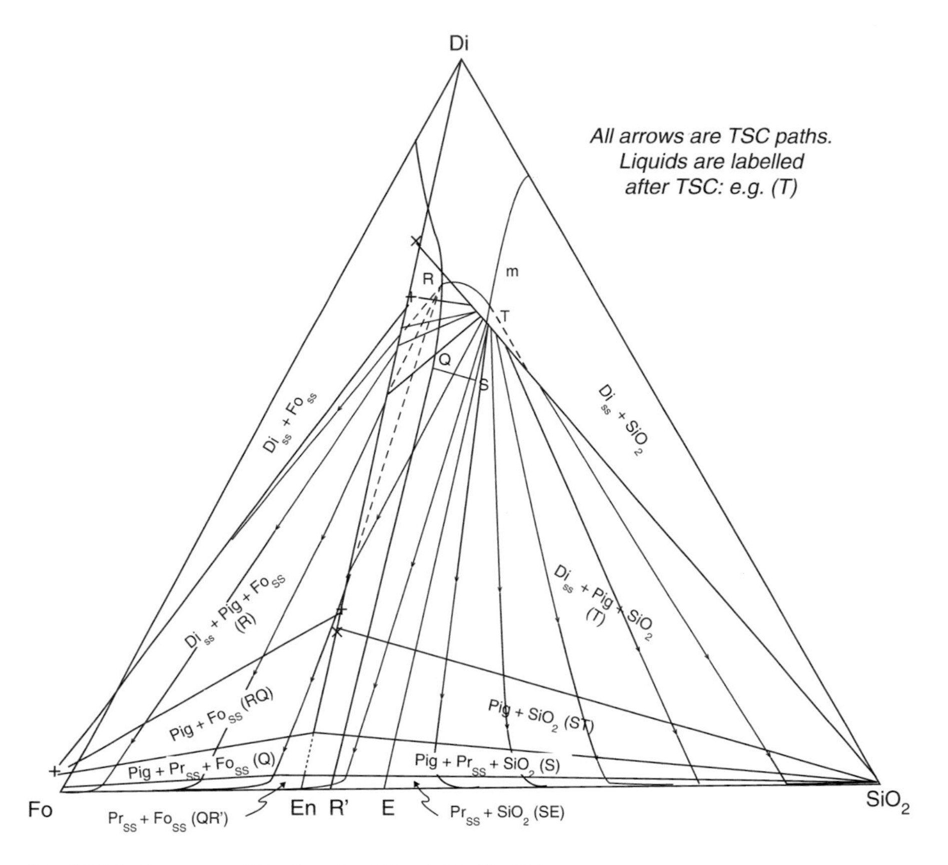

Fig. 12.18 Summary of fractional melting in Fo-Di-SiO_2

made, partly because a great deal happens in this part of the system over a very small temperature range. The interpretation to be offered here is speculative with respect to the existence of the minimum *m* and the maximum M' on the Di_{ss}-Tr boundary, but there are sound reasons for accepting this interpretation as the most probable given the data and the geologic knowledge which we have at hand.

It will be recalled from Fig. 12.1 that Di_{ss} shows a relatively large range of composition at 1390°C. It is also known that pure $CaMgSi_2O_6$ melts incongruently, 1391.5°C being the liquidus point (Kushiro 1972). This phenomenon is indicated in Fig. 12.1, where it will be seen that a tie line from Di_{25} to a $CaSiO_3$-rich liquid passes through pure $CaMgSi_2O_6$ at 1390°C. The lozenge-shaped field of Di_{ss} at 1390°C may be conceived as the map of an elongate island at a "water level" (temperature) of 1390°C. Assuming that the island has a smooth shape, as suggested by the map outline, then raising the temperature should have the effect of shrinking the lozenge proportionately in all directions. The last bit of the island to become submerged should be a point about in the centre of the lozenge. Such a point would be defined as the composition of Di_{ss} of *maximum thermal stability*. Its stable temperature limit is probably appreciably above the liquidus of pure $CaMgSi_2O_6$ itself, perhaps around 1395°C.

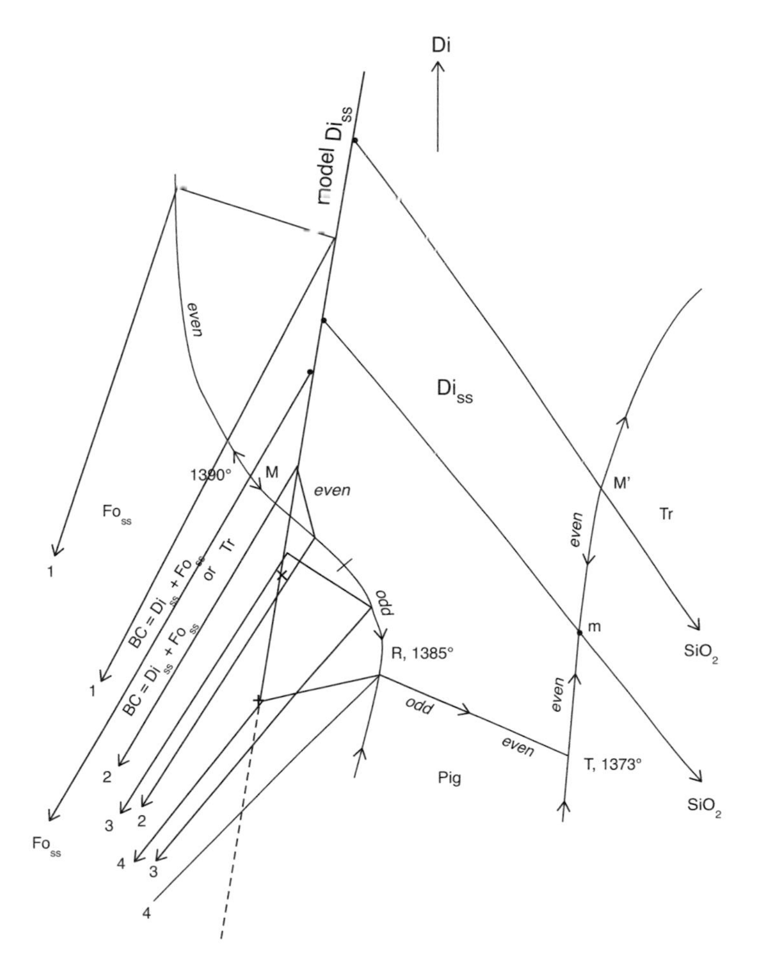

Fig. 12.19 Cartoon of liquid curves and tie lines in the Di-rich portion of the system. *M* and *M′* are maxima; *m* is a minimum. The symbols *X* and + have their conventional significance, as labelled in Fig. 12.3 et. seq. Four 3-phase triangles with Fo_{ss} are depicted. The labels "odd" and "even" refer to the type of reaction along segments of curves

Now if Di_{ss} by itself has a special composition of maximum thermal stability, it follows that there must exist special Fo_{ss}-Di_{ss} and Di_{ss}-SiO_2 tie lines of maximum thermal stability. These will not terminate at the Di_{ss} maximum, but on shoulders of that maximum in line with the saturating phase (Fo_{ss} or SiO_2). The arrangement of crystal-liquid tie lines found by Kushiro (1972), including many not shown in Fig. 12.1, tends to confirm these deductions. An interpretation of these relations is shown in Fig. 12.19. In this figure, the line labelled "model Di_{ss}" is meant to represent the axis of the lozenge in Fig. 12.1; it lies off the join Di-En. A most stable Di_{ss} (Fo_{ss}) occurs near this model line, very near or even at the tip of the lozenge in Fig. 12.1 at 1390°C. The tie line joining this composition of Di_{ss} to Fo_{ss} is a unique tie line of maximum thermal stability. If precisely these two compositions of Fo_{ss} and Di_{ss} are heated together, they will melt at 1390°C to a liquid which lies exactly on the tie line joining them. This is known as a *degenerate equilibrium*, when three ternary phases are collinear so that in discussing them the ternary system has degenerated to a special binary system composed of the two phases at the ends of the tie line. In the present case, no other tie lines between Fo_{ss} and Di_{ss} can occur at this high temperature. The composition of the unique liquid produced on melting this special assemblage of maximum thermal stability is itself a *maximum* (M in Fig. 12.19) on the L (Fo_{ss}, Di_{ss}) curve. All other liquids on this curve move away from M on crystallization. The temperature of M is taken as 1390°C; the composition lies off the join Di-En, as shown in the figure.

Compositions in the vicinity of M crystallize with even reaction, the liquids moving away from both Di_{ss} and Fo_{ss} together. Kushiro's data, however, show that Di_{ss} compositions move rapidly towards En with falling temperature, reaching the + symbol at 1385°C, in equilibrium with liquid R. Because of the chemographic arrangement of +, R, Pig, and Fo_{ss}, it is clear that the reaction of liquid R is $L + Fo_{ss} = Di_{ss} + Pig$; the reaction is odd, since Fo_{ss} has a negative sign when considered as a product of liquid on cooling. This equilibrium is denoted by the three-phase triangle labelled 4 in the figure. Triangle 3 also shows an equilibrium in which the reaction is odd, i.e. $L = Di_{ss} - Fo_{ss}$, because the tangent to the liquid path cuts the extension of the tie line Di_{ss}-Fo_{ss}. As the temperature rises towards 1390°C, the composition of Di_{ss} moves so rapidly towards Di that the reaction becomes even, the tangent to the liquid path now cutting the Di_{ss}-Fo_{52} tie line itself. A small tick mark on the liquid path indicates the transition between odd and even reactions; at this point the tangent to the liquid path falls exactly on the Di_{ss} end of the Di_{ss}-Fo_{ss} tie line. Liquids moving from M to R first produce and then consume Fo_{ss}, as the reaction changes from even to odd. This reaction sequence may occur for both silica saturated and oversaturated bulk compositions.

Liquid R is in equilibrium with both Di_{ss} and Pig, the former having the most En-rich limit permitted by the solvus, at the composition denoted by the + sign. At temperatures below R, the En-saturated composition of Di_{ss} moves *away* from En, down the limb of the solvus (recall Fig. 12.2), thus reaching X in the figure when the liquid is at T, 1373°C. The tangents to the curve RT do not all fall to the En side of this set of Di_{ss} solvus compositions for reasons to be explored shortly; hence the reaction is partly odd and partly even along the curve RT. The reaction at T is odd, namely $L + Pig = Di_{ss} + Tr$. This is because the tie line Di_{ss}-SiO_2 from composition X falls below point T. Once pigeonite is used up in reaction, the liquid is free to proceed along Tm in an even reaction, $L = Di_{ss} + Tr$.

Analytical data indicate that the Di_{ss} of maximum thermal stability in equilibrium with SiO_2 lies well toward Di from that in equilibrium with Fo_{ss}. The tie line from this unique maximum Di_{ss} (SiO_2) must generate a maximum (M' in Fig. 12.19) on the curve L (Di_{ss}, Tr). If such a maximum exists, as seems probable, then the minimum m must also exist, because as we have just seen, liquids move northward from T along the Di_{ss}-Tr curve, and must move southward as well as northward from M'. The properties of a ternary minimum are like those of a maximum in that liquid m lies collinear with a special composition of Di_{ss} and Tr; the system is degenerate, and m is invariant by restriction ($c = 2$, ϕ, $= 3$, and $W_P = 0$). Further discussion of crystallization at a minimum is given below.

Nature of the Diopside-Pigeonite Boundary

The experimental evidence indicates that the boundary curve *RT* is nearly straight, or at least not strongly curved. Because of this and the locations of Di_{ss} in equilibrium with liquids on *RT*, crystallization processes involving this curve are special. The tie lines from diopside solid solution to *R*, *T*, and two points on *RT* are shown schematically in Fig. 12.20. For simplicity, it is assumed here that *RT* is rigorously straight. Each of the tie lines is part of a three-phase triangle Di-Pig-*L*. Forsterite and silica occur as additional saturating phases for liquids at *R* and *T*, respectively. Consider the bulk composition *B* which, when fully crystallized, will consist of Di_{ss}, Pig, and Tr. In order to achieve this TSC, the liquid will need to reach *T* and then produce Tr by reaction. At present, bulk composition *B* consists only of $Di_{ss} + L$. The reader may verify for herself that the liquid got from *B* to its present location via the reactions $L = \text{Fo}$, $L = \text{Di} - \text{Fo}$, and $L = \text{Di}$. The scene depicted in Fig. 12.20 occurs just as the liquid has reached *RT*. In order for crystallization to proceed any further, the liquid must move towards *T*. In doing so, it will encounter tie lines to diopside that fall progressively above *B*, and therefore pigeonite must be continually added to the TSC. The *reaction* along this part of curve *RT* is therefore undeniably $L = Di_{ss} + \text{Pig}$, an *even reaction*. Nevertheless, the curve itself here is *odd*, because its tangent lies on Di − Pig.

How can there be an even reaction on an odd curve? First, note that this can happen only during equilibrium crystallization; in fractional crystallization the liquid would leave the curve *RT*. Secondly, we note that the curve is odd because its tangent falls to the negative side of one of the equilibrium crystal compositions. The part of

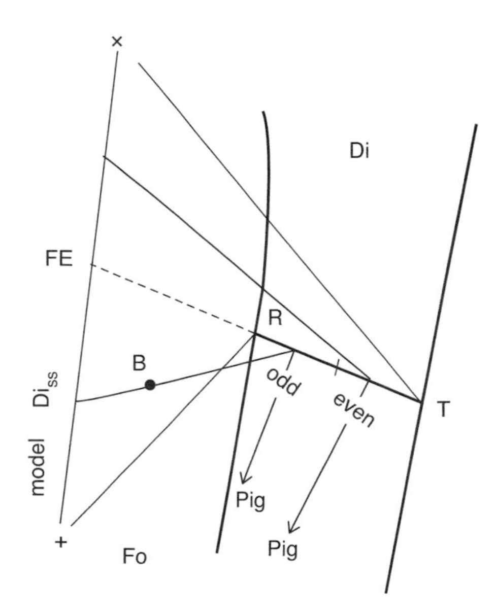

Fig. 12.20 Later stages of equilibrium crystallization of bulk composition *B*. The liquid lies on *RT* and is connected by a lever through *B* to the TSC, which consists of diopside solid solution. The liquid has just reached *RT* via paths described in the text. It now proceeds down *RT* towards *T* while pigeonite (Pig) is added. The composition of Di_{ss} will move towards *X* in the figure. Point FE is the fictive extract, on the tangent to *RT*

the curve nearer *T* is an even curve, because its tangent now falls *between* Di and Pig, the composition of Di_{ss} having moved upward.

Liquids must always move away from the material being extracted from them. In fractional crystallization, this material is the tangible ISC. However, in equilibrium crystallization, with solid solutions, the material being extracted from the liquid is not identical with the equilibrium crystal composition, but is instead the material *needed to convert the crystals to their new compositions*. In this book, this material is called the *fictive extract*; it always lies on the tangent to the liquid path. For the case being considered, the fictive extract lies at the point labelled FE in Fig. 12.20. Point FE must contain a positive quantity of Pig, since pigeonite is crystallizing. It must also contain a positive quantity of the fictive Di_{ss} component being removed from the liquid to convert Di_{ss} towards the top of the diagram. This fictive Di component, therefore, must simply lie above FE on the Di line. A retrospective analysis of the equilibrium liquid path $L =$ Di across the corner of the diopside field will show that this fictive material, always on the tangent to the *L* path, has lain above FE all the time while diopside crystallized alone. It is thus possible to map the location of the fictive extract and to show that when the liquid first touches the *L* (Di, Pig) boundary curve, the fictive extract jumps discontinuously from some location near *X* to the point FE.

Another example of the fictive extract is shown in Chap. 15, in connection with Fig. 15.21.

The Diopside Field

Bulk composition 8, Fig. 12.21, lies in the solid field of Di_{ss} + Tr. Crystallization begins with the production of Di_{ss}, which moves towards Di with falling temperature, causing the liquid to follow a concave-down path. The liquid path is shown in the figure; it reaches a point on *Tm* such that the Di_{ss}-*L* tie line passes through the bulk composition. This arrival initiates the first three-phase triangle Di_{ss}-*L*-Tr. The liquid now moves along the curve *Tm* until the Di_{ss}-Tr leg of the three-phase triangle cuts BC, when crystallization ceases. The principles are exactly the same as those in Di-An-Ab, where a solid solution (plagioclase) occurs with a phase of fixed composition (Di). If Fig. 12.21 is turned on its left side, a resemblance to the liquid path in Di-An-Ab will be discerned. The behaviour of Di_{ss} is that of an ordinary solid solution loop, although the liquid in this case does not lie on the Di_{ss}-En join. The minimum *m* has nothing to do with this particular crystallization path.

The general properties of minima may as well be introduced here, in any event. Figure 12.22 shows the hypothetical system A-B, with a minimum melting relationship. The diagram is to a first approximation equivalent to two binary loops butted together at *m*, and equilibrium crystallization in either loop follows the same principles as for plagioclase. The point *m* occurs at the point of tangency of solidus and liquidus; here, a special composition of solid AB and a special composition of liquid are identical, and lie at the same temperature. The tangent to *m* must be

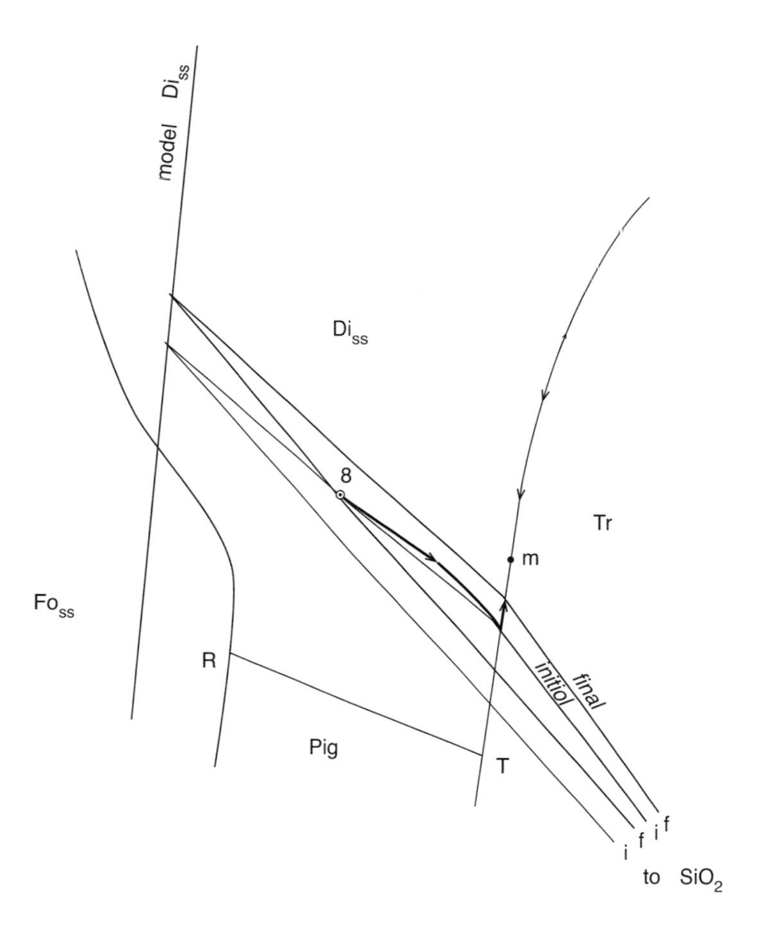

Fig. 12.21 Equilibrium crystallization of bulk composition 8. Initial tie lines to Di_{ss} are not shown

isothermal, i.e. horizontal. A liquid of composition *m* would crystallize isothermally at *Tm* to a homogeneous crystal of composition *m*. If such a minimum-melting system is embedded in a solvus, a eutectic system such as that shown in Fig. 5.5 results, in which the crest of the solvus and the bottom of the minimum loops are metastable and not shown. With this general and sketchy introduction to melting minima, we may now turn to a depiction of Di_{ss} relations in Fo-Di-SiO_2, by means of two *projections*.

Projections of pictures may be achieved in various ways; for example, with the light behind the observer, or shining towards the observer from behind the screen. We are already familiar with projections along a temperature axis onto a composition plane, as seen in all our polythermal ternary diagrams. What we now propose to do is to project liquid and solid compositions related to Di_{ss} from one or another saturating phase, Fo_{ss} or SiO_2. Figure 12.23 illustrates how such a projection from Fo_{ss} is derived. The aim is to visualize the *T-X* relationships of liquid and Di_{ss} when these

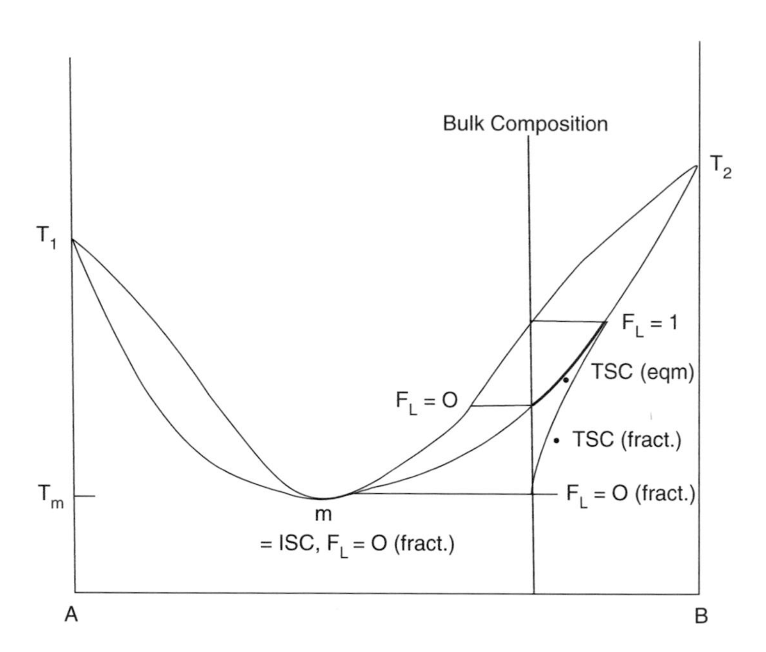

Fig. 12.22 Hypothetical binary system with a minimum melting composition (*m*). This feature with an internal minimum falls into the class of an azeotrope

two phases do not lie in the same *T-X* plane. The right-hand side of Fig. 12.23 is the desired *T-X* projection, separated from the parent *X-X* plot by a *fold line*. The parent *X-X* plot will be recognized, it is hoped, from Fig. 12.19. The line Di_{ss}-En is chosen as the base of the projection plane T-D_{ss}-En. We wish to map onto this *T-X* plane all compositions of Di_{ss} and *L* that are saturated with Fo_{ss} and with each other. This is done by means of rays from Fo_{ss} which either pass through the point to be projected and then on to the reference plane, or pass through the reference plane to the point. In either case, the intersection of the ray with the reference plane is the desired projection point. Those of the second type are indicated with arrows backward along the ray, whereas those of the first type are indicated with arrows forward along the ray. The fold line is set off an arbitrary distance from the reference plane, and the projection points are carried from the reference plane to the fold line by normals to both. The fold line marks the transition from *X-X* to *T-X* plot.

In the *X-X* plot, tie lines are shown between Di_{ss} and liquid. One of these tie lines is the unique line Fo_{ss}-M-Di_{ss}, and this plots in projection as a point, a maximum, in the *T-X* diagram. Three other tie lines are projected, one of them being $+\ -R$ at the bottom of the *X-X* diagram. Such tie lines are isothermal lines in the *T-X* diagram. From the disposition of such tie lines, properly assigned to their respective temperatures, the full *T-X* projection from Fo_{ss} can be constructed. It consists of a maximum *M* joining a pair of loops, one of which terminates in the Pig-Di_{ss} solvus at an apparent eutectic, *R*. This point is, in fact, of the eutectic type with respect to Di_{ss} + Pig, but as we know, it is a peritectic with respect to *L* + Fo_{ss} (this peritectic

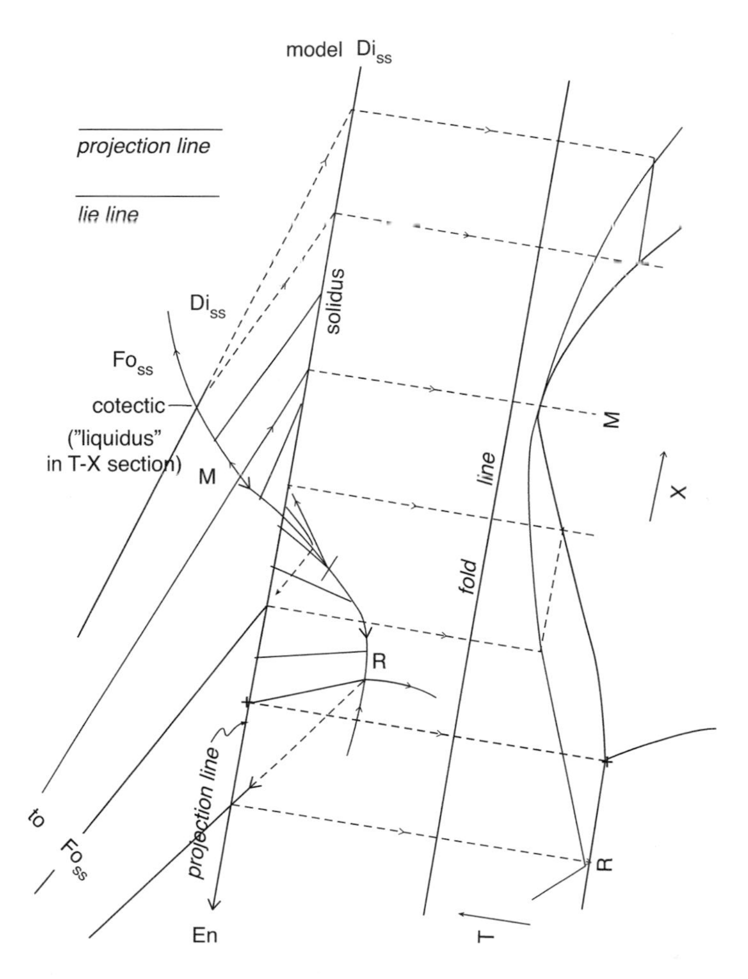

Fig. 12.23 Scheme of projecting liquid and crystal compositions saturated with Fo_{ss} onto a *T-X* plane. All crystal compositions are assumed to lie on the model Di_{ss}-En line, which is also taken as the base of the *T-X* plane. A number of crystal-liquid tie lines are shown, four of which are chosen for projection

relationship cannot be shown in the projection, but it would be apparent in a projection from Di onto the Fo-SiO_2 sideline). The significant property of the maximum is that the projected solidus and "liquidus" become tangent there. The term "liquidus" is put in quotation marks because this curve is in fact the cotectic (elsewhere peritectic) equilibrium L (Fo_{ss}, Di_{ss}).

The projection of Fig. 12.23 is repeated at another scale in Fig. 12.24, along with a projection from SiO_2 of the compositions of phases saturated with SiO_2. This projection is also made onto the Di-En-*T* plane. In the Fo_{ss} projection, Fig. 12.24a, Fo_{ss} is to be understood as an added phase in all regions, since this is a projection from Fo_{ss}. Similarly, Cr or Tr is to be understood in all labels on Fig. 12.24b. In the

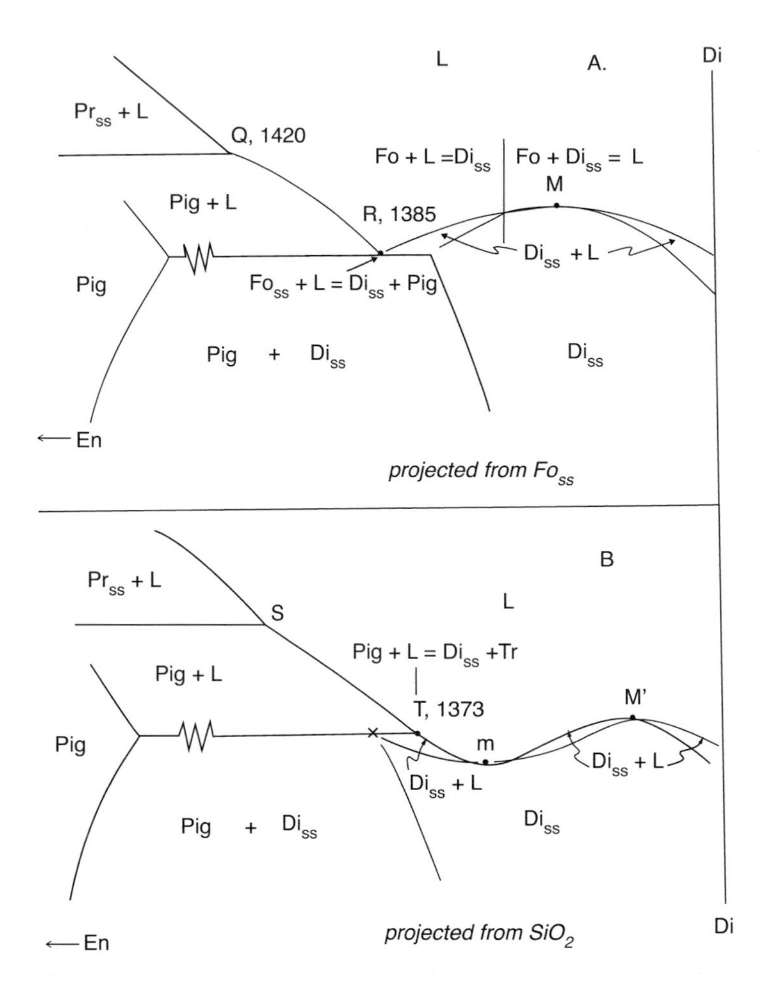

Fig. 12.24 Schematic *T-X* relations of Di_{ss}-liquid equilibria. (**a**) projected from Fo_{ss}; liquids are those along *R′QRM* in *M* in Fig. 12.3. (**b**) projected from SiO_2; liquids are those along *ESTmM′* in Fig. 12.3

upper figure, *Q* is seen to be a reaction point where Pr_{ss} is converted to Pig; compare Fig. 12.2, which shows the *T-X section* along Di-En rather than the *projection*. Point *R* shows even reaction relative to Di_{ss} + Pig, as mentioned above. The change from odd to even reaction along the curve *RM* cannot be discerned in the *T-X* diagram, as it is a property of ternary relations involving Fo_{ss}, the projecting phase. In Fig. 12.24b, the Di_{ss} composition labelled *X* is the same as in previous diagrams, e.g., Fig. 12.19, and the projection from SiO_2 emphasizes that this composition has a peritectic relationship to liquid *T* when tied to SiO_2. The isothermal tie line Pig-*X*-*T* truncates the Di_{ss}-*L* loop to the left of the minimum *m*. It is on this segment of the loop that the final stages of crystallization of bulk composition 8, Fig. 12.21, take place. The remainder of the right-hand side of the diagram illustrates the supposed maximum on the L (Di_{ss}, Tr) curve. In both figures, the behaviour of Di_{ss} near pure $CaMgSi_2O_6$ is left unspecified; further discussion of these relationships is given by Kushiro (1972).

Application to Basalts and "Granites"

The minimum m prevents liquids from leaving the Fo-Di-SiO_2 system by fractionation, and is consistent with natural fractionation sequences which lead to pyroxene-quartz-feldspar rocks rather than to wollastonite-bearing siliceous rocks which would result from fractionation out of the system, beyond the maximum M'. Although the natural systems are usually iron-rich, thus making comparison with the system FDS rather tenuous, the analogy is probably valid as a provisional consideration.

The main features of the system FDS as a model of natural rocks occur in the basaltic region. As already discussed, the reaction of olivine to both Pr_{ss} and Pig is a phenomenon for which much evidence exists in natural rocks. The reaction relationship between Pr_{ss} and Pig, as modelled by the equilibria along QS, is a matter which has been hotly debated among field workers. Again, the natural sequence is the same, but involves iron enrichment leading to the stabilization of Pig instead of Pr_{ss}. However, iron enrichment occurs with falling temperature, and may simply be a parallel effect rather than a causal one. In any event, it is a common experience to find, in tholeiitic layered intrusions, sills, and lava flows the sequence olivine-hypersthene-pigeonite, and the present system serves as a good starting model for this sequence. An added feature of the natural rocks is that this sequence commonly occurs in the constant presence of augite, rather than leading to late crystallization of a clinopyroxene, as would be implied by the system FDS. Reasons for this will not be rigorously explored here, but it can at least be said that the lowering of crystallization temperature by Fe^{2+} and other components of natural magmas serves to embed the liquid-crystal equilibria more deeply into the diopside – Ca-poor pyroxene solvus, and to stabilize diopside in equilibrium with liquid and the other mafic phases. In this way, it is possible to imagine a cotectic relation among Ca-poor phases (including olivine), liquid, and diopside, in company with peritectic relations between the Ca-poor phases. Imagine, for example, the composition of plagioclase being affected by the co-existence of monoclinic pyroxene. We shall perhaps see.

The reaction relation of olivine plus liquid to orthopyroxene is the hallmark of tholeiitic basalts. This reaction occurs for all liquids along $R'Q$ (Fig. 12.3 et. seq.), hence for all bulk compositions within the triangle Fo_{ss}-Q-R', whether they are olivine tholeiites in Fo-Di-En or oversaturated tholeiites in Di-En-SiO_2. The reaction of olivine plus liquid to pigeonite is an entirely equivalent reaction on QR. The long reach from R' on Fo-SiO_2 to R in the ternary system practically guarantees that the reaction will occur in any situation remotely approaching equilibrium crystallization, for almost any liquid of tholeiitic affinity. In pure fractional crystallization, the sequence olivine – Ca-poor pyroxene would be encountered instead, but in natural processes the theoretical limit of fractional crystallization can never be pervasively maintained, so natural rocks are bound to show, somewhere, evidence of the reaction, commonly as a reaction rim along the corroded edge of an olivine crystal.

The system Fo-Di-SiO_2 has been used extensively in the interpretation of tholeiitic layered intrusions formed by large-scale fractional crystallization and crystal settling, particularly by Irvine and Smith (1967) and Irvine (1970). Although there have been important modifications in the system since that time, the general conclusions are interesting and valid. In brief, these are that in the Muskox layered intrusion, Canadian Northwest Territories, fractional crystallization of a magma analogous to bulk composition 60-25-15 in Fig. 12.16 produced a series of ISCs very closely modelled by the experimental system, namely peridotite, orthopyroxenite, and websterite (Cpx-Opx rock). Modal analyses of the actual rocks, when plotted in the FDS system, fall in distinct concentrations almost exactly where they would be predicted from the experimental ISC. The experimental diagram is therefore shown to be a quantitatively rather accurate model of natural processes, and with the benefit of such hindsight, it might be possible to use modern analytical methods to construct slightly modified versions of the experimental system, accurate in subtle details of pyroxene compositions and reactions, for each natural body of rock. (People actually do these things.)

The system Fo-Di-SiO_2 also serves as an excellent starting model for the generation of basaltic magma in the mantle. We shall eventually examine the high-pressure relationships of this system, but the low-pressure version serves to illustrate several important principles. One is that no matter how Ca-poor or silica-poor the model mantle composition is in the triangle Fo-R-R', the initial liquid produced on heating is moderately to very Ca-rich, and silica oversaturated at low pressures. For example, if the bulk composition falls in the field of Fo_{ss} + Pig + Pr_{ss}, the initial liquid is Q, with more than 50% diopside component relative to orthopyroxene. This fact no doubt explains why tholeiitic basalt is almost invariably saturated or nearly saturated with augite, of which it produces large quantities during crystallization. Moreover, it is clear, as already mentioned, that liquids saturated with several phases, such as Q, S, and liquids along QR', are much more likely to be generated by either fractional or equilibrium melting than by fractional crystallization from ultramafic liquids. This principle follows from the extensive presence of odd reaction curves in the system, which shed liquids on fractional crystallization but hold them on melting.

G-X Diagrams for Systems with Maxima or Minima

Melting at a maximum or minimum has been introduced in this chapter because such behaviour occurs in the system Fo-Di-SiO_2. Here we have used projected liquid and solid compositions, because the system is ternary. However, the principles apply to simple binary systems, and, in turn, the *G-X* diagrams for a binary system can be used to illustrate the principles operating in a ternary system.

Figure 12.25a, b shows two related types of system displaying, respectively, melting at a maximum and melting at a minimum. They can be analysed essentially

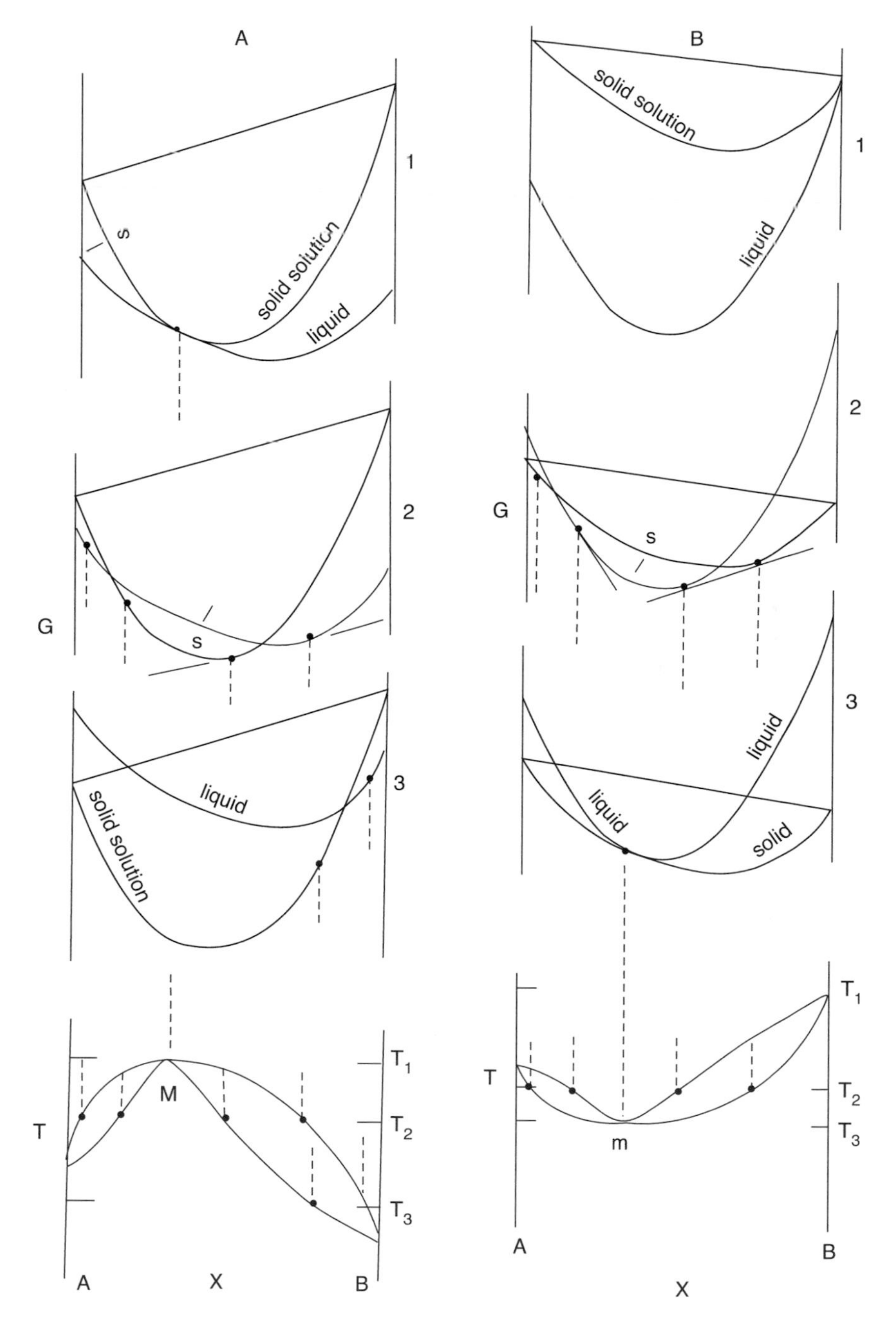

Fig. 12.25 *G*- and *T-X* diagrams for systems showing melting at a maximum (**a**) and at a minimum (**b**)

as for the plagioclase system, but with the observation that at either a maximum or a minimum, both solidus and liquidus *must* coincide on an *isothermal* tangent in *T-X* space, and in *G-X* space the solid and liquid *G-X* curves simply have a common tangent point. The common tangent has zero slope in *T-X* but not in *G-X*. It is common to see diagrams which violate this principle, especially in having liquidus curves entering a minimum at a high angle. This is incorrect. It may happen that the liquidus curves approach closely before swinging in a tight arc to isothermal tangency, but that would imply a very large difference in the radius of curvature of the liquid and solid *G-X* curves (see section 3 of Fig. 12.25b), and this is unlikely, to say the least, in silicate systems in which the melts are highly polymerized and structurally analogous to the crystal-line solutions from which they form. Such are the insights one may gain from an appreciation of *G-X* diagrams, which constantly rescue from error the practising petrologist who remembers to use them.

Chapter 13
Layered Intrusions

Introduction

A layered intrusion is a set of igneous rocks emplaced into the crust of the Earth in such a way as to be exposed to eventual uplift and weathering where we can see it. Such a body is ideally defined by oldest rocks at the base and youngest rocks at the top, with the exception of any roof rocks preserved that will meet the upward layering rocks in what has been pleasantly named a sandwich horizon. Layered intrusions have important ways of revealing to us examples of phase equilibria that are fundamental. We only have to discover what these principles are.

It appears that every layered intrusion is unique (although some may appear more unique than others!) and that small details are likely to define a solidification history that is special. There are on record at least 30 exposed layered intrusions on Earth, of ages running from 4406 Ma (Ujaraaluk, Quebec) to 56 Ma (Skaergaard, Greenland). They are described compactly by Scoates and Wall in the valuable book "Layered Intrusions" by Charlier et al. (2015).

Our objective here is to find and illustrate the principles of phase equilibria that are illuminated and in some cases defined by a small set of layered intrusions. We can do this comfortably in large part because the Charlier book contains nine general chapters describing many aspects of layered rocks and then follows with six specific descriptions of individual intrusions including the 1307 Ma Kiglapait intrusion of Labrador. Experiments on the origin of the Kiglapait magma deep in the Earth's upper mantle and its assembly at the Moho have been described by Morse et al. (2020) in the *Canadian Mineralogist*.

S. A. Morse, *Basalts and Phase Diagrams*,
https://doi.org/10.1007/978-3-030-97882-2_13

Stillwater Complex, Montana, USA

We may start our selective wanderings with the intuitive studies of the Stillwater Complex by Harry Hammond Hess (1939, 1960). This large body lies in southern Montana and has been dated at 2704 $\pm$ 1 Ma (Scoates and Wall, pp. 3–74 *in* Charlier et al. 2015). Hess's observations lead us to the birth of the Adcumulate theory that eventually found its way into glory in the first year of the *Journal of Petrology* under the authorship of Wager, Brown, and Wadsworth. Hess (p. 149) had discovered an important paper by F. F. Grout (1918) that invoked convection to explain the crystallization of large basaltic magmas. Hess noted that distinct layers of Stillwater rock contained only settled minerals of constant composition. If each layer had been a closed system, then surely the mineral compositions would have evolved in place. That result could *only* be avoided if the lower-temperature components had all *escaped*, leaving only the liquidus compositions.

This might sound like a special case, but only consider that the ambient temperature in the accumulated mafic rock would tend to be constant, because any cooling front would be far away through the overlying magma.

Then, the resident crystals must keep on growing until the whole local mass was crystallized—at constant temperature—hence *isothermally*. Hess recognized that this was the "two-phase convection" of Grout. The lighter phase is rising up, while the denser phase is continuing to crystallize in place.

So what happens to the lower-temperature component? What we now may call the *rejected solute*, being lighter than the isothermally growing crystals, will rise into the main magma: "phase 1". And in turn—this is the fun part—the resident crystals, being of a single composition, will grow by adding only the *isothermal* component of the main magma, hence at constant composition.

Grout (1918) wrote that "A consideration of the above figure suggests that convection is inevitable in large basaltic magma".

The principles of *solute rejection* are well known in the metallurgical literature and were explicitly applied by Wager et al. in the first year of the *Journal of Petrology* (1960) where it was named *adcumulus growth*. This process implies isothermal solute rejection and accounts for crystal growth at constant composition. Such isothermal growth requires magmatic exchange of incoming fresh magma and simultaneous rejection of depleted magma.

Skaergaard Intrusion, Greenland

Introduction

This relatively small intrusion in East Greenland occurs within and about a glacial sheet that has exposed many kilometres of outcrop in the last two decades especially. Although "small" and difficult of access, it is the perhaps the most studied, most important, and most gifted body of layered igneous rocks in the world. It was discovered in 1930–1931 by L. R. Wager and studied with new colleagues, especially W. A. Deer, over several seasons, leading to the publication of the 1939 "Memoir" by Wager and Deer (1939). It has been dated at 55.960 Ma (see Scoates and Wall in Charlier et al. 2015). An annotated bibliography of this intrusion has been made by Brooks (2018) and covers 777 papers in 60 pages. Since then, two substantial papers by Nielsen et al. have been published in the *Journal of Petrology* (2019).

The Skaergaard intrusion has an oval shape, longest at ~7 km in WSW-ENE directions and with a total original depth near 4 km. The shape and internal conditions are usefully shown in the book by Brooks as richly detailed by T. N. Irvine. The plagioclase composition ranges from about An_{66} at the base to An_{25} at the sandwich horizon where the Upper Zone meets the inverted Upper Border Zone. The mineralogy evolves from augite–olivine–plagioclase at the base through the arrival of Ti-magnetite within the Lower Zone, then the loss of olivine and its return in the Middle Zone, now at An_{50}, Fo_{40}, and thence through the Upper Zone to the sandwich horizon where the plagioclase reaches An_{25} and the olivine Fo_0. Note that the plagioclase does not reach an evolved albite content, whereas the olivine does reach a Fo limit. The pyroxene does also reach the limit of Fo_0.

A study by Morse et al. (1980) noted that the "EFS" reaction $En = Fo + SiO_2$ could be used to estimate silica activity in the Skaergaard intrusion, which when plotted against *1/T* appears thoroughly consistent with the disappearance of olivine in the Middle Zone. The activity of silica in the system governs the relationship between olivine and orthopyroxene: if the silica activity is high, olivine becomes unstable and pyroxene is stable. If the silica activity is low, then olivine is stable. Figures in the cited paper illustrate these results in terms of silica *activity* and oxygen *fugacity*. *Activity* is formally treated as a thermodynamic property related to chemical potential in Chap. 17. A further relationship of thermodynamic interest, *fugacity*, is discussed in Chap. 16.

Temperatures and Compositions

The liquidus temperatures of the Skaergaard intrusion have been determined experimentally in several laboratories, and they are plotted here in Fig. 13.1. They cover the region from the base of the Lower Zone at 1191°C with a systematic drop that contains more than a bit of subtlety. Clearly, from LZc to somewhere between the Middle Zone and Upper Zone (a) there is a tendency to flattening, followed by the return of a steeper slope. This flatter slope coincides significantly with the disappearance of olivine in the cumulate, as discussed above. It implies that the cooling of the magma is slower with fewer phases to crystallize. The more phases, the faster the cooling. This effect involves the release of the fractional latent heat, a matter to which we shall return.

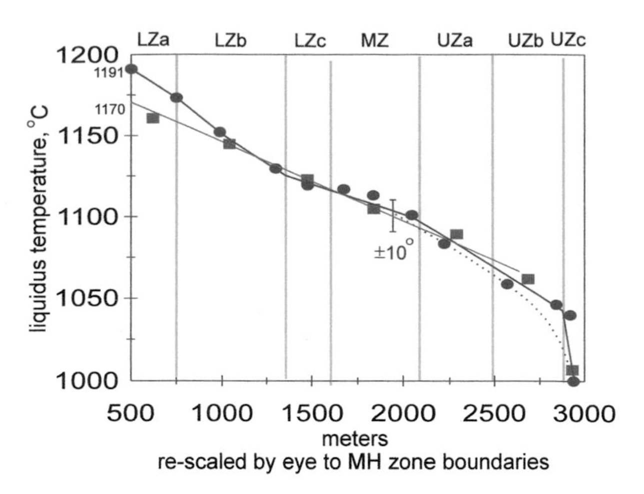

Fig. 13.1 Experimental temperatures of the Skaergaard intrusion deduced from studies of McBirney and Naslund (1990, red squares) and blue dots and line from Toplis et al. (2007) as revised by Morse (2008)

The roof of the Skaergaard intrusion is a mirror called the Upper Border Series that scales inversely down to the sandwich horizon. That means that it records the cooling history in miniature. Although the end of crystallization does not reach beyond An_{25}, the general end point plots near many other evolved, syenitic systems near the triple point in the system olivine–feldspar–augite.

New Advances

Studies of the Skaergaard intrusion flourished before and after the turn of this century with renewed access by ship and air, leading to new sampling, drilling, and a flood of publications. Studies by petrologists and geochemists such as Alex McBirney, Neil Irvine, Christian Tegner, Kent Brooks, Richard Naslund, Troels Nielsen, and Marian Holness produced a prodigious amount of data, particularly for mineral compositions and field relations. As noted above, Nielsen perhaps has the last word about the mineral compositions and field relations. The booklet by Brooks covers much of the old and new territory.

Here, we record a novel and prolific approach by Holness and colleagues that has enlightened the physics of crystallization. Steeped in the microscopic study of minerals, she turned her techniques to the investigation of textures in the Skaergaard intrusion. Her main tool was the three-axis universal stage of optical mineralogy, a tool often preached but rarely loved among petrologists. The gift of this tool is that one may move the contacts of adjacent mineral species until one is looking down the exact boundary common to three crystals. Her trick was to study the dihedral angle among coexisting augite–augite–plagioclase, or in her notation cpx-cpx-plag. To do this with desired precision, one must examine very many instances in each thin section. She did this for each of the eight Skaergaard zones and plotted the results as median dihedral angles against the standard stratigraphic height in metres.

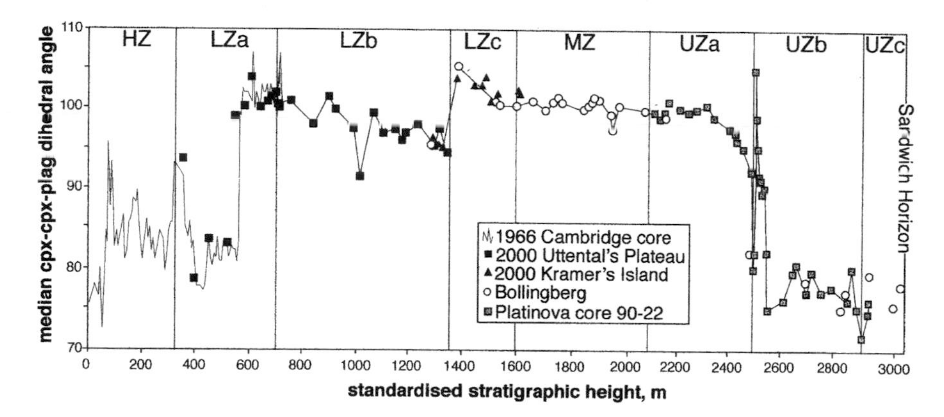

Fig. 13.2 Dihedral angles of cpx-cpx-plag in the Skaergaard intrusion. (Holness et al. 2007)

It turns out that there is a jump or some other visual effect at each zone boundary. The data are messy, of course, but definitive (Fig. 13.2). This result immediately implies that there is something inherent in the cooling system that stalls or enhances the removal of heat from the system. The literature became more and more explicit in a string of papers by Holness, Nielsen, and Tegner (2007) and the same group plus Holness, Tegner, Nielsen, Stripp, and Morse (2007). The latter is described by Brooks (2018, corrected here) as "The clinopyroxene-clinopyroxene-plagioclase dihedral angle in gabbroic cumulates records the time-integrated thermal history in the subsolidus and provides a measure of textural maturity".

As might be imagined, this author soon examined many slides from the Kiglapait intrusion, and so did Professor Holness, but in vain. The huge size and million-year cooling history of this intrusion have erased any evidence of such an immature cumulate angle. However, it has other games to play, as we shall see.

More Progress: The Latent Heat

The Holness group advanced to a further discussion of "A textural record of solidification and cooling in the Skaergaard Intrusion, East Greenland" in the *Journal of Petrology* (2007). In this study, we paid attention to the fractional latent heat of crystallizing magmas, which led to a much longer treatment of that subject by this author. It appears sufficient and appropriate here to present the abstract.

The Fractional Latent Heat of Crystallizing Magmas, by S. A. Morse, American Mineralogist 96, pp. 682–689 (2011).

Abstract

The fractional latent heat of a crystallizing system is simply the latent heat of fusion divided by the total enthalpy. When plotted against temperature, this function displays robust pluses with the successive saturation of each incoming crystal phase. Each pulse endures for tens of degrees, corresponding to hundreds of thousands of years in large magma bodies. From an original 1963 development by P. J. Wylie using a synthetic system, I show that the liquidus slopes (degrees per gram of solid produced) decrease discontinuously at the arrival of a new phase, and their inverse, the crystal productivity, undergoes sharp upward pulses at the same time. The overall liquidus slope increases continuously, interrupted by small downward jumps, with the evolution of the multicomponent melt. The crystal productivity pulses feed the fractional latent heat pulses, which dominate the crystallization history of the melt. These elementary relationships govern the near-solidus growth of dihedral angles in cumulates as they relate to the liquidus events. The latent heat ordinarily approximates to about 80% of the total enthalpy budget, but jumps to 100% (by definition) when solidification occurs by isothermal adcumulus growth and approaches 100% when mafic phases such as augite and Fe-Ti oxides are over-produced (relative to their equilibrium saturation) in layered intrusions. The feedback of latent heat to a self-regulating cooling history of large magma bodies is deduced here in principle. The overall results help to clarify the crystallization history of mafic magmas. In particular, they support the solidification of floor cumulates by interchange with parent magma and without the help of compaction.

The secret to the success of this work lies in the development of the MELTS spreadsheet by Mark Ghiorso and the help in using it provided by Mark and John Brady. The result surprised all of us, and the essence of the latent heat history is presented in Figs. 13.3 and 13.4 here.

In both figures, one interesting feature is that the MORB composition, in other words most of the Earth's magmatic composition, crystallizes in order of decreasing temperature the minerals plagioclase, olivine, augite, magnetite, and apatite.

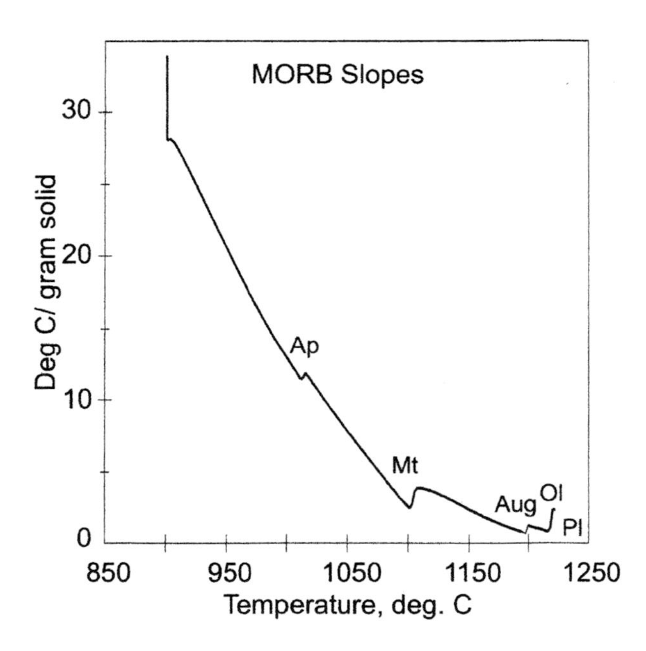

Fig. 13.3 Changes in slope during the crystallization of a damp MORB composition at 1.0 kbar as modelled by the Melts programme of Ghiorso and Sack (1995). (Figure 3 of Morse 2011)

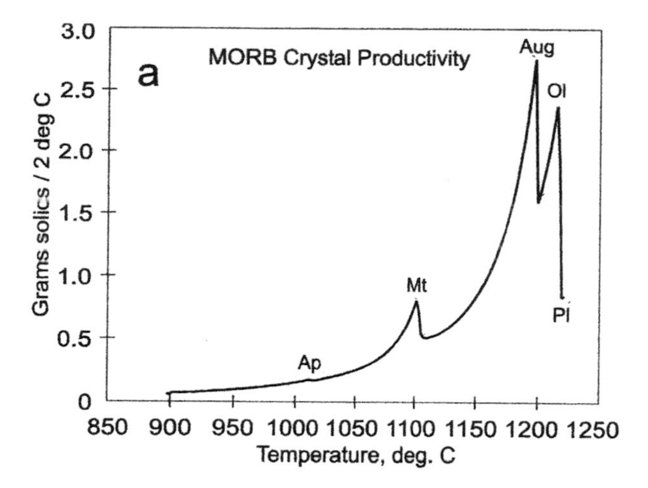

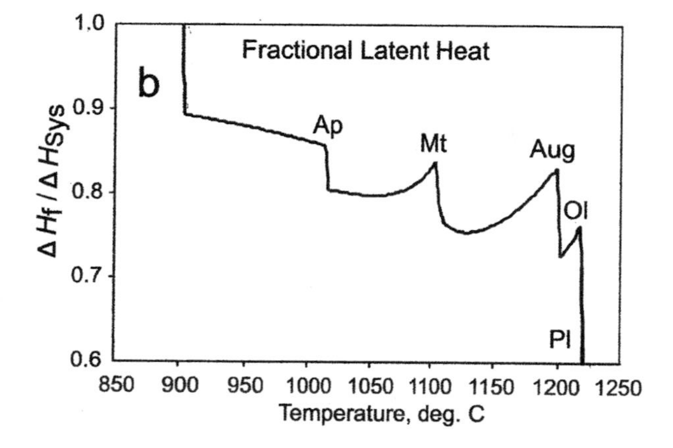

Fig. 13.4 (**a**, **b**) Crystal productivity in the MORB experiment of Ghiorso (1997) and spreadsheet furnished to Morse in 2008, stated in terms of grams per solid per 2°C, the step size at which the thermodynamic properties are calculated in the numerical experiment

Interestingly, this is the order found in many if not most layered intrusions. And here in Fig. 13.3, we find the principle that as the crystallization temperature falls, the rate of cooling (°C) per gram crystallized rises dramatically to the end of crystallization. This feature may be imagined as an increasing efficiency of crystallization.

This feature is repeated in Fig. 13.4a, which is, clearly, simply a detailed inversion of Fig. 13.3. This has the effect of sharpening the peaks of the incoming crystal species. The treasure of the experiment is in Fig. 13.4b, in which the whole exercise is converted to the fractional latent heat, i.e. the momentary latent heat divided by the latent heat of the system. Here, we find very abrupt peaks of plagioclase and olivine, followed by a distinctly curved approach to magnetite. In the field, we may frequently observe such a feature as the local saturation of magnetite (at any scale from millimetres perhaps to centimetres) suddenly discontinued. This is simply a local overstepping of the cooling system that relates to the relative ease of nucleation in the phases involved. A similar but more subdued effect occurs as the saturation of apatite is approached.

Finally, as crystallization ends, the fractional latent heat goes to 1.0 and the system now cools without regard to liquid, which is gone.

That the sequence Pl-Ol-Aug-Mt-Ap is found in that order in this exercise on the MORB system, and then again in the Kiglapait intrusion, shows that the latter body has a bulk composition close to that of MORB.

Solute Paths

Aside from this geochemical-thermal approach to the crystallization of magma, there are local features that simply describe the local physics of crystallization. It can be helpful to think of the interface between growing crystals and their parent magma, and for this purpose, we may use some of the slang of metallurgical systems. In Fig. 13.5, we see the interface between crystals and magma rendered vertically for convenience. The system is that of plagioclase feldspar crystals and related magma. In order for the crystals to grow at constant composition, the albite portion of the liquid must go back into the magma, while the anorthite portion replaces departing albite. This process may be thought of as a replenishment of the anorthite component accompanied by a rejection of the albite component. This is the principle of solute rejection on crystallization, leading to crystals of constant composition. This is the principle of *adcumulus growth* as described by Wager et al. (1960).

For cumulates growing on a sloping floor, we may imagine two end members: dense minerals such as olivine and light minerals such as plagioclase. A cartoon of

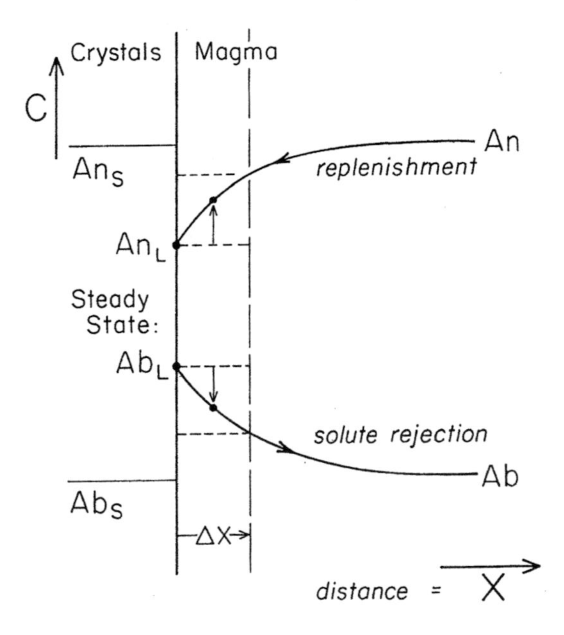

Fig. 13.5 Cartoon of dense and light phases crystallizing from magma, such that the denser phase component represented by pure anorthite (An) becomes crystallized by replenishment, while the pure albite (Ab) phase component returns to the magma by solute rejection. Subscripts: S for solid, and L for liquid. C for crystals; *X* for distance. Vertical image becomes more realistic as it rotates to the left

these cases is shown in Fig. 13.5. If the cumulate is dense, the rejected solute rises up and away. If the cumulate is light, the rejected solute is dense and will sink.

Kiglapait Intrusion, Labrador and Newfoundland, Canada

Highlights

We take occasion to report a few specific details of the Kiglapait intrusion in northern Labrador that may help to illustrate its breadth of igneous principles. Details may be found in abundance in the Kiglapait chapter of the book on layered intrusions by Charlier et al. (2015). Figure 13.6 here is a photograph from space showing the intrusion and its surroundings in a spring snow. The faint circular features approximate the igneous layering in the intrusion. The black spots are lakes, already thawed. The present area of the intrusion is 560 km^2; the original volume is estimated at 3500 km^2. The location is about 2/3 the northern extent of the Labrador Coast, about 30 miles north of the town of Nain, which is now the northernmost town in Labrador.

As shown in Fig. 13.7, the intrusion is divided into a Lower Zone of troctolite (plagioclase–olivine rock) that accounts for 84% of the estimated volume of the intrusion, and an Upper Zone that includes, in sequence, augite, Ti-magnetite, sulphides dominated by pyrrhotite, apatite, with feldspar changing to antiperthite, all from 84 to 96 per cent solidified ("PCS"). The rest of the mineralogy retains all the foregoing list, while the feldspar evolves to mesoperthite and the An content reaches a minimum of 10% and the olivine reaches pure Fe.

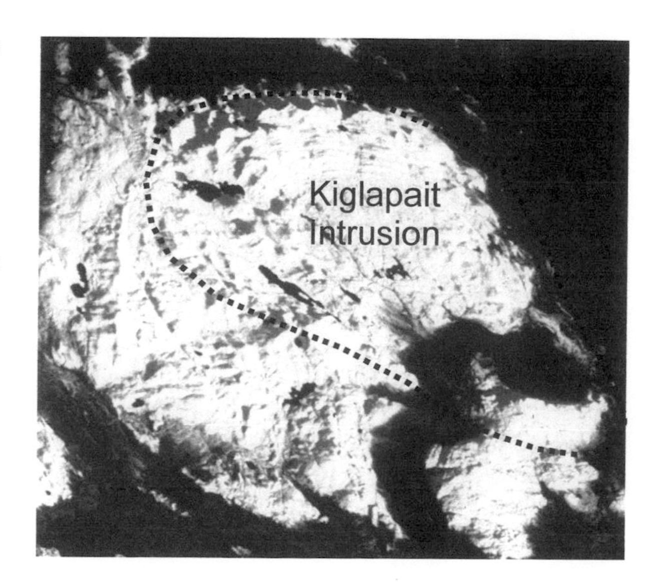

Fig. 13.6 US Shuttle image from space showing the shape and topology of the Kiglapait intrusion on the coast of Labrador (Province of Newfoundland and Labrador, Canada). Hasselblad Image courtesy of Prof Lew Ashwal while at the US Lunar and Planetary Institute. The length of the intrusion is 34 km; width 27 km. Black is water, showing that the time of year is near summer

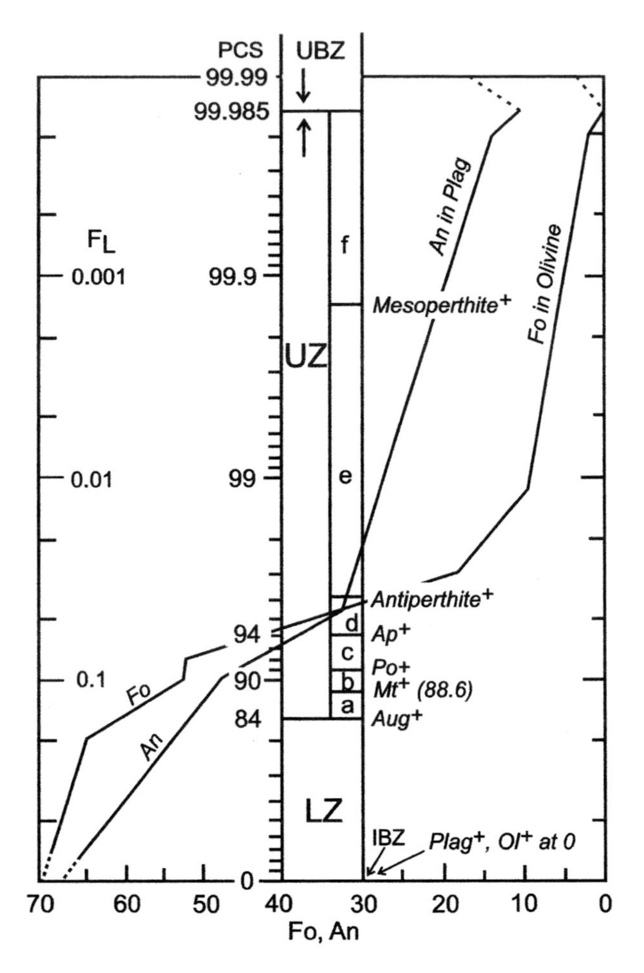

Fig. 13.7 Kiglapait stratigraphic column. The *x*-axis denotes compositions of olivine and plagioclase as plotted in the "Fo" and "An" lines in the vertical part. Left margin: fractionation progress given as the fraction of liquid remaining "F_L". Left of column: PCS = per cent solidified. Centre of column: list of zones (Lower, Upper, and Sub-zones a–f given by the notes alongside). At base: IBZ = Inner Border Zone. "*Aug*", augite; "*Mt*", titanomagnetite ± ilmenite; "*Po*", pyrrhotite (near troilite) and other sulphides; "*Ap*", fluorapatite. The vertical scale is required to be logarithmic in order to plot all the samples at high stratigraphic levels, given that the section is 8400 m thick and there are still 266 m of rocks above 99.9 PCS. The Upper Border Zone (UBZ) records the entire stratigraphy from the roof downwards

All this news is collapsed into a more vivid sequence based on calculated magma density (Fig. 13.8), which shows the rejected solute originating (we think) from olivine alone to the addition of plagioclase and then all the previously listed phases to a maximum at Ap+ and 94 PCS.

Yet another feature that warrants attention is that of Fig. 13.9, which shows the calculated residual porosity of the individual plagioclase compositions. This

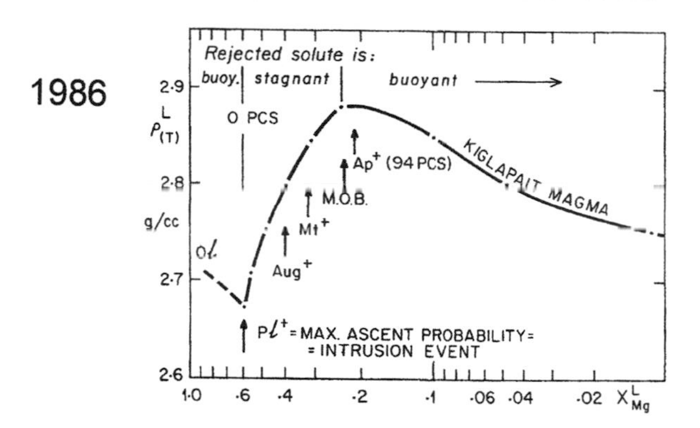

Fig. 13.8 Calculated density (g/cm^3) of the Kiglapait magma (Morse 1979a) against molar MgO/(MgO + FeO) (Morse 1981). The arrival of each cumulate phase is indicated by the mineral abbreviation and a plus sign: *01*, Pi (Intrusion event) through augite, magnetite, the main ore band (MOB), and apatite. The diagram serves as a record of the status of the rejected solute, as indicated. The RS only becomes buoyant after the crystallization of a major level of Fe-Ti oxide minerals in the MOB. (Image after Morse 1986)

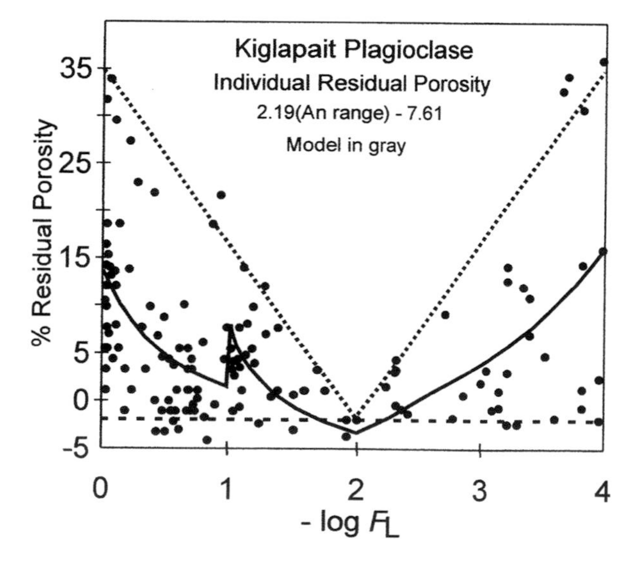

Fig. 13.9 Residual porosity of individual samples of the Kiglapait intrusion as calculated from the equations of Morse (2012), and with the model (solid curve) derived from the map of the An range given in Fig. 17 of Morse (2015). The dotted lines approximate the upper limits of the data. The data show, first, that the very lowest part of the intrusion (left edge) has the most variable range of plagioclase compositions, after which the main variation runs from less than zero (error 2–3% An) to 7. Next, there is a distinct jump near 1 unit that occurs where the feldspar content falls and mafic minerals take charge near 90 PCS. The dotted lines merely characterize the limits that fall to zero at 99 PCS. The residual porosity is simply a measure of the local variation in the An range of plagioclase

mysterious quantity is derived from the An range of individual plagioclase grains, which being the variation in An content for a given ten measurements among randomly chosen plagioclase grains in immersion mounts of ground and sieved crystals using the polarized light microscope. The message here is that a completely uniform plagioclase composition range in a given slide would signify a complete connection with the main magma for the entire history of crystallization. Note that the total error runs (in the figure) from −2% to +2% relative to zero. In this case, all values from −2 to +4 may be considered essentially zero, which then approximately defines the majority of the data. Values 5 and above are also commonly found in the Lower Zone, whereas values up to 35 occur near the base of the intrusion (−log F_L 0–0.5) and (rarely) at the end of crystallization.

The black line in Fig. 13.9 highlights the nominal results from the beginning of crystallization to about 90 PCS, where it takes a kick. It then goes down to zero at 99 PCS and rises again to a nominal value of negative 15 at the end of crystallization. What is so special about 99 PCS? If you find out, let me know.

The other stuff is just normal noise as the magma volume becomes near zero and varies from place to place without being well stirred (Fig. 13.10).

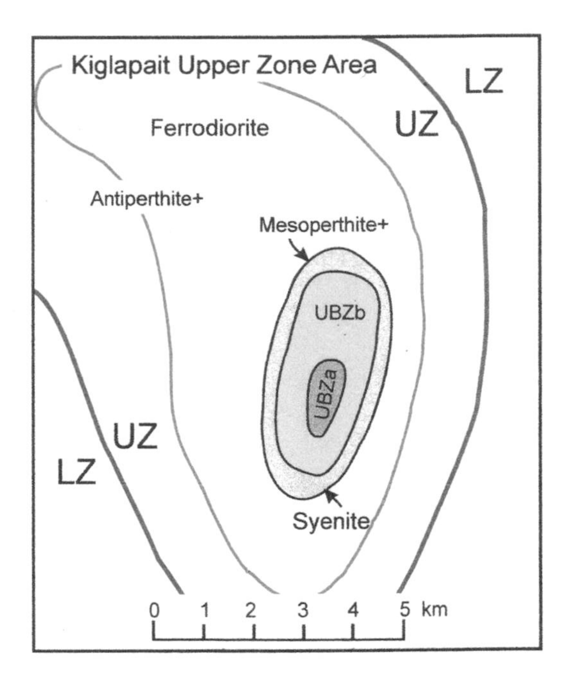

Fig. 13.10 Sketch map of the Kiglapait Upper Border Zone to show the relationships near and at the end of crystallization. The boundary between the Lower Zone (LZ) and Upper Zone (UZ) is shown near the margins. (Recall from Fig. 13.7 that the LZ-UZ boundary occurs at the arrival of Augite in the layered sequence.) The feldspars become antiperthite at ~96 PCS and that occurrence is marked by a red line on the map. The feldspar becomes mesoperthite at 99.85 PCS until the end of crystallization. The oldest part of the Upper Zone is given the name UBZa (for Upper Border Zone a) and that landscape contains the entire history of the intrusion upside-down, the part labelled UBZa is the Lower Zone equivalent, and the UBZb is the inverse of the Upper Zone in composition

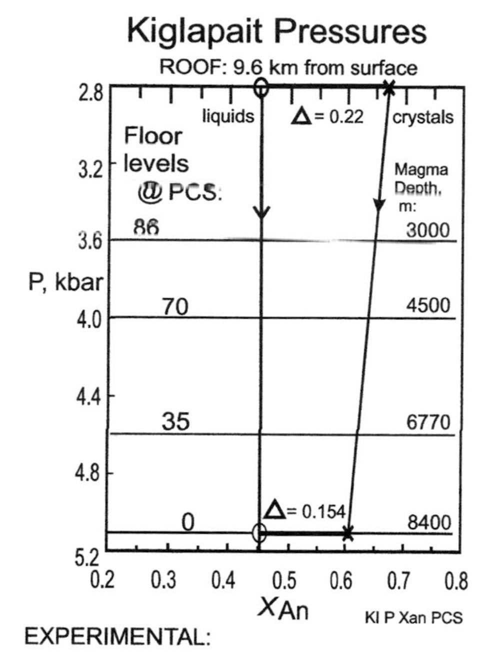

Fig. 13.11 Pressures at depth in the Kiglapait intrusion during crystallization. The results are experimental, using the piston–cylinder apparatus and therefore confined to $\geq$5 kbar, which is appropriate for the Lower Zone depth at the beginning of crystallization. Higher depths are corrected for pressure at the floor level. The plagioclase compositions in the liquid and crystals are shown to vary from a width of 0.154 X_{An} at the start (high pressure) to a nominal 0.22 X_{An} at the roof, which is calculated to be 9.6 km from the surface at the end of crystallization

As the pool of magma becomes smaller the cooling rate may speed up, but the volume is still in the range of 10 × 20 km, as shown in Fig. 13.11. This region is called the Upper Zone (UZ), and this is the place where new components arrive according to the lists in Figs. 13.7 and 13.8. Augite begins to crystallize, and then titanomagnetite, first in dribbles and eventually in a continuous Main Ore Band that has a thickness of at least half a metre. This represents a kind of relief, where cooling can now proceed more effectively, and apatite soon appears on the liquidus and stays there. The feldspar begins to show blebs of antiperthite, which remains increasingly abundant until overcome as *mesoperthite* that fills the feldspar with very fine threads that are so coherent that they may not coarsen unless encountered by a fluid.

The central part of the Upper Border Zone (UBZa) is an inverse assemblage of the Lower Zone, and the next outer part (UBZb) is the inverse of the Upper Zone. Thus, the total UBZ is a compressed recorder of the main part of the intrusion. Its intersection with the upward-growing Upper Zone tends to contain the most evolved fractions of the intrusion. It is to be supposed that the original roof rocks extended geographically far beyond the small sample left behind for us to see.

Geochemical Features: Intensive Parameters

For intrusions like Kiglapait and Skaergaard, there is enough information to reconstruct certain parameters like oxygen fugacity and silica activity. Here, we mention a few results that are given more detailed treatment in Charlier et al. (2015).

In the Skaergaard intrusion, it was quickly noticed that there was an "olivine hiatus" in the Middle Zone where olivine ceased to crystallize, only to return at the Upper Zone. The absence of olivine was soon seen to occur because the oxygen fugacity had increased, so that orthopyroxene was still stable, but olivine was not stable. This feature allowed the linear plotting of oxygen fugacity against the 5-kbar Kelvin temperature for both intrusions, with a linear Lower Zone trend where the oxide–mineral components accreted in the liquid, thereby eventually elevating the oxygen fugacity to the loss of olivine in Skaergaard but not in Kiglapait. Then, after the oxide phases crystallized, the f_{O_2} became once again adequate for olivine to crystalize in Skaergaard (Morse 1980).

In the Kiglapait intrusion, it became important to have a reasonably accurate estimate of the pressure at the time of crystallization, because it turns out that pressure affects plagioclase behaviour. We have experimental information that helps us along in this quest. Figure 13.12 is a plot of pressure and magma depth

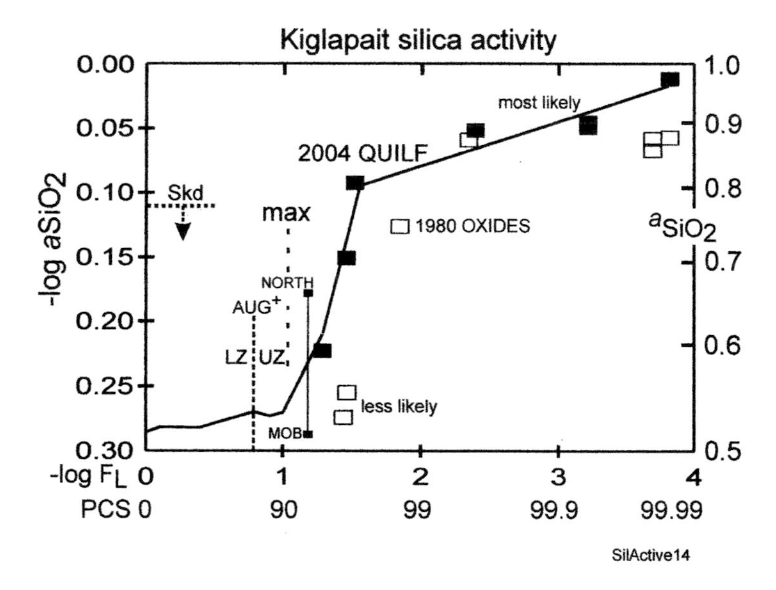

Fig. 13.12 Silica activity in the Kiglapait liquids calculated from coexisting pairs of augite and olivine at high temperature. The activity is scaled as log on the left and directly on the right. The *black rectangles* are from the *QUILF* equilibrium of Anderson et al. (1993) calculated by Donald Lindsley (Pers. Comm. 2004). The LZ-UZ region was calculated by Morse (2014). The vertical dashed line is located at the maximum modal abundance of Augite (Morse 1979b). Open symbols represent augite with higher Ca, considered to have gained most Ca from their initial condition and hence less likely to give relevant results

plotted against the An behaviour in the Kiglapait feldspars. The diagram shows that increasing pressure reduces the difference between the natural crystals and their parent liquids (i.e. the width of the plagioclase loop). The loop width is found experimentally to vary from 0.22 A (An) at low pressure (2.8 kbar) to 0.15 A (An) at the highest pressure—5 kbar—and greatest magma depth. The Upper Zone pressure is taken to begin at 3.6 kbar and run to 2.8 kbar at the roof.

The upshot of the feldspar study was that troctolitic magma is *silica-poor* and hence it reduces the activity of the *NaSi* albite component of plagioclase, hence favouring the An component of the liquid and crystals and thereby weakening the fractionation process.

Then, as the normative augite component in the magma rises from the base of the Lower Zone to the base of the Upper Zone, the activity of silica also rises slightly and its consequent effect on plagioclase composition tends to diminish.

In a more general statement, the type and amount of pyroxene being crystallized have a non-trivial influence on the silica activity and hence on the fractionation of plagioclase.

In the course of this exercise, it became feasible to plot a complete map of the silica activity in the Kiglapait magma, with the help of Professor Donald Lindsley, and that is now shown here as Fig. 13.12. The left margin shows the negative log of the silica activity, which is a linear function. The right margin shows the activity itself, which is not linear.

The earliest stages of Kiglapait crystallization, rich in olivine and plagioclase, have almost no effect on silica activity, which stays low in the figure. Something important happens: the magma starts to crystallize titanomagnetite and eventually reaches the half-metre thick Main Ore Band. This of course sends the silica activity flying upward. The succeeding black rectangles are determined from pairs of coexisting olivine and augite using the QUILF equilibrium of Andersen et al. (1993) as calculated by Professor Lindsley.

A map of the coexisting crystals of Kiglapait olivine and augite in the pyroxene quadrilateral (Fig. 13.13) illustrates the complete fractionation of these crystals to Mg-free, Fe compositions, along with their calculated liquid compositions. The "PCS" labels mean "per cent solidified" as given in Fig. 13.7. One may suppose that this is the most complete mafic history as yet found in nature.

As to liquids and their history in the Kiglapait intrusion, the entire crystallization history as derived experimentally is shown here in Fig. 13.14, as determined by Morse and Brady (2017) using the piston–cylinder apparatus at 5 and 3 kbar.

Final Kiglapait Crystallization History

The end point of crystallization in the Kiglapait intrusion was a seven-phase eutectic that is possibly unique in petrologic history. When told in detail, it runs to many pages and individual papers, especially on the complicated and intriguing feldspar exsolution history. Our goal here is to present the basic story without specialized

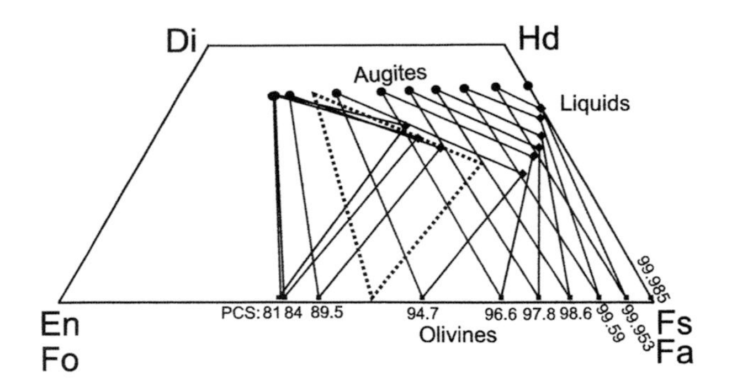

Fig. 13.13 Calculated Kiglapait liquid compositions in equilibrium with augite and olivine. The stratigraphic "PCS" values are shown along the base of the quadrilateral for reference to the liquid compositions, which are taken from Morse (1981). However, the most Mg-rich *triangle* at 81 PCS represents an experimental determination of the liquid composition at the augite saturation point at 5 kb in graphite (Morse et al. 2004). (Figure from Morse and Ross 2004)

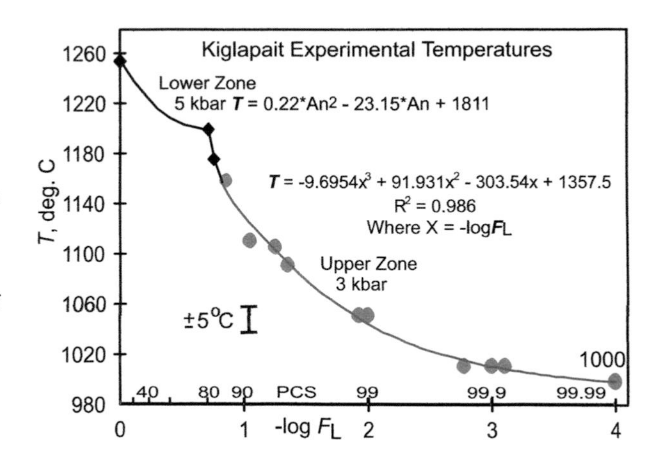

Fig. 13.14 Experimental temperatures of the Kiglapait magma compositions referred to 5 and 3 kbar as indicated. Determinations made in graphite at 5 and 3 kbar. The UZ data are based on the solidus of adcumulates, taken as the magma temperature at the closure of reactive exchange (Morse et al. 1980)

detail. That goal is accomplished in Fig. 13.15, which represents the path from the start of crystallization to the last fragments of liquid at a pressure of 3 kbar and temperature of syenite crystallization involved in a seven-phase eutectic containing a feldspar written as calcic sanidine (before exsolution to much lower temperatures and too complicated to discuss here), fayalite, ferrohedenbergite, fluroapatite, ilmenite, troilite (FeS), and melt.

Ternary Feldspars

Beyond the feldspar end point cited above, there was a continual reconstruction of the ternary feldspars leading to intergrowths (symplectites) that expanded to within ~3-4% Or at the Ab-An sideline and the counterpoint at Or_{80}, An_0. There remained a

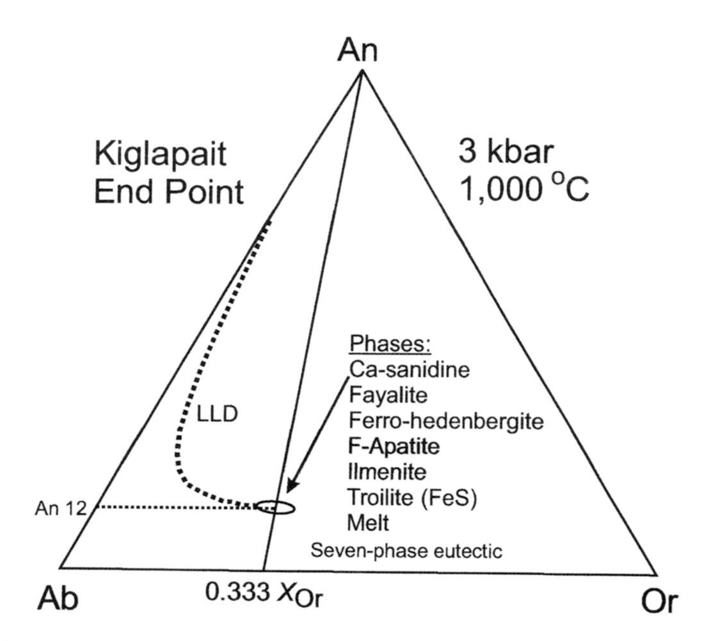

Fig. 13.15 Crystallization path of the Kiglapait liquid referred to 3 kbar at 1000°C. "LLD": liquid line of descent generalized from a swarm of observed data, and ending at a multi-phase eutectic as listed. This is a simplification of data that spread out on cooling, leaving a resistant cluster of quenched melt near the final oval shown here, but also cooled to final equilibrium reaching almost from the left sideline near An 12 to an An-free point at Ab_{20}-Or_{80}

cluster at the magmatic end point at An_9 Or_{34} Ab_{46}. After this point was reached, some feldspars with high energy stayed there, and others began to exsolve. From observations and theory, it then became possible to calculate the entire solvus, now illustrated as Fig. 13.16.

This figure maps the liquidus, solidus, and solvus of the Kiglapait alkali feldspars adjusted to 3 kbar. The projected Ab-Or azeotrope is generated using linear partitioning (Morse 2000) from observed Kiglapait tie lines. The binodal is generated from experiments by Morse (1969b, 1970) adjusted for temperature and Ba content. The eutectic temperature at 1000°C is experimental from studies of selected Kiglapait pairs. The spinodal is estimated from the literature; the coherent solvus (dotted) is also estimated from the literature. The dotted centre line is the rectilinear diameter (r.d.). Liquidus experiments require that the solidus and solvus are embedded. Details are further discussed in the paper by Morse (2017).

It is not a common thing to find data enough from natural rocks that can be translated to such a complete result. The binodal curves have also been extrapolated to much lower temperatures that account for random feldspars that coexist near the An-Ab sideline and also at the Or_{80}-An_0 sideline as mentioned above.

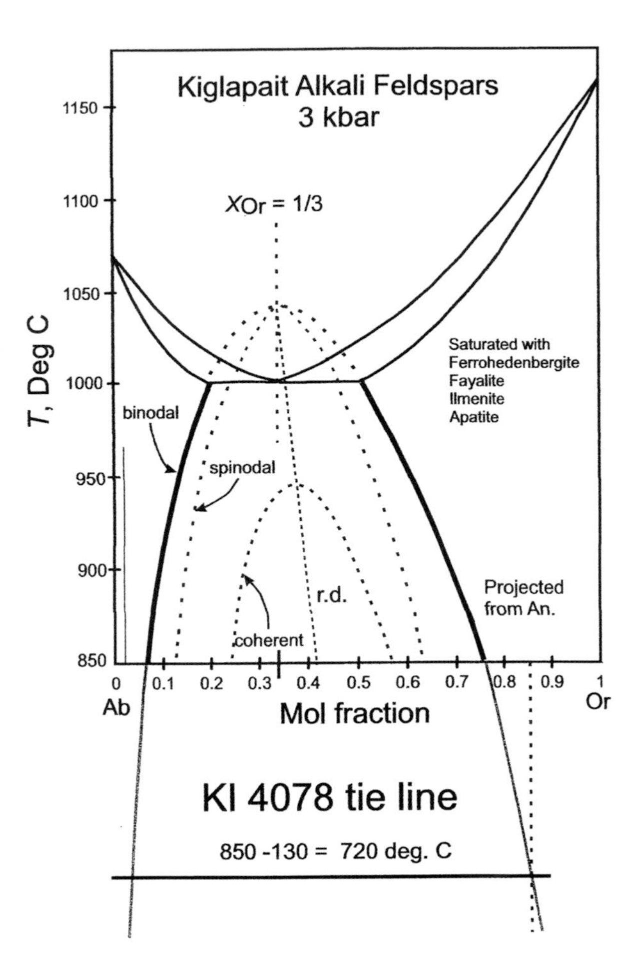

Fig. 13.16 Late exsolution history of the Kiglapait alkali feldspars at 3 kbar, based on experimental and natural data. The light upper curves represent the liquidus compositions extended to their liquidus end members and terminating at the experimental temperature of 1000°C. Intermediate temperatures of the actual liquid equilibria are listed from 1050 to 1000°C. Hypersolvus and solvus relations are shown as binodal, spinodal, and coherent solvi. "r.d.": rectilinear diameter. The binodal has been extended to as low as 350°C

The Thermal Structure of Crystallizing Magmas

Many studies of layered intrusions deal with such matters as where the heat goes and whether, for example, mushy zones of crystals and liquid might have existed and, if so, how deep might they have gone. The answer has been adequately discussed in the literature and the conclusion is that there might have been mushy zones in small intrusions (that have cooled fast) but little if any in large bodies of magma. In order to evaluate this sort of question, it helps to have a clear understanding of the thermal structure of the roof and floor of a crystallizing body.

Heat Flow

A test: What happens to the tar road in Hawaii when hot lava flows over the road? Your answers are either (a) it melts or (b) nothing. Obviously, this is a catch, so you sensibly choose (a).

Wrong.

The dominant flux of heat from a hot magma is up into the atmosphere, and the flux is so intense that it overwhelms any tendency to heat the floor. (There is another test of this principle involving boiling tea water in a thin metal pan just removed from a hot fire: with care it can be held, briefly, by an up-held hand. This test is not recommended unless you are a seasoned and wise canoe man in the wilds of Quebec.)

These mundane thoughts have relevance to the cooling history of large magma bodies. For example, the Kiglapait magma was so very large that it might have been as deep as 10 km under a roof at least half that thick, and it is reckoned that it took a million years to solidify (e.g. Morse 2015).

Roof Heat

Figures 13.17 show the general nature of heat arrangement near the top of a deep troctolitic magma body, the Upper Border Zone, in a depth–temperature diagram. The ROOF is far away up to the sky. There has been freezing or deposition of melt, and the upper level of that is now a (mostly) solidified cumulate. Below a solidification front, there is a mixture of crystals and trapped liquid. Heat is escaping upward through that cumulate layer. Lower down, there exists a space within which crystals may form owing to loss of heat upward. That space now becomes a thermal buffer, through which heat must pass upward, thus causing some crystals to form. These are denser than the melt, and they start to sink and may be accompanied by some cooling melt. Mostly they will stay at or near the liquidus, sinking with (in this case of troctolite) plagioclase plus olivine. After leaving the thermal boundary layer, they remain strictly at the liquidus. On occasion, there may be downward plumes carrying cooler crystals. The nominal liquidus may carry heat at about 6.7°C/kbar.

For all this to happen, there must be a thermal feature that returns the heat of crystallization upward towards the roof. We note that this *return adiabat* is very much hotter than the thermally active layer, because it brings the latent heat from the bottom of the crystallizing magma, which we shall soon see! As indicated, the return adiabat may be as different from the liquidus as perhaps 6–18°C. It is superheated.

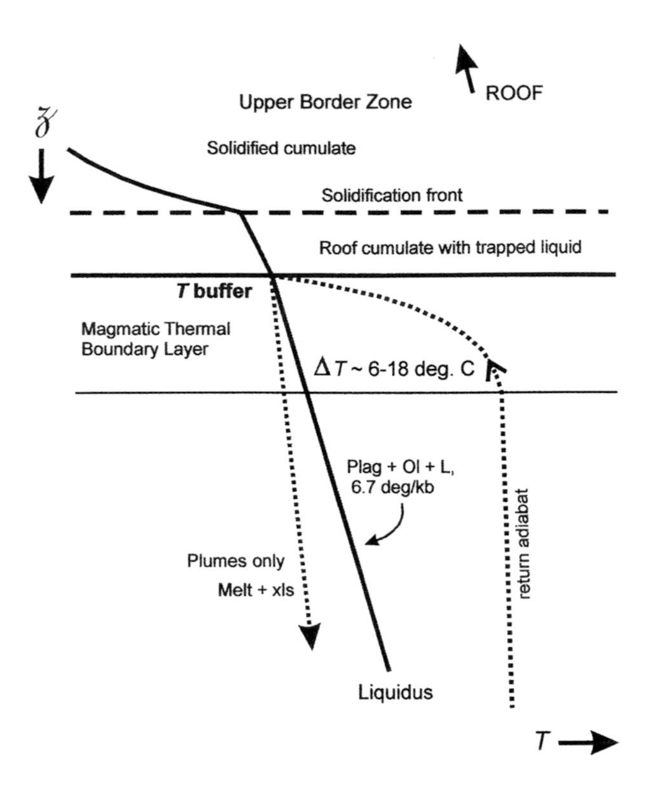

Fig. 13.17 Heat arrangement near the top of a troctolitic magma body. Heat escapes upward through the cumulate. The lower space becomes a thermal buffer, through which heat must pass, causing crystals to form. The crystals are dense, so they sink along the liquidus. The "return adiabat" arises from the floor and carries the latent heat of crystallization into a sharp cooling region

Floor Heat

We continue with the floor region (Fig. 13.18). The temperature may be rising at a rate of 6.7°C/kbar. The scales are as before, depth and temperature. The liquidus is an adiabat, to the right of which no crystals exist. At a slightly cooler temperature, there is a body of melt plus crystals with downward flux that must transfer through any mushy zone and end at a point below which there is a cumulate plus trapped liquid.

Through this cumulate, there must be a floor below which no melt occurs. Below this "solidification front", all are solid. The cumulate region is buffered at the top to the ambient temperature of crystallization, called the latent heat hump (ΔH_f). This gives way downward along a gradient that ends at complete solidification.

Here ends the general treatment of crystallizing magmas. We now turn to a mystery in the Kiglapait intrusion that has never been discussed. It is fitting, having exercised our minds with the rigours of Earth physics, to entertain a problem that we cannot even attack without consensus from the local polar bears.

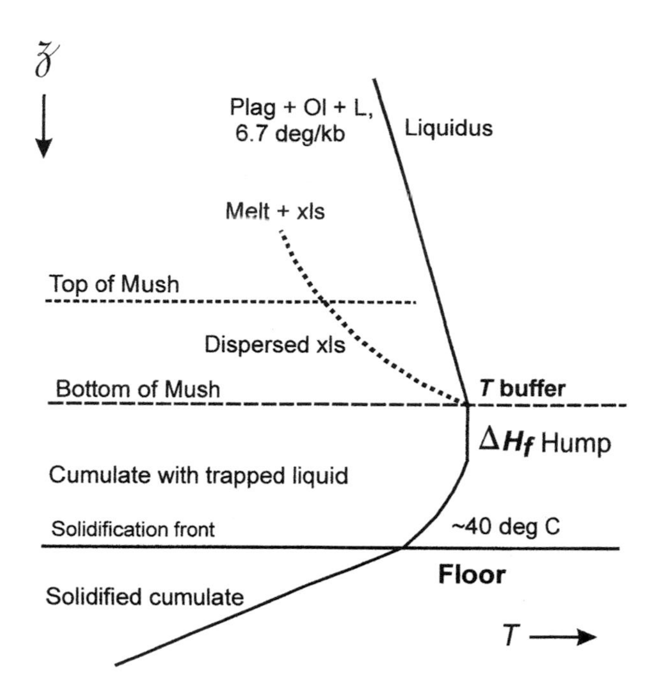

Fig. 13.18 Thermal constraints near the floor of a steady-state igneous cumulate. Vertical depth **is** indicated by the down arrow and temperature rises to the right. It is assumed that a dilute crystal-bearing packet arrives from above (heavy dotted line), such that it is metastably supercooled with respect to the *liquidus*. A narrow mushy zone contains dispersed crystals and reaches a crystal-bridging criterion of—50%, leading to a porosity of—50%. This assemblage locks the ambient temperature to the crystal–liquid maximum, which is nominally the highest temperature in the intrusion assuming perfect stirring. For adcumulates, the solidification front lies at the base of the latent heat hump. For extreme orthocumulates, the temperature may range be as great as 60°C (Morse 2013). The trapped liquid cumulate must reach a solidification front (the floor) below which no melt exists

The Grand Mystery

The Kiglapait Mountains, for which the intrusion is named, form an east–west line at about the 20–40 PCS level, making a strongly ridged region in the northern part of the intrusion. A similar range in the south of the intrusion was probably present but now exists mainly as the single Mount Thoresby, near 50 PCS. (Climbers on that mountain have been puzzled by its magnetic field, which runs all over the place—at least during lunch hour.)

The northern mountains consist of at least eight major peaks and as many that are slightly lower, mostly near 1000 m (3000–3400 ft) high. They have not been sampled, except possibly for one sample containing pyroxene at the east coast edge of the northern mountains, and a view of a pyroxenite slab near the west end of the range. The mountains have no vegetation beyond mosses. The northern parts of the intrusion running to the south are heavily ridged to about the 60 PCS contour.

Why are these mountains there? If they were precursors to the intrusion, they should have been outside. And further, why are they mountains at all, sitting on well-weathered lowlands? Clearly, they are resistant to weathering and erosion, so must be of a singularly different composition from the ordinary troctolite that forms the main regions sampled systematically by drilling.

I have a hunch, but do not expect it to be explored very soon. I suspect that in the mountains we may be seeing pyroxene-rich edifices and that they came from the mantle region of harzburgite that was a precursor to the Kiglapait liquidus near 1250°C. Our experimental data show Kiglapait-like compositions at 13 kbar pressure and 1290°C and have shown a plausible evolution of the Kiglapait magma from harzburgite to troctolite (Morse et al. 2020).

Not that we have yet seen harzburgite or nepheline in this intrusion, but we have not looked hard at the mountain regions.

The closest we came to sampling the mountain region was when my assistant and I were camped in the northern region and there was an amazing rockfall from somewhere in the mountain region. Upon inspection, we found a piece of troctolite about the size of a locomotive with white dust still falling off it. The plagioclase was white, and the olivine was a glorious glassy bright green in 5-cm long pieces. That is what fresh olivine looks like, until in a few years it oxidizes to olive. (See Appendix: Kiglapait Intrusion, Labrador Coast: A Record of Discovery.)

Chapter 14
Nepheline–Silica and the Rest of the Basalt Tetrahedron

Introduction

In this chapter, we return to the basalt tetrahedron as a model for the crystallization of basalt magma and conclude the discussion of phase diagrams related to this system at 1 bar pressure. A brief discussion of the system nepheline–silica ($NaAlSiO_4$-SiO_2) serves to introduce the important principle of the thermal barrier (or thermal divide), which tends to prevent silica-undersaturated liquids from producing critically undersaturated residua and vice versa. A review of the rest of the tetrahedron serves as a summary of earlier chapters and as a conclusion regarding the lessons of the fundamental basalt tetrahedron. This review will serve as the background for examination, in succeeding chapters, of the effects of other components and of pressure on the phase relations of basaltic systems.

It is to be emphasized that in this chapter no attempt is made to provide a thorough discussion of basalt genesis, but instead to focus on some of the theoretical constraints and principles provided by phase diagrams.

Nepheline–Silica

Nepheline, $NaAlSiO_4$, or 0.5 ($Na_2O \cdot Al_2O_3 \cdot 2SiO_2$), is the most important of the feldspathoid group of minerals from the standpoint of rock genesis. As we have seen, the standard CIPW norm calculation reflects the central importance of this mineral among the critically undersaturated natural rocks. By the convention of this calculation, silica deficiency is first remedied by converting albite to nepheline, and the appearance of *ne* in the norm is always sufficient evidence of a critically undersaturated rock warranting names such as alkali basalt, basanite, and olivine nephelinite. Nepheline may occur in such rocks as part of a late mesostasis, intergrown with alkali feldspar, or more rarely as independent crystals. The natural mineral always

S. A. Morse, *Basalts and Phase Diagrams*,
https://doi.org/10.1007/978-3-030-97882-2_14

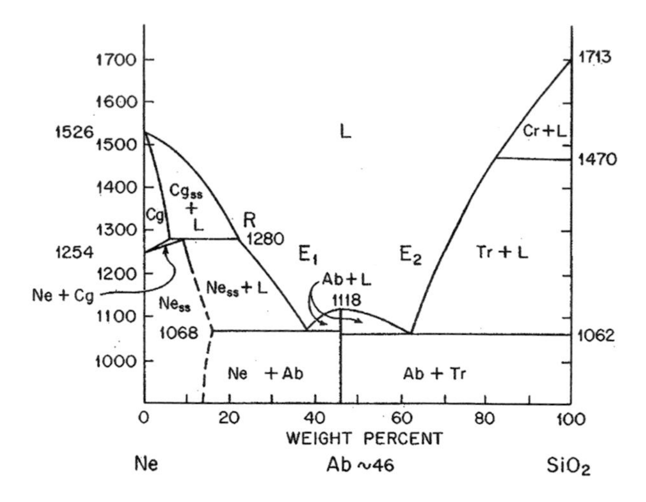

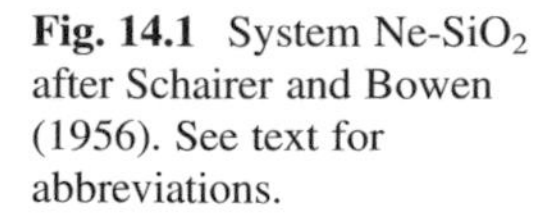
Fig. 14.1 System Ne-SiO_2 after Schairer and Bowen (1956). See text for abbreviations.

contains “excess” silica (deficient alkali and aluminium) when it is soda-rich, the amount of this “excess” decreasing as the composition of Ne_{ss} varies towards $KAlSiO_4$.

Nepheline is synthesized, as is albite, from sodium disilicate ($Na_2O \cdot 2SiO_2$), which serves to fix the volatile alkali metal, and Al_2O_3. The procedure is described in detail by Schairer and Bowen (1956). In the laboratory, a high-temperature polymorph is encountered above 1254°C, persisting to the melting point at 1526°C; this polymorph was named carnegieite in honour of Andrew Carnegie, the founder of the Carnegie Institution of Washington.

The system Ne-SiO_2 (Greig and Barth 1938; Schairer and Bowen 1956) is an excellent example of a binary system with an intermediate compound, albite, which melts at a maximum and generates two eutectics, one with each of the end member phases (Fig. 14.1). The Ne-rich portion of the diagram presents certain complexities, which may be disposed of in a brief analysis.

As Fig. 14.1 shows, solid solution towards SiO_2 is a feature common to both polymorphs of $NaAlSiO_2$, the sole exception being at the melting point of the substance, where the highest thermal stability is shown by the pure compound. The Ne-rich side of the diagram therefore shows large regions of solid Ne_{ss} and Cg_{ss} where no liquid is involved. Each of these regions is isobarically divariant ($c = 2$, $\phi = 1$, $W_p = 3 - 1 = 2$); T and X can be varied without changing the state of the system within each region. A small univariant phase transition loop originates at the inversion point, 1254°C, on the Ne sideline, and is truncated by the melting loop $Cg_{ss} + L$ at 1280°C. This small two-phase field represents the mutual saturation limits of relatively silica-poor Cg_{ss} and relatively silica-rich Ne_{ss} at any given temperature in the range 1254–1280°C. At any given temperature in this range, any bulk composition within the loop is represented by compositions of Cg_{ss} and Ne_{ss} lying on the solid saturation curves at the ends of an isothermal tie line. The lever rule may be applied to find the ratio of the solids to one another. As the temperature rises towards 1280°C, both solids become richer in SiO_2, but Ne_{ss} is always the more silica-rich of the two. When bulk compositions near the $NaAlSiO_4$ composition are heated, a complete conversion to Cg_{ss} occurs over a temperature interval (as the Ne_{ss} lever becomes shorter and finally reaches length zero), and Cg_{ss} then begins to melt at the solidus somewhere between 1280 and 1526°C. The $Cg_{ss} + L$ loop is (except for its lower truncation) like any simple binary melting loop (the plagioclase loop may serve as an example) and melting takes place with continuous conversion of the solids towards more silica-poor compositions.

If a bulk composition corresponding to Ne_{ss}, say at 12% SiO_2, is heated, a similar melting loop is encountered, and Ne_{ss} also changes composition continuously away from SiO_2 as liquid is formed. The $Ne_{ss} + L$ loop is a close counterpart of the $Cg_{ss} + L$ loop, the two being joined at a discontinuity where, upon addition of heat, Ne_{ss} is converted to Cg_{ss} in the presence of liquid R. The coexistence of three phases marks this assemblage as an isobarically invariant one; the three phases are collinear, and R is evidently a peritectic (reaction) point for which the heat-removing reaction can be written $L = Ne_{ss} - Cg_{ss}$. The reaction, which is odd, can also be written $L + Cg_{ss} = Ne_{ss}$.

The field of Ne_{ss} reaches its maximum SiO_2 content at 1068°C, the eutectic temperature of Ne_{ss}–Ab mixtures. The limits of the Ne_{ss} field in equilibrium with L or Ab are dashed because of experimental uncertainties, but follow the form expected for such a system, as discussed on an earlier page (Fig. 5.5) where the properties of an idealized binary system are displayed in connection with the system diopside–anorthite. The dashed curve of Ne_{ss} (Ab) (read "Ne_{ss} saturated with albite") is in fact the solvus of these two phases, and it is intersected by the solidus Ne_{ss} (L) at 1068°C.

The isobaric eutectic E_1 involving Ne_{ss}, Ab, and L can most easily be analysed by ignoring all of the systems to the SiO_2-rich side of Ab, whereby it becomes evident that Ne_{ss}-Ab is a simple binary eutectic subsystem of the larger system. The crystallization of any liquid lying between the albite composition and about 20% SiO_2 and 80% Ne is a straightforward process typical of that in any binary eutectic system, ending with the isothermal crystallization of eutectic liquid E_1 at 1068°C. Fractional crystallization has no effect on the composition of the final liquid, and only a minor effect on the amount of the final liquid if Ne_{ss} is the fractionally removed phase. Fractional crystallization of pure Ab has no effect on the amount of eutectic liquid. Equilibrium and fractional fusion are both straightforward processes, the main result being the isothermal production of liquid E_1 until one of the solid phases is exhausted.

The SiO_2-rich side of Ne-SiO_2 (Fig. 14.1) is also by itself a simple binary eutectic E_2 between Ab and tridymite at about 62% SiO_2, 1062°C.

Figure 14.2 shows an alternative possibility for the phase diagram in the region of albite. In this figure, albite shows limited solid solution towards both Ne and SiO_2, and attains the pure composition $NaAlSi_3O_8$ with liquid only at its melting point, 1118°C. A moderate degree of solid solution towards Ne was thought by Greig and Barth (1938) to occur, but was not verified by Schairer and Bowen (1956). Since that time, numerous authors have claimed evidence for such an Ab_{ss} field as shown in Fig. 14.2, based on classical wet chemical analyses of natural feldspars and on

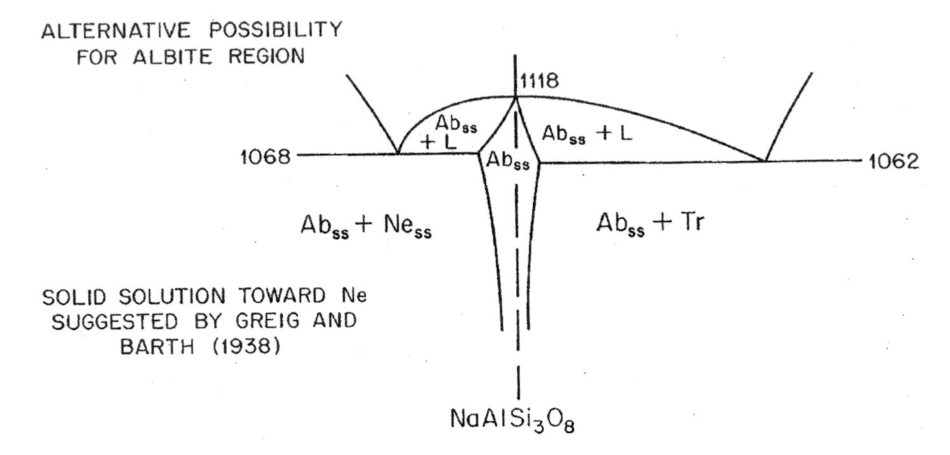

Fig. 14.2 An alternative interpretation of the albite composition region in $NeSiO_2$, assuming some solid solution of albite towards both Ne and SiO_2. For discussion, see text

experimental studies. Most or all of these claims have been inconclusive because of well-known sources of error inherent in analytical procedures, or because of suspected kinetic difficulties (metastability) in experimental runs. A convincing resolution of the question awaits further work. There can be no doubt that such an Ab_{ss} region exists, but whether it is wide enough to have geologic significance is the question at hand. The nature of what might constitute "geologic significance" will be discussed shortly.

Assuming either configuration near Ab, it can be seen that albite melts at a local maximum; all liquids in the primary phase field of feldspar will move away from feldspar on crystallization, towards either E_1 or E_2. The compound albite thus generates a thermal barrier or divide; no liquid can cross the divide, although two liquids lying infinitesimally near to each other (and to Ab) may move in opposite directions, to E_1 and E_2, thus appearing to yield either residual product from the "same" parent. Moreover, fractional or equilibrium fusion on either side of the divide will yield Ab as the final solid, and there is no way to drive the residual solid composition across the barrier.

The eutectic E_1 is an approximate model of the natural rock phonolite, which is the residual end member of the commonly observed natural alkali basalt fractionation series. Similarly, eutectic E_2 is a synthetic analogue of rhyolite, the common acid end member of tholeiitic suites. The composition of albite itself is a model of the natural rock trachyte, and one of the mysteries of petrology is why some fractionation trends terminate in trachyte (or the intrusive equivalent syenite), at an apparent maximum, instead of sliding off towards natural versions of E_1 or E_2. Explorations of this so-called syenite problem to date have shown that the relative geometric relations are not changed by the addition of K_2O, H_2O, and pressure (Morse 1969a). It is possible that an intricate interplay among Fe oxides and silicates, feldspar, and oxidation controls the fate of such liquids, at least in part. Another contributing factor may be the ability of feldspar to tolerate some deviation from the stoichiometric composition, as in Fig. 14.2, so that with complex magma compositions the thermal "barrier" has finite width, like a plateau rather than a sharp peak. A plateau would be more likely to hold residual trachytic liquids than a peak. This is the main reason for being interested in the extent to which Fig. 14.2 is correct.

The Thermal Barrier in More Complex Systems

We have already seen, in Chap. 7, that the join diopside–albite is nonbinary and cannot itself be a thermal barrier, because liquids leave the system towards Na_2O- and SiO_2-rich compositions. Moreover, since Di-Ab is one edge of the critical plane Di-Ab-Fo, it is obvious that the critical plane itself is not a thermal barrier. Nevertheless, natural liquids do appear to fractionate away from augite, feldspar, and olivine towards either tholeiitic or alkali basaltic residua, and there are good grounds to suppose that a thermal barrier exists, even though it is not exactly the critical plane in composition. It is appropriate to examine the extent to which the concept does hold up in experimental studies.

Yoder and Tilley (1962) discussed this question extensively, and the ternary diagrams of Fig. 14.3 are taken directly from their Fig. 11. These diagrams show the effect of adding, in turn, leucite (Lc), anorthite, forsterite, fayalite (Fa), diopside, and spinel to the system Ne-SiO_2. Diagrams (c) and (e) are, respectively, the base and the rear face of the basalt tetrahedron (Fig. 2.1). For present purposes, we are concerned mainly with the arrows on boundaries crossing the join connecting albite to each of the other principal phases. Nevertheless, it is possible to recognize familiar relations, particularly in Fig. 14.3c, where the incongruent melting of En (labelled Pr) is observed on the sideline Fo-SiO_2 and inside the ternary system as well. The field of Fo overwhelms this diagram and covers most of the join Pr-Ab and most of the join Fo-Ab. The cotectic boundary *L* (Fo-Ab) crosses the Fo-Ab join very close to the Ab composition. Note that arrows point away from the line Fo-Ab on either side, showing that liquids tend either to the isobaric ternary eutectic *L* (Fo, Ab, Ne) or to its silica-saturated counterpart, *L* (Pr, Ab, Tr). The join Fo-Ab is a thermal barrier. Fractional removal of Fo from bulk compositions on either side of Fo-Ab can only magnify the effect of the barrier. The iron end member olivine, fayalite, shows a similar relationship in Fig. 14.3d. Evidently, all members of the olivine series should maintain the thermal barrier relationship with albite.

Figure 14.3e shows that liquids move away from the join Di-Ab towards eutectics *L* (Di, Pl, Ne) and *L* (Di, Pl, Tr). As we have seen, the plagioclase effect causes liquids initially on the join to become enriched in Na_2O and SiO_2; therefore, the join itself is not a thermal barrier, but despite this, compositions sufficiently rich in Ne will yield critically undersaturated residua. Evidently, there is a thermal barrier in this system, although it does not precisely coincide with Di-Ab. In the absence of other components, the plagioclase effect will lead to silica-saturated sodic rocks (pantellerites) with molar $Al_2O_3 < (Na_2O + K_2O)$.

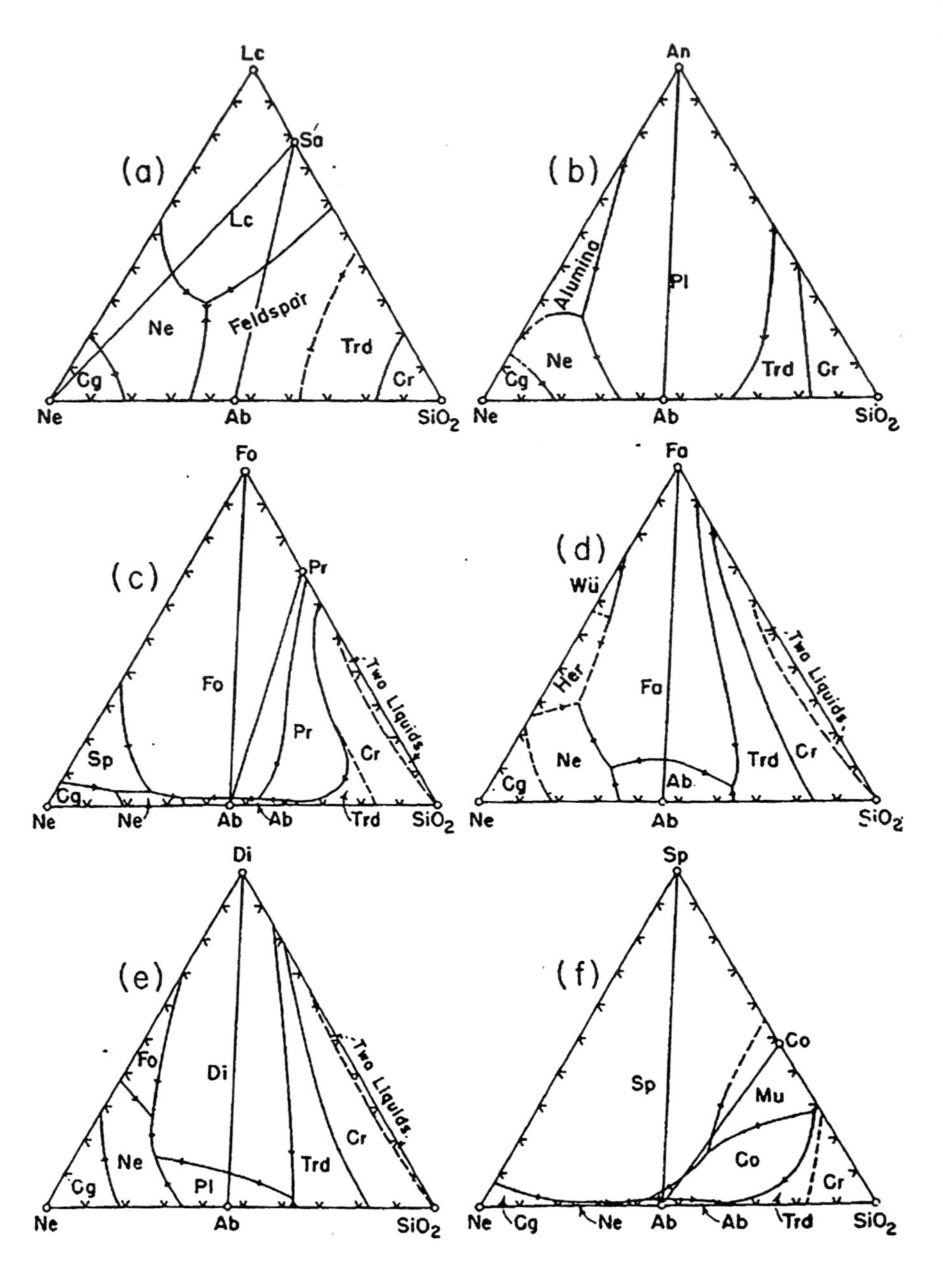

Fig. 14.3 Phase diagrams illustrating the nature of the equilibrium thermal divides in relationship with the principal join Ne-SiO_2. (**a**) Lc-Ne-SiO_2: Schairer and Bowen (1935), Schairer (1950). (**b**) An-Ne-SiO_2: Schairer (1954). (**c**) Fo-Ne-SiO_2: Schairer and Yoder (1961). (**d**) Fa-Ne-SiO_2: Bowen and Schairer (1938). (**e**) Di-Ne-SiO_2: Schairer and Yoder (1960). (**f**) Sp-Ne-SiO_2: Schairer and Yoder (1958). (From Yoder and Tilley 1962, by permission)

In Fig. 14.3f, it may be seen that equilibria among spinel, cordierite (Co), albite, and liquid will lead to a transit of the thermal barrier towards Ne. However, such equilibria cannot be of general importance for basaltic magmas, although spinel and cordierite might conceivably be produced in gabbros by assimilation of pelitic sedimentary rocks. The plagioclase join, Fig. 14.3b, acts as a thermal barrier in the system Ne-An-SiO_2, driving liquids constantly towards either the Ne-Ab or Ab-Tr eutectic, possibly via either of the univariant equilibria *L* (Ne, Pl) or *L* (Tr, Pl). Whether a liquid first reaches one of these univariant curves, or the sideline Ne-SiO_2, depends on how far the bulk composition lies from the join An-Ab and whether crystallization is of the fractional or equilibrium type. However, when another calcic mineral such as diopside is present, the plagioclase effect will probably insure that the liquid reaches either the Ne or the Tr field, because the plagioclase will not reach the composition of pure Ab. By comparing diagrams (b) and (e) of Fig. 14.3, one may reach the conclusion that diagram (b) closely resembles the diopside-saturated surface in the quaternary system Di-An-Ne-SiO_2. It may be an instructive exercise for readers to sketch for themselves the tetrahedron representing this system, starting with the base Ne-An-SiO_2, and placing Di at the upper apex so that the front face is diagram (e), Ne-Di-SiO_2. The plane Di-An-Ab approximately bisects the tetrahedral volume, and the cotectic in this system serves as a guide for locating the *L* (Di, Pl) saturation surface. The Di-Pl cotectic is the thermal barrier in the quaternary system. It is the *L* (Di, Pl) saturation surface, projected onto the Ne-An-SiO_2 base, which shows approximately the same geometry as the Pl field in diagram (b), and therefore, saturation with diopside will not appreciably alter the thermal barrier nature of the plagioclase join in that diagram.

Although we have not yet discussed potassium-bearing systems, it is possible to conclude from diagram (a) that introduction of a potassium feldspar component in small quantities will not alter the thermal barrier relationship. Diagram (a) is the system Ne-Lc ($KAlSi_2O_6$)-SiO_2. The orthoclase composition (Sa, $KAlSi_3O_8$) lies on the sideline Lc-SiO_2, and at high temperatures, a continuous alkali feldspar series exists between sanidine and albite. Sanidine itself, and sanidine-rich feldspar generally, melts incongruently to Lc + *L*, and the liquids in equilibrium with these two phases may cross the barrier towards critically undersaturated compositions under conditions of equilibrium crystallization (they may not do so, however, under conditions of fractional crystallization, although by fractional removal of leucite, K-rich liquids may make the reverse transit from critically undersaturated to oversaturated compositions). The important feature of diagram (a), for all but the very special potassic basalts which need not concern us here, is that the Ab-rich part of the alkali feldspar join is in fact a thermal divide, and therefore, the entire ternary feldspar join acts as a thermal barrier when the feldspars involved are plagioclases showing moderate enrichment in K as they reach lower contents of An.

We may therefore conclude from the diagrams of Fig.14.3 that the plagioclase feldspars (including varieties as K-rich as anorthoclase, or calcic sanidine) constitute a thermal divide between merely undersaturated and critically undersaturated liquids, that this divide lies very near the critical plane in the basalt tetrahedron, and that the properties of the divide persist in the presence of iron-bearing silicates such as fayalite.

In a thorough discussion of this thermal barrier and the various mechanisms alleged to breach it, Yoder and Tilley (1962, pp. 401–403) cite only one, other than those cited above in connection with Fig. 14.3, which may reasonably be considered to cause liquids to cross the critical plane at low pressure. This mechanism is oxidation or reduction in iron in the magma, a subject so important that it deserves treatment in a separate chapter. The effect of oxidation on silica saturation may be simply stated; however, by noting that if the ferrous iron normally incorporated in silicates (e.g. fayalite) is oxidized enough to cause formation of magnetite ($FeO \cdot Fe_2O_3$), the crystallization of this non-silicate mineral will "release" silica to the residual liquid which otherwise would have been bound up in ferrous silicates. The same result can be seen in the norm calculation; if enough FeO is converted to Fe_2O_3 in an alkali basalt analysis, so much magnetite may be formed in the norm calculation that the analysis becomes hypersthene- or quartz-normative. This points up the great importance of accurate FeO/Fe_2O_3 determinations when a correct classification of a basaltic rock is desired.

Apart from the effects of pressure and oxidation–reduction reactions, the concept of the thermal barrier appears to be a durable one. We may therefore state that at low pressures and within certain ranges of oxygen partial pressures, no normal basaltic liquid can make a transit of the natural critical plane of silica undersaturation; alkali basalts cannot be parental to olivine tholeiites and vice versa. It is true that two parents on either side of the critical plane may be indistinguishably close together and yield very different residua, which might appear to have a common parentage. However, in principle, no single parent can yield both trends.

The Critical Plane as a Model System

The critical plane in the basalt tetrahedron is the ternary system Fo-Di-Ab. Although pure Ab generates the thermal barrier discussed above, it is an unrealistic component by itself for basalt discussions; intermediate plagioclase would be far better. The discussion of critically undersaturated basalts can be improved by adding the component An to form the *critically undersaturated basalt system* (Yoder and Tilley 1962, p. 394). This is the system Fo-Di-An-Ab, containing the principal silicate phases of basalt: plagioclase, clinopyroxene, and olivine. It can be sketched as a tetrahedron (an instructive exercise for energetic students) from the systems Di-An-Ab and Fo-Di-An, which we have previously examined, and some straightforward assumptions about Fo-Di-Ab. The liquidus surfaces in the tetrahedron are shown in Fig. 14.4, after Yoder and Tilley, who note several important features of the system and of the natural basalts which it represents. One of these is that the four-phase curve *L* (Pl, Fo, Di) extends across the system with a temperature range from 1270 to 1135°C; both the curve itself and the temperatures are very close to those in the olivine-free system Di-An-Ab, upon which we have already remarked (Chap. 8) as

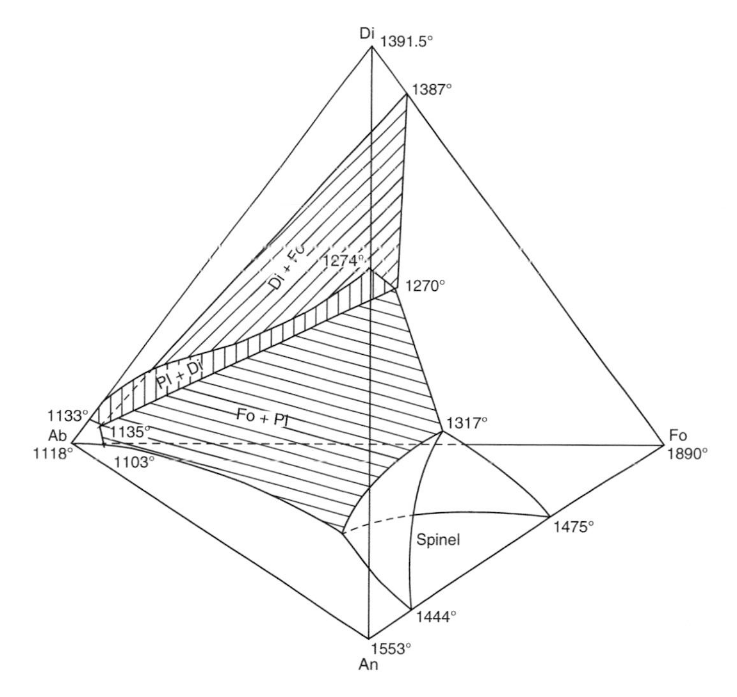

Fig. 14.4 System Fo-Di-An-Ab, the expanded critical plane of the basalt tetrahedron. (From Yoder and Tilley 1962. Reproduced with permission)

to the compositions and temperatures of liquids (like basalts). Saturation with olivine hardly changes the results. In fact, the quaternary system bears out what we have said all along about the success of Di-An-Ab and Fo-Di-An as models of basalt crystallization.

Yoder and Tilley used this system as a basis for discussing experimental studies on natural basalts. They had hoped to deduce parent–daughter relationships from the relative temperatures at which the natural rocks, when melted and then crystallized, reached the four-phase *L* (Pl, Fo, Di) saturation condition. However, they found that all the basalts studied reached this condition in a very small range of temperature (1155–1170°C), implying that they all lie very near the same fractionation path, or a set of equivalent fractionation paths in the several primary phase regions. This coincidence of thermal behaviour and of composition suggested to Yoder and Tilley that "most basalts are themselves a product of fractional melting or crystallization" (1962, p. 397). This is somewhat understated, for it is almost obvious that, given an essentially peridotitic mantle, basalts must be the product of *partial* melting (not necessarily fractional in our current usage, i.e., that of Presnall 1969), but more than that, they must be products of similar *controls* on partial melting in terms of the phases melted and left as solid residue, and of temperature–pressure regimes during ascent to the surface of the earth. It may happen that nearly all basalt magmas undergo fractional crystallization of the primary (i.e. liquidus) phase during ascent;

commonly this phase will be olivine, and when magmas arise so rapidly as to fail of fractional crystallization en route, they may then be termed picritic or, indeed, peridotitic. The aphyric basalts carry no phenocrysts of the primary phase, and if their parent magmas have reached saturation during ascent, they have passed through or resided in a region of *T-P* such that saturation in a primary phase no longer occurs.

The critically undersaturated basalt system (Fig. 14.4) may provisionally explain why some basaltic magmas having compositions in or very near the critical plane produce trachytic (syenitic) residua instead of their Ne- or SiO_2-bearing counterparts, phonolite, or rhyolite. A bulk composition in or sufficiently close to the critical plane should yield residua, which remain within this system upon fractional crystallization, and should not become enriched in either Ne or SiO_2. If this is the only reason why trachytes occur, then it merely transfers the trachyte–syenite problem to the problem of why critically undersaturated basalts occur, because trachytes would then be the products only of critically undersaturated basalts (i.e. those on the critical plane). The abundance of trachytes in the oceanic association, and of syenites in the continental association, is high enough to engender the suspicion that controls during fractionation of basaltic magma are also important, although closeness of the parent magma to the critically undersaturated condition undoubtedly also plays a part.

The Rest of the Basalt Tetrahedron

All the principal elements of the basalt tetrahedron have now been covered in our discussion and can be summarized very briefly as follows.

- The critical plane divides the tetrahedron into three fundamentally different regions, the undersaturated basalts which fractionate away from the plane and towards Ne, the critically undersaturated basalts which always remain critically undersaturated in the trachyte trend, and the tholeiitic basalts which fractionate away from the plane and towards hypersthene or silica saturation.
- No common parent of both alkali basalt and tholeiite appears to exist at low pressures.
- Alkali basalt cannot, at low pressures, fractionate to produce tholeiitic basalt (unless oxidation is involved), nor can tholeiitic basalt fractionate to produce alkali basalt.
- Within the tholeiitic volume to the silica-rich side of the critical plane, olivine tholeiites may fractionate to produce hypersthene basalt or quartz basalt; the reverse process cannot occur. Olivine tholeiites can be and no doubt often are parental to any of the suite of rocks having higher degrees of silica saturation, up to and including rhyolite (in small amounts controlled by the amount of alkalies, Na_2O and K_2O, in the initial magma).

- The system Fo-Di-SiO_2 is a satisfactory model for that part of the basalt tetrahedron to the right of the critical plane. Addition of An or plagioclase to this system (see, for example, Fig. 13.10) may improve our detailed knowledge of crystallization processes, but does not alter the fundamental parent–daughter relations among tholeiitic liquids as deduced from Fo-Di-SiO_2.
- All major basalt types can be explained by the production in the mantle source region of three major primary magma types: undersaturated, critically undersaturated, and olivine tholeiitic. This is not to say that some magmas which could be daughters by fractionation may not also originate directly by partial melting in the mantle, but it does imply that genetic knowledge of the three primary magmas is necessary and perhaps sufficient for an understanding of the origin of most basalts.

Chapter 15
Potassium: Petrogeny's Residua System and Ternary Feldspars

Introduction

The main effect of potassium when present in the small amounts typical of most basalt is to lower temperatures slightly and enter into the feldspar as small amounts of the orthoclase molecule, $KAlSi_3O_8$. Potassium is a large alkali ion (1.46 Å radius if we use the compilation of radii by Whittaker and Muntus 1970), considerably larger than sodium (1.10 Å). Because of its large size, it is almost entirely excluded from the common mafic minerals of basalt, and it occurs in only limited amounts in calcic plagioclase. For this reason, K is conserved in residual liquids upon fractional crystallization of most basaltic compositions, because it finds a happier home in the liquid structure than in the structure of the average crystals. This effect is called partitioning. The result of the fractionation process can often be seen in the micropegmatite mesostasis (commonly sanidine + quartz) of some tholeiitic basalts. Although we may rightly say the effect of potassium is minor in such a role, we must nevertheless strive to consider its behaviour in phase diagrams if we wish to follow the question of what basaltic liquids may become as they crystallize. Moreover, despite its small, seemingly trivial abundance in basalts, it would be a grave mistake indeed to treat potassium lightly in any discussion of basalt genesis, for even rather small variations in potassium content appear to be very important indicators of basalt type, source, or even mode of origin. And finally, there is a whole class of strange potassic igneous rocks, whose basic members may include leucite basalt. Although we shall not attempt here to explain why some suites of rocks are potassic, we shall at least explore some of the consequences.

We begin with the melting behaviour of $KAlSi_3O_8$, potassium feldspar. This chemical formula is the orthoclase component (or CIPW molecule), but it is usually represented by the monoclinic, disordered polymorph sanidine at magmatic temperatures.

S. A. Morse, *Basalts and Phase Diagrams*,
https://doi.org/10.1007/978-3-030-97882-2_15

Sanidine: The System Kalsilite–Silica

Sanidine is a bit tricky to prepare from scratch. Synthesis usually involves addition of potassium as the carbonate, and when the CO_2 is sintered off, it commonly carries away some of the volatile K, which must be restored by adding a little extra potassium carbonate. The first component made is $K_2O \cdot 6SiO_2$, to which is then added Al_2O_3. The preparation of such starting materials is elegantly described by Schairer and Bowen (1955). A much more satisfactory way of making a starting material is to take natural microcline, readily available as large single crystals, and subject it to alkali chloride exchange (Waldbaum 1969). The ground feldspar is held for perhaps a day in a KCl melt, into which impurities such as Ba, Rb, and Na will diffuse out of the feldspar, being replaced there by K from the chloride melt. A second exchange may be made if desired. The KCl with impurities is simply dissolved in water after each exchange, leaving the feldspar intact. This will still be microcline in structure, but it may be convened to sanidine, if desired, by heat treatment near 1000°C. Such a procedure cannot guarantee a stoichiometric feldspar composition in terms of Al and Si, unless the starting material is stoichiometric.

If one now attempts to melt pure sanidine, one finds in the run, instead of glass alone, crystals of *leucite* ($KAlSi_2O_6$, Lc) plus glass. The crystals persist to very high temperatures. Obviously, the melting is incongruent. It must be described in terms of the binary system leucite–SiO_2 just as the incongruent melting of enstatite must be described in terms of the system Fo-SiO_2 (see Chap. 10). And as with the case of enstatite, there is nothing really special about the incongruent melting of sanidine; interested readers may construct *G-X* diagrams to illustrate this in the manner done for Fo-SiO_2.

The phase diagram, after Schairer and Bowen (1955), is shown in Fig. 15.1. This is actually the system $KAlSiO_4$-SiO_2 (kalsilite, Ks-SiO_2), for reasons that will later

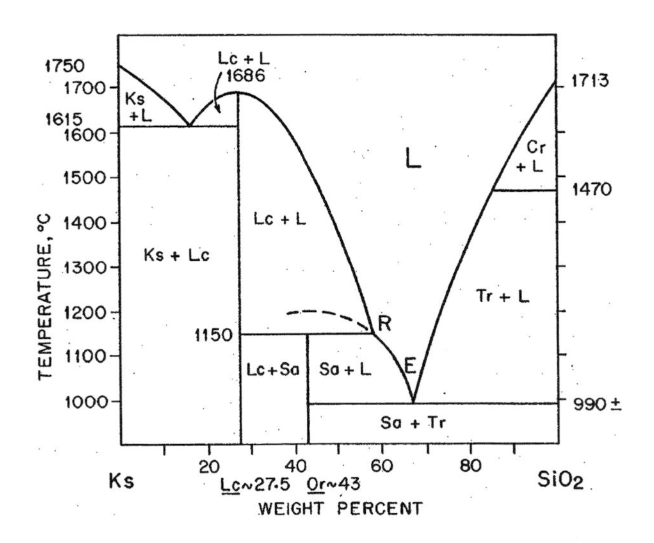

Fig. 15.1 System kalsilite (Ks)–silica (SiO_2) (after Schairer and Bowen 1955). *Lc*, leucite; *Sa*, sanidine; *Tr*, tridymite; *Cr*, cristobalite; *Or*, orthoclase (component)

become apparent. The melting of sanidine involves only the portion Lc-SiO_2. Leucite itself melts congruently at 1686°C and cristobalite at 1713°C. The system is strongly reminiscent of Fo-SiO_2. Sanidine melts incongruently at 1150°C to Lc and silica-rich liquid *R*. Sanidine and tridymite melt together at a eutectic at about 990°C. Crystallization and melting paths may be analysed for Fo-SiO_2. *R* is a peritectic reaction point of isobarically zero variance: $c = 2$, $\phi = 3$,

$$\begin{aligned} W_p &= c + 1 - \phi \\ &= 3 - 3 \\ &= 0 \end{aligned} \quad . \tag{15.1}$$

Fractional crystallization must always yield liquid at *E*. Equilibrium processes must, on the other hand, involve a thermal arrest at 1150°C, while leucite is destroyed (or made, on melting) in the reaction Lc + $L \rightleftharpoons$ Sa. If the bulk composition lies to the left of Or, crystallization will end at 1150°C unless fractionation occurs.

As in the case of the other alkali feldspar, albite, we are not sure whether sanidine is always strictly stoichiometric with respect to SiO_2, or whether on the other hand it may show limited solubility towards and away from SiO_2. Figure 15.2 indicates how the phase diagram might look in the latter case; it is presented mainly in order that we do not lose sight of this possibility, for which there is some tenuous experimental evidence.

In contrast to albite, sanidine provides no thermal barrier. Clearly, the removal of leucite from early liquids can cause a transit from silica-undersaturated to silica-oversaturated liquids. Just as clearly, this can happen only for potassic rocks. But where does the thermal barrier represented by albite give way to the incongruent relation shown by sanidine? Since we know of extensive solid solution in nature between albite and sanidine, it would appear to be a critical matter to know where the incongruent melting behaviour takes over as the crystals become richer in orthoclase component. This means we should look at the system Ab-Or, which turns out to be novel to us in two ways: it is in part an example of a binary system with a *minimum*

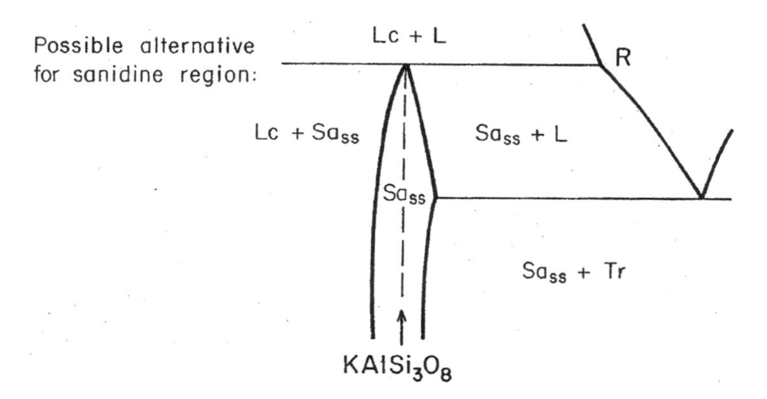

Fig. 15.2 Possible solid solution of sanidine towards silica and leucite

melting relationship and in part a *nonbinary section* showing incongruent melting. In order to understand completely the so-called join Ab-Or, we shall eventually need to look at the ternary system nepheline–kalsilite–SiO_2, called by Bowen "Petrogeny's Residua System". We turn first to the alkali feldspar join Ab-Or.

Albite–Orthoclase

This join was studied by Schairer (1950), who found the crystallization behaviour of the feldspars so sluggish that he was unable to determine a feldspar solidus. By 1969, James B. Thompson Jr. and David R. Waldbaum had completed a series of studies on the thermodynamic properties of sanidine–high albite crystalline solutions, and were able to calculate a phase diagram entirely consistent with Schairer's liquidus data (Waldbaum and Thompson 1969). Although they suggested a method of testing their calculated solidus experimentally, this has not yet been done explicitly. Instead, it is feasible to use the linear partitioning principle to reconstruct the Ab wing of the azeotropic system Ab-Or as done here in Fig. 15.4. The procedure follows that used for An-Ab as reconstructed from Bowen by Morse (2015) with the Ab melting corrected to 1100°C as discussed in that paper. The trick is to use the Waldbaum–Thompson data in a linear partitioning diagram to constrain the Ab limb to the derived value of $K_D = 0.320$, draw the solidus, and fit the liquidus to conform to the K_D value. This approach is shown in the lower part of Fig. 15.4.

The series of papers by Thompson and Waldbaum is a landmark in the theoretical and experimental treatment of crystalline solutions, and its principles find frequent use in advanced studies.

The join is shown in Fig. 15.3. On the Or sideline, the temperature 1150°C marks the incongruent melting of pure sanidine, and the temperature 1530°C marks the limit of the Lc + L field; these temperatures occur in Fig. 15.1 along the Or composition line. The rest of the diagram is too compressed to comprehend easily, so we turn instead to an expanded temperature scale in Fig. 15.4a. This is a complete azeotropic melting diagram for sanidine crystalline solutions. The metastable feature from Fig. 15.3 is removed for clarity and merely identified by the dotted line. The "*m*" in Figs. 15.3 and 15.4 is the azeotropic point, here (Fig. 15.4) given its correct location at 0.331 mol fraction Or. More details are given in caption (a) to the figure. A more detailed assessment of the sanidine solutions is given in the lower Fig. 15.4b, which is a linear partitioning diagram (Morse 1996) that shows the experimental data for each "leg": albite and orthoclase, along with their tie lines. A valuable feature of such diagrams is the presentation of data with their equations and the impressive precision of the results (note the azeotropic point at 0.331 ± 0.006!).

In Fig. 15.4a, we see right away the answer to our main question: the incongruent melting leaves off about halfway across the diagram towards albite. (This is no "magic number", it is merely fortuitous that the melting behaviour changes at about $Or_{50}Ab_{50}$.) There is a reaction point at 1080°C, below which the alkali feldspar join is binary and above which are found equilibria involving leucite. Leucite does not lie

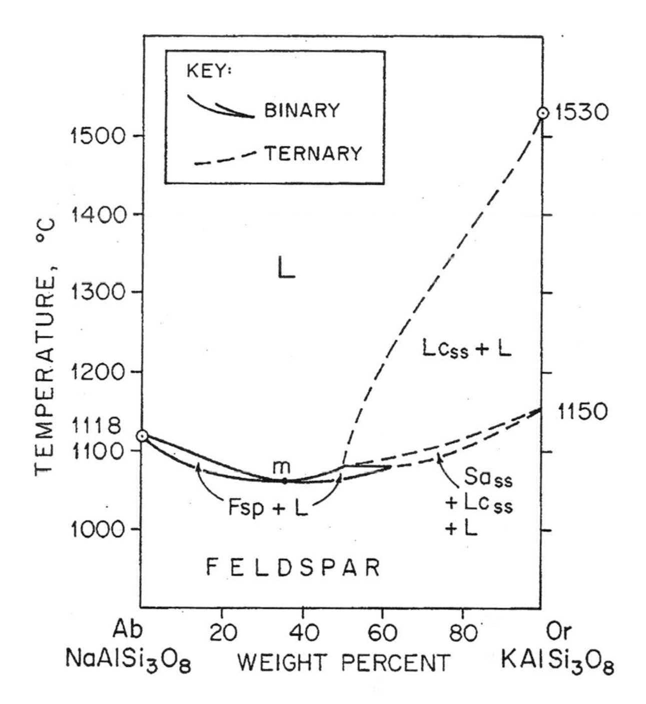

Fig. 15.3 System albite (Ab)-orthoclase (Or) (after Schairer 1950). *Fsp*, feldspar solid solution; *Lc*, leucite. All equilibria involving leucite, i.e., from 1080°C, are ternary, not binary

in the join Ab-Or (see Fig. 15.1 for a reminder), so all liquids in equilibrium with leucite must also lie off the join, towards SiO_2. It is for this reason that we call this part of the diagram "nonbinary", which means ternary or higher order. In this case, as we shall find, ternary will do. It is a bit hard to understand such "pseudobinary" representations of higher-order equilibria, but we have encountered these before, as in the spinel field of Fo-An, and one can learn to read them as road maps without worrying too much about the missing dimensions. One region that is still binary in the right half of the diagram is the region below the solidus, labelled "sanidine solid solution". We see that this field is continuous to albite solid solution and that the whole region represents an arbitrarily variable solid solution from pure Ab to pure Or.

Turning now to the less bizarre part of the system, we find a binary minimum subsystem of alkali feldspars. This looks a little bit like two plagioclase loops joined at their lower ends and indeed ignorant people sometimes draw minimum diagrams that way. But we shall learn better, namely that at the minimum, the solidus and liquidus curves *must coincide*, and their common tangent *must be isothermal*, i.e. parallel to the composition axis. This means that the minimum composition is very special indeed—it might at first seem to be a "univariant point" (!), but of course there can be no such thing. The point is *invariant by restriction* because for that particular composition the system can be described as a one-component system (the single component being a feldspar or liquid of the minimum composition), so for $c = 1$ and $\phi = 2$:

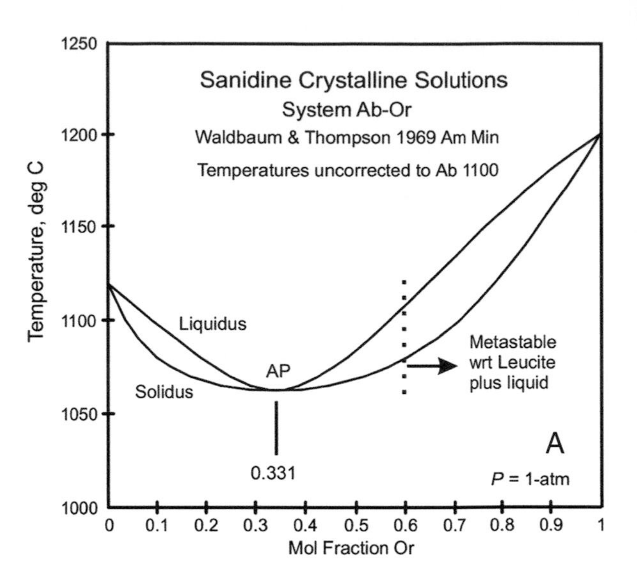

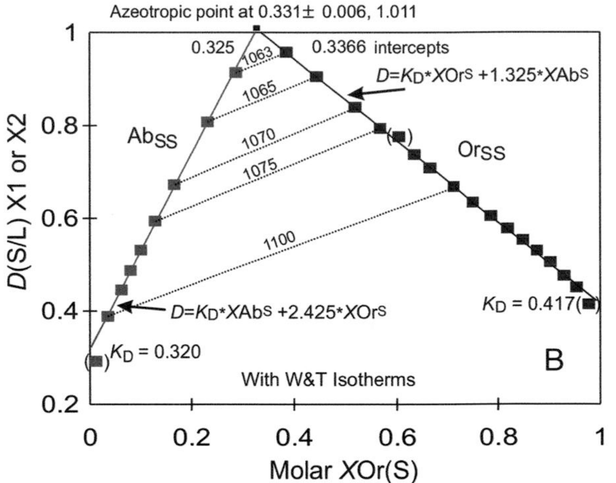

Fig. 15.4 (**a**) System Ab-Or with expanded temperature scale. Nonbinary above 1080°C beyond Or_{50}, except for the metastable feldspar melting loop, which is binary (after Waldbaum and Thompson 1969). The temperatures are uncorrected for the adjustment to 1100°C for albite melting (Anovitz and Blencoe 1999). (**b**) Linear partitioning diagram of the Ab-Or binary constructed from the tables of Waldbaum and Thompson and with K_D values used to fit the liquidus curves in (**a**). (From Morse 2015)

$$\begin{aligned} W_p &= c + 1 - \phi \\ &= 2 - 2 \\ &= 0 \end{aligned} \quad . \tag{15.2}$$

Such points, *invariant by restriction*, are also a type of *singular point*; here, singularity means the identity of feldspar and liquid compositions.

Otherwise, we treat crystallization in such a system just as we would in a plagioclase loop. To continue with our example, $Or_{60}Ab_{40}$, the liquid may now cool below 1080°C, maintaining equilibrium with Sa_{ss} whose composition lies on the solidus. At a temperature near 1078°C, this solidus cuts the bulk composition, and the system is now entirely crystalline. The last liquid has a composition near $Or_{48}Ab_{52}$. Note that the liquid will not reach the minimum in equilibrium crystallization unless the bulk composition is the minimum.

Crystallization of Ab-rich liquids proceeds in a straightforward way, as may safely be left to the reader to ascertain.

Fractional crystallization in the truly binary part of the system will always yield liquid and a few crystals of the minimum composition. Only an infinitesimally small amount of minimum liquid will result, however. TSC paths will terminate at 1063°C and the bulk composition, having originated at the solidus. Melting paths can be deduced from the principles of the plagioclase system, except that one should not attempt fractional melting into the incongruent region, but should instead wait to see the ternary system. Note!!! that on melting, *no* liquid of minimum composition will be produced except for a bulk composition at the minimum itself.

"Petrogeny's Residua System", Ne-Ks-SiO_2

Liquidus Diagram

The ternary system Ne-Ks-SiO_2 is shown in Figs. 15.5 and 15.6. The diagrams are based on extensive work by Schairer (1950), who was able only with difficulty to

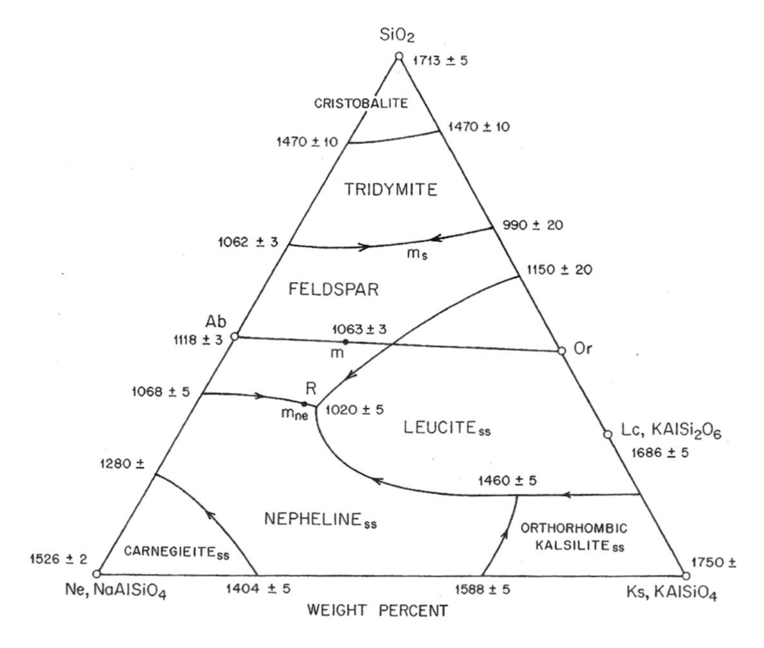

Fig. 15.5 System Ne-Ks-SiO_2 (after Schairer 1950). Letter *m* signifies a minimum: m_s, silica-saturated; m_{ne}, nepheline-saturated. Location of m_s is approximately 960 °C (Johannes and Holtz, 1996)

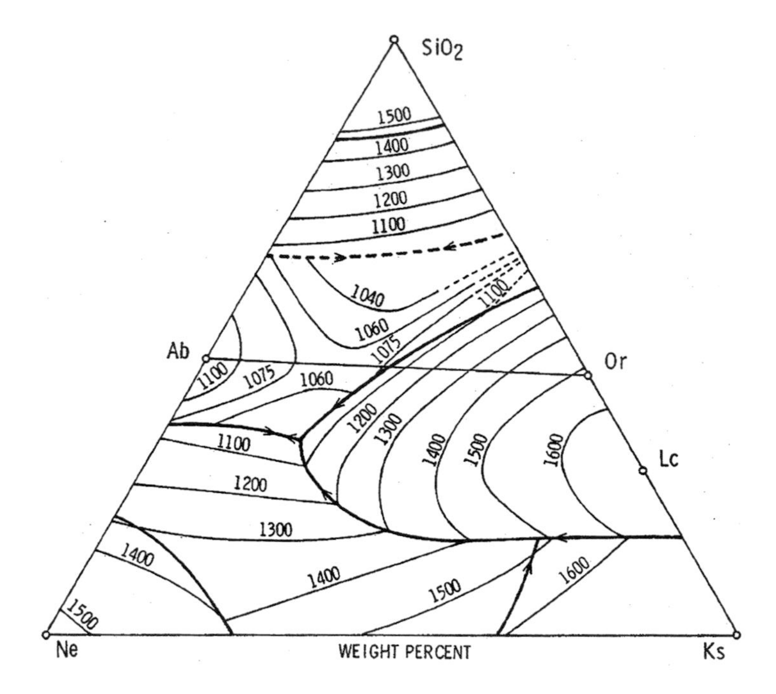

Fig. 15.6 Isotherms in Ne-Ks-SiO_2. (After Schairer 1950)

determine liquidus temperatures and the approximate locations of minima. Work on this system at 1-atm pressure was to some degree supplanted by the revolutionary studies of Tuttle and Bowen (e.g. 1950, 1958) and others using elevated steam pressures to facilitate reaction and to correspond more closely to conditions at some depth in the crust of the earth. As a result, there are many details of the 1-bar system that we do not know, but can more or less adequately infer from studies at higher pressure. We do not, for example, know any tie lines in the dry system. It will suffice to analyse the diagram with inferred tie lines, and this has the added benefit that we can illustrate the principles without worrying excessively about the details.

Of the bounding binary systems, we have studied Ne-SiO_2 and Ks-SiO_2 already. Ne-Ks has been studied (Tuttle and Smith 1958), although not with complete success; its main feature of importance to us is that it is dominated by a large range of solid solutions from kalsilite towards nepheline and a smaller range of solid solutions from carnegieite towards kalsilite. The junction of these fields is not satisfactorily resolved with respect to the liquidus diagram of Fig. 15.5, so we shall be forced to make some arbitrary assumptions about relations near the base of the ternary diagram.

The features of chief interest in Fig. 15.5 are the large primary field of leucite solid solution, which overlaps the alkali feldspar join; the large fields of nepheline and feldspar solid solution; and the three *ternary minima* that lie in what Bowen called a "thermal valley" and approximate the compositions of rhyolite (m_s), trachyte (m), and phonolite (m_{ne}). (The point m is actually a binary minimum in Ab-Or.)

These minima closely represent the natural residua from fractionation of, respectively, tholeiites, critical plane basalts, and alkali basalts, and it is for this reason that Bowen aptly called the system Ne-Ks-SiO_2 "petrogeny's residua system". The minimum m on the alkali feldspar join is now known with good accuracy, but compositions 5% to each side of Or_{35} gave the same liquidus temperature, being bracketed by 1060°C (some crystals in glass) and 1065°C (all glass) in 3-week experiments. The minimum therefore could lie anywhere from Or_{30} to Or_{40}, but Or_{33} is the best estimate for the binary join Ab-Or, to be discussed.

The silica-saturated ternary minimum, here denoted m_s, is taken from Johannes and Holtz (1996) as 960°C. The position and temperature are compatible with experiments at higher pressure.

The nepheline-saturated minimum, here denoted m_{ne}, is not precisely located here, but its position and temperature are much more tightly constrained by the liquidus data. It must lie to the left of the triple junction R labelled 1020°C, which is a reaction (peritectic) point, and yet close to that point, from liquidus data.

Another reaction point occurs at 1460°C, involving orthorhombic kalsilite, nepheline, leucite, and liquid. As with the several peritectics in Fo-Di-SiO_2 (Chap. 12), one can deduce that this point and R are peritectics because arrows flow away from them and towards them. This could not happen with an even (eutectic) reaction.

Of the *five* solid solutions feldspar, nepheline, leucite, carnegieite, and kalsilite in this system, only feldspar and leucite are known with any accuracy. We know the feldspar series to be complete at these high temperatures. Fudali (1963) showed that leucite contains up to about 40% by weight of the component $NaAlSi_2O_6$ ("soda leucite", not jadeite!), and we shall interpolate compositions up to that limit in succeeding diagrams. Nepheline is known to contain vacancies, which amount to solid solution towards albite, and we will depict it essentially as determined at 5 kbar by Morse (1969a), although the solid solution limits must be somewhat different at the higher temperatures of the 1-bar diagram. Carnegieite solid solution is limited by the presence of nepheline and can be adequately deduced from the liquidus diagram and from the work of Tuttle and Smith. Kalsilite solid solution is poorly known and complicated by the possible presence on the liquidus of a hexagonal polymorph. We shall be able to find suitable constraints on kalsilite composition and on leucite–nepheline tie lines near the 1460°C reaction point. Given these generalities, we now turn to a series of isobaric, isothermal sections to illustrate what can be inferred about this ternary system from the liquidus diagram and a few basic principles.

Isothermal Sections

The isotherms as determined by Schairer are shown in Fig. 15.6. These are based on a great density of points and contain much detail not apparent at first glance.

Ten isothermal sections are shown in Fig. 15.7. These depict, in a down-temperature series, the instantaneous equilibria that aid in constructing crystallization or melting paths. Brief comments on each are given below.

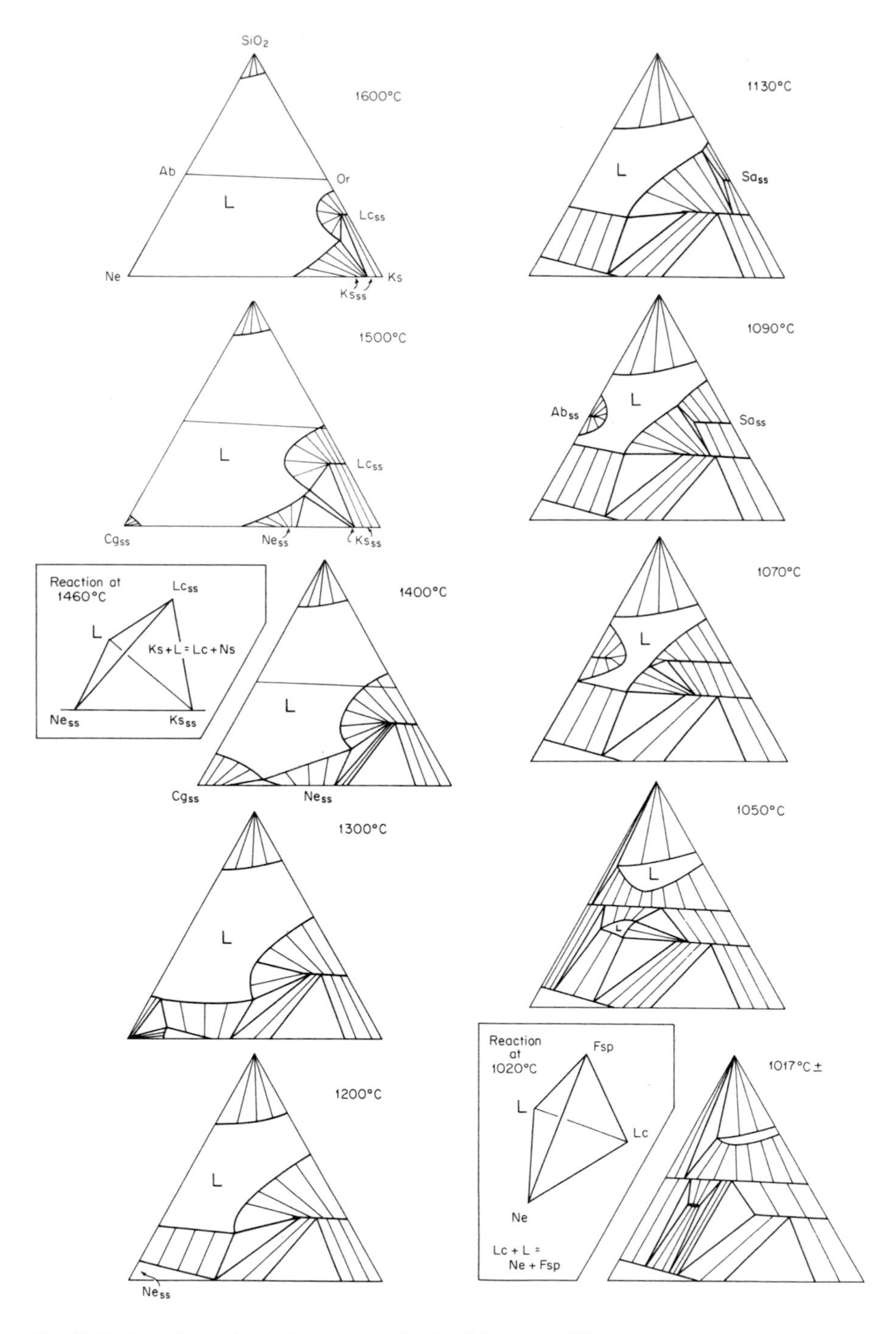

Fig. 15.7 Isobaric, isothermal sections in Ne-Ks-SiO_2 at ten different temperatures

1600. Small field of cristobalite + liquid near SiO_2 corner. An extensive field of Ks_{ss} + L is already established, and this was joined at 1686°C by the beginnings of Lc_{ss} crystallization. Since Ks crystallized first, we know that its solid solution towards Na-bearing compositions must be more advanced than that of Lc_{ss}. The invariant ($W_{p,t} = 0$) three-phase triangle Lc-Ks-L reflects this condition; it is bounded by fields of Lc_{ss} + L, + Lc_{ss} + Ks_{ss}, and Ks_{ss} + L.

1500. Cr + L field has grown. Cg + L field is new. Ks + L field has shrunk drastically at the expense of Ne_{ss} + L, newly arrived. The limits on Ne must be about as shown because the curve L (Ne, Ks) represents an odd reaction (L = Ne − Ks). The topology around the three-phase triangle Lc-Ks-L remains unchanged.

1400. The inset shows the isothermal destruction of Ks at 1460°C by the reaction Ks + L = Lc + Ne. Note that in this reaction four phases coexist at equilibrium, and the system is isobarically invariant without an external imposition of fixed temperature ($W_p = 0$, $W_{p,t} = -1$, which is irrelevant). At 1400°C, the silica polymorph is now tridymite. A larger field of Cg_{ss} + L now occurs, as well as a 3-phase equilibrium Cg + Ne + L. The equilibrium Lc_{ss} + Ne_{ss} is well established, isolating Ks from further interaction with liquid. Both Lc_{ss} and Ne_{ss} have become much more sodic. The 3-phase solid triangle Lc-Ks-Ne is arbitrarily located; its limits are unknown except for certain logical geometric constraints.

1300. Much the same picture as before. The field of Cg_{ss} + L is now nearly gone. Ne_{ss} + L is approaching a configuration that will remain essentially unchanged to lower temperatures. Note the increased solution of Ne towards albite, and the enlarged field of Cg + Ne.

1200. Cg is now gone altogether. The liquid pool continues to shrink at the expense of Lc_{ss}, Ne_{ss}, and tridymite, each in equilibrium with liquid.

1110. Sanidine makes its appearance for the first time at 1150°C. We see it breaking into the Lc + L field as a 3-phase triangle with apex on Sa_{ss}. The tangent to the liquid boundary falls drastically outside the Lc + Sa line, hence on the negative extension (Sa − Lc), and the reaction is odd, L = Sa − Lc. Lc_{ss} has advanced further towards soda leucite.

1090. Albite has entered from the left at 1100°C and now forms an arc of Ab_{ss} + L. Note that, as with Lc_{ss} before, the maximum enrichment of the opposite component occurs on the join, whereas liquids to either side of the join are tied to less advanced solutions. The 3-phase triangle Sa + Lc + L has swept rapidly towards more sodic compositions, as dictated by the location of liquid on the field boundary where intersected by the isotherm. The liquid pool is now being constricted mainly from the sides, since the tridymite and nepheline liquidus surfaces are very steep.

1070. Recall that at 1080°C in the Ab-Or join, the leucite field disappeared. This event corresponds in the ternary system to the coincidence of the Sa_{ss} + L edge of the 3-phase triangle with the Ab-Or join. By 1070°C, this edge has fallen below the join, and the triangle has radically changed shape from the initial one at 1130°C. Despite this, there has been no change in the nature of the reaction, which is still odd. The extent of sanidine solid solution is probably somewhat exaggerated in the

1070 section. Note that the Ab_{ss} + L field has grown considerably and that Lc_{ss} + L has shrunk considerably. Clearly, this field will soon be wiped out.

1050. The alkali feldspar join has closed off at 1063°C, leaving a continuum of solid solutions in equilibrium with various other phases. From the left, we see the encroachment of a 3-phase triangle with Tr + Ab_{ss} + L; this will sweep to the right leaving a solid triangle with Tr + Ab_{ss}, which has swept in from the left also. Two liquid pools (rhyolitic and phonolitic) are separated by the feldspar join. An almost complete range of Ab-Or solution coexists with rhyolitic liquid. A more limited range coexists with phonolitic liquid, bounded to the right by the 3-phase triangle Fsp + Lc_{ss} + L. (We can no longer distinguish the feldspar compositionally as sanidine, because of its intermediate composition, but it is undoubtedly of the sanidine structure.) The Lc_{ss} + L field is smaller than ever.

1017±. The peritectic reaction at 1020°C was L + Lc = Fsp + Ne, as shown in the inset. Now, the Fsp + Ne field has grown to isolate Lc_{ss} from liquid. The extremal Lc_{ss} composition has now begun to withdraw, generating Ne + Fsp. Only a tiny pool of undersaturated liquid remains (presuming the minimum temperature to be 1015°C), bounded on each side by opposed 3-phase triangles each having the identity Ne + Fsp + L. These will, at 1015°C, meet and simultaneously become a single line Fsp + L + Ne_{ss}, just as the last trace of liquid disappears. The feldspar and nepheline coexisting with this last liquid will have singular compositions, dictated by the collinearity, but note that the feldspar composition will not lie (except by chance) at the feldspar minimum on the join Ab-Or. It is important to understand the independence of the singular feldspar composition tied to Ne + L and the minimum feldspar composition in equilibrium with its own liquid. This principle will be stressed again below.

Below 1015°C, not shown, the undersaturated portion of the residua system will consist of the following fields of solids: Ne_{ss} alone, Ne_{ss} + Ab-rich feldspar, with tie lines as in the 1017 section, a 3-phase field Ne_{ss} + Fsp + Lc, two 2-phase fields Fsp + Lc_{ss} and Ne_{ss} + Lc_{ss}, the 3-phase field Ne_{ss} + Lc_{ss} + Ks_{ss}, and the field Lc_{ss} + Ks_{ss}.

The siliceous part of the system is still dominated by liquid equilibria at 1017°C. We presume that at 990°C a new 3-phase triangle Sa_{ss} + Tr + L will enter from the right and that the two triangles will meet at 980°C in a singular line through the ternary rhyolite minimum.

Equilibrium Crystallization

Two examples, both originating in the leucite field, will serve for now to illustrate equilibrium crystallization. The first bulk composition, BC-1 in Fig. 15.8, yields only sanidine and leucite. The liquid first moves in equilibrium through the leucite field, tied through the bulk composition to the TSC on the leucite join. It eventually reaches the field boundary, whereupon reaction ensues, using up leucite to make

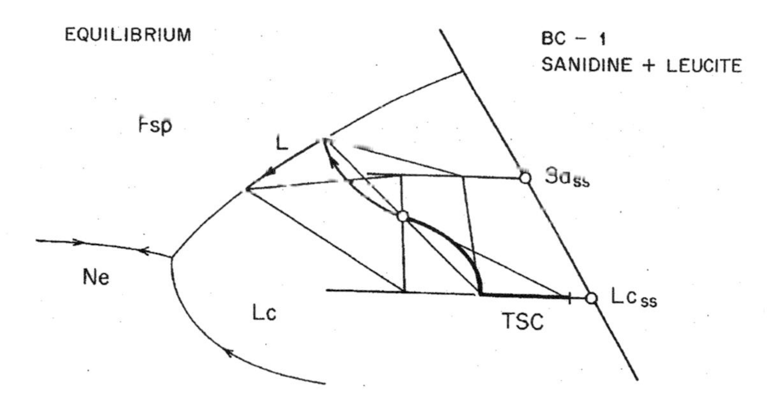

Fig. 15.8 Equilibrium crystallization of BC-1

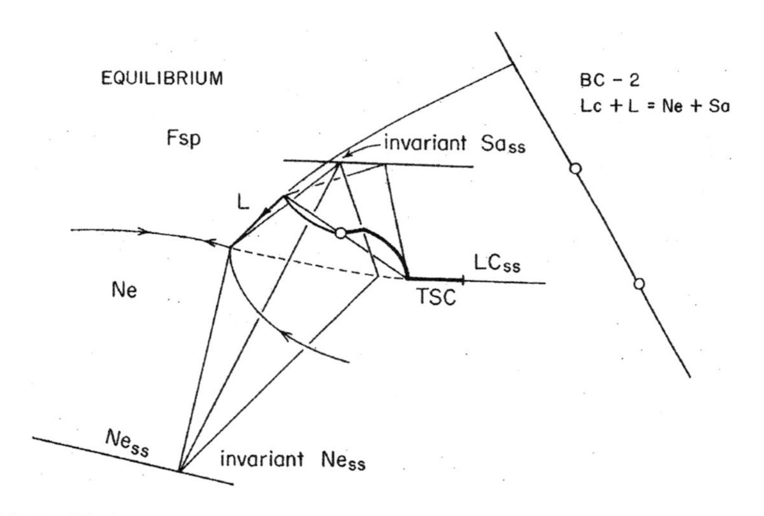

Fig. 15.9 Equilibrium crystallization of BC-2

sanidine. The TSC advances towards the bulk composition along successive solid-solid legs of 3-phase triangles, as shown by the TSC path in the figure. Crystallization ceases when the Sa + Lc tie line cuts the bulk composition.

The crystallization of BC-2 is shown in Fig. 15.9. This composition lies within the terminal, invariant 3-phase triangle Ne + Lc + Fsp at 1020°C, and hence consists of these three phases when solid. The liquid initially moves along a path with small curvature while Lc_{ss} crystallizes. It then moves along the field boundary, first dissolving a small amount of Lc_{ss} as feldspar forms, and then, for reasons discussed later in this chapter, producing leucite again. The TSC rises along a curved path of relative enrichment in feldspar. The liquid then reaches the reaction point *R* at 1020°C and remains there while the TSC moves directly towards *R* and to the bulk composition. This is accomplished by the crystallization of invariant Fsp + Ne at the expense of Lc and liquid.

Other equilibrium crystallization exercises, particularly with bulk compositions in the silica-oversaturated region of the leucite field and on the feldspar join itself, can be undertaken by the reader for herself. Certain bulk compositions just below the feldspar join yield complex paths, and the nature of reaction along the leucite-feldspar boundary changes from odd to even for certain bulk compositions. These complexities are discussed in a section near the end of this chapter, to which the reader should refer for a rigorously accurate understanding of equilibrium crystallization. The present examples will suffice for an elementary understanding.

Pseudoleucite

In some natural rocks, there occur patchy or stringy intergrowths of alkali feldspar and nepheline, often having the external outline of euhedral leucite phenocrysts. Such objects are called "pseudoleucite". Figure 15.9 leaves little doubt that these may originate by reaction of liquid *R* with leucite phenocrysts, much in the manner of BC-2. Conceivably, the reaction may run rapidly enough to convert all the leucite to Ne + Fsp in some rocks. The reaction need not be confined to such bulk compositions as BC-2, within the invariant quadrangle, for with some degree of fractional crystallization, a large range of Lc_{ss} crystals could come into contact with peritectic liquid. Ideally, at equilibrium, these crystals would first be converted to the maximum soda leucite content before breaking down to Ne + Fsp, but in nature it is possible that the nepheline and feldspar will nucleate and grow rapidly from metastable Lc_{ss} which is not at the limiting composition. Any liquid remaining after the conversion will then fractionate towards the minimum, enriching both Ne and Fsp in soda components.

While the peritectic reaction may theoretically account for pseudoleucite, Fudali (1963) has argued persuasively that most natural examples have originated by a simpler mechanism, namely the *subsolidus* breakdown of soda-rich leucite itself to nepheline + feldspar. At subsolidus conditions, soda leucite becomes unstable and spontaneously reacts to Ne + Fsp. In other words, the 3-phase triangle Ne + Fsp + Lc shifts towards potassic compositions with falling temperature. Given such behaviour, there is no need to call upon crystal–liquid reaction to account for pseudoleucite, which can form in the solid state. Correct interpretation of natural occurrences must therefore depend to large degree on the chemical and textural evidence bearing on the reaction.

Fractional Crystallization

Using the same bulk compositions as treated above for the equilibrium case, we may examine two contrasting examples of fractional crystallization. These are shown in Fig. 15.10. In the case of BC-1, the liquid follows a shallow curve away from ISC-1,

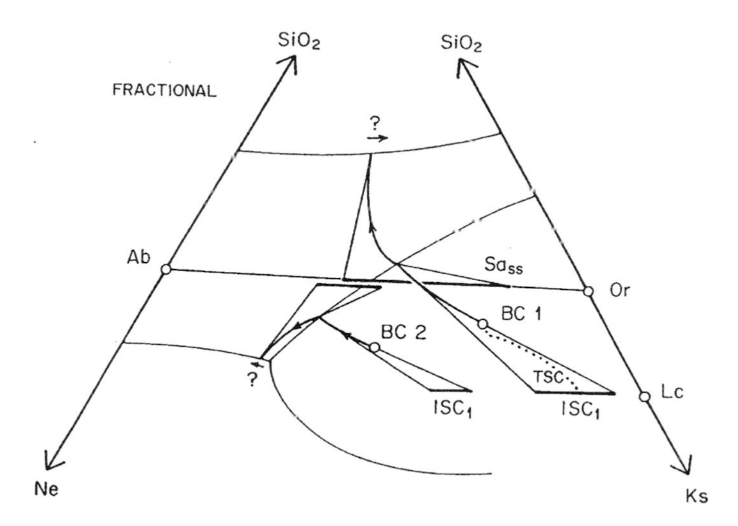

Fig. 15.10 Fractional crystallization of BC-1 and BC-2

which lies on the leucite join. Upon reaching the field boundary, the liquid immediately ceases to produce leucite and instead begins to produce sanidine along a wide range of compositions. The liquid follows a fractionation path that must be determined by experiment, but that no doubt has the general form shown in the figure. The Sa_{ss} formed during this time constitutes ISC-2 (not labelled), and the TSC migrates towards ISC-2 as shown by the dotted path in the figure. When the liquid reaches the silica field boundary, tridymite crystallizes along with feldspar until the liquid is used up just as it reaches the ternary minimum. The minimum may lie either to the right or left of where the liquid hits the field boundary, not at that point except by chance. In Fig. 15.10, it is assumed that the ternary minimum lies to the right of the liquid path, and it should be noted that this will produce *reverse zoning* in the feldspar: successive layers will now be more Or-rich, whereas before they were more Ab-rich.

Bulk composition 2 (Fig. 15.10) yields undersaturated residual liquids. Again, the liquid leaves the Lc + Fsp field boundary the instant it arrives when the ISC changes from Lc_{ss} to Fsp. The feldspar presumably lies on or very near the join, but is offset for clarity in the figure. The liquid now follows a curved fractionation path across the feldspar field until it reaches the cotectic field boundary with nepheline. It then follows the cotectic to the undersaturated ternary minimum, which may lie to either side of the incoming liquid path. If the minimum lies to the left, normal feldspar zoning will continue to be produced; if to the right, reverse zoning will be produced. The final ISC (number 3) will lie on the Ne + Fsp leg of a three-phase triangle, on the tangent to the cotectic, and will end up at the ternary minimum. The TSC path is not shown, but can readily be deduced by passing levers from liquid through the BC.

The two illustrated bulk compositions typify the contrasting trends of fractionation from the leucite field. In particular, they show that it is easy to breach the critical plane (strictly, the feldspar join) towards silica saturation by fractionation of

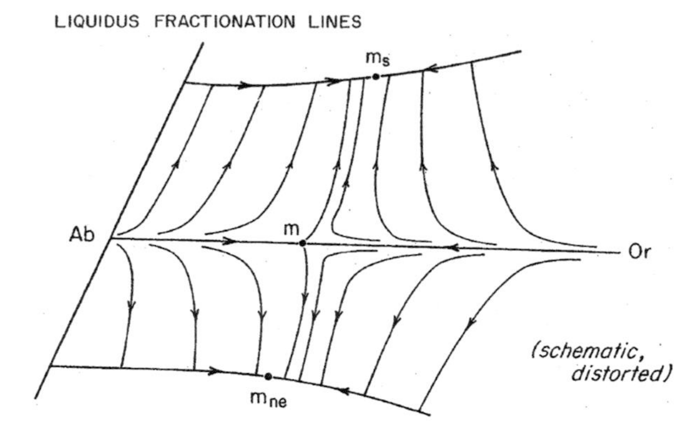

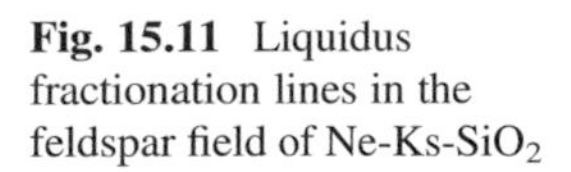
Fig. 15.11 Liquidus fractionation lines in the feldspar field of Ne-Ks-SiO_2

leucite, but only for potassic compositions. Intermediate undersaturated compositions will fractionate towards the undersaturated minimum. Observe that between BC-1 and BC-2 there must exist a unique fractionation curve that arrives at the feldspar join just at its crossing with the field boundary. Liquids lying on this unique curve will theoretically stay on the feldspar join and go to the trachyte minimum at 1063°C. Given a slight amount of reaction with previously formed leucite, however, they will again become undersaturated and move towards the undersaturated cotectic. On the other hand, given the slightest delay in feldspar nucleation, they will become oversaturated and move towards the oversaturated cotectic. Thus, with slight imperfections in fractionation processes, an undersaturated parent liquid may produce either type of residuum, or conceivably both, in closely related areas.

Figure 15.11 summarizes the *feldspar fractionation paths* or more precisely, the *liquid fractionation lines from feldspar* in the residua system. The diagram is distorted, to show relations away from the leucite field, but it corresponds qualitatively to relations observed at moderate pressure, where such fractionation is more likely to occur than at the low pressures of lava extrusion. There are several features of importance in this stylized diagram. First, fractionation paths originating rigorously on the feldspar join remain on that join and terminate at m. All others terminate eventually at either m_s or m_{ne}, respectively, the oversaturated and undersaturated ternary minima. All but two paths originate (either directly or in extension) at either pure Ab or pure Or. Two paths originate at m, the feldspar minimum. These are unique liquidus fractionation lines, and they have the interesting property of curvature, so they do not lead straight away from m. That this must be so was discussed in an elegant exposition by Bowen in 1941 and again by Tuttle and Bowen (1958). While this may at first seem surprising, it follows logically from the fact that the feldspar composition in equilibrium with liquid at either ternary minimum is not the composition m, and this in turn arises from the fact that a silica-rich liquid, for example, has different thermodynamic mixing properties from those of a pure feldspar liquid. Accordingly, the G-X surfaces for liquid solutions off the feldspar join have points of tangency with feldspar solid solutions that are different from the points of tangency to liquids lying on the feldspar join itself. The curvature of the unique fractionation lines has the interesting consequence that liquid paths close to the join and close to the unique fractionation lines may undergo a reversal of curvature, and at the inflection point, the feldspar zoning will become reversed. According to Fig. 15.11 (which may be incorrect in detail), silica-oversaturated liquids approaching m_s from near the unique fractionation curve may produce still more strongly reversed zoning as they move along the cotectic towards m_s, and undersaturated liquids may cause feldspars to change from reverse to normal zoning as they move along the cotectic towards m_{ne}.

Note that the unique fractionation lines do not lead directly to either m_s or m_{ne}, but only to the cotectic somewhere near these points. This is because the addition of a saturating phase, tridymite or nepheline, generates a new criterion for equilibrium, namely a singularity wherein the saturating phase, ternary minimum liquid, and feldspar must all be collinear in the composition diagram. This can only be realized if the *G-X* surfaces for liquid and feldspar, projected from the saturating phase, have a single common tangent at some special value of *G* and *X*. There are, of course, special fractionation lines that do lead directly to each of the ternary minima, but these do not originate at *m*. They originate at Or or Ab (note arrows in Fig. 15.11) or another cotectic.

Fractional Melting

We may safely dispense with exercises in equilibrium melting, which as usual simply reverses the process of equilibrium crystallization. Turning to fractional melting and Fig. 15.12, we are reminded by the 3-phase triangle *L*-Fsp-SiO_2 that solidus fractionation lines are found by constructing a family of short tangents away from liquid on the solid–solid legs of several such triangles. One triangle is shown, with a dashed construction line leading from the liquid apex to a circled solid composition, at which the construction line is tangent to a solidus fractionation line. The figure emphasizes those fractionation lines leading to the feldspar join and to the undersaturated regions. In the oversaturated regions, the repeated construction soon shows that all solidus fractionation lines must radiate from m_s. One of these is straight, because it is radial from SiO_2 through m_s. The fractional removal of liquid m_s merely chases the solid composition towards feldspar, which must have the same composition irrespective of the tridymite content along this singular line joining tridymite, invariant liquid, and invariant feldspar. Such unique fractionation lines are characteristic of systems with a ternary minimum. Note the fundamental contrast between such straight, unique *solidus* fractionation lines and the curved, unique *liquidus* fractionation lines in Fig. 15.11; it may require a little reflection on the differences between fractional melting and crystallization to come to terms with the difference in unique paths.

In the undersaturated portion of Fig. 15.12, the construction leads to the further conclusion that all solidus fractionation lines towards nepheline, as well as towards feldspar, also originate at the minimum, here m_{ne}. Again, the unique fractionation lines run straight away from m_{ne} in both directions.

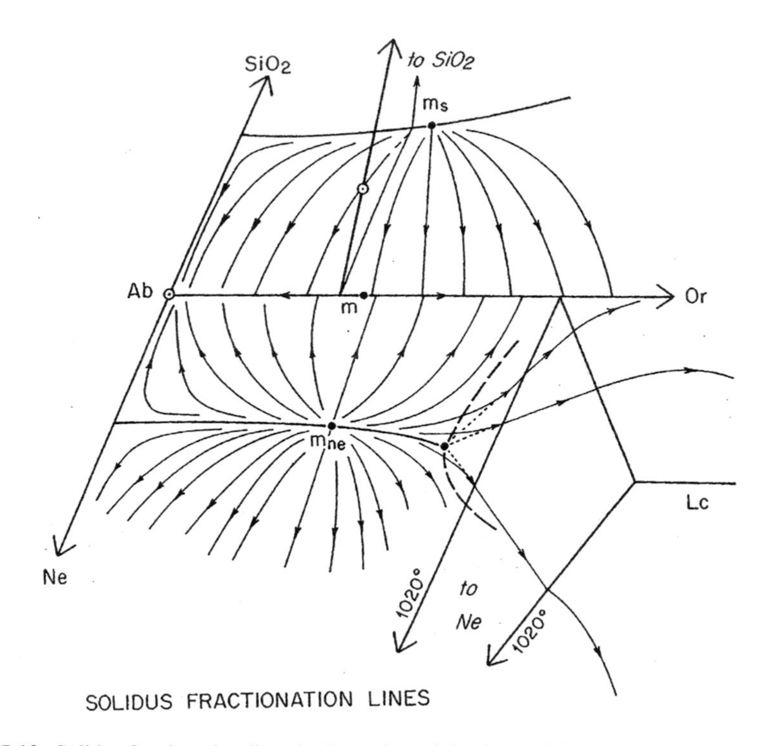

Fig. 15.12 Solidus fractionation lines in the region of the three minima

Note that in no case does the unique solidus fractionation line lead to the feldspar minimum *m*, which has no control over solidus fractionation towards the feldspar join. For the special, singular compositions that lie on the unique lines, fractional melting resembles that in eutectic systems. The initial liquid has the composition of a ternary minimum, and it is extracted isothermally until the saturating solid phase is exhausted. No further melting occurs until the temperature of the feldspar solidus is reached. Liquid is thereafter fractionally extracted over a continuous range of temperature.

For all other bulk compositions, not lying on singular lines, the initial melting temperature is a function of the particular bulk composition, rising with distance from the unique solidus fractionation line. Note that for a given 3-phase triangle, the initial melting temperature is indifferent to distance from the cotectic.

The distorted form of Fig. 15.12 serves to illustrate the interesting behaviour of solidus fractionation lines involving the peritectic point *R*. Lines originating at m_{ne}, for bulk compositions sufficiently undersaturated, eventually encounter the invariant solid triangle Ne-Fsp-Lc. When they do so, the temperature has increased to 1020°C, and the liquid has moved continuously along the cotectic from the *direction of* m_{ne} to the reaction point *R*. It then remains at *R*, with isothermal removal of liquid, while the TSC moves in a straight line across the invariant solid triangle. When the TSC reaches the limiting side of the triangle where either nepheline or feldspar is exhausted, the temperature must now rise continuously from 1020°C along one of

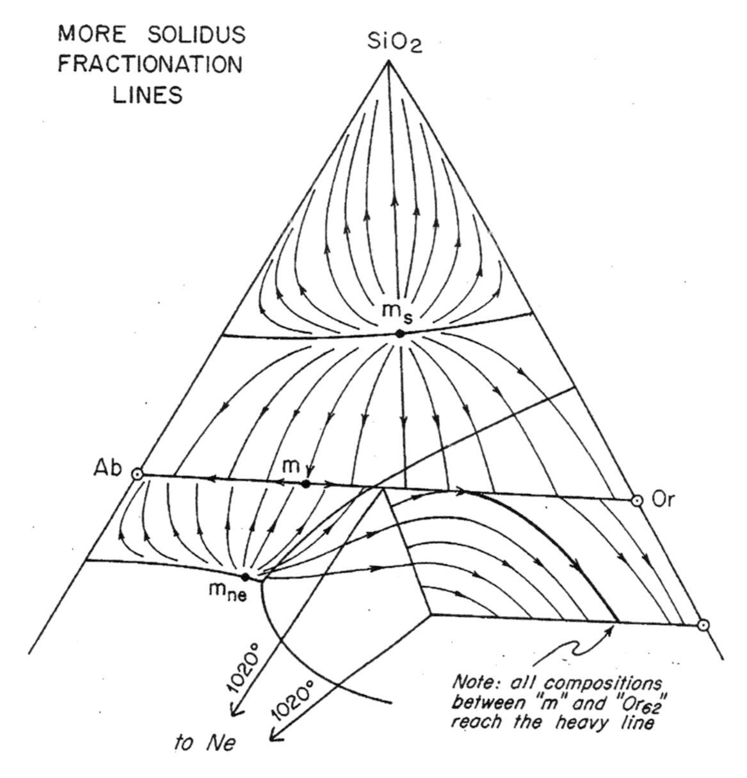

Fig. 15.13 Solidus fractionation lines extended to the regions of leucite and silica mineral

the leucite field boundaries, and the solidus fractionation line now curves continuously away from the straight segment. Note the sharp contrast with systems lacking solid solution (e.g. Fo-SiO_2), in which extraction of peritectic liquid is followed by a large hiatus without melting.

Solidus fractionation lines are summarized at proper scale in Fig. 15.13, which permits some further comment. First, note that silica-oversaturated bulk compositions follow fractionation paths originating at m_s and terminating at SiO_2. Second, note that one line leads, by chance, directly to the feldspar minimum m. Third, note that all lines in the feldspar field originating at m_s terminate on the feldspar join; the liquidus field boundary with leucite plays no role here, since all liquids lie on the silica-saturated cotectic. If we assume that Or_{62} is the solidus composition where feldspar begins to melt incongruently at 1080°C to leucite plus liquid, then all feldspars to the right of m (up to Or_{62}) yield a series of continuous liquids on the join, eventually followed by a continuous series of liquids along the leucite field boundary as the total solid composition follows the heavy solidus fractionation line towards Lc_{ss}. This heavy line is common to all compositions lying between m and Or_{62}. Feldspars farther towards Or follow their own fractionation lines to the Lc_{ss} join and eventually to soda-free leucite along that join.

One peculiarity remains to be resolved and that is the behaviour of compositions close to the feldspar join but in the Lc + Fsp solid field. Observe that these are driven towards the feldspar join by extraction of liquids along the field boundary *L* (Lc, Fsp) near *R*. But then, the liquid moves towards the feldspar join and eventually across it. No TSC can, therefore, remain on the feldspar join for more than an instant, at Or_{62}. Fractionation lines approach infinitesimally close to the join, but touch it only at Or_{62}, whereupon they immediately leave it along the heavy line. Clearly, fractional melting of a leucite-bearing composition will not yield an exactly saturated solid residuum, although it may eventually yield an oversaturated liquid. By contrast, many nepheline-bearing original compositions will yield solid residua strictly on the feldspar join.

Using Fig. 15.13, we may draw certain conclusions about fractional melting in this system at 1 bar. Melts of strictly minimum composition will be rare, restricted to singular bulk compositions lying on special lines. All other melts will be of variable initial temperature and composition on a cotectic or peritectic boundary. They will be extracted only over a continually rising temperature range except for those that reach *R* and stay there for a while. No melting process can yield undersaturated liquids from initially oversaturated compositions. A very restricted range of bulk compositions may yield undersaturated solid residua from initially saturated solid compositions.

Petrologic Summary

Once again, we return to the theme of what basaltic liquids may become on fractionation. In general, as Bowen long ago pointed out, they yield residua closely modelled by the system Ne-Ks-SiO_2, and more specifically, these residua tend to lie in a relatively limited region, which Bowen called a "thermal valley", encompassing the three minima, m_{ne}, m, and m_s (Fig. 15.14). Natural rocks closely corresponding to

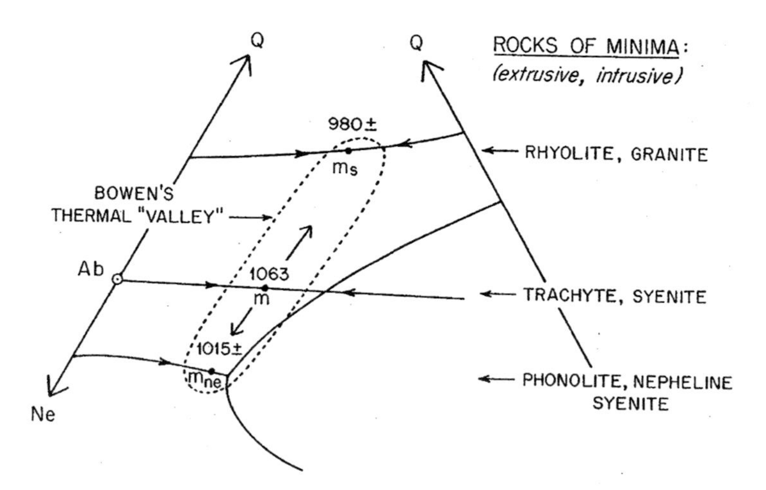

Fig. 15.14 Bowen's thermal "valley" in the residua system

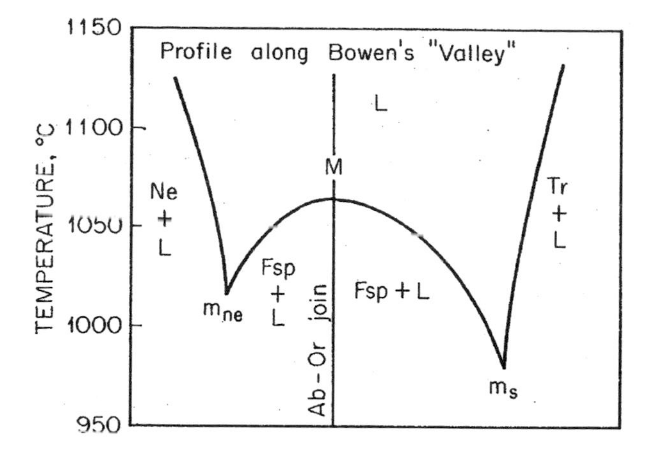

Fig. 15.15 Scale drawing of liquidus temperatures along Bowen's thermal "valley". The line of section connects the three minima, but may not be straight through *M*

these three compositions occur in abundance, and their extrusive equivalents are called, respectively, *phonolite*, *trachyte*, and *rhyolite*. We may therefore speak informally, and quite appropriately, of the *phonolite minimum*, *trachyte minimum*, and *rhyolite minimum*. These are the logical and quite often defensible residua from fractionation of alkali basalt, critical plane basalt, and tholeiite, respectively.

In terms of actual liquidus temperatures, Bowen's thermal "valley" is not a valley at all, but a saddle, or negatively curved surface, with a culmination at the trachyte (syenite) minimum at 1063°C. If this is a saddle, the Ab-Or join is the horse, and cantle and pommel slope towards m along the feldspar join itself. The culmination, or seat, then slopes off away from the join towards the stirrups at m_s and m_{ne}. A cross section through the liquidus surface along the general line (transverse to the horse) through or near the three minima is shown in Fig. 15.15. This section is drawn to scale, and it leaves no doubt that m is a pronounced *maximum* relative to m_{ne} and m_s. It shows, furthermore, the very steep liquidus slopes leading to the two "stirrup minima", especially from the directions of nepheline and silica. The feldspar minimum m is simultaneously a minimum in Ab-Or and a maximum relative to m_{ne} and m_s.

The general configuration of Fig. 15.15 is maintained at high pressure (to at least 10 kbar) in the presence of steam (Morse 1969a). It is therefore somewhat surprising that trachytes exist at all, for one would suppose that the slightest deviation of a bulk composition from the critical plane (or feldspar plane) would lead to phonolitic or rhyolitic residual liquids. The common existence of trachytes and syenites requires, then, some mechanism for keeping liquids in the critical condition during fractionation, as well as a non-unique source mechanism for generating critical plane liquids in the mantle. The fractionation control need only be minor if initial liquids are *strictly* of the critical plane type, for then they will have no inherent tendency to fall away from the feldspar join. But the source controls for generating such liquids repeatedly in the mantle are essentially unknown. Moreover, the syenitic rocks derived from mafic magmas are typically saturated with those mafic phases such as Fe-rich pyroxene and olivine, whereas almost all of the well-known syenites of southwest Greenland and elsewhere are largely or extremely felsic, far from saturation with mafics (Morse and Brady 2017). The "syenite problem" remains a thorn in the side of igneous petrology.

Introduction to Ternary Feldspars

We shall take a very simplistic view of ternary feldspars, which can be quite complex in the plutonic environment. Our purpose here is simply to see how feldspars behave with fractionation in basaltic systems and how well the join Ab-Or proxies for feldspars in general.

We begin unconventionally by introducing a ternary plot of feldspar analyses from one thin section of a basalt from Picture Gorge, Oregon, studied by D. H. Lindsley and Douglas Smith (1971). This is shown in Fig. 15.16. The feldspar composition ranges all the way from An_{84}, $Or_{0.5}$ to $An_{0.3}$, $Or_{59.8}$. The Or content increases moderately, while the An content moves from An_{80} to An_{40}, after which it begins to increase strongly, following a smooth curve that is convex towards Ab and nearly symmetrical about the bisector of the Ab corner angle. This extreme range of compositions can only be due to fractional crystallization, and from studies of many other basaltic rocks, it is apparent that the trend found by Lindsley and Smith is the normal one.

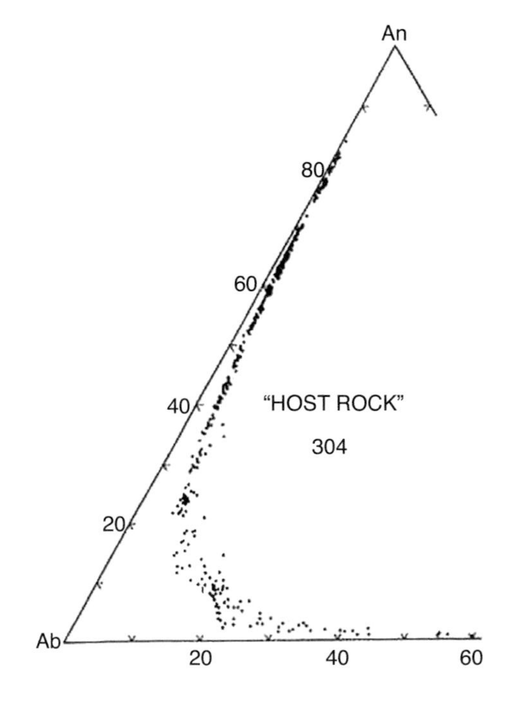

Fig. 15.16 Plot of 304 feldspar analyses from a single thin section of Picture Gorge basalt, Oregon, by Lindsley and Smith (1971). Reproduced with permission of the authors. Or lies at the missing corner of the triangle

The trend shown in Fig. 15.16 is, in effect, an ISC path of fractional crystallization plotted in the ternary feldspar diagram. We do not know what role was played by equilibrium reaction of crystals and liquid, but from the extreme range of compositions we may safely assume that it was very small. We must recall that the *liquids* from which these feldspars crystallized were *not* in the ternary feldspar plane, but basaltic to rhyolitic. Nevertheless, we may view Fig. 15.16 effectively as a phase diagram if we imagine liquids to be projected onto the feldspar plane from all other components, and imagine the feldspars themselves to lie in the plane. Suppose we assume that the bulk feldspar composition lies somewhere near An_{60}. (We have no simple way of retrieving this from the plot, but a reasonably calcic composition is required by the calcic nature of the initial ISC.) Then, the initial liquid must have contained slightly more Or than the ISC at An_{60}, for otherwise the ISC would equal the TSC at An_{60} and crystallization would be complete there. The liquid path, when plotted on Fig. 15.16, will need to lie very close to the ISC path, with tie lines reaching backwards to more An-rich ISCs. It appears as though the liquid must contain more Or than the normal coexisting feldspar, because otherwise it could never become Or-enriched. On the other hand, the difference in Or must be very small at first, because otherwise Or enrichment in the liquid would proceed very rapidly and lead, as other experience shows, to crystallization of two feldspars. The diagram shows that we are concerned essentially with a single-feldspar trend, which sweeps through the plagioclase series and into the sanidine structural series (which reaches over to the $Ab_{60}Or_{40}$ region). We can therefore postulate the *qualitative* liquid path, shown in Fig. 15.17, as a generalized example of basaltic feldspar fractionation. In this summary diagram, crystallization begins with the establishment of a tie line from liquid "BC" to the first feldspar and continues with feldspars tied by tangents to the liquid path. The feldspars fall along and define the ISC path. The end of the process is assumed to occur when the liquid and feldspar both reach a minimum on or near the Ab-Or join.

A very different result would arise from equilibrium crystallization, which is qualitatively displayed in Fig. 15.18. The initial BC feldspar tie line must be the same as in the fractional case, but all others are different, since they must pass through the bulk composition. The liquid and TSC paths are much more strongly curved than in the fractional case.

It is helpful to consider the geometry of each type of crystallization in a *T-X* diagram as projected from An onto Ab-Or, shown in Fig. 15.19. It should be stressed that both liquids and crystals are projected; none lie in the section except possibly those at *m*. The initial difference in Or content between crystal and liquid has been greatly exaggerated for clarity. This initial isothermal tie line supposedly connects a liquid bulk composition near An_{60} with a feldspar near An_{80}. As fractional crystallization proceeds, both crystal and liquid become enriched greatly in Or with a large drop in temperature. Liquid and ISC paths are in principle the same as those in Fig. 15.17. Crystallization is assumed to terminate at a minimum. The fractional TSC path begins at the solidus and reaches the bulk composition at the temperature of the minimum. Equilibrium crystallization, assumed to have the paths shown in Fig. 15.18, terminates at a much higher temperature. The TSC path reaches the bulk

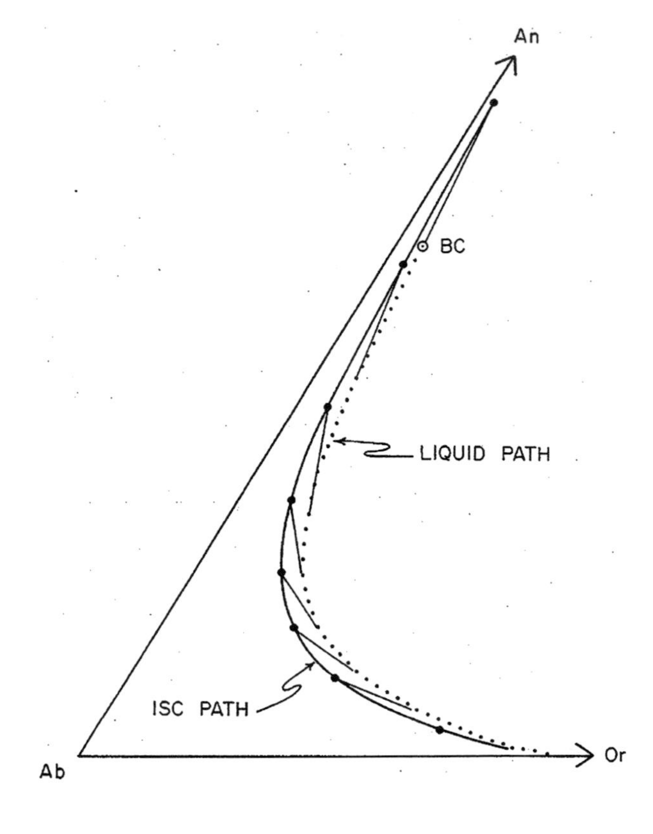

Fig. 15.17 ISC path for fractional crystallization corresponding to the feldspar trend in Fig. 15.16, with inferred liquid path projected from all other components onto the feldspar plane

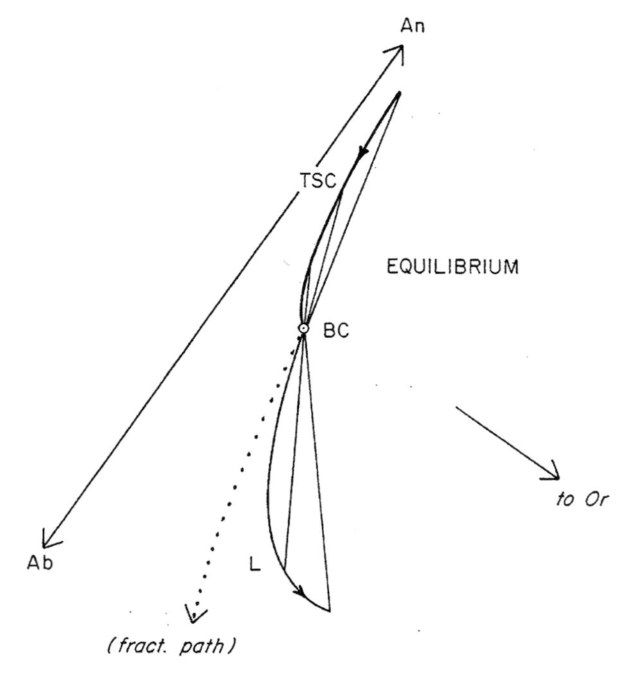

Fig. 15.18 Scheme of equilibrium crystallization for Or-poor ternary plagioclase

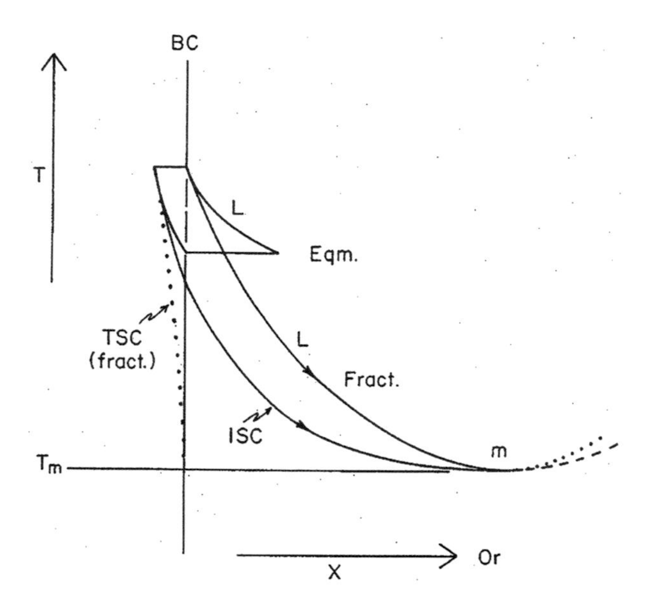

Fig. 15.19 Comparison of equilibrium and fractional crystallization paths, in a *T-X* section projected from An onto the Ab-Or join

composition when the liquid has made a considerable (but exaggerated) excursion towards Or enrichment. Figure 15.19 emphasizes the difference between paths seen in projection and those lying in a *T-X* plane. In the latter case (i.e. in a binary system), there is only one liquidus and one solidus, whatever the process. In projection, and of course in the parent ternary diagram of Fig. 15.18, the liquid paths differ depending on the process, and so do the crystal paths. Each bulk composition falls on a unique equilibrium path. Families of bulk compositions lie along unique liquidus fractionation lines like the liquid path in Fig. 15.17.

The principles governing crystallization and melting paths in systems with ternary solid solution, like the feldspar system, are described by Roeder (1974).

Concluding Remarks on Feldspar

The discussion above is intended to illustrate the familiar fact that in basaltic systems, the liquid is slightly enriched in potassium relative to feldspar. Fractionation of feldspar *and associated mafics* causes the feldspar fractionation trend typified by Fig. 15.16. We must now draw a very careful distinction between the foregoing illustrations and the pure ternary feldspar system itself. The assumed ISC paths result from the total fractionation of all phases, including, but not limited to, feldspar. In general, the feldspar contains almost all the K and the mafic phases almost none. The crystallization of these K-poor mafic phases therefore must contribute greatly to the liquid fractionation trend: in particular, to its enrichment in Or. Fractional crystallization of a pure ternary feldspar liquid would certainly yield a trend falling much closer to the Ab corner of the diagram.

There is by now an enormous literature on feldspars, including large books by Smith and Brown (1988) and Deer et al. (2001), and many papers including that of Parsons et al. (2015). The experimental range has been usefully extended by the use of graphite containers in piston–cylinder apparatus (Chap. 18) that permit studies at high temperature and dry conditions. One such study that includes feldspars is that of Morse et al. (2004) on the Kiglapait intrusion. A comprehensive treatment of Kiglapait feldspars and their fractionation paths and solvus relations is given in Morse (2017) with thermal history by Morse and Brady (2017). Of particular interest is the effect of silica activity on the fractionation of plagioclase in the Kiglapait Lower Zone (Morse 2014), in which it is shown that fractionation progress is somewhat impeded by low-silica activity in the Kiglapait Lower Zone.

Finally, it must be noted that the fractionation path observed by Lindsley and Smith (Fig. 15.16) arose from relatively rapid cooling of a thick flow and therefore quite surely represents a large degree of supercooling. This phenomenon produces crystals, which do not lie on the equilibrium solidus, but instead lie nearer to the liquid. (Some of the scatter around the bend of the trend may result from varied degrees of supercooling.) The illustrated trend need not, therefore, pertain in detail to deep-seated fractionation processes that involve much less supercooling.

Complex Crystallization of Solid Solutions

Curious things may happen during equilibrium crystallization when two solid solutions are present in a ternary system, for example leucite and feldspar in the system Ne-Ks-SiO_2. Furthermore, curious things may happen whenever a liquidus field boundary crosses a solid solution join (such as the feldspar join) in such a system. Some analysis is required to know when the reaction on such a boundary is odd or even. The system Ne-Ks-SiO_2 offers a good opportunity to comment on such complex paths. The analysis offered here derives from that of Fudali (1963) and ultimately from the remarkable pioneering study of another system by Bowen and Schairer (1935).

Odd and Even Reactions

Consider Fig. 15.20, which shows part of the crystallization of a bulk composition BC located in the leucite primary phase field. Leucite has crystallized while the liquid has moved from BC to 1, at which time feldspar appears. Now leucite dissolves as feldspar forms, and the liquid moves along the field boundary in odd reaction. Eventually, the liquid reaches *R* where the invariant reaction will set in, but we are not concerned with that. By means of the lever rule, we may calculate the fraction of leucite present in the system, first by finding F_s from the levers through BC to the TSC and then multiplying by F_{lc} (fraction of leucite in the solids) obtained

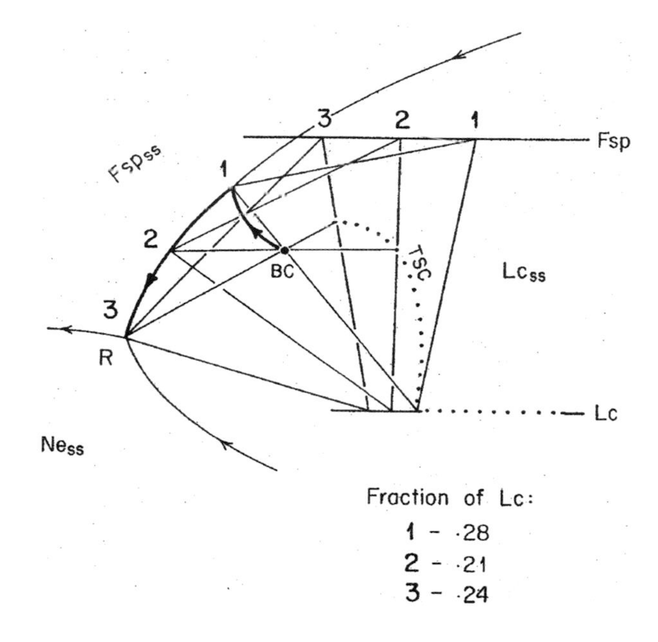

Fig. 15.20 Partial equilibrium crystallization path from a bulk composition in the leucite field, showing first the loss and then gain of leucite. (After Fudali 1963)

from the solid–solid leg of triangles 2 and 3. We find the sequence 0.28, 0.21, and 0.24. Clearly, leucite has been dissolved between 1 and 2, but added between 2 and 3. This can only mean that the reaction has changed from odd to even somewhere in the vicinity of 2.

This may at first seem an amazing result, apparently in contradiction with the positions of equilibrium feldspars and the tangent to the field boundary. But in equilibrium crystallization, the material being removed from the liquid at any instant is far in advance of the equilibrium crystal composition, and it is this material from which the liquid must always flee. We shall find that our tangent rule for odd and even reactions still holds true when we consider the material being removed from the liquid rather than the equilibrium compositions of solid solutions.

In Fig. 15.21, we see equilibrium compositions F and L tied to each other and to liquid. For the solid solution, let us suppose that fictive feldspar material of composition f_1 is being removed from liquid l to *convert* F_1 towards F_2. Similarly, let us suppose that fictive leucite material of composition l_1 is being removed from liquid l to convert L_1 towards L_2. The liquid must always move away from what we may call the *fictive extract*, a summation of f and l. This fictive extract must therefore lie on the tangent to the field boundary, and its position on f_1l_1 is defined by the intersection of this tangent with f_1l_1. In the case of liquid l, it is clear that the tangent lies on the *extension* of f_1l_1, hence the fictive extract consists of $f_1 - l_1$, and the reaction is odd: liquid equals fictive feldspar minus fictive leucite. In the case of liquid 2, the fictive compositions are so far advanced that the fictive extract is now a positive combination of feldspar and leucite, i.e. $f_2 + l_2$. Now the reaction is even. The rule is that the liquid must always move away from the fictive extract, not from the equilibrium compositions.

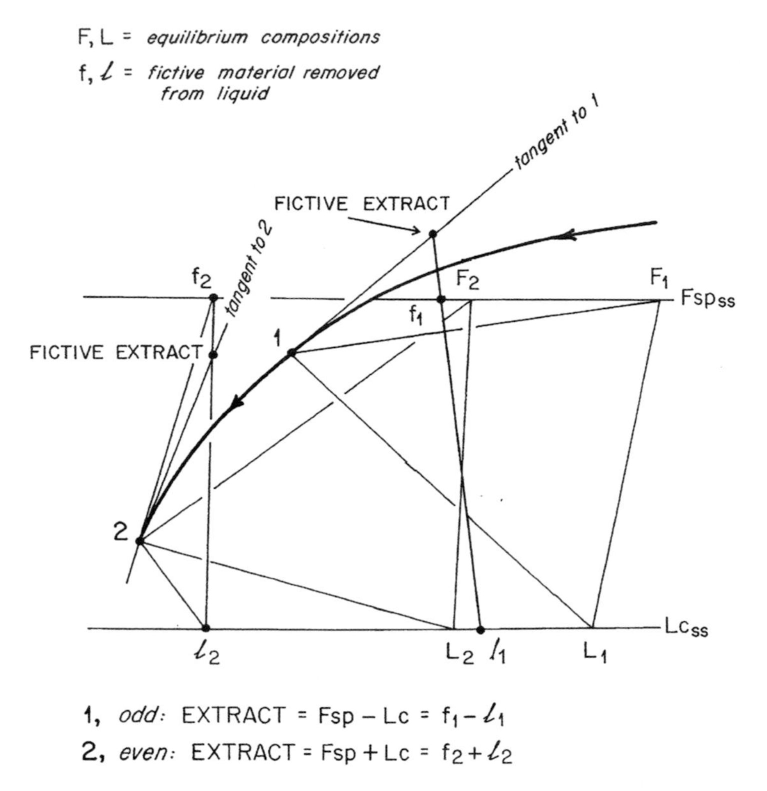

Fig. 15.21 Illustration of the fictive extract and how it determines whether equilibrium crystallization is odd (case 1) or even (case 2)

It is sometimes erroneously said or implied that the reaction changes from odd to even when the fictive feldspar composition crosses the field boundary, but this is not the case (unless the boundary is straight). The change in sign occurs only when the *fictive extract* reaches the feldspar join from a negative position. One may construct on Fig. 15.21 a fictive tie line halfway between f_1l_1 and f_2l_2 and observe that the tangent to the liquid path halfway between 1 and 2 still cuts the fictive tie line at negative leucite. The correct location of the sign change is to be found by constructing the entire path of fictive extracts by such means and finding where this path cuts the feldspar join.

It should be noted that a fictive composition such as l_2 may lie far beyond any actually attainable leucite crystal composition. Moreover, the fictive extract must be farther advanced for a large fraction of solids than for a small fraction, so it depends on the bulk composition.

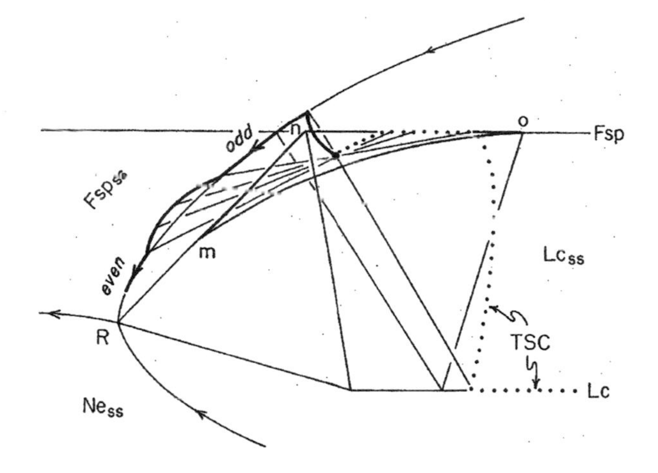

Fig. 15.22 Equilibrium crystallization from region *m o n* showing divariant excursion when the bulk composition lies on feldspar–liquid tie lines. (After Fudali 1963)

Complex Variance Changes

A restricted group of bulk compositions in the leucite primary phase field can be expressed by equilibrium tie lines from feldspar to liquid. This means that in the course of crystallization, the liquid will leave the field boundary and then return to it. This is a general property of systems in which a peritectic field boundary crosses a solid solution join. An analysis is given in Fig. 15.22. The liquid first crystallizes leucite until it reaches the field boundary. Now feldspar arrives at the expense of leucite, and the reaction is odd. Construction of the TSC path allows one to discover a point when the TSC equals feldspar alone; all the leucite has been dissolved. Now, the liquid leaves the boundary and proceeds in divariance across the feldspar field. But continual reaction with feldspar pulls it back once again to univariance, on the field boundary. Observe that at this moment, the fictive extract changes from pure feldspar on the tangent to the *divariant* curve, to a positive combination of fictive feldspar and leucite on the tangent to the *univariant* curve. Now the reaction is even, and both leucite and feldspar are produced from liquid. The TSC now approaches the bulk composition along the solid–solid legs of three-phase equilibrium triangles, until crystallization is complete.

We therefore see an equilibrium process in which leucite is created, destroyed, and created again. This explains how it is possible, in principle, to find two stable generations of the same phase separated by an interval of resorption. Such features are quite common in complex ternary feldspars.

Fudali (1963) has pointed out that the region *m o n* contains all bulk compositions that may exhibit this complex change in variance. This region is bounded by (1) the feldspar join, (2) the tie line from R to invariant feldspar, and (3) the locus of all the intersections of all feldspar–liquid edges of three-phase triangles between *o* and *n*.

Fractional Crystallization and Melting

In fractional crystallization, the fictive extract is always identical to the instantaneous *equilibrium* composition (the ISC) of the separating solids, and none of the above complexities arise. The liquid always moves away from the real ISC. Likewise in fractional melting, the TSC always moves away from the real ILC, and questions involving the geometry of fictive material do not arise. The complexities discussed in this section arise from the *equilibrium* reaction of crystals and liquid. The leucite–feldspar field boundary is peritectic along its entire length for fractional crystallization, but cotectic in places for equilibrium crystallization.

G-X Diagrams for the Albite–Orthoclase Join

The studies of Waldbaum and Thompson have led (1969) to an elegant presentation of *G-X* diagrams and calculated phase diagrams. Although these topics are a side issue to the matter of basalts, they are so fundamental to the advanced treatment of phase diagrams that we shall digress briefly to review them here. Four isothermal *G-X* diagrams illustrating the melting of sanidine-high albite are shown in Fig. 15.23. In contrast to the *G-X* diagrams shown earlier in this volume, which were drawn *a posteriori* from the phase diagrams, these more nearly represent *a priori* calculations from which the phase diagram was derived. This is not quite the case, because

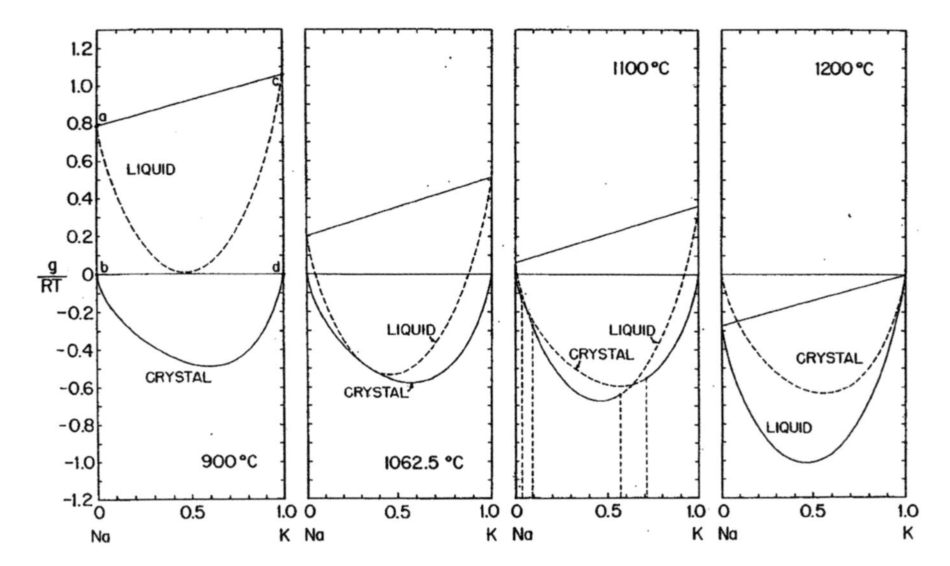

Fig. 15.23 Calculated *G-X* diagrams for the melting of sanidine at 1 atm (from Waldbaum and Thompson 1969, reproduced by permission). Ordinate scale is described in the text. Abscissa is in mole units

Schairer's (1950) determination of the minimum composition and temperature was used as input, along with the (actually current) melting point of albite and the metastable congruent melting point of sanidine. For the rest, the calculation makes use of the *mixing properties* of the crystalline solution series, which is to say the non-ideal ways in which the entropy, internal energy, molar volume, and Gibbs energy are additive over the series from pure Ab to pure Or. These properties were calculated from a variety of thermodynamic data by Thompson and Waldbaum. Then, the mixing properties of the melts themselves were calculated from Schairer's minimum determination alone, and finally, the assumption was made, for lack of any relevant data, that the so-called excess entropy of mixing along the liquid solution series is zero. This is an assumption of ideal mixing, very likely not true in detail but not seriously in error for the purpose at hand. The procedure of calculation is beyond the scope of this book, and in fact it required what was then a large amount of computer time because it involves an iterative solution to some complicated equations. The procedure is thoroughly described in the set of papers cited in Waldbaum and Thompson (1969).

Samples of the results are shown in Fig. 15.23. The figure is a particularly convenient representation of *G-X* relations in that the motion (with temperature) of the *G-X* curve for the solid solution is held fixed, while the curve for the liquid moves past it. This is done by the clever artifice of reducing the Gibbs energy to a difference, such that the quantity on the ordinate for either pure Ab crystal or pure Or crystal or any mechanical mixture of them is zero. The quantity plotted (divided by the product of the gas constant and the Kelvin temperature) is given by the relation:

$$g \equiv \overline{G}(\text{crystal or liquid}) - N_1\mu_1^{\text{o}},\ \text{crystal} - N_2\mu_2^{\text{o}},\ \text{crystal}, \qquad (15.3)$$

where

$\overline{G}$ = molar Gibbs energy
N = mole fraction
μ^{o} = the chemical potential $\partial G/\partial n$ at standard state (n = No. of moles)
1, 2 = subscripts denoting components Ab and Or, respectively.

For pure substance 1, $N_1 = 1.0$, $N_2 = 0$, and $\mu_1^{\text{o}} = \overline{G}$. Similarly for pure substance 2. This is why the value of g is zero for the mechanical mixture of pure solids. The quantity $\overline{G}$, however, includes the mixing effect for crystalline and liquid solutions and is always less than a linear combination of end points. Without the reduction, the Gibbs energy of Ab would plot higher than that of Or, and without the further reduction of dividing by RT, these points and the crystal *G-X* curve would move down along the ordinate with temperature.

Figure 15.23 may be compared with the *T-X* diagram, Fig. 15.26 (or Figs. 15.3 and 15.4). In the figure, the 1100°C diagram illustrates the determination of solidus and liquidus compositions by double tangents to the curves. Numerical solutions for these double tangents at many closely spaced temperatures are given in the original reference, and it is these values from which the *T-X* diagrams have been constructed.

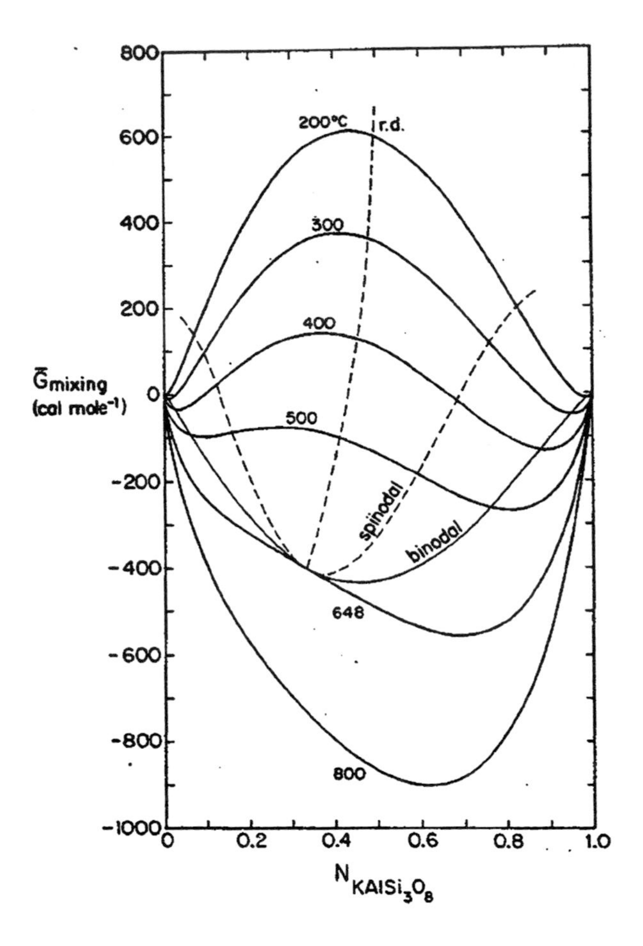

Fig. 15.24 Calculated *G-X* diagram for the sanidine solvus at 1 atm. The curves are isothermal sections through the Gibbs energy of mixing surface. (From Waldbaum and Thompson 1969, by permission)

Note that in Figs. 15.23, 15.24, 15.25, and 15.26 the abscissa is in mole fraction rather than weight units, since mole units must be used in, and emerge from, all calculations.

Figure 15.24 treats the alkali feldspar solvus, which is the curve bounding the miscibility gap in crystalline solutions at temperatures below the 1-atm solidus. The *T-X* diagram of this solvus is shown in Fig. 15.26. Both diagrams illustrate features of solvi that we have not yet discussed, so we shall do so briefly here.

Figure 15.24 is a composite diagram showing isobaric (1 atm) slices through the *G-X* surface at six different temperatures. The ordinate is the *molar excess Gibbs energy of mixing*, in calories per mole. The reference state is a mechanical mixture of the pure end member phases at 1 atm and the temperature of interest, and hence, it is zero. The excess (positive or negative) molar Gibbs energy of mixing is a summation of the mixing properties of the molar internal energy, the molar volume, and the molar entropy. These quantities all change systematically with composition: the internal energy and volume deviate positively from a mechanical mixture of end members, while the entropy shows a negative deviation. The fundamental thermodynamic relationship is as follows:

$$\begin{aligned} G &= H - TS, \\ H &= E + PV, \\ G &= E + PV - TS, \end{aligned} \tag{15.4}$$

where G is the Gibbs energy, E is the internal energy, P is pressure, V is volume, T is Kelvin temperature, and S is entropy. These symbols represent the absolute quantities for a chemical system. By writing the mass-dependent (extensive) quantities G, E, V, and S with a bar over them, we denote the reduced or molar quantities, i.e. $\overline{G} = G/n$, and the reduced quantities are inter-related mathematically in the same way as the absolute quantities. Figure 15.25 shows the additivity of the reduced mixing quantities in the alkali feldspar system (sanidine structure) at 650°C and 10 kbar, relative to a reference state of a mechanical mixture of pure end members. Note here that while $\overline{EPV}$ and $T\overline{S}$ appear almost perfectly symmetrical with composition, their summation in the heavy $\overline{G}$ curve is highly asymmetrical, and the alkali feldspars belong to the class of *asymmetrical solutions*. This is further shown by the trace of the rectilinear diameter (r.d.) in Figs. 15.24 and 15.26. The rectilinear diameter is the mean composition of the coexisting phases on the binodal solvus, which is now to be discussed.

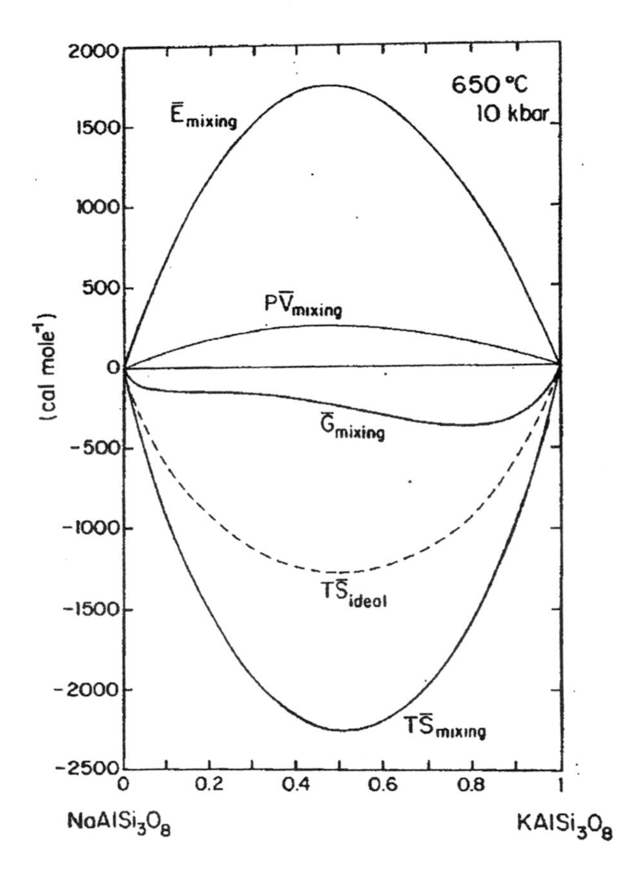

Fig. 15.25 Mixing properties of sanidine solutions at 650°C and 10 kbar. (From Thompson and Waldbaum 1969, by permission)

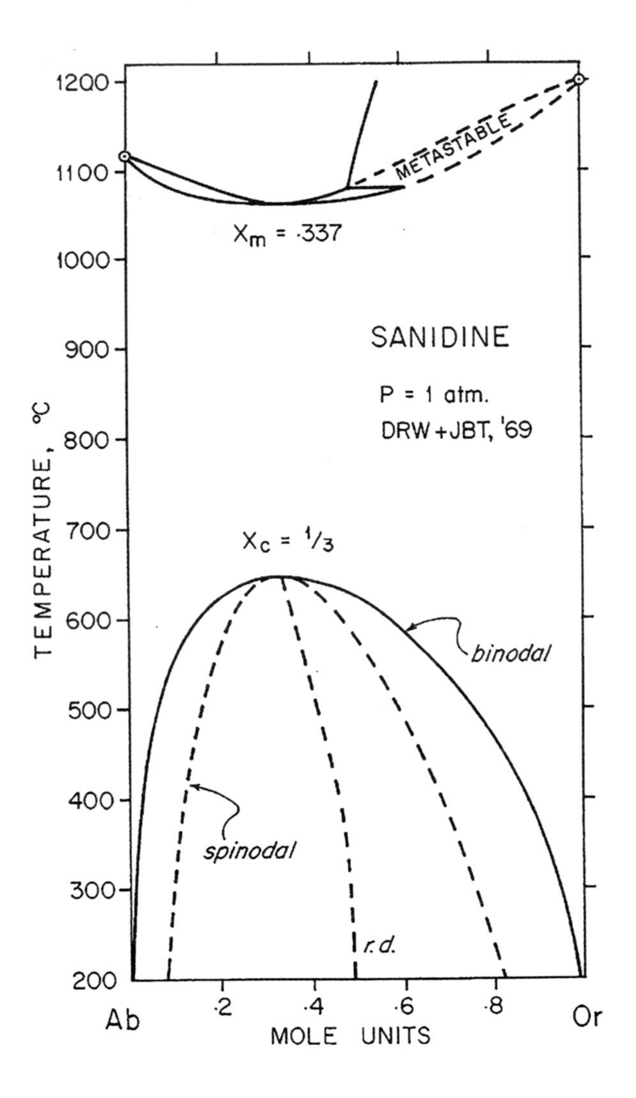

Fig. 15.26 One-atmosphere phase diagram for Ab-Or, sanidine solutions (after Waldbaum and Thompson 1969). X_{m} is the melting minimum and X_c is the critical composition of the solvus. Binodal and spinodal are explained in the text, as is r.d., the rectilinear diameter

Two of the *G-X* curves in Fig. 15.24 are characterized by a single curvature; these are for solutions at temperatures above (800°C) or equal to (648°C) the critical temperature of the solvus. The other four curves show both upward and downward curvature, characteristic of *G-X* curves for temperatures below the critical temperature of a solvus. The *stable* equilibrium compositions of the solvus limbs are given, for any selected temperature, by the common tangent points to the two *nodes* of the *G-X* curve. The solvus curve in *T-X* space, which describes the locus of such double tangent points, is called the *binodal* curve or binodal solvus. This curve defines the stable equilibrium composition of the coexisting feldspars. A mechanical mixture of two feldspars external to the binodal will lie at higher *G* than the common tangent and will therefore be metastable relative to the binodal pair; such a mechanical mixture will tend to come to the equilibrium compositions by diffusive exchange of alkalies. The crystals will do this along portions of the *G-X* curve having the second derivative >0 (concave up), and this means that the rate of approach to minimum *G* will become less and less as the crystals approach the binodal compositions. In fact, kinetics being involved, such crystal pairs may closely approach but not attain the binodal compositions, and this causes problems for the experimentalist.

The same upward curvature extends inward past the binodal points; here also the feldspars are metastable with respect to diffusion. But along the *G-X* curve, there are two *inflection points* where the second derivative $\partial^2 G/\partial N^2$ changes from positive to negative. These are the so-called *spinodal points*, and the curve describing their locus with temperature is called the spinodal curve or spinodal solvus. A homogeneous phase lying within the region bounded by this curve is inherently unstable with respect to internal diffusion. One way of appreciating this is to note that the isothermal *G-X* slope increases towards the spinodal points in this region, and hence, the change in *G* increases with composition. The spinodal curve may thus be regarded as the *limit of metastability* on going away from the binodal compositions. At the spinodal, metastability gives way to true instability as long as diffusion processes can act fast enough. A feldspar in this region of instability will tend to unmix (exsolve) spontaneously. Only if internal diffusion is too slow (as in quenching) will this fail to occur.

Figure 15.24 shows the origin of the spinodal curve as the locus of inflections in the *G-X* curves at various temperatures. Figure 15.26 shows the *T-X* trace of the spinodal. Note that it unites with the binodal at the critical point.

There is yet another type of solvus, called the *coherent* solvus, which takes account of the strain energy developed at the physical interface between two intergrown phases (phases that cohere). The coherent solvus does not unite with the other two at the critical point, but lies everywhere inside the spinodal. A valuable discussion of this phenomenon is given by Yund (1975). There is some question as to how far the concept of coherency applies to intergrowths of feldspars, pyroxenes, and amphiboles, because the exsolution directions in these minerals tend to be non-rational crystal planes (optimal phase boundaries) along which the lattice strain is minimized, possibly even to zero or a value very close to it, at the temperature of formation of the intergrowth (Bollmann and Nissen 1968; Robinson et al. 1971). This feature may help to account for the fact that some natural feldspars can maintain such fine-scale periodic structures as mesoperthite even where other crystals in the same sample have exsolved to their limiting low-temperature compositions, as in the uppermost syenites of the Kiglapait intrusion (Morse 2017).

Several properties of the alkali feldspar solvus shown in Fig. 15.26 deserve further comment. The figure pertains only to sanidine structures. For other feldspars with more ordered distributions of tetrahedral Al and Si, the solvus rises, perhaps as much as 230°C for microcline-low albite (Bachinski and Muller 1971). The solvus also rises with pressure, at a rate of at least 14°C/kbar. Both these factors must be considered when thinking about melting of crustal rocks.

Chapter 16
Iron and Oxygen

Introduction

In the systems heretofore discussed, we have dealt mainly with the six components SiO_2, Al_2O_3, MgO, CaO, Na_2O, and K_2O, among which may be found the principal phases or mineral components of the basalt tetrahedron and of basalts themselves: nepheline, albite, orthoclase, quartz, anorthite, plagioclase in general, forsterite, diopside, enstatite, and Fe-free pigeonite. Missing from this list are the Fe-Ti oxides, which require consideration of FeO, Fe_2O_3, and TiO_2 as components. FeO is required also for discussion of the ferrous silicate components of olivine and pyroxene. Phosphorus is also a constituent of importance in basalts, reporting either in groundmass glass or in apatite, which crystallizes during late fractionation stages when the concentration of phosphorus becomes high enough to saturate the melt in apatite component. Present generally in minor quantities, phosphorus can be ignored for present purposes, always bearing in mind that its abundance level in basaltic magmas may contain important information, which we may or may not be able to analyse correctly.

The above list of neglected components—oxides of Fe, Ti, and P—all have one very important property in common. They all tend to be conserved in residual liquids upon fractionation of basaltic magma, or conversely, to be incorporated into relatively low temperature melts on partial fusion of basalt. P commonly has the status of a minor element except after several-fold concentration by a fractionation or fusion process. Fe is a major element of basaltic rocks, commonly reaching levels of 10% (expressed as FeO), and Ti commonly lies between 1% and 2%, expressed as TiO_2. We shall ignore Ti for the present. It is both necessary and convenient to discuss Fe in some detail.

S. A. Morse, *Basalts and Phase Diagrams*,
https://doi.org/10.1007/978-3-030-97882-2_16

Iron in Silicates

In common basaltic rocks, most of the iron in silicate minerals is ferrous iron, Fe^{2+}. We shall treat these minerals throughout as ferrous silicates, ignoring small amounts of Fe^{3+} which may be present but which do not, in large degree, affect the fractionation or genesis of basaltic magmas. The minor iron in leucocratic minerals such as plagioclase feldspar will also be ignored. Most of the ferrous iron in basaltic silicates can be accounted for in olivine and pyroxene, and for all practical purposes, we may assert that ferrous iron occurs in these mafic minerals as a proxy for Mg, in three major solid solution series. The first of these is easily and satisfactorily modelled as the olivine series forsterite (Fo)–fayalite (Fa) and Mg_2SiO_4-Fe_2SiO_4. The pyroxenes are less clear-cut, but are composed of a Ca-rich (clinopyroxene) series and two Ca-poor (orthopyroxene, pigeonite) series in the pyroxene quadrilateral, diopside (Di)–hedenbergite (Hd)–enstatite (En)–ferrosilite (Fs), in which we emphasize the Si_2O_6 suffix:

$$\begin{aligned} \mathrm{Di} &= \mathrm{CaMgSi_2O_6} \quad \mathrm{En} = \mathrm{MgMgSi_2O_6} \\ \mathrm{Hd} &= \mathrm{CaFeSi_2O_6} \quad \mathrm{Fs} = \mathrm{FeFeSi_2O_6} \end{aligned} \tag{16.1}$$

The Ca-rich clinopyroxene series commonly contains up to 1/5 En-Fs in solid solution (hence lying below the Di-Hd line in the pyroxene quadrilateral) and is known as the augite–ferroaugite–ferrohedenbergite series. The Ca-poor series commonly carries less than 1/10 Di-Hd in solution except when occurring as pigeonite. All three series, unsurprisingly, share the property that Fe lowers the melting temperatures.

Fayalite

Fayalite (Fa) is a well-known product of blast furnace slags, and a common method of laboratory synthesis involves an analogous reaction of hematite plus quartz in a reducing atmosphere. This can be done by sintering pure Fe_2O_3 and SiO_2 in appropriate proportions to yield the compound $2FeO \cdot SiO_2$. However, a simpler recipe is to heat in a sealed silica glass tube an equimolar mix of FeO and hematite with half again as many moles of silica. This produces the reaction $2FeO + 2Fe_2O_3 + 3SiO_2 = 3Fe_2SiO_4$ (recipe courtesy D. H. Lindsley 2016).

All experimental work involving a compound of FeO presents difficulties, for it is difficult or impractical to control the partial pressure of oxygen to just that value (or small range of values) which maintains Fe^{2+} without excess Fe or Fe^{3+}. Experimental techniques with ferrous minerals have commonly involved a compromise between what may be theoretically desirable and what may be expeditiously achieved. From the early work of Bowen and Schairer (1932), it has been a common practice to maintain most of the iron in the ferrous state during melting experiments

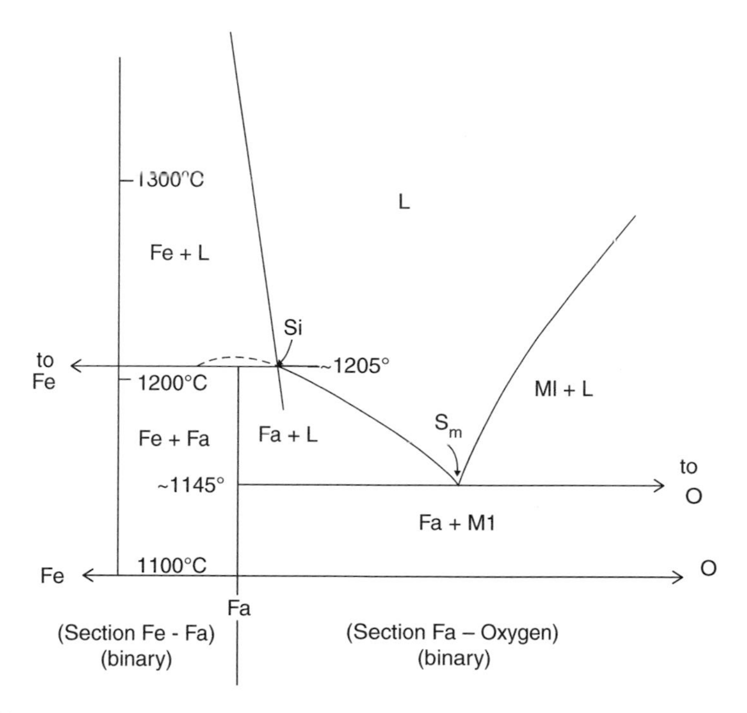

Fig. 16.1 Two binary *T-X* sections joined at fayalite (Fa) to illustrate its melting relations. The sections are Fe-Fa and Fa-oxygen, and are taken from Fig. 16.4

by establishing equilibrium with an iron crucible in a nitrogen or other gas atmosphere, sometimes with iron wool within the charge as well. Experiments performed in nitrogen under these conditions will, amazingly, contain *all three* oxidation states of iron, but FeO predominates. Bowen and Schairer reported the melting point of fayalite as 1205°C, but this actually represents the incongruent melting of Fa to iron plus liquid in the system Fe-Si-O. Figure 16.1 illustrates the nature of this reaction in a strategically chosen *T-X* section through the ternary subsystem Fe-SiO_2-O. There are actually two sections, Fe-Fa and Fa-O, joined together at Fa. Each of these sections is binary by restriction, being compositionally degenerate so that Fe-Fa, extended slightly beyond the Fa composition, would rigorously express the reaction Fe + L = Fa, and Fa-O expresses the binary reactions Fa + O = L in the presence of iron and Fa + O = L in the presence of magnetite. The derivation of these sections will become apparent on a later page, where the full ternary system is discussed. Figure 16.1 is used here to show the general similarity of the melting behaviour of fayalite to that of enstatite in the system MgO-SiO_2. Note, however, that the incongruency involves separation of a metal phase and an oxygen-rich liquid, rather than a silica-poor solid and a silica-rich liquid.

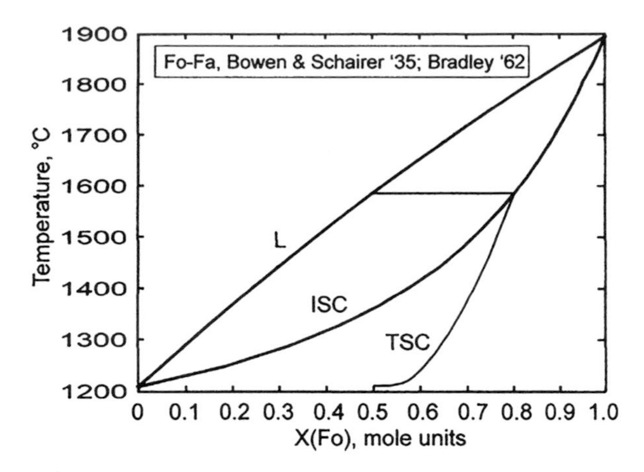

Fig. 16.2 Melting of olivine as determined by Bowen and Schairer (1935) in the presence of iron, and as computed by Bradley (1962)

The Fo-Fa Series

The system Fo-Fa as determined by Bowen and Schairer (1932) and calculated by Bradley (1962) is shown in Fig. 16.2, taken from Fig. 1 of Morse (1997). This is not strictly a binary system because in the 1932 determination the olivines and liquids are in equilibrium with metallic iron, and the liquids are enriched in oxygen relative to the stoichiometric olivine composition. This incongruency disappears, of course, at the melting point of pure Fo, and is probably of minor importance except possibly at Fa-rich compositions. It is not clear that the Fe saturation is carried over into the Bradley calculation, and in any event, there are other matters that complicate the olivine loop, in particular, the effects of the liquid structure.

The main geological impact of the system is that the olivines constitute a binary loop in which Fa is enriched in liquids relative to solids, and in which temperatures are drastically lowered as the Fo content falls. The diagram accounts for the enrichment in iron relative to magnesium commonly observed in olivines of basalts and their daughters, including the appearance of fayalite in rhyolites and trachytes and their plutonic counterparts, granite and syenite. Pyroxenes undergo similar iron enrichment with falling temperature, and the olivine system furnishes an adequate preliminary example for the generalization that the main effect of FeO in magmas is to reduce temperatures of crystallization and promote enrichment of FeO at the expense of MgO in residual liquids, in parallel with enrichment of Ab as a result of plagioclase crystallization.[1] The story is not quite that simple, however, and we

[1] It has been noted from time to time that olivine actually contains more FeO than the basaltic liquids from which it crystallizes. The phase diagram tells us that olivine crystals are enriched in the Mg component relative to the Fe component, and we find that this is still true in basaltic systems. Basaltic systems also fractionate Mg–olivine, enriching the liquid in components of Fe–olivine, as the phase diagram predicts. Of course, basalt consists of much more than olivine, and it should occasion no surprise that a metal-rich silicate such as olivine should contain more FeO than the liquid, which also carries potential plagioclase and other components. Magnesian olivine also

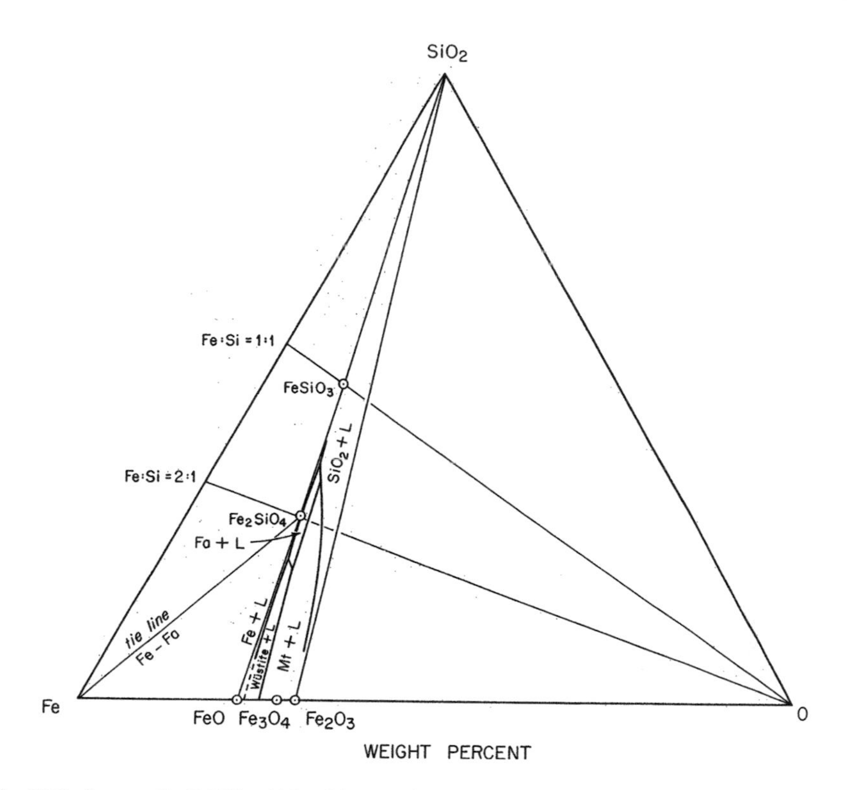

Fig. 16.3 System Fe-O-SiO_2. (After Muan and Osborn 1956)

should soon turn to a more detailed analysis of olivine melting as an introduction to the subject of oxidation and reduction in magmas.

Olivine Melting, Revisited

Both Figs. 16.1 and 16.2 contain elements as yet unexplained, and we now undertake some explanations. Figure 16.3 depicts the system Fe-SiO_2-O, showing the location of several phases important to basaltic rocks. Attention is called to the join FeO-SiO_2, which contains fayalite, Fe_2SiO_4, and the end member orthopyroxene ferrosilite (Fs), $FeSiO_3$. Ferrosilite is not stable at low pressures, a fact which makes the iron-rich orthopyroxenes of great importance as "natural barometers"—

---

contains more MgO than the liquid. It is also true that plagioclase contains more Na_2O than basaltic liquid. The *relative* crystal/liquid partitions illustrated by phase diagrams of simple systems should never be misconstrued to imply anything about *total abundances* of a single end member or component in a complex magma.

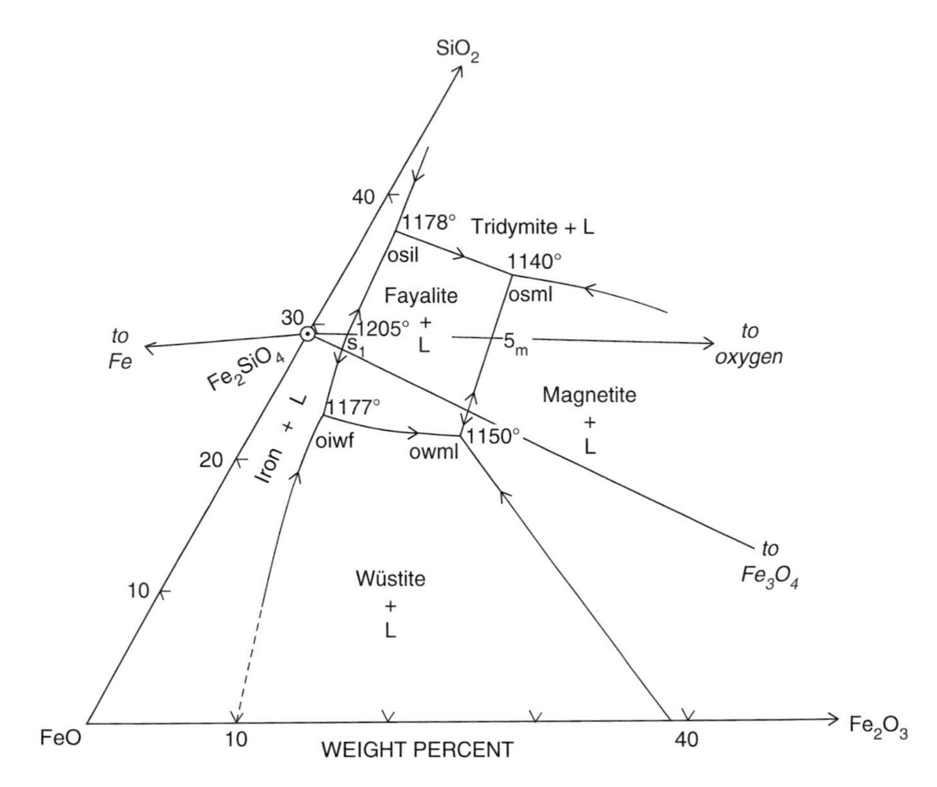

Fig. 16.4 System $FeO\text{-}Fe_2O_3\text{-}SiO_2$. (After Muan and Osborn 1956)

indicators of pressure during crystallization. Owing to this fact, we shall reserve discussion of ferrosilite to a later chapter where the effects of high pressure are considered. At 1 bar, the ferrosilite composition is represented by fayalite plus quartz or tridymite. Also shown in Fig. 16.3 are the compositions of magnetite (Mt, Fe_3O_4, or better, $FeO \cdot Fe_2O_3$) and hematite (Fe_2O_3), on the join Fe-O. Not shown (for lack of room and present geologic interest) is the mineral wüstite (German *Wüste* = desert), $Fe_{1\text{-}x}O$, a variable solid solution occupying part of the range between FeO and Fe_3O_4.

Two tie lines of importance are shown in Fig. 16.3: that from Fe to Fa and that from O to Fa. These are actually the section lines of Fig. 16.1, but we shall see them to better advantage in the next figure. In Fig. 16.3 are sketched the liquidus boundaries, highly compressed and difficult to appreciate, of SiO_2, magnetite, wüstite, and iron, which collectively enclose the field of Fa + *L*. These boundaries are more readily seen in the system $FeO\text{-}Fe_2O_3\text{-}SiO_2$, part of which is shown in Fig. 16.4. This figure is the lower left quarter of the full system $FeO\text{-}Fe_2O_3\text{-}SiO_2$, and in it are depicted the same two tie lines Fe-Fa and Fa-O, now with much flatter slopes, as are shown in Fig. 16.3. Of special note is the squarish field of Fa + *L*, bounded by fields of Tr, Mt, Wü, and Fe. Ignore for now the lower case abbreviations *osil*, etc. The boundary *L* (Fa, Fe) contains the incongruent melting liquid at 1205°C, which results when fayalite melts to iron plus liquid. This liquid composition must lie on the extended tie line from Fe to Fa (not the tick labelled S_1), and it lies at a maximum on the curve L (Fa, Fe), which runs to 1178°C at a eutectic with SiO_2 and 1177°C at a eutectic with wüstite. The field of iron + L covers the fayalite composition. The position of the 1205°C liquid is a ternary expression of the familiar incongruent melting relationship. Figure 16.1 is a section along the Fe-Fa line, abutted to a section along Fa-O. The 1205°C maximum does not quite lie in the plane of the Fa-O section in Fig. 16.1 (hence the approximation ~1205°C in the figure); the points S_l and S_m do lie in that section, and they represent the melting reaction Fa + O = L in the presence of iron and magnetite, respectively. These are so-called *singular points* arising from the collinearity of Fa, L, and O, which causes the *degeneracy* of the ternary system into the binary system Fa-O. These terms and points will be discussed at greater length in a subsequent analysis.

Having now seen the ternary geometry of the incongruent melting of fayalite, we may add MgO to the system to represent the melting of all olivines in the Fo-Fa series. The quaternary system MgO-Fe-O-SiO_2 is sketched in Fig. 16.5a, the base of which is the ternary system just discussed (Figs. 16.3 and 16.4). Fo occurs on the MgO-SiO_2 edge of the tetrahedron, and the triangular plane Fe-Fo-(SO = 65:35) contains the series Fo-Fa. This plane is extracted in Fig. 16.5b as an equilateral triangle. Since the olivine series lies in the plane, and the crystals and liquids all coexist with metallic iron (the crucible), which lies at one apex, it is evident that the plane may be treated as a degenerate ternary system for the purposes of discussing the melting of olivine; that is, iron, olivine, and liquid all lie in the plane. The liquid is shown an exaggerated distance away from Ol and Fe (for clarity), and a few sample tie lines from liquid to crystal are shown across the liquid-crystal field. One sample three-phase triangle Fe-Ol-L is also shown. This gusset-shaped field of Fa-Fo-liquid is shown (distorted) in Fig. 16.2b, where the X-X relationships of Fig. 16.5b are shown in relation to the T-X diagram, Fig. 16.2a. One sample tie line is shown in both the T-X and the X-X sections. Together, the diagrams of Fig. 16.2 illustrate that the apparently binary loop is really a *projection* along the ternary plane of Fig. 16.5b onto a T-X section arising from the Fo-Fa join. The *liquids* are projected into the T-X section, whereas the crystals already lie in the section. The projection is made from (or to, depending on the arrangements of lamp, viewer, and screen) the corner Fe. It should be noted that the actual shape of the "liquidus" in the X-X plot is not known to me and is arbitrarily represented as a straight line. Furthermore, it is not a liquidus in the strict sense, but a peritectic curve denoting an odd equilibrium with two solid phases. Here endeth Part I of the olivine story. Now we go on to consider fayalite and its relations to magnetite, silica, and oxygen, among other things.

Fayalite and Its Oxidation

A hundred years ago, one would hardly have dreamed that one might discuss, let alone measure, the partial pressure of oxygen in a crystallizing (or long crystallized) magma. Even now one might ask, why bother? One answer has already been

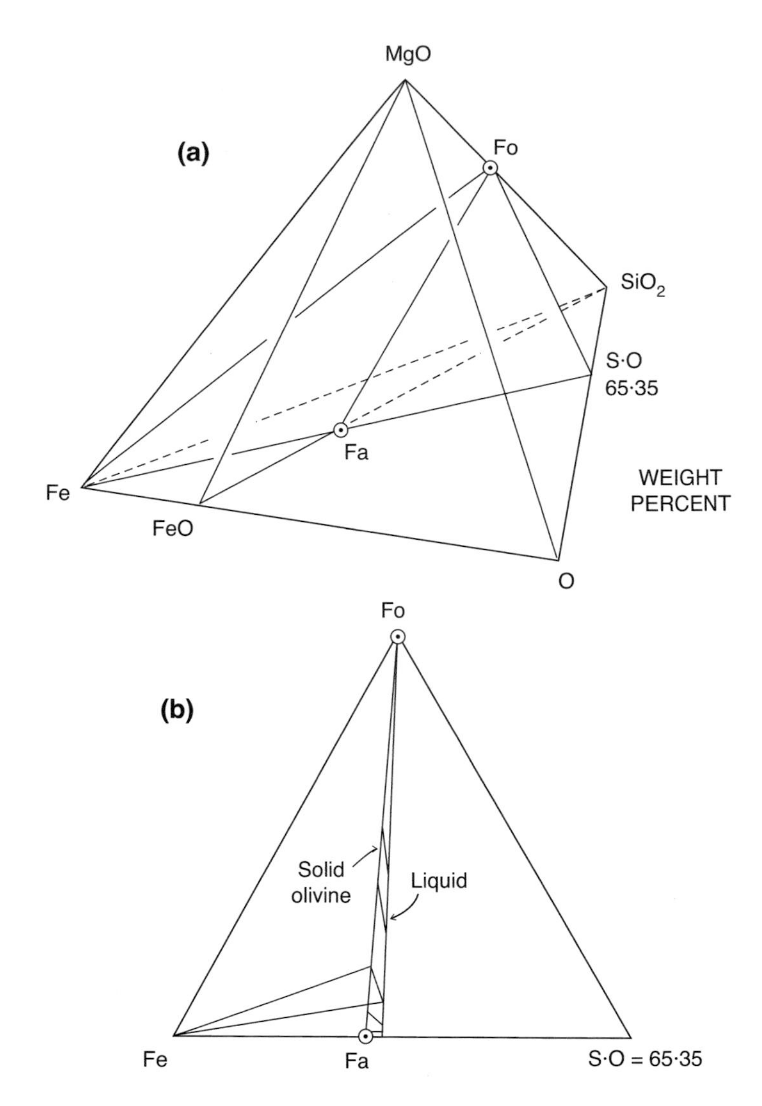

Fig. 16.5 (**a**) System MgO-Fe-O-SiO_2. (**b**) Fe-Fo-(65:35) plane taken out of the above tetrahedron

hinted on an earlier page, but it bears a brief amplification. Perhaps a historic context will help.

Bowen and Fenner long argued about whether the normal course of fractional crystallization led to iron-rich or alkali, silica-rich end products. Bowen (e.g. 1928, p. 110) favoured the alkali, silica enrichment typical of the calc-alkali series: basalt–andesite–dacite–rhyolite, claiming no enrichment of iron. Fenner (e.g. 1929, p. 238) held that iron enrichment was the normal case. That these two able men were unconsciously comparing separate sets of conditions began to be evident with the work of G. C. Kennedy (1955), who showed that a higher state of oxidation favoured the Bowen trend and more reducing conditions favoured the Fenner trend. [The report in 1939 of strong iron enrichment in the Skaergaard intrusion (Wager and Deer 1939) must surely have suggested such a dichotomy of process almost at once.] Osborn (1959), in a landmark study which for the first time put the matter of oxidation in magmas on a rigorous experimental basis, amply confirmed the Kennedy conclusions and discussed in detail the various mechanisms by which oxygen might influence the course of magmatic fractionation. It is now clear that an appreciation of oxidation–reduction reactions is essential for an enlightened approach to the problems of magmatic fractionation and perhaps even genesis. During the latter half of the last century, much progress was made by Osborn and his colleagues at the Pennsylvania State University, using controlled gas atmospheres at 1 bar, by H. P. Eugster and his colleagues at the Johns Hopkins University, and especially by the school of Donald F. Lindsley at Stony-Brook. Eugster invented the oxygen buffer which has been essential to studies at any pressure.[2]

The resolution of the Bowen–Fenner uproar can be appreciated very quickly with the aid of a single and very important reaction:

$$\begin{array}{llcllcl} 3Fe_2SiO_4 + O_2 & & \rightleftharpoons & 2FeO \cdot Fe_2O_3 & + & 3SiO_2 & \\ \text{Fa} & \text{gas} & & \text{Mt} & & Q & . \\ \text{F} & & \rightleftharpoons & \text{M} & + & Q & \end{array} \tag{16.2}$$

This is the famous FMQ (QFM)[3] oxygen buffer reaction; it is also a genuine magmatic reaction by means of which a high degree of silication may result in the residual liquid (right-hand arrow) or a high degree of iron (fayalite) enrichment may result in the residual liquid (left-hand arrow). In effect, the right-hand arrow leads to the Bowen trend and the left-hand arrow to the Fenner trend. The reactants may be considered normative components of the magma; the products are crystals produced by the magma. The reaction does not necessarily go to completion. Mechanistically,

[2] The oxygen buffer consists of a fixed mineral assemblage surrounding the experimental charge and ordinarily requires a semi-permeable membrane for diffusion of hydrogen between the buffer assemblage and the charge: the hydrogen acts as the messenger of oxygen fugacity from the buffer to the charge. The invention is a fine example of necessity being the mother. Eugster attempted to make the iron mica annite [$KFe_3AlSi_3O_{10}(OH)_2$] in a sealed platinum tube, using ferrous oxalate as the source of Fe^{2+}. But he kept getting magnetite + sanidine instead, so he realized that hydrogen was escaping through the platinum and leaving excess oxygen behind which oxidized the iron. He first cured this by surrounding the platinum capsule with Zn + HCl, the classical Kipp generator, to create a back pressure of hydrogen, and it worked. This led to Fe + HCl, then Fe + H_2O (the iron–wüstite buffer), and finally a large variety of solid assemblages, with H_2O as the source of messenger hydrogen. The oxygen buffer is not only a beautiful experimental discovery, but also a powerful concept when applied to geological systems. I am grateful to the late Hans-Peter Eugster for providing a brief account of the discovery. Early descriptions of the technique may be found in Eugster (1959) and Eugster and Wones (1962).

[3] The code "FMQ" realizes the actual expression "Fayalite [+ Oxygen] = Magnetite + Quartz". The less informative code "QFM" implies the strange expression "Quartz = Fayalite – Magnetite [+ Oxygen]". It is helpful to use codes that make the most sense to the widest audience.

what happens is that if the initial partial pressure of oxygen is sufficiently high, magnetite will crystallize early in the fractionation history, thus locking up iron as oxide and silicating the residual liquid. If the initial partial pressure of oxygen is low, magnetite may not crystallize at all, so that the iron will continue to crystallize as silicate, using maximum silica from the magma. Conditions intermediate to these two extremes may well be imagined.

It must be emphasized that it is the early-, late-, or never-crystallization of magnetite which has the effect on silica saturation in the residue, assuming a closed system. Of course, it is also possible to imagine a contamination process in which respectably large quantities of oxygen are introduced to the magma (say as H_2O vapour), causing massive precipitation of magnetite. A variety of such a mechanism was in fact invoked by Osborn (1959) as a model for the calc-alkali series.

Oxygen in Melts

Before considering the buffering qualities of the FMQ reaction, we should perhaps consider the source and nature of oxygen in the natural magmatic environment. First, it should be recalled that oxygen composes about 90% by volume of the Earth's crust and most magmas. It must not be assumed that a gas phase, particularly one composed of oxygen, occurs with every magma. Gases that do occur with lavas are composed overwhelmingly of H_2O and CO_2, along with many other species in major or minor amount. Dissociation of H_2O to $H_2 + 0.5\ O_2$ occurs to some degree at all magmatic conditions, and selective leakage of the highly mobile H^+ ion is a favourite means of increasing the partial pressure of oxygen in a magma. An influx of H_2 leading to reduction is less likely, but conceivable.

Many magmas do not possess a gas phase at all, particularly at high confining pressures which tend to force gaseous components into solution with the melt, to achieve reduction in volume. However, even with H and O in solution in a silicate melt, diffusion of H as momentary H^+ is likely to occur, and an effective increase in the oxidation potential of the magma can thereby be achieved. Dissociation of CO_2 into $CO + 0.5\ O_2$ is another possible mechanism of influencing oxidation. Once again, whether these species are present in a gas phase or merely as dissolved components is relatively immaterial. The initial ferrous: ferric ratio of a magma is an indication of its potential for oxidation and reduction.

Even dry magmas contain some ferric iron, except perhaps on the moon, where equilibrium with metallic iron or ilmenite is the rule. Lunar magmas may then be characterized as having all their oxygen bonded to silicate and metal oxide structures (polymers in the liquid, crystals in the solid state). Lunar metal equilibria are therefore dominated by the species Fe + FeO, or in other words, the iron–wüstite assemblage. Terrestrial basaltic magmas have enough oxygen to satisfy the silicate polymers and crystals and still have some left over for metal oxides. Therefore, Fe metal is absent (except in rare, perhaps freakish occurrences like Disko Island,

Greenland), all iron is at least as oxidized as FeO, and some (or perhaps much) reports as Fe_2O_3.

Melts and rocks may be considered as dominantly oxygen networks, and the amount and distribution of cations in polymers, crystals, and dissolved gases controls how much oxygen may be available for affecting the valence state of transition metals. To speak of oxidation and reduction in magmas is thus to imply something far more subtle and intricate than mere infiltration with oxidizing or reducing gases.

One might suppose (indeed one did, some years ago) that the potential interactions among cations and oxygen would be so numerous and interlocking as to defy conceptual analysis. Fortunately, while this may be true some of the time, there are important special cases where some kind of oxidation potential can be rigorously specified, and there are enough such special cases in nature to yield, potentially, a rather detailed and specific picture of the influence of oxygen. There are even broadly limiting conditions which tell us something; for example, that the ferrous: ferric ratio is almost always greater than one already tells us that most magmas are far from being strongly oxidized. The most potent sort of information, however, arises from such reactions as the FMQ reaction, which uniquely specifies the oxidation state of a system as long as all reactants and products are present. It is now fitting to take a closer look at this oxygen buffer reaction and to introduce a parameter to describe things related to oxygen without continually having to resort to such vague or inappropriate terms as oxidation potential, which really relates more appropriately to electrochemistry.

Oxygen Fugacity and the Concept of the Oxygen Buffer

There are many ways to describe the behaviour and amount of oxygen in a system, but for magmatic processes there are two functions which are most useful and conceptually helpful. These are the *partial pressure* and the *fugacity*. The partial pressure is a helpful concept because it is perhaps the most amenable to a mechanistic visualization. The air we breathe is a complex gas, at a pressure of about 1 bar (1 atm), of which about 21% is oxygen. We therefore thrive under a partial pressure of oxygen of 0.21 bar. If we breathe into a paper bag, we gradually replace the oxygen with CO_2, thus reducing the partial pressure of oxygen to values well below 0.21 bar until we become unhappy or unconscious. The partial pressure is easily evaluated, since the sum of the mole fractions times the partial pressures equals the total pressure, and (at 1 atm) one therefore needs only the analysis of the gas. The partial pressure of oxygen reflects the amount of oxygen present among other components of a gas phase and is a convenient parameter as long as a gas phase exists, but it becomes undefined for the case where no gas phase exists, for example at high pressure in the presence of a splendid solvent such as a melt. *Fugacity* is a more general parameter which works well whether or not a gas phase is present. It is in fact an alternative expression of the chemical potential μ, which is the partial molar Gibbs energy of a component, $\partial G/\partial n$. The relation between the chemical potential of oxygen and the fugacity (f) is as follows:

$$\mu O_2 = \mu O_2{}^{\circ} + RT \ln f_{O_2}, \quad (16.3)$$

where $\mu O_2{}^{\circ}$ refers to the chemical potential of oxygen in some standard state, and R and T are the gas constant and the Kelvin temperature, respectively. For an ideal gas, $f_{O_2} = p_{O_2}$, and $f/p \rightarrow 1$ as ideality is approached. Fugacity has the dimension of pressure and is measured in bars or atmospheres or pascals. The f_{O_2} of a reaction is a continuous function of temperature, and for solid: vapour equilibria, the $\log f_{O_2}$ of a reaction has the happy property of being linear with respect to $1/T$, the reciprocal of the Kelvin temperature. This falls out of the Clausius–Clapeyron equation, which may be reduced to:

$$\log P = -\Delta H/2.303RT + C, \quad (16.4)$$

where P is pressure or partial pressure, H is the enthalpy, R and T have their usual meaning, and C is a constant. Figure 16.6 compares the f_{O_2}-T plot for the FMQ reaction with the f_{O_2}-$1/T\,^{\circ}$K plot.

The FMQ reaction is an oxygen buffer, which means that it is a regulator of f_{O_2}. As an analogy, a eutectic melting reaction is a temperature buffer in which the addition or subtraction of heat to or from the system has no effect on the temperature, as long as all reactants and products are present. The assemblage Di + An + L, for example, may be subjected to heat from a furnace without changing the temperature from 1274°C until such time as Di or An is consumed. Heat may also be extracted isothermally, until L disappears, at which time the assemblage loses its thermal buffering capacity. The calories in the thermal analogy may be likened to atoms of oxygen in the oxygen buffer reaction. How many calories need to be added or subtracted, and how many oxygens, both depend on the initial state of the system and on its mass. As long as the specified buffer condition holds, the T or the f_{O_2} is constant and unrelated to numbers of heat or oxygen units involved in a flux.

We may place pure fayalite in a furnace and pass a stream of oxygen-rich gas over it, thereby beginning to oxidize it by reaction (16.2) to magnetite plus quartz. At constant temperature, a specific f_{O_2} will now be established and maintained as long as Fa, Mt, and Q remain in the assemblage. If the flow of gas continues, all fayalite will eventually be converted to Mt + Q, and the f_{O_2} will then rise without inhibition. If the gas is changed to a reducing mixture, Fa will be generated at the expense of Mt + Q, again at a specific f_{O_2}. The buffering capacity persists as long as there is one grain of each phase present—when the last crystal of any species goes, there goes the buffer.

The great beauty of the oxygen buffer principle is that the buffer is insensitive to other phases in the rock. If pure Fa, Mt, and Q coexist in a rock, the f_{O_2} of the rock system lies on the buffer curve in Fig. 16.6, no matter what liquid or other solid phases may also be present. Of course in the natural environment, the phases are not

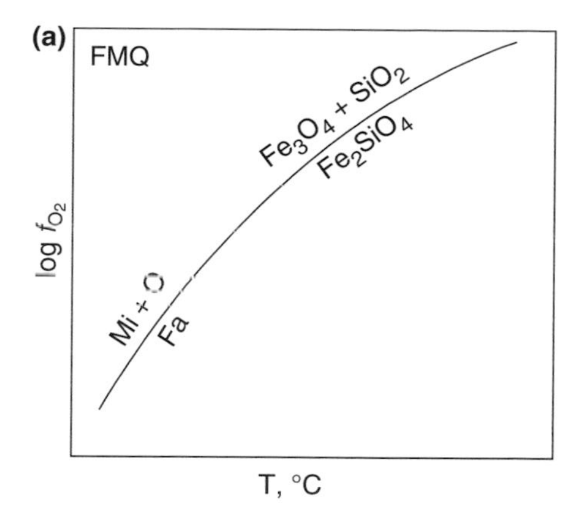

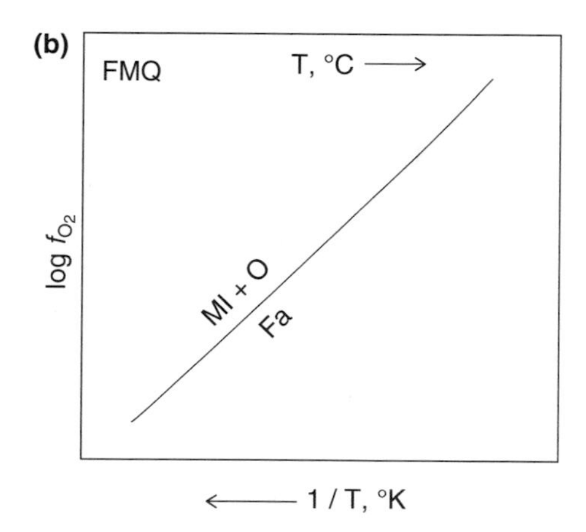

Fig. 16.6 FMQ buffer reaction plotted as a function of oxygen fugacity against (**a**) T °C, and (**b**) reciprocal Kelvin temperature. In (**b**), T °C is nonlinear but $1/T$ °K is linear

pure members of the Fe-Si-O system, but they do often closely approach such purity, and the principle holds, with appropriate changes in calibration, even for "dirty" phases, which usually contain Mg and Mn in some quantity, for example. The presence of Fa, Mt, and Q in a rock is *prima facie* evidence that the rock crystallized or equilibrated under f_{O_2}-T conditions corresponding to the buffer curve or its "dirty" analogue. Whether or not liquid partook of this equilibrium depends on the textural evidence for mutual crystallization *ex* liquid, or subsolidus reaction. As it turns out, plutonic rocks cool so slowly that the apparent temperatures recorded are rarely over 800°C, and commonly as low as 500°C, but many volcanic rocks record temperatures very close to their crystallization temperatures.

Here endeth Part II of the olivine story. We now undertake to describe the whole melting story at 1 bar pressure in terms of T-f_{O_2} relations.

Olivine Melting: A T-f_{O_2} Analysis

We have described the incongruent melting of olivine in the presence of iron metal at 1 bar, and we have described the solid-state breakdown of fayalite to magnetite plus quartz as a function of T and f_{O_2}. In terrestrial rocks, we shall not generally expect to find metallic iron involved with olivine and liquid, whereas we do see evidence (in some granites) that fayalite, magnetite, and quartz have crystallized from a silicate melt. What are the melting relations of olivine along the FMQ buffer curve? How relevant is the iron-saturated olivine melting diagram to natural conditions? Are there other oxygen buffer reactions relevant to olivine melting? One might perhaps suppose that a great deal of hit-or-miss experimentation with controlled f_{O_2} would be required to answer such questions. This is not the case. By means of a very powerful geometrical analysis due to the physical chemist F. A. H. Schreinemakers (published over the period 1915–1925), it is possible to organize information about coexisting phases in such a way as to write the set of all possible reactions among the phases and to deduce all the geometric relations among these reactions in P-T space, or, for example, f_{O_2}-T space.

By means of such a Schreinemakers analysis, D. H. Lindsley was able to sit down and work out in a few hours all the relevant relationships among the phases fayalite, iron, silica, wüstite, magnetite, and liquid in the system Fe-Si-O and plot them as a function of T and f_{O_2}. (And then present the results to an evening seminar for young petrologists, one of whom helpfully identified a particular graphical feature as a "ruled surface".)

The only experimental data needed to fix the arrangement of 13 reaction curves were the previously known f_{O_2}-T slopes of four oxygen buffer curves and the locations of two out of six invariant points. We shall now review the procedure followed by Lindsley, Speidel, and Nafziger (1968), because not only does the result answer the questions raised above, but also the system provides an elegant and satisfying introduction to the methods of Schreinemakers analysis, which is a powerful tool in petrology, both as a guide to laboratory experimentation and to the interpretation of field relations among minerals, particularly in metamorphic rocks.

We revert to the shorthand notation of the original paper, in which the phases considered are as follows:

o= fayalite
s = tridymite or quartz
i = iron
w = wüstite
m = magnetite
l = liquid

The liquidus relationships of these phases at 1 bar are shown in Fig. 16.4. The boundaries of the fayalite field are reproduced in Fig. 16.7. These are labelled according to the phases in equilibrium along each curve; for example, the curve of

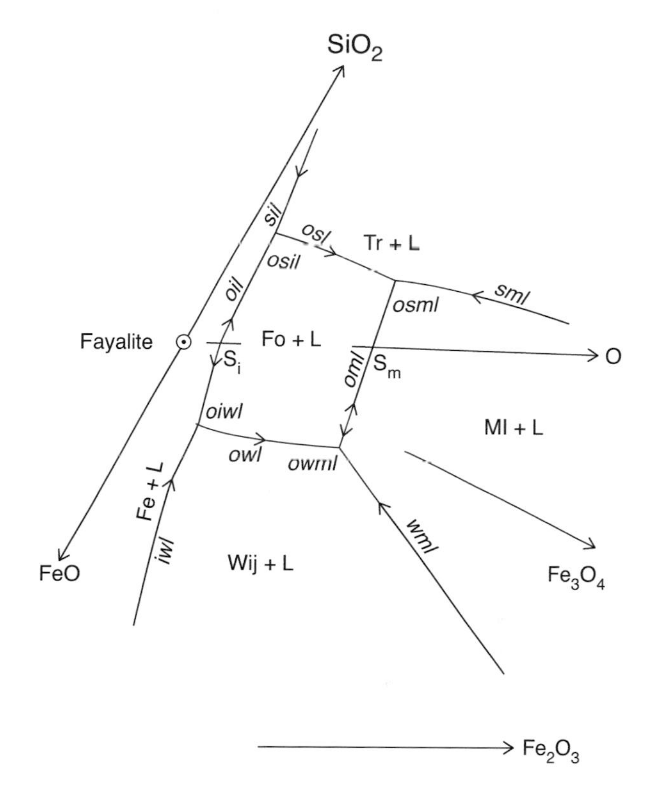

Fig. 16.7 Part of the system Fe-SiO_2-O, taken from Fig. 16.4. Labels are clarified in the text

the reaction *olivine* → *iron* + *liquid* is labelled *oil*. This would be *L* (Fe, Fa) in our usual notation. The system is ternary, Fe-SiO_2-O. Along each of these three-phase curves, the isobaric variance W_p is equal to $3 + 1 - \phi = 1$. The four four-phase points *osil*, *osml*, *oiwl*, and *owml* are, by the same reasoning, isobarically invariant ($W_p = 0$). Each of these invariant points is generated by the intersection of three univariant curves in the liquidus diagram. For example, the point *osml* involving fayalite, tridymite, magnetite, and liquid (i.e. essentially FMQ + *L*) is formed by the three curves *osl*, *sml*, and *oml*. In Fig. 16.8, these are given the labels (*m*), (*o*), and (*s*), respectively, where the *label in parentheses signifies the phase which is absent from the equilibrium* along that curve.[4] This is a conventional and convenient shorthand for purposes of analysing the arrangement of univariant curves around an invariant point. In this case, the label *osml* signifies the point itself, and (*m*) therefore designates a univariant curve involving equilibria among the other three phases, *o*, *s*, and *l*. (After a little experience, this notation becomes second nature.) One of the curves, (*l*), cannot be shown in the *T-X* projection of Fig. 16.8, because this is a liquidus diagram (map of the liquidus surface) and the curve (*l*) involves the solids *o*, *s*, and *m* in the *absence* of liquid. To study this curve, we shall need to plot

[4]The "missing" phase may be present in the system but it does not take part in the reaction.

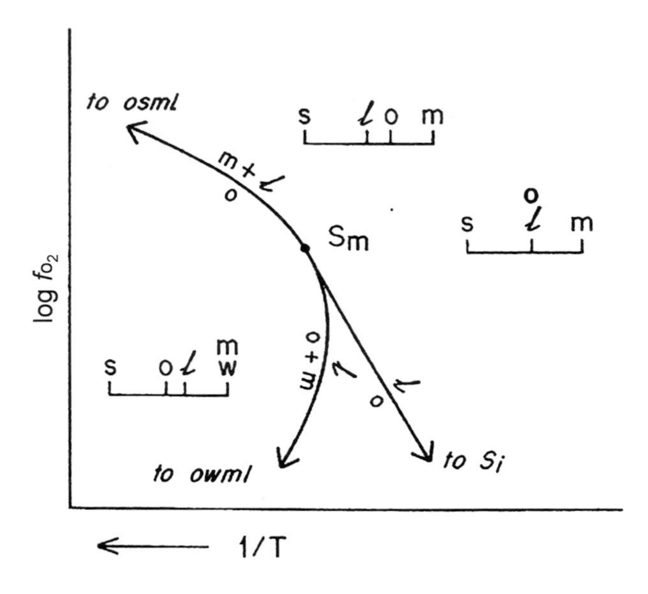

Fig. 16.8 Liquidus curves in the vicinity of the point *osml*, showing labels used in the "missing phase" notation. Coordinates of the figure are the same as in the ternary diagram, Fig. 16.7

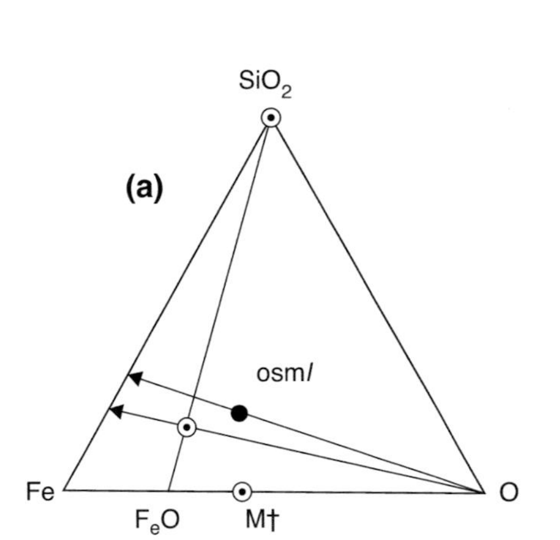

Fig. 16.9 Chemographic diagram for the point *osml*. (**a**) Ternary diagram showing projections from oxygen through fayalite (circle on FeO-SiO_2 line) and liquid (at point *osml*) onto the join Fe-SiO_2. (**b**) Chemographic line Fe-SiO_2 showing projected positions of magnetite (*m*), olivine (*o*), liquid (*l*), and the position of silica (*s*) at the end of the line

an f_{O_2}-T diagram. First, we need to examine the compositions of the phases relative to each other. We shall use for this purpose an oxygen projection onto the SiO_2-Fe join, and in so doing, we will never need to worry about whether oxygen is evolved or used in a reaction, since addition or subtraction of oxygen will occur normal to the page. The derivation of such a projection for the point *osml* is shown in Fig. 16.9, along with a linear scale (chemographic diagram) showing the projected compositions. Only the relative compositions are needed; no matter if the distances are incorrect in Fig. 16.9b.

In Fig. 16.9b, we note that the projected composition of olivine lies to the right (Fe-rich) side of liquid; this is readily seen in Fig. 16.4 or 16.7, where it is clear that the composition of liquid at *osml* lies towards SiO_2 with respect to the projection line Fa-O. By using the chemographic diagram, we may judge the nature of the reaction (o) involving the phases s, m, and l. The rule is that the reaction must represent the correct linear combination of phases. Among s, m, and l, the following reactions are geometrically impossible:

$$\begin{aligned} & s + l = m (m \text{ lies outside the pair } s, l \text{ and would have to be } - m) \\ & m + l = s (s \text{ lies outside } m - l) \end{aligned} \tag{16.5}$$

The correct reaction is as follows:

$$s + m = l \ (\text{or } s + l = -m) \tag{16.6}$$

because l lies between s and m. The reaction (o) is therefore the congruent melting reaction Tr + Mt = L.

Our immediate goal is to plot the complete sequence of univariant reaction curves around the invariant point *osml* in T-f_{O_2} space. For the moment, we are unconcerned with the calibration of T and f_{O_2} and concerned only with the relative slopes and positions of curves. We shall use a plot similar to Fig. 16.6b of $\log f_{O_2}$, versus $1/T$ K, with the abscissa scale increasing to the left so that T increases to the right. In Fig. 16.10, the invariant point *osml* is arbitrarily located, and the reaction curve (o) is plotted with a positive slope above the invariant point. There are two rules that

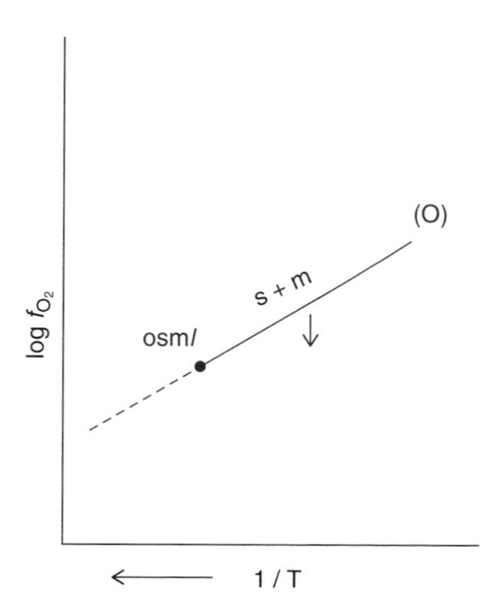

Fig. 16.10 Arbitrary initial position of the univariant curve *sml* in $1/T$-f_{O_2} coordinates

govern the plotting of this univariant reaction curve. One is the *fundamental axiom* (our Rule 1) which says that the curve separates the region where the assemblage $s + m$ is more stable from the region where l is more stable, each *relative to* the other assemblage. For "more stable", one can also substitute "less metastable". (Note that nothing is implied about the stability of either of these assemblages relative to some other assemblage not directly involved in this reaction. Our analysis is concerned for now only with reactions in the immediate vicinity of the invariant point and of the univariant curve.) The second rule (our Rule 2) is that the *metastable extension of (o)* must lie in a region where the phase o is stable (hence the usefulness of the missing phase notation). Obedience to this rule ensures that *no divariant assemblage can occupy an angular sector larger than 180°* about the invariant point. Both the use and the logic of these rules will become apparent after some experience is gained.

A third, informal rule (Rule 3) is that the slope and labelling of the reaction curve make sense from a physical, experimental, or geological standpoint. This is not essential for an eventually correct topological analysis, but it is an efficient way to proceed.

In the current case, we have a melting reaction, $s + m = l$. It would be silly to plot l on the low-temperature side of the curve. As to the slope, we have three basic choices, as suggested in Fig. 16.11: positive, negative, or infinite (slope zero is also a possibility, but one which we can usually ignore for a melting reaction.) An infinite slope would imply that f_{O_2} has no effect on the temperature of melting. We would expect this to be the case for the reaction cristobalite → liquid, for example, but not for the melting of an iron-bearing mineral, especially one like magnetite containing two oxidation states of iron. A negative slope would imply that higher f_{O_2} favours liquid relative to magnetite, but this seems unreasonable, as liquid has no special

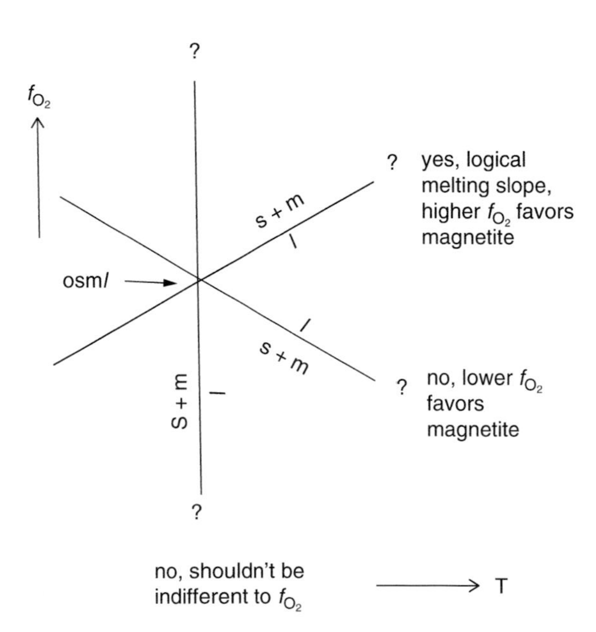

Fig. 16.11 Choices of slope for the melting reaction $s + m = l$

need for Fe^{3+}, whereas magnetite certainly does. A positive slope is consistent with this need and appears to represent a reasonable melting curve.

Which segment of the positive slope curve in Fig. 16.11 should we choose as the stable segment, the one above the invariant point or the one below? The curve label is (o), so if we choose the lower segment as stable, that implies that olivine is stable in the region above and to the right of the invariant point. We do not yet "know" that this is unreasonable, but since the invariant point involves *osml* it presumably will involve an olivine melting reaction, and to have olivine stable at higher T will mean that liquid will be stable at lower T, an unlikely event. It will be prudent to assume that the stable segment is the one above the $W_p = 0$ point, as in Fig. 16.10.

It should be pointed out at this stage that we are using a short-cut analysis in trying to guess the most sensible configuration by the informal Rule 3. The procedure of Schreinemakers yields only the correct *sequence* of curves about an invariant point, and when the whole sequence is established, you know it is right because it is internally consistent. However, without additional information you do not know whether you have the physically correct configuration of the *whole set* of curves, or its *enantiomorph*. Reversing the stable and metastable segments of the (o) curve in Fig. 16.10 would in fact yield the enantiomorph, which would turn out to be geometrically correct but physically unlikely. A strictly formal analysis would require that an arbitrary choice be made for the first curve plotted and that the test for enantiomorphism be applied after all the curves were plotted. Although potentially hazardous, our method of trying to make physically meaningful guesses early in the game is a worthwhile short cut as long as we recognize the danger of becoming wedded to an incorrect notion.

We now wish to build onto Fig. 16.10 by adding the curves (s), (m), and (l). Starting with (s), the assemblage is *oml*, and the reaction can be deduced by means of the bar graph in Fig. 16.9b. The compositional relations do not permit a congruent melting reaction $o + m = l$, because l is not a positive linear combination of o and m. The reaction is odd, i.e. $o - l = m$ or $o = m + l$. Again, we choose among the slopes represented in Fig. 16.11. We wish to have liquid on the high-temperature side, and magnetite on the high f_{O_2} side, so this requires $m + l$ to lie on the right side of a curve with negative slope. If the metastable extension of the curve (s) is allowed to fall on the low-T side of the invariant point, it lies in a region where silica is, so far as we know, stable, because it occurs nearby with m on the curve (o). Hence, the configuration in Fig. 16.12 seems logical.

The reaction (m) involves o, s, l, and the composition diagram, and Fig. 16.9b indicates the congruent melting reaction $o + s = l$. A curve of negative or nearly infinite slope, with l on the high-temperature side, would be satisfactory. A curve of positive slope would put the ferrous mineral olivine on the high f_{O_2} side, which is unreasonable. We cannot put the curve to the right of the curve (s) (try it), because then we have olivine partaking of a reaction ($o + s = l$) in a region beyond the breakdown of olivine itself ($o = m + l$). This is impossible. An obvious rule is that in a correct sequence of reactions, the breakdown of a pure single phase comes last of

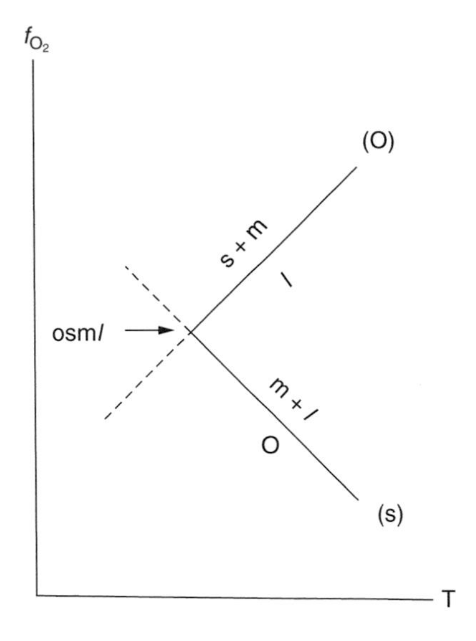

Fig. 16.12 Addition of curve (s)

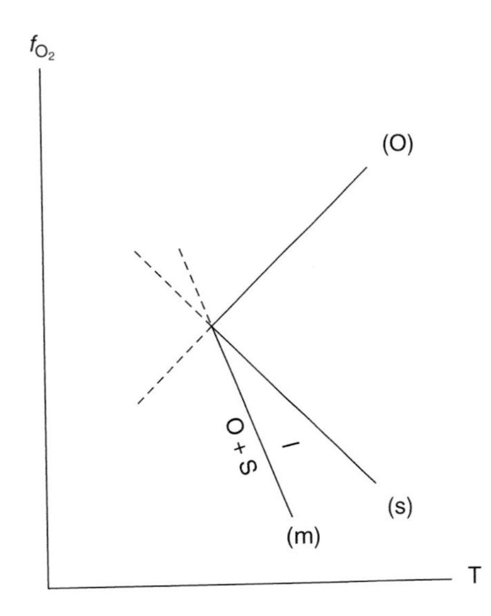

Fig. 16.13 Addition of curve (m)

all the reactions involving that phase. We therefore get the configuration of Fig. 16.13.

The reaction remaining to be plotted is (l), involving o, s, and m. The reaction is clearly $o = s + m$, the FMQ oxygen buffer reaction. Obviously, the $s + m$ side should lie towards higher f_{O_2}. Applying our second rule, this reaction (l) must lie with its metastable extension in the region where liquid (the absent phase) is stable. This means that the stable part of (l) must lie above the metastable extension of (o). That

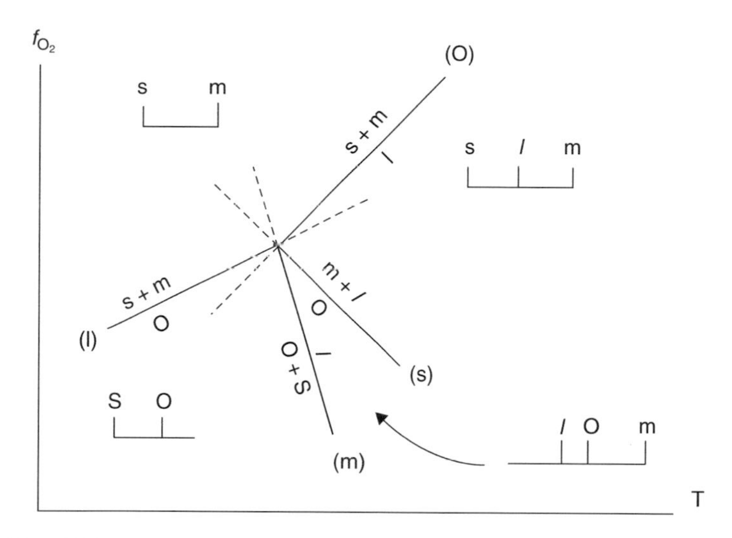

Fig. 16.14 Addition of curve (l) and final array of curves about invariant point *osml*. Chemographic diagrams show compatibilities of phases in each of the divariant regions between the curves

the curve should have a positive slope, as we have already learned in Fig. 16.6b, is perhaps less obvious. If it had a negative slope, that would mean that it is easier to oxidize olivine to silica plus magnetite at higher temperatures than at lower temperatures. Olivine would then be stable at relatively high f_{O_2} at low temperatures. We should commonly expect to see silica plus magnetite as crystallization products of magma, and we should expect to see evidence for subsolidus reaction of $s + m$ to olivine (olivine reaction rims between magnetite and quartz). Instead, we commonly see olivine as a crystallization product and magnetite + quartz as subsolidus reaction products. Actually, Fe tends to gain an electron more easily at higher temperature, and it is always true that at a given f_{O_2}, the ferrous Fe^{2+} state is favoured at high T and the ferric Fe^{3+} state is favoured at low T. In consequence, *all* oxygen buffer curves involving the element iron have positive slopes in f_{O_2}-T space. The curve (l) is therefore constrained to lie in a small region between slope zero and the metastable extension of (o), as in Fig. 16.14, which is the complete set of curves about the invariant point *osml*.

We now verify that Fig. 16.14 is correct insofar as the rules and logic are concerned. Three "assemblages", $s + m$, l, and o, occupy angular sectors no greater than 180°, which is correct according to Rule 2. We may also go around the sequence and follow the reactions in turn, to see if they make sense. Start with the $s + m$ field. Crossing the (o) curve, we see liquid coming in between s and m. Crossing the (s) curve, olivine now intervenes between l and m. Crossing the (m) curve, l becomes unstable and $o + s$ can now coexist. Crossing the (l) curve, o becomes unstable relative to $s + m$ which is the assemblage begun at. The reactions appear to be in proper sequence. To see what an improper sequence would be like, try reversing the positions of (*s*) and (*m*), and you will see that the assemblage demanded by the lower side of the (*m*) curve is not the same as that required by the higher side of the (*s*) curve. *The entire region between any two curves must be occupied by the same assemblage*, whether or not that assemblage partakes of a reaction at one of the curves.

The advantages of using a chemographic diagram projected from oxygen should now be apparent. The reaction $o = s + m$ can be deduced from the bar graph without concern for the reaction coefficients or the consumption of oxygen. From the reaction $o = s + m$, the full FMQ reaction

$$O_2 + 3Fe_2SiO_4 \rightleftarrows 3SiO_2 + 2Fe_3O_4 \tag{16.7}$$

can readily be written when desired, but need not be for purposes of graphical analysis.

We have derived a valid sequence of univariant reaction curves about the invariant point *osml*, and it now remains to do the same for each of the other three invariant points in Fig. 16.7, independently, and then finally to put all the points and curves together in f_{O_2}-T space, by using identical reactions which emanate from different invariant points. We may now turn our attention to the invariant point *osil*.

The phases *o*, *s*, *i*, and *l* are shown in the bar graph of Fig. 16.15a, which is to be compared with the ternary diagram (Figs. 16.4 and 16.7). The composition relations are much the same as for *osml*, except that the liquid lies slightly closer to *s*. The curve (*l*) involves *o*, *s*, and *i* and must represent the reaction $o = s + i$ in the absence of liquid. The stable part of this curve must therefore lie below the invariant point, as in Fig. 16.15a, so that its metastable extension lies in the field where liquid is stable. The reaction is akin to the FMQ reaction $o = s + m$, except that in this case it represents a *reduction* of Fe^{2+} to Fe°; it is an oxygen buffer reaction

$$Fe_2SiO_4 = SiO_2 + 2Fe + O_2 \tag{16.8}$$

and clearly *o* must lie on the high-f_{O_2} side. The curve is plotted with a positive slope in Fig. 16.15a.

The reaction (s) represents equilibrium among *o*, *i*, and *l* and represents the familiar incongruent melting of olivine, $o = i + l$. Raising the f_{O_2} should favour olivine over iron, and liquid should lie to the high-*T* side of the curve, so a positive slope is indicated (try a negative slope to see if you agree). The stable part of the curve must lie above (*l*), because reaction (*s*) also limits the stability of olivine by itself. The stable segment then must lie above the temperature of *osil*, and it must lie 180° or less from curve (*l*) so that the angular sector occupied by olivine is no greater than 180° (Rule 2). A solution meeting these conditions is shown in Fig. 16.15b.

Reaction (*i*), involving *o*, *s*, and *l*, must be the melting reaction $o + s = l$; the curve (*i*) cannot lie above (*s*) in temperature, since the disappearance of the pure phase *o* on curve (*s*) prevents its subsequent appearance in a reaction at higher *T*. The curve

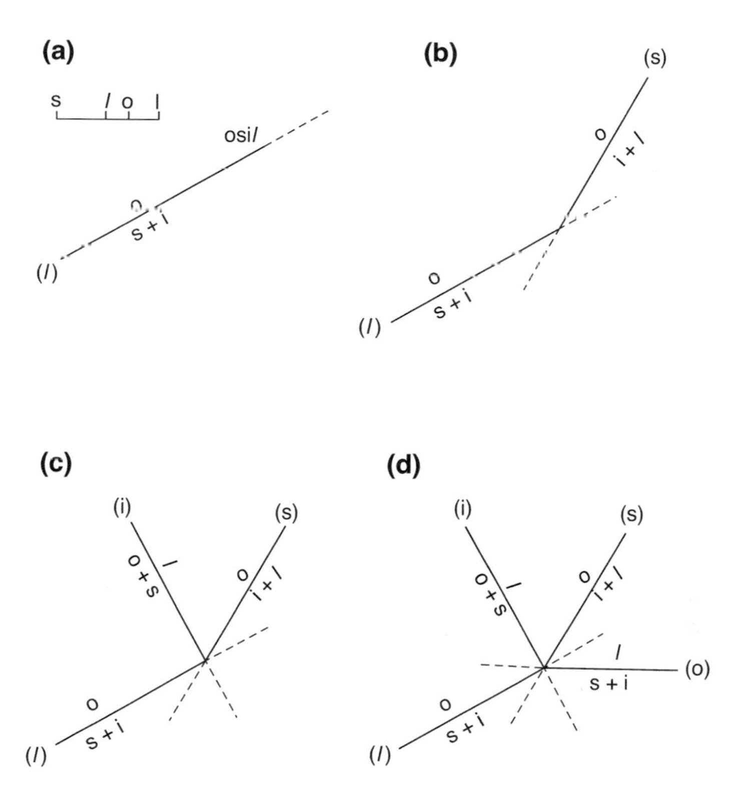

Fig. 16.15 (**a–d**) Development of the array of univariant curves about invariant point *osil*. Coordinates are f_{O_2} and $1/T$

should be fairly insensitive to f_{O_2}, neither *o* nor *l* requiring any large change in oxidation state. On the other hand, it is clear from Fig. 16.7 that *l* has the higher oxygen content and should be slightly favoured by f_{O_2} and *T*. Note that the metastable extension lies where *i* is stable, which would not be the case if the curve were plotted below the invariant point in f_{O_2}.

The reaction curve (*o*) must now be plotted with its stable segment to the high-*T* side of the invariant point, so that its metastable extension lies where *o* is stable. It must also lie so that the angular sector occupied by *l* alone is $\leq 180°$. The assemblage *s* + *i* must lie on the low-f_{O_2}, side of the curve, to be contiguous with the same *s* + *i* assemblage on curve (*l*), and also to leave metallic iron in the low-f_{O_2} region. A flat or low positive slope is possible; a negative slope would imply that it is easier to oxidize iron at high *T* than at lower *T* which is contrary to our experience. Therefore, the curve (*o*) is plotted with a nearly flat slope in Fig. 16.15d. This is evidently a retrograde melting curve, in which the solid → liquid reaction can occur with falling temperature *but with the addition of energy in the form of oxygen*. (There is still no such thing as a free lunch, and retrograde melting reactions always involve pumping in something else instead of heat.)

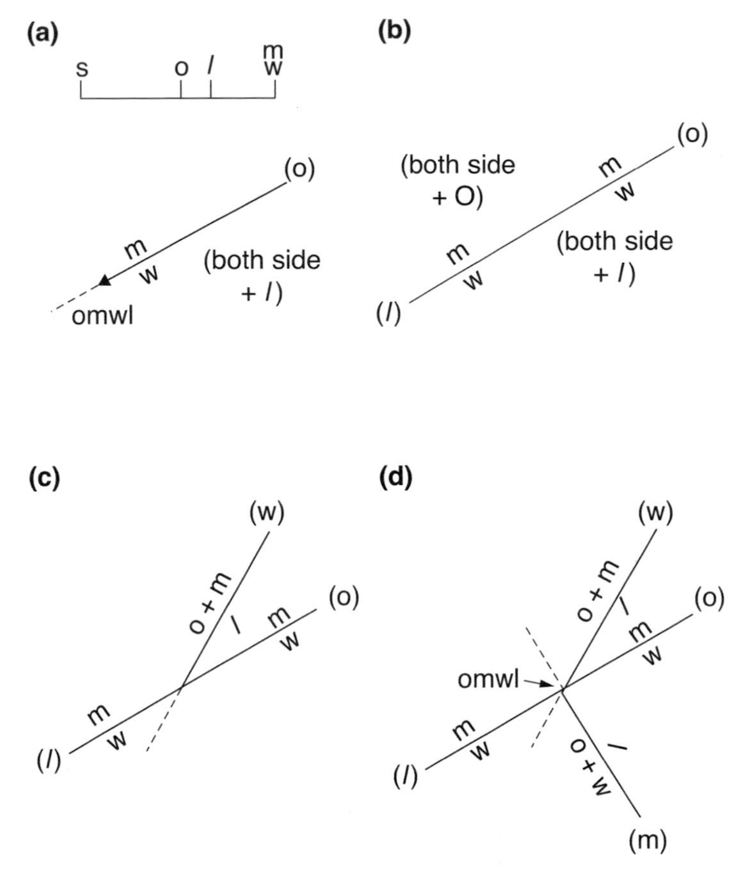

Fig. 16.16 (**a**–**d**) Development of invariant point *omwl* in f_{O_2}-$1/T$ coordinates

Figure 16.15d may now be verified as to the correct sequence of reactions, as was done with invariant point *osml*. It is left to the reader to do this. Attention is called, however, to Fig. 16.14, curve (*m*), which is the identical reaction to curve (*i*) in Fig. 16.15d, namely $o + s = l$. This common reaction will later be used to tie these two invariant points together in f_{O_2}-T space.

Invariant point *omwl* offers a special treat, a degeneracy caused by the collinearity of *w* and *m* with oxygen. The composition diagram, Fig. 16.16a, shows this as a superposition of *m* and *w* at the Fe terminus. Note also that *l* now plots slightly to the right of *o*, as may be verified with Fig. 16.4 or 16.7. The reaction (*o*) involves equilibrium among the phases *m*, *w*, and *l*, but there is no way to write a reaction involving *l* as a reactant or a product. The phase *l* is here an *indifferent* phase, and the reaction must be $m = w$, which is the magnetite–wüstite buffer reaction (MW):

$$\underset{\text{(wüstite)}}{3\text{“FeO”} + 0.5O_2} \rightleftharpoons \underset{\text{(magnetite)}}{Fe_3O_4}. \quad (16.9)$$

Since this is a buffer reaction, the curve should have a positive slope similar to the other buffers, and since liquid is present, the stable part of the curve must lie above the invariant point. Obviously, magnetite must lie on the high-f_{O_2} side. The curve may be plotted as in Fig. 16.16a.

Consider now the reaction (l), involving equilibrium among o, m, and w. Again, we have an indifferent phase, o, and again the reaction must be $m = w$. In this case, the metastable extension must lie where liquid is stable and the stable part below the invariant point. We now note that if the curve (l) makes some angle with the curve (o), Rule 2 will be violated, because either m or w will occupy an angular sector >180°. The only solution is for these two curves to lie collinear, so that their stable segments and metastable extensions and their sector angles are equal to 180°. The same buffer reaction, $m = w$, occurs on both sides of the invariant point, but in the presence of olivine on one side and of liquid on the other side. The phases o and l do not participate in the reaction, but coexist stable with the equilibrium assemblage. The curve $m = w$ is *univariant by restriction* because the system required to express these phases has degenerated to the binary system Fe-O, so $c = 2$, $\phi = 2$, and $W_p = 1$. The curves (l) and (o) are plotted in Fig. 16.16b.

The reaction (w) is evidently $o + m = l$. Liquid should be favoured by high temperature, and magnetite should be rather strongly favoured by oxygen, so the curve should have a positive slope as shown in Fig. 16.16c, where it will be noted that the metastable extension lies in the field where w is stable.

The reaction (m) is also evidently a melting reaction, $o + w = l$. Its metastable extension must lie above the $m = w$ curve and to the left of the stable part of curve (w). Liquid should be favoured by high T. A negative slope, while not obviously required by the relative oxygen contents of l and w, is reasonable, and the curve is plotted in Fig. 16.16d. If the negative slope is wrong, we shall be able to correct it later.

The sequence of reactions about the point *omwl* may now be verified in the usual way. No assemblage occupies a sector >180°, and the sequence of reactions is orderly in that no curve falls outside the stability field of the reactants.

The invariant point *oiwl* will be seen to mimic *omwl* very closely, with a degeneracy of iron and wüstite, an $i = w$ buffer reaction (Fe + O = “FeO”) with indifferent phases o and l, and two melting reactions: (i), $o + w = l$ and (w), $o + i = l$. The complete array of curves about *oiwl* is shown in Fig. 16.17 and can be verified in the usual way. The slope of (i) must be compatible with the slope of (m) leading to the invariant point *omwl* (Fig. 16.16d), since the reactions are the same, $o + w = l$. It now becomes apparent that these last two invariant points will be linked together by this reaction in f_{O_2}-T space.

We now turn briefly to two special *singular points*, S_m and S_i, which are labelled in Figs. 16.4 and 16.7. These are points collinear with fayalite and oxygen, and they each represent the singular event which occurs when a moving phase composition

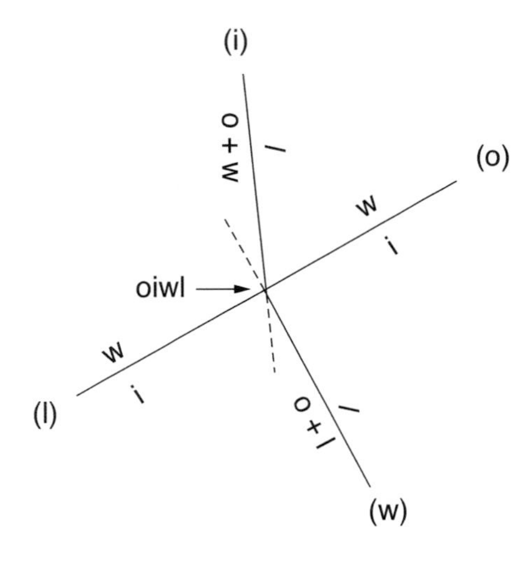

Fig. 16.17 Invariant point *oiwl* in f_{O_2}-$1/T$ coordinates

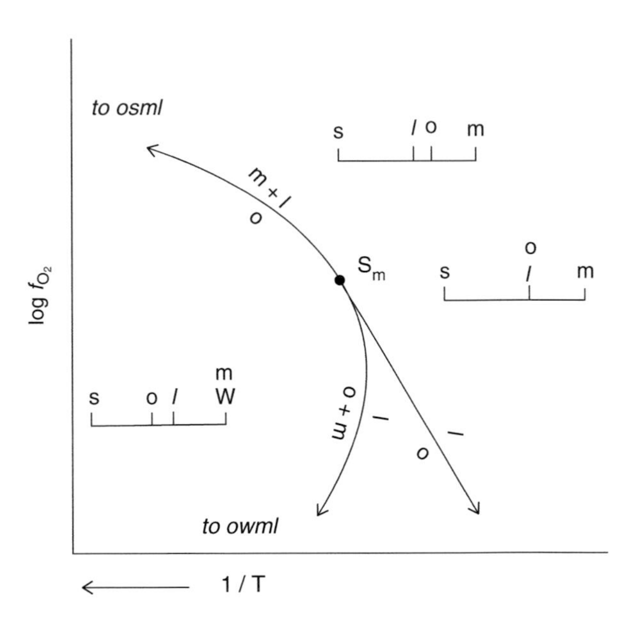

Fig. 16.18 Relations about singular point S_m

(in this case liquid) by coincidence reaches collinearity with two other phases. As mentioned on an earlier page, when *o*, *l*, and the projecting component oxygen are all collinear, the system is degenerate, no other phases need enter the reaction, and the melting reaction is the *congruent* reaction $o = l$. The variance of the singular point is zero, because the third phase (*m* or *i*) is not indifferent at each side of the singular point. We shall examine first the point S_m to see why this is so.

S_m lies on the curve *oml* (Fig. 16.7), and this assemblage has appeared in two of our diagrams, as the curve (*s*) at *osml* (Fig. 16.14) and the curve (*w*) at *owml* (Fig. 16.16d). These two curves are shown in Fig. 16.18, along with the composition

Fig. 16.19 Relations about singular point S_i

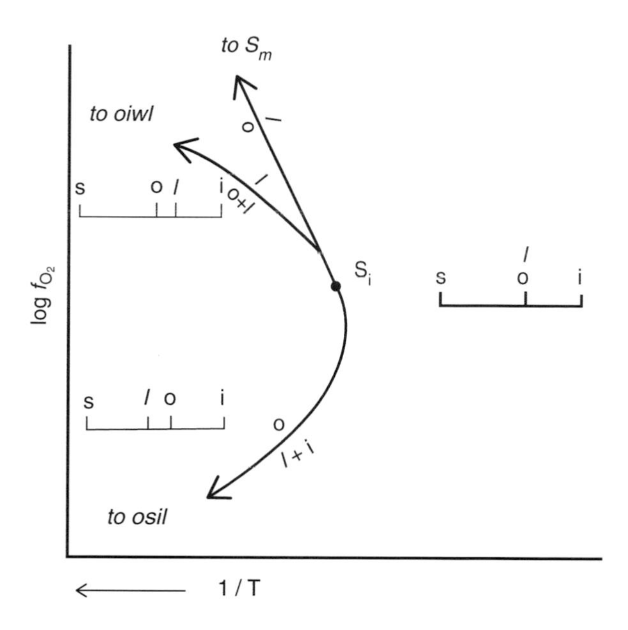

graphs illustrating the nature of the reaction. The curve *oml* emanating from *osml* is the odd melting reaction $o = m + l$; the one emanating from *owml* is the even reaction $o + m = l$. Magnetite is necessary for the reaction on both sides of S_m and so while the reaction at S_m is simply $o = l$, magnetite is a necessary third phase; $c = 2$, $\phi = 3$, and $W_p = 0$.

As Fig. 16.7 indicates, the reaction $o = l$ is not confined to S_m and S_i, but occurs all the way across the olivine field along the line Fa-O. Thus, in Fig. 16.18, S_m is the terminus of a reaction curve $o = l$ which runs across T-f_{O_2} space to S_i, the other terminus. The singular point S_i occurs on the curve *oil*, which is the standard fayalite melting reaction for iron crucibles. This curve connects the invariant points *osil* and *oiwl* (Figs. 16.7, 16.15d, and 16.17) and is reproduced in Fig. 16.19, which may be analysed in the same way as Fig. 16.18.

We now have all the information needed to plot the complete 1-bar T-f_{O_2} section for olivine equilibria in Fe-SiO_2-O, namely the curves about the four invariant points *osml*, *omwl*, *oiwl*, and *osil*, and the singular points S_m and S_i. The four buffer curves *osm*, *mw*, *wi*, and *osi* are experimentally known, but their relative positions are dictated in any event by the univariant curves emanating from invariant points, and the arrangement shown in Fig. 16.20 can be deduced without reference to experiment. It is easiest to start by plotting the curve $o + s = l$ between *osml* and *osil*, and then the curve $o + w = l$ between *omwl* and *oiwl*. Note that the temperatures of the invariant points are given in Fig. 16.4. The singular point curves of Figs. 16.18 and 16.19 are then plotted, and the singular points connected by single line $o = l$. This curve marks the absolute limit of fayalite stability as a function of f_{O_2} and T, but note that it is not likely to be achieved in nature.

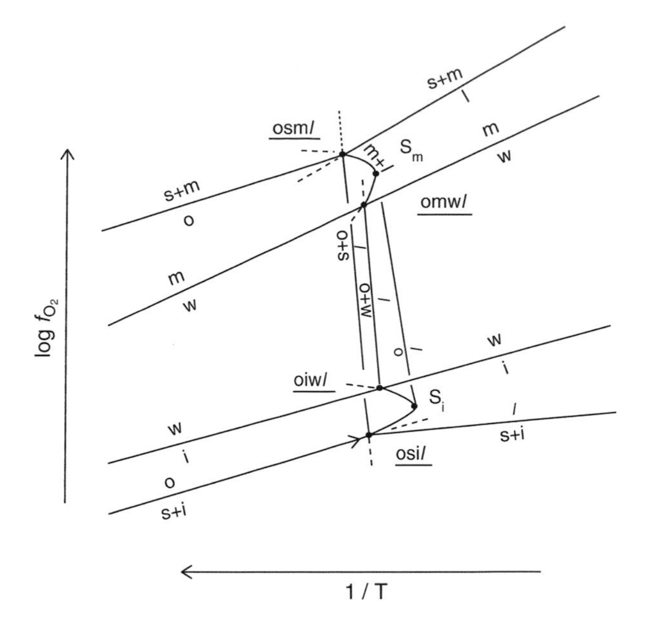

Fig. 16.20 T-f_{O_2} section through the system Fe-SiO₂-O at 1-atm pressure. (After Lindsley, Speidel, and Nafziger 1968)

We can now answer the question, how relevant is the early determination of fayalite melting in iron crucibles? The melting occurs along the line between *oiwl* and S_i in Fig. 16.20. The answer is not very relevant in terms of f_{O_2}, because equilibrium with metallic iron is rare in terrestrial magmas, but not bad from the standpoint of temperature, since all the olivine melting curves have steep slopes. It may be seen that the not uncommon rhyolite–granite equilibrium *osml* and its relatives *sml*, *oml*, and *osm* are quite restricted in f_{O_2}; a relatively small drop in f_{O_2} will lead to wüstite equilibria. On the other hand, the reaction $o + s = l$ may occur over the whole range of f_{O_2} from the stability field of magnetite to that of iron. The reaction $o + w = l$ also occupies a large range of f_{O_2}. The FMQ oxygen buffer is destroyed by the melting of olivine, although the liquid on $s + m = l$ is constrained to a somewhat similar T-f_{O_2} path. The MW buffer, on the other hand, is indifferent to olivine melting, and any liquids that should happen to be saturated in $m = w$ would be controlled by that buffer, no matter what other phases should appear or disappear. From direct measurements in lavas and inferences from natural plutonic rocks, it appears probable that the MW buffer curve represents the lower f_{O_2} range of common terrestrial magmatic T-f_{O_2} paths and that equilibria in the range between MW and FMQ are very common. Equilibria outside the range of olivine, above the *osm* and *sml* curves, also occur in nature at least up to the hematite-magnetite (HM) buffer.

Figure 16.20 delineates the entire f_{O_2}-T stability field of pure fayalite, which is bounded by *osm*, S_m, $o = l$, S_1, and *osi*. The effect of adding MgO to the system will be to increase both the T and the f_{O_2} limits of the olivine field to the limit of pure Mg_2SiO_4, which is of course f_{O_2}-independent at 1898°C. Another effect of MgO is to stabilize orthopyroxene at 1 bar pressure, for pyroxene compositions with Mg/(Mg + Fe^{2+}) greater than about 0.25. Orthopyroxene will be stabilized, in part, by the release of silica from olivine at the *osm* breakdown, so it will persist to higher f_{O_2} than the FMQ buffer curve. The effect of MgO on the system has been summarized in detail by Speidel and Nafziger (1968).

Richardson's Rule

The treatment of Schreinemakers' rules in this chapter is based on that of Lindsley et al. (1968) and on the thorough exposition by Zen (1966), which should be consulted for a rigorous and general understanding. The version of Rule 2 used here is a useful shortcut to the correct sequence of curves about an invariant point. An even more useful shortcut was suggested to me by Dr. S. W. Richardson in Oxford, October 1977. His version may be stated as follows, using the missing phase notation:

> The curve denoted by the missing phase lies with its metastable extension in the sector where the missing phase is generated by all other reactions, *stable or metastable*.

In almost no time, this lovely shortcut became routine among petrologists interested in phase equilibria. (But of course, nobody does that anymore, leaving it up to some code or other.)

In using this rule, it is convenient to place small arrows on stable curves *and* their metastable extensions, pointing towards the sector where the missing phase is relatively more stable than its reaction products. The new curve is then placed so that its metastable extension lies between the arrows. This practice speeds up the correct placement of the curve.

As an example, consider the point *osil* (Fig. 16.15d). According to our version of Rule 2, the curve (*i*) could be placed between curves (*s*) and (*o*). Using Richardson's rule, the two closest arrows pointing into the desired sector stabilizing *i* would lie on the stable segment of curve (*o*) and on the *metastable* segment of curve (*s*). The correct placement of (*i*) would then immediately result from putting its metastable extension between the arrows. Further examples appear in the following chapter.

The value of Richardson's rule lies in the fact that correct placement of the new curve can be made immediately, before the reactants and products are written on the curve. In the case cited above, using the old rule, the incorrect position of (*i*) would have become clear only in attempting to write the full label. Richardson's rule prevents this type of unsuccessful trial. This seems to be the last word in the refinement of Schreinemakers' analysis to a straightforward and rewarding exercise (unless you have the code).

Chapter 17
Iron-Bearing Olivines and Pyroxenes

Introduction

In this chapter, we continue to examine the effects of adding iron to simple magnesian systems. As expected, we shall find that Fe lowers the melting temperatures appreciably, but in addition we find that fundamental changes occur in the reaction relation of olivine to pyroxene. The chapter opens with a review of the system MgO-FeO-SiO_2, taken in part from the original research paper. It then continues with an iron-bearing version of Fo-An-SiO_2, the effect of adding oxygen to MgO-FeO-SiO_2, a discussion of quadrilateral pyroxenes and their evolution, and silica activity in magmatic systems.

The System MgO-FeO-SiO_2

The olivines and Ca-poor pyroxenes of natural basaltic rocks are Mg-Fe solid solutions. We have seen the incongruent melting of enstatite in Chaps. 10–12 and briefly discussed the olivine diagram in Chap. 16. But what happens to the incongruent melting of orthopyroxene as iron is added? The orthopyroxene series forms one corner of the plane of silica saturation, and we have concluded that liquids may penetrate this plane by the removal of olivine, because of the incongruent melting of pure En. If the melting becomes congruent as iron is added, however, liquids may or may not penetrate the plane depending on the iron content. These matters are illuminated by the relations among olivine, pyroxene, and silica minerals in the system MgO-FeO-SiO_2 (Bowen and Schairer 1935).

A simplified liquidus diagram of the system is shown in Fig. 17.1. There is a wedge-shaped field of pyroxene between the fields of olivine and silica minerals, but this wedge tapers to a line within the diagram, allowing the coexistence of iron-rich olivines and tridymite. We see immediately, therefore, that the olivine reaction

S. A. Morse, *Basalts and Phase Diagrams*,
https://doi.org/10.1007/978-3-030-97882-2_17

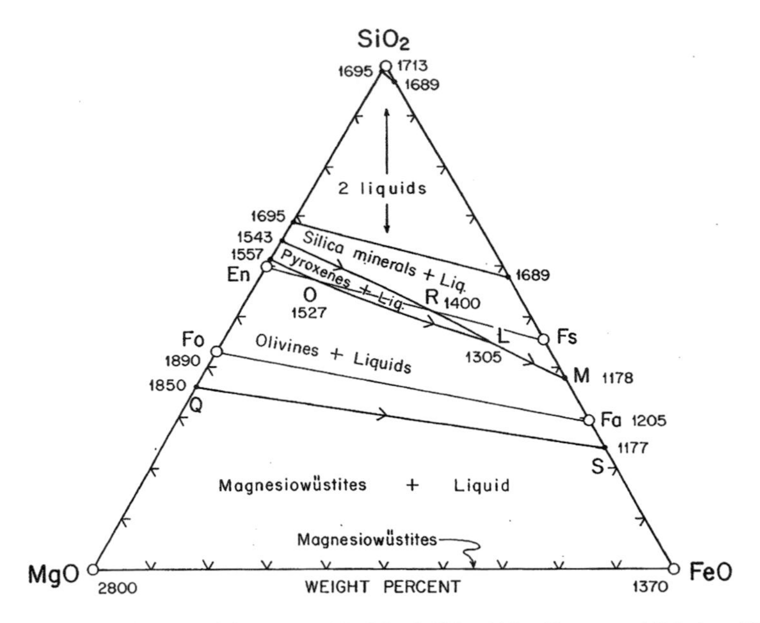

Fig. 17.1 Liquidus diagram of the system MgO-FeO-SiO_2. (After Bowen and Schairer 1935)

relationship common to tholeiitic basalts gives way to olivine + silica compatibility, as we would expect from the existence of fayalite granites. Clearly, at some value of iron enrichment, the melting of pyroxene becomes congruent, but this does *not* occur at point O in Fig. 17.1 because liquid O is in equilibrium with a pyroxene between O and En. Figure 17.1 represents a ternary liquidus map covering several binary solid solutions of the continuous type (like plagioclase), and a correct interpretation of the diagram requires careful analysis. It also explains many features found in real rocks, as we shall see.

The paper by Bowen and Schairer (1935) is a classical account of a powerful experimental attack, and it is elegantly expressed in a style characteristic of the authors. Their descriptions of crystallization behaviour in this system are thorough and lucid. The paper itself is frequently used as a direct source in teaching.

Perhaps the foremost petrologic result of the study by Bowen and Schairer was to illuminate the limits on the incongruent melting of Ca-free pyroxene. The systems Fo-An-SiO_2 (Chap. 11) and Fo-Di-SiO_2 (Chap. 12) predict that olivine is always in reaction relation to Ca-free pyroxene and that the two species cannot be produced together for any significant time during fractional crystallization. On the other hand, many basic layered intrusions show extensive olivine–bronzite cumulates that appear to suggest cotectic crystallization. We now see that for liquids sufficiently rich in iron that is the expected relationship. In part of the paper omitted here, Bowen and Schairer correctly predicted that high pressure would enlarge the field of pyroxene relative to olivine, and as we shall see, that effect eliminates the incongruent melting of pure En at low pressures corresponding to shallow depths in the crust. Hence, to summarize, the reaction relation holds for shallow, low-pressure magmas (such as basalts) when they are relatively Mg-rich. It disappears at high pressure or for Fe-rich magmas.

Another classical problem illuminated by the experimental study is the *olivine hiatus* of fractionated tholeiitic-layered intrusions, sills, and lava lakes. This is a hiatus wherein olivine ceases to crystallize at some magnesian composition and then reappears at a more iron-rich composition. As Bowen and Schairer emphasize (p. 210), such behaviour is readily explained by the phase diagram, in which a liquid may leave the olivine–pyroxene field boundary but remain below the pyroxene join and eventually return to the field boundary, on fractional crystallization. This behaviour owes much to the fact that the pyroxene join lies at a very low angle to the field boundary.

The olivine join represents the critical plane in the basalt tetrahedron, and the pyroxene join, the plane of silica saturation. Note that according to the phase diagram, liquids in equilibrium with olivine may not cross the metasilicate join towards silica unless they are very magnesian. Therefore, it should be common experience to find liquids "trapped" between the two joins, and this is realized in the common occurrence of olivine tholeiites. Note further that late-stage, Fe-rich liquids will crystallize a silica mineral even though they lie to the silica-poor side of the pyroxene join. This result is ultimately due to the instability of ferrosilite at low pressure.

Oxygen Content of Iron Silicate Liquids

One of the most startling results of the experimental study was the demonstration that appreciable amounts of Fe_2O_3 (or dissolved oxygen) occur in liquids in contact with metallic iron. This implies that natural terrestrial liquids crystallizing ferrous silicates contain even more dissolved oxygen than found by Bowen and Schairer in the presence of iron and under a nitrogen atmosphere. D. H. Lindsley (personal communication, January 1979) has pointed out that this phenomenon may explain the very common late-stage oxidation of titanomagnetite to ilmenite plus magnetite found in basic plutonic rocks. If Fe-rich residual liquids contain excess oxygen relative to the crystals separating from them, as suggested also by the liquid compositions in the system Fe-SiO_2-O, Chap. 16, then the inevitable result must be either a late-stage liquid + crystal oxidation reaction or the release of an oxygen-rich vapour phase at the end of crystallization. Such a vapour would be capable of "recharging" through the hot rock and oxidizing those minerals most susceptible to oxidation. These latter appear from petrographic evidence to be the Fe-Ti oxide minerals. A late, oxygen-rich liquid phase would explain the "oxysymplectites" of magnetite + hypersthene locally formed at the expense of olivine in some troctolitic rocks (e.g. Morse 1969, Plate 34; Morse 2015b, Fig. 13.19). These intergrowths represent in large part the local oxidation of the fayalite component of olivine to magnetite and silica (the FMQ reaction!), which then combines with the remaining magnesian olivine to form hypersthene (Presnall 1966, p. 785).

The System Olivine–Anorthite–Silica

How does the effect of adding iron to pyroxene carry over into other systems? Is the incongruent melting of pyroxene influenced by adding plagioclase? Useful answers to these questions can be obtained by studying an iron-bearing version of the system Fo-An-SiO_2 (Chap. 11). Figure 17.2 shows the system Fo-An-SiO_2 itself, re-plotted in cation equivalents. The reason for using cation units is that pyroxenes of varying Mg/Fe ratio will then plot at a single point, rather than over a range as with weight units, and thus, the iron-bearing system can be compared directly with the Mg system and rigorously analysed by itself. Figure 17.2 is very similar to Fig. 11.1 because cation plots closely resemble plots in weight units.

The effect of adding iron is shown in Fig. 17.3, based on the work of Lipin (1978). This is a *liquidus map* of part of the plane: molar $MgO_{59}FeO_{41}$, An-SiO_2 in the quaternary system MgO-FeO-An-SiO_2. The ratio of MgO to FeO is here expressed in *molar percent, not weight* percent. The Ol-SiO_2 sideline of Fig. 17.3 is taken directly from the work of Bowen and Schairer (1935) and corresponds to a line from the SiO_2 apex to $MgO_{45}FeO_{55}$ (weight units) in Fig. 17.1. Lipin (1978) determined experimentally the locations of the piercing points *D, R,* and *E* and reported their temperatures in the International Practical Temperature Scale of 1968. The reported values have been reduced to conform to the geophysical laboratory scale used throughout this book.

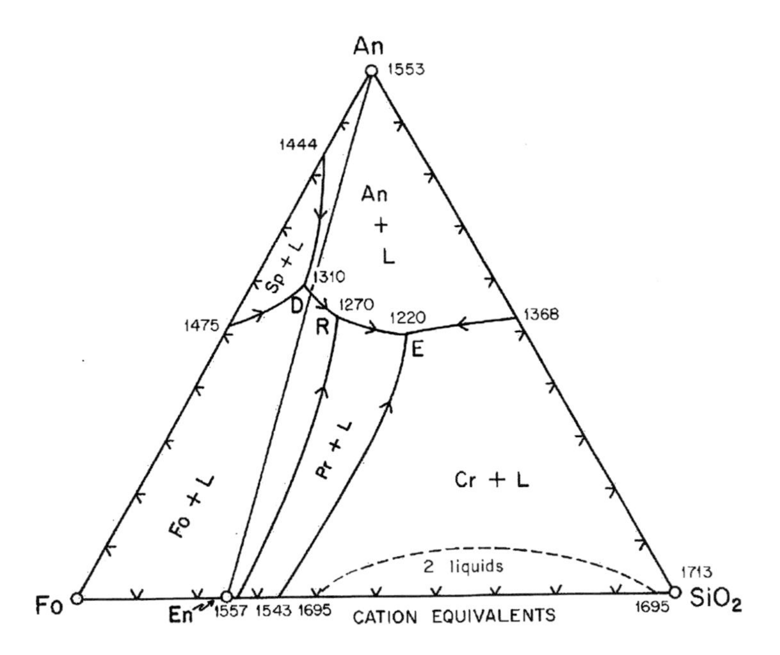

Fig. 17.2 Liquidus diagram of the system Fo-An-SiO_2, re-plotted in cation units. (Based on data of Andersen 1915, Osborn and Tait 1952, and Irvine 1974)

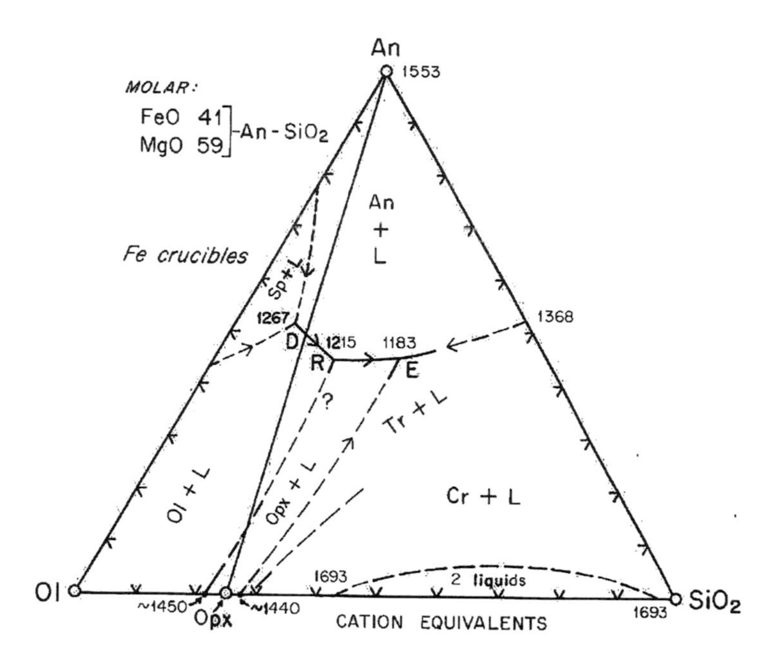

Fig. 17.3 Liquidus relations in the system olivine–anorthite–silica for a molar ratio of $MgO_{59}FeO_{41}$ (after Lipin 1978). Dashed lines represent inferred boundaries. Side line Ol-SiO_2 is from Bowen and Schairer (1935). Query refers to curvature of boundary *L* (Ol, Opx)

The field boundaries near the piercing points are known from Lipin's work, but elsewhere they have not been precisely located, hence are shown dashed in Fig. 17.3.

On the whole, there is a striking similarity between Figs. 17.2 and 17.3. Clearly, the addition of Fe does not greatly alter the geometry of Fo-An-SiO_2. The relative positions of *D, R,* and *E* are little changed. The primary phase field of anorthite is somewhat enlarged because the temperatures of the olivine and pyroxene fields are lowered with iron content. Note, however, that in nature the addition of iron would be accompanied by addition of albite, so the plagioclase field would be smaller than the An + *L* field in Fig. 17.3; it might instead look much like the An + *L* field in Fig. 17.2. It is important to note that no liquid to the left of *E* will remain in the plane of Fig. 17.3 with crystallization, because the separation of olivine or pyroxene crystals will drive the liquids towards lower Mg/Fe.

As before, *R* is a peritectic point representing the reaction Ol + *L* = An + Opx. It is of interest to examine the nature of the curve *L* (Opx, Ol) between the Ol-SiO_2 sideline and point *R*. On the sideline itself, the reaction is even (cotectic), because the MgO_{45} (weight percent) line lies to the right of point *b* in Fig. 22 of Bowen and Schairer, and hence, the pyroxene composition lies to the right of point *O* in that figure. In the quaternary system (pseudoternary plane of Fig. 17.3), the liquid moves sharply away from the sideline, hence away from olivine and pyroxene collectively, in such a way that the tangent to the liquid path clearly lies between Opx and Ol. The reaction along the curve is therefore also cotectic or even. Unfortunately, however, we have no knowledge of the curvature of the *L* (Opx, Ol) boundary as it approaches the piercing point *R*; this uncertainty is indicated by a query on Fig. 17.3. If the curvature is slight, as drawn, the boundary may be even all the way to *R*. If the curvature is stronger, as in the iron-free system (Fig. 17.2), then the boundary becomes odd on the way to *R*. In the latter case, equilibrium liquids will resorb olivine before reaching *R* and continue to do so at *R*; fractionating liquids will leave the boundary and cross the pyroxene field to the boundary *RE*, avoiding *R* altogether. If the boundary changes from even to odd, cotectic fractions of olivine + orthopyroxene will be succeeded by an interval of orthopyroxene alone before plagioclase joins the assemblage, and the sequence harzburgite–bronzite–norite will result, with fractional crystallization. If the boundary remains cotectic to *R*, no bronzite event will occur between harzburgite and norite. In the Stillwater Complex (see Chap. 13), olivine disappears before norite appears, suggesting that the *L* (Opx, Ol) boundary becomes odd near *R*. A complete evaluation would require not only data on the curvature in Fig. 17.3, but also data for other planes of different Mg/Fe ratio.

Figure 17.3 describes a plane of constant bulk composition, having atomic Mg 59 Fe 41, and ignores the minor amounts of Fe_2O_3 present in the liquid. However, crystals coexisting with liquids in this plane are more magnesian than the plane itself, and in any process of crystallization, liquids will leave the plane towards more iron-rich compositions. This may not be a serious handicap to understanding if the geometry changes but little with increasing Fe, as seems likely. Lipin (1978) furnishes analysed compositions of some liquids and crystals lying out of the plane.

In summary, the iron-bearing system olivine–anorthite–silica retains most of the features of Fo-An-SiO_2 except that the olivine–orthopyroxene boundary is cotectic at least in part. The boundary may be peritectic near the reaction point *R*. It is to be emphasized that Fig. 17.3 refers to equilibrium with metallic iron and cannot be used rigorously for terrestrial magmas crystallizing at higher fugacities of oxygen. It should also be noted again, anticipating a later chapter, that increased pressure tends to move the pyroxene field towards olivine, thus strengthening the likelihood that the *L* (Opx, Ol) boundary will be even if the oxygen fugacity is sufficiently low.

T-f_{O_2} Relations in Mg-Fe-Si-O

The system MgO-FeO-SiO_2 discussed earlier in this chapter was, as we have said, studied in the presence of iron. For terrestrial rocks, we are more interested in the phase relations at higher oxygen fugacities than dictated by equilibria with iron. Figure 17.4 summarizes the effect of adding oxygen (as Fe_2O_3) to the "ternary" system MgO-FeO-SiO_2; the figure is similar to one compiled by Speidel and Nafziger (1968). Two triangular faces of the tetrahedron are familiar to us: the base, from Bowen and Schairer (1935), and the right face, from Muan and Osborn (1956). The interior of the diagram is also due to the latter authors. The figure is not to scale, and the equilibria with wüstite are incompletely shown. The main message of Fig. 17.4 is that there is a *magnetite saturation surface* (defined by labels OPML,

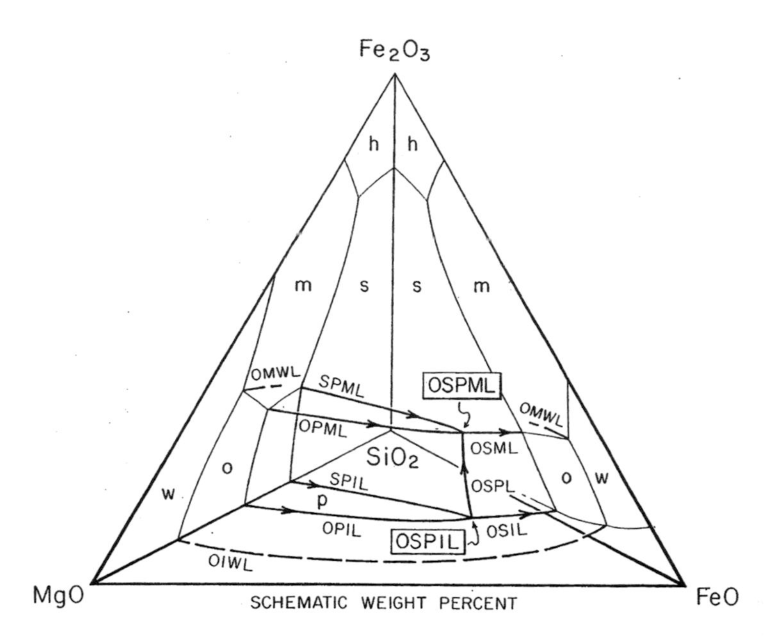

Fig. 17.4 System MgO-FeO-Fe_2O_3-SiO_2, showing primary phase volumes. *o*, olivine; *s*, silica mineral; *p*, pyroxene; *m*, magnetite; *L*, liquid; *h*, hematite; *w*, magnesiowüstite. (After Speidel and Nafziger 1968)

SPML, OSPML, OSML) having the same essential geometry as the familiar curves of olivine, pyroxene, and tridymite with liquid on or near the base of the tetrahedron. The invariant point called "*L*" by Bowen and Schairer is now fully labelled OSPIL to show the presence of iron; it is a 5-phase point in a quaternary ($c = 4$) system, and hence for the 1-atm isobaric restriction has $W_p = 0$; it is an isobaric invariant point. A univariant curve, OSPL, runs from the iron-saturated invariant point to its magnetite-saturated analogue, OPSML. This curve OSPL and the four heavy lines leading to it from the left define the primary phase volume of pyroxene plus liquid, bounded by equilibria with iron, olivine, magnetite, or silica. The phase abbreviations follow the style of Chap. 16.

The liquidus relations of Fig. 17.4 and the associated reaction curves may be rearranged in T-f_{O_2} space with the aid of a Schreinemakers analysis, extending the analysis of Chap. 16 to the Mg-bearing system. Such an analysis can be done very simply with the aid of a correct compositional projection and the use of Richardson's rule. We begin with an analysis of the univariant curves in the *immediate vicinity* of the invariant point OSPIL, noting that our analysis will not necessarily describe correctly the univariant reactions farther away from the invariant point.

Where shall we plot the phase compositions? To do so within the tetrahedron would require a feat of three-dimensional visualization worthy of the most fanatical structural geologist; it would not be a rigorously quantitative exercise, and it would fail anyway because iron does not lie in the tetrahedron. We shall instead make use

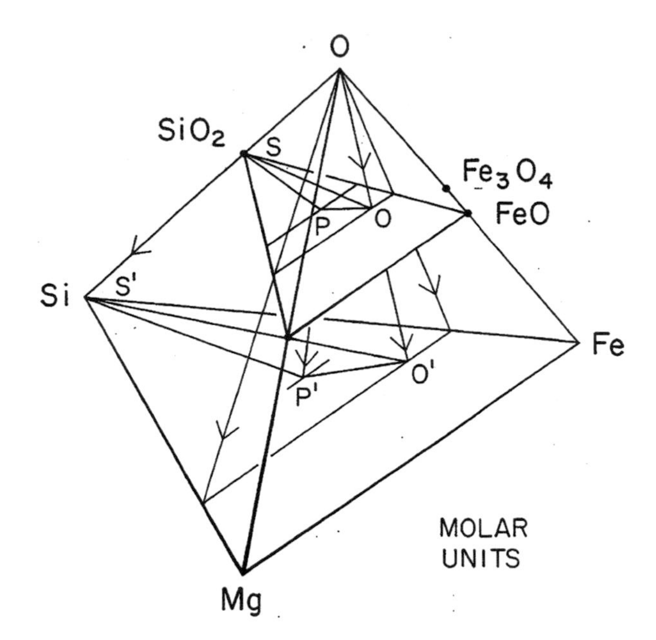

Fig. 17.5 Projection from oxygen of a triangle *ops*, in the plane MgO-FeO-SiO_2, onto the base Mg-Fe-Si of the quaternary system Mg-Fe-Si-O. Letters *o*, *p*, and *s* refer to olivine, pyroxene, and silica; the same letters primed refer to the projected compositions of these phases on the base. Pure magnetite plots at the point labelled Fe_3O_4 and projects to Fe

(as before in Chap. 16) of a very powerful method, a projection from oxygen of all the phases, this time onto the triangle Mg-Fe-Si. Such a projection is shown in Fig. 17.5; in the figure, the triangle *osp* is projected from MgO-FeO-$5iO_2$ onto the base, where it is labelled o' s' p' in projection. Now, the projected compositions all lie in the same plane as iron, Fe, and saturation with iron could be indicated by drawing lines from o', s', and p' to Fe. Note for future reference the position of magnetite, Fe_3O_4, in Fig. 17.5; this will project to the Fe corner when we get around to it.

Figure 17.6 summarizes the composition space as projected onto the base Mg-Fe-Si. The exact positions of *p* (pyroxene), *o* (olivine), and *l* (liquid) are of no concern as long as the *relative* positions are correct. Liquid is shown over a range of compositions merely as a reminder that it, as well as pyroxene and olivine, will have somewhat different compositions depending on the temperature and the nature of the reaction. We should recall that *l* lies towards oxygen, above the plane MgO-FeO-SiO_2 in Fig. 17.5; but this knowledge is not required for correct chemographic analysis in projection. It should be mentioned that the projected compositions of *o*, *p*, and *l can* be plotted exactly in their experimentally determined positions if desired; the projection is quantitatively rigorous.

The Schreinemakers analysis of invariant point OSPIL is illustrated in Figs. 17.7 and 17.8. If you start with the curve *ospl* and refer to Fig. 17.4, you will see at once that the curve must lie sub-parallel to the oxygen axis and terminate with iron at the lower end (and with magnetite at the upper end, as we shall see). We now write all the reactions about the invariant point, referring to Fig. 17.6 to see the chemographic relations that dictate the reactions. In shorthand form, the five reactions are as follows.

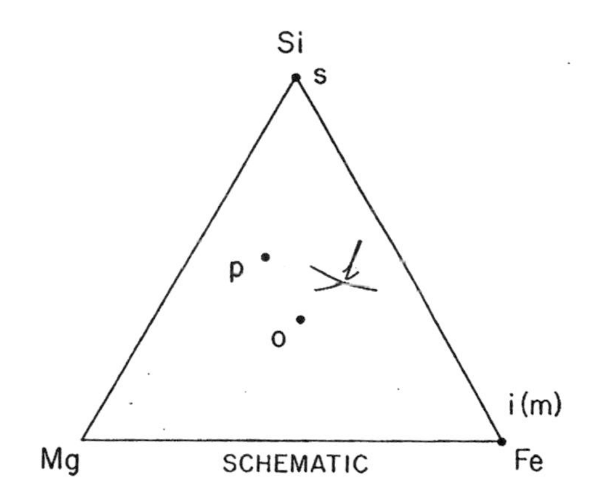

Fig. 17.6 Chemographic triangle Mg-Fe-Si, showing the projected compositions of olivine (o), pyroxene (p), silica (s), magnetite (m), and liquid (l). Iron (i) lies on the base at the Fe corner and is therefore not projected

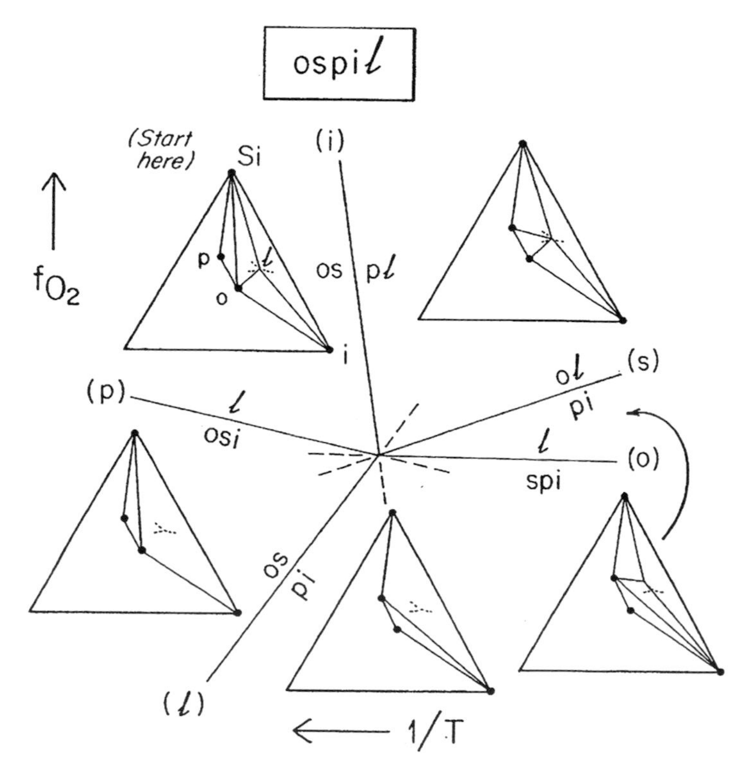

Fig. 17.7 Invariant point OSPIL, with chemographic diagrams based on Fig. 17.6

ospl (*i*)	$os = pl$
osil (*p*)	$osi = l$
ospi (*l*)	$os = pi$
opil (*s*)	$ol = pi$
spil (*o*)	$spi = l$

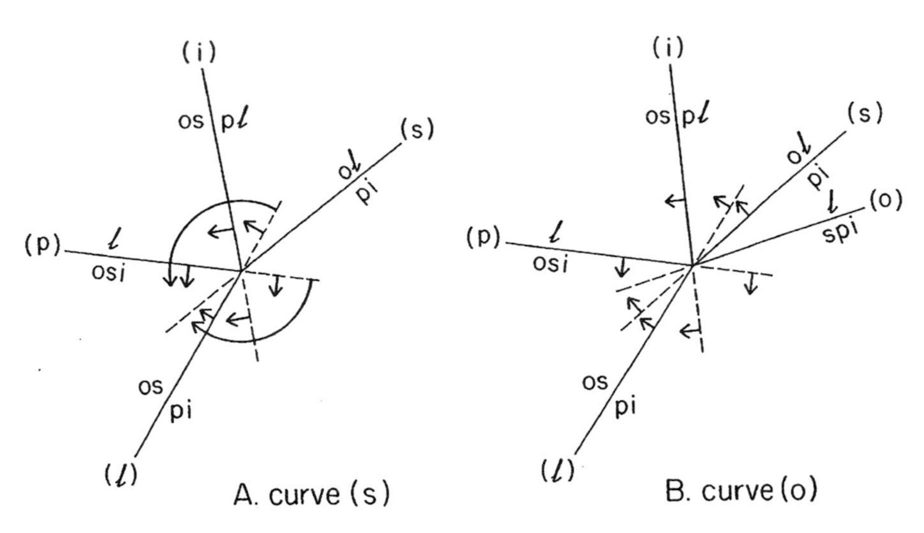

Fig. 17.8 Use of Richardson's rule to locate the curves (*s*) and (*o*) about the invariant point OSPIL

Curve (*p*) is known from Fig. 17.4 to lie at low temperature compared to the invariant point, and the same must obviously be true of curve *l*, a subsolidus reaction. The first three curves are therefore easily plotted. Curves (*s*) and (*o*) appear to offer problems at first, as their metastable extensions might lie anywhere between curves (*i*) and (*l*). But Richardson's rule solves the problem immediately, as shown in Fig. 17.8. For curve (*s*), merely draw arrows pointing towards the sector where *s* is generated by all the other reactions, stable *and* metastable (Fig. 17.8a). The result shows that the metastable extension of (*s*) must lie between curves (*p*) and (*l*). Note that the arrows are symmetrically disposed about the correct position of curve (*s*) and that two arrows point away from the stable segment of (*s*). The location of curve (*o*) is likewise determined with ease, as shown in Fig. 17.8b. We now have the complete geometry around invariant point *ospl*, and can verify from the chemographic diagrams that the sequence is correct because the tie lines are broken one at a time or appear or disappear three at a time where liquid is involved in a triangular field (Fig. 17.7). The configuration of the tie lines in the sector between (*s*) and (*i*) is actually that of the invariant point itself; within the sector the composition of liquid changes with T and f_{O_2}.

With the benefit of Fig. 17.7, it is now child's play to construct the sequence of curves about invariant point OSPML. First, referring to Figs. 17.5 and 17.6, we note that *m* plots at the Fe corner where *i* plotted before. The list of reactions about OSPML is therefore identical to the list for OSPIL (see above) if *m* is everywhere substituted for *i* in the old list. This means, of course, that the sequence of curves must be the same about the two invariant points; however, the two bundles of curves are *enantiomorphous* to one another. This can easily be seen from the fact that OSPML is the upper terminus of the stable portion of curve *ospl*; hence, it is a mirror image of OSPIL. The new bundle may therefore be obtained without further ado by reflecting across the normal to *ospl*; this has been done to obtain Fig. 17.9.

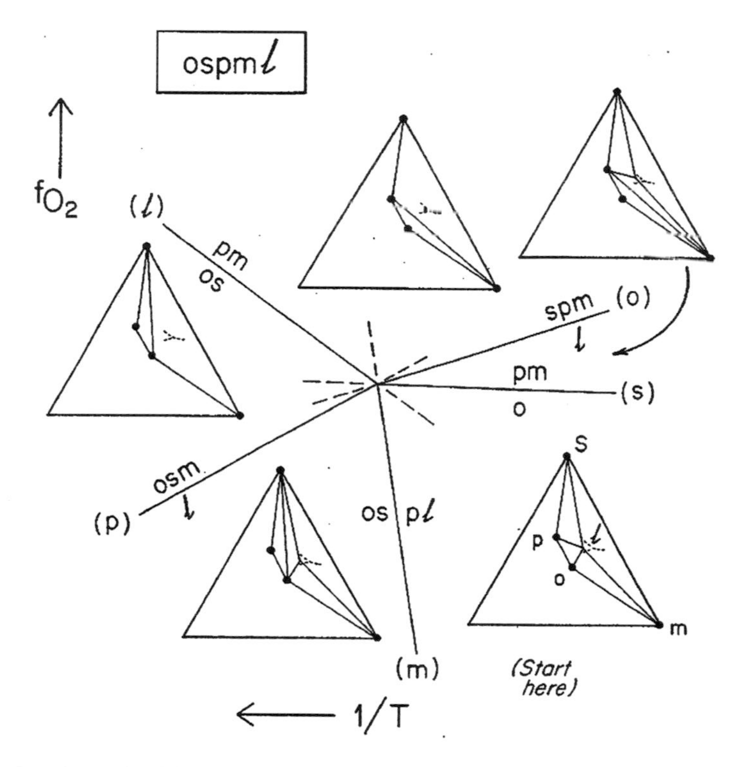

Fig. 17.9 Invariant point OSPML with chemographic diagrams

The two quaternary invariant points are united by curve *ospl*, as shown in Fig. 17.10. The figure is approximately scale-true, after Speidel and Nafziger (1968), who located curve *ospl* by experiment. Their version of this figure is quantitative and somewhat more complete than Fig. 17.10. The light lines near the left edge of the figure refer to the ternary equilibria in Fe-SiO_2-O (Chap. 16) and involve several buffer reactions for the pure Fe system, such as *osm* (i.e. FMQ). In Fig. 17.10, quaternary equilibria are denoted by heavy lines and capital letters, ternary equilibria by light lines and small letters. The divariant regions between the univariant curves correspond to the ones shown (in composition space only) in Fig. 17.4; OPL and OSM are examples. OPL is the surface between OPIL and OPML; OSM is the surface between OSIL and OSML. It is worth some effort to compare Figs. 17.4 and 17.10 until the relation between them becomes clear. Figure 17.10 can be characterized in two major ways: It shows the rise in T accompanying the addition of Mg to the Fe system, and it describes the incoming of a new phase, pyroxene. Note that the ternary *in*variant *points* (Fig. 17.10) generate quaternary *uni*variant *lines* having the same label. It will be seen that the known coordinates of the curves about OSPML require a sharp curvature of OSML near the invariant point in order for the metastable extension to lie in the correct sector (see

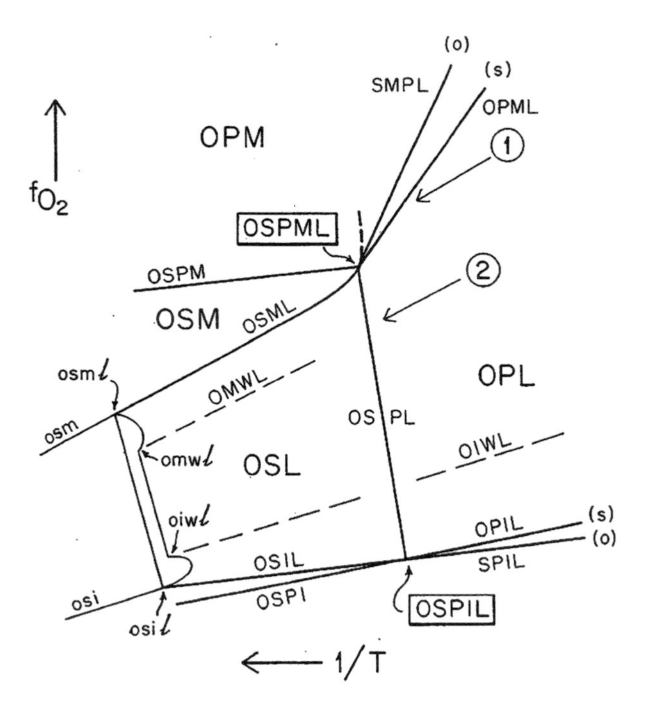

Fig. 17.10 Arrangement of invariant points, univariant curves, and divariant regions in the system MgO-FeO-Fe_2O_3-SiO_2. Heavy lines denote quaternary equilibria. Light lines denote ternary equilibria in the system Fe-SiO_2-O, as discussed in Chap. 16. Points labelled (1) and (2) are bulk compositions discussed in the text. (After Speidel and Nafziger 1968)

Fig. 17.9). Given the experimental data on the curves, without the Schreinemakers analyses, one would very likely draw an incorrect phase diagram at OSPML.

Crystallization paths can be followed qualitatively in both Figs. 17.4 and 17.10. Bulk composition (1) in the latter figure represents a liquid saturated with both olivine and pyroxene. It can be shown from external evidence (Morse 1979; Morse, Lindsley, and Williams 1980) that in a closed system such liquids follow a slope in T-f_{O_2} space roughly like that of the FMQ buffer (OSM). Such a path would be concave upward in Fig. 17.4. The liquid will eventually encounter OPML and then run along that curve as long as olivine and liquid survive. If olivine is lost on fractionation, the liquid will leave OPML but return to it as pyroxene and magnetite crystallize. The liquid may reach OSPML. At this peritectic point, pyroxene would be consumed by reaction, or with fractionation, the liquid would proceed along OSML. The sequence of rocks produced on fractionation would be harzburgite, magnetite-harzburgite, and then a rock consisting of fayalitic olivine, tridymite, and magnetite. Bulk composition (2), of lower oxygen content, would encounter OSPL and there resorb pyroxene if equilibrium crystallization occurred, conceivably travelling along OSPL all the way to OSPML. With fractionation, such a liquid would blithely cross OSPL and continue along OSL to OSML or to the Mg-free system. In all such cases, the extraction of ferrous silicates must raise the oxygen content of the liquid and magnetite must crystallize in order to account for the oxygen in the liquid. It is very important to note that the oxygen *content* of the liquid *must rise* along paths (1) and (2) even though the oxygen *fugacity falls*. The two properties are not necessarily correlated directly.

If we saturate the entire system MFSO with an approximately inert phase like plagioclase, the MFSO equilibria will persist largely unchanged but the crystallization temperatures will all be lowered. Thus, the whole array of Fig. 17.10 will shift to the left, but *down* and to the left approximately along the paths of OSML and OSIL. The diagram is qualitatively applicable, therefore, to basalts, gabbros, diorites, and related rocks. Assuming feldspar saturation, the rocks fractionally produced by bulk composition (1) would then be olivine norite, magnetite–olivine norite, and perhaps quartz–magnetite–fayalite–monzonite. In any case, at low pressures, Ca-poor pyroxene would disappear. The sequence wherein magnetite appears and Ca-poor pyroxene disappears is common in tholeiitic-layered intrusions, but the disappearance of orthopyroxene is not, in general, related to a silica-saturated reaction point like OSPML, as we shall see.

Role of Oxygen Fugacity in Complex Systems

Most of the principles of crystallization and fractionation pertaining to the system MgO-FeO-Fe_2O_3-SiO_2 carry over into more complex systems, such as the diopside-bearing system obtained by adding CaO to the above list. Presnall (1966) has studied this system at various oxygen fugacities and given a useful review of its bearing on the crystallization of basaltic magmas; his review includes a summary and elaboration of earlier work by Osborn (1959) which first showed quantitatively the effect of oxygen fugacity on fractionation of basic magma. Osborn distinguished two limiting cases: crystallization at constant f_{O_2} and crystallization at constant bulk composition. The former case requires membrane equilibrium with an external reservoir to supply oxygen and can be regarded as a kind of open-system model. The second case requires isolation from the surroundings, as in the example we have considered (Fig. 17.10), and constitutes a closed system model. One may readily see from Fig. 17.10 that imposing a constant oxygen fugacity on a crystallizing liquid would drastically alter the results, leading to strong silica enrichment caused by the abundant crystallization of magnetite; less extreme external regulation of oxygen fugacity would have a similar but less drastic effect. In either case, the Bowen trend of alkali and silica enrichment would result. In the case of constant bulk composition, the Fenner trend of iron enrichment results.

Conditions approaching constant oxygen fugacity during fractionation would require substantial addition of free oxygen to the system, an event likely to be rare in nature. However, the Bowen trend does not require such an extreme condition as constant oxygen fugacity. The favourite mechanism for introducing oxygen has been the dissociation of H_2O, under the assumption that the differential escape of hydrogen (as a proton) from the magma system would occur with ease. Osborn (1959) suggested that the Bowen trend, calc-alkaline series of volcanic rocks could be accounted for by contamination of magmas with water as fractionation proceeded, a case strengthened by the common association of calc-alkaline magmas with orogenic zones where water would be readily available from the voluminous sedimentary pile. Further discussion of the contamination hypothesis is given by Presnall (1966) and Czamanske and Wones (1973).

The Pyroxene Quadrilateral

In the system Fo-Di-SiO_2 (Chap. 12), we saw the coexistence of pigeonite with either diopside or protoenstatite. In Chap. 13, we deduced a relationship for plagioclase- and iron-bearing systems such that the assemblage Di + Pr_{ss} would give way to Di + Pigeonite with fractionation, to account for the sequence seen in tholeiitic-layered intrusions. We have seen in this chapter the evidence for iron enrichment in Ca-poor pyroxenes, but of course pigeonite cannot be obtained in the Ca-free system. In order to understand the crystallization of all three pyroxenes: augites (or diopsides), Ca-poor orthopyroxenes, and pigeonites, we should have to consider the system CaO-MgO-FeO-SiO_2 (CMFS). However, pyroxene relations in this quaternary system are exceedingly complicated and difficult to study experimentally, for several reasons. Miscibility gaps (i.e. solvi) and polymorphic transitions occur in the pyroxene solid solutions, and these chain silicates may nucleate or persist metastably in experimental runs (and evidently in nature also). As in the case of the ternary feldspars, it will suit our purpose here to fall back on the data from natural rocks as a guide to the salient phase relations.

The plane $CaSiO_3$ (Wo)–$MgSiO_3$ (En)–$FeSiO_3$ (Fs) lies at 50 mol% SiO_2 in the system CMFS, and it contains the pure CMFS analogues of the major rock-forming pyroxenes of basaltic rocks. None of these pyroxenes occur with compositions more calcic than 50 mol% Wo, so it is common practice to pay attention only to the Wo-poor half of the ternary plane, namely the *pyroxene quadrilateral* defined by the truncated triangle: Di-Hd-En-Fs. Figure 17.11 shows this composition region.

On Fig. 17.11 are plotted the pyroxene trends for two layered intrusions, the Skaergaard and the Kiglapait. Olivine occurs in both intrusions, except during the olivine hiatus in the Skaergaard intrusion, for which the compositional range in Mg/(Mg + Fe) is marked on the diagram. Tie lines from olivine to Ca-poor pyroxene in the Skaergaard intrusion are shown for the limits of the olivine hiatus. Tie lines from olivine to augite are omitted for clarity, but can be interpolated as parts of three-phase triangles from the other two types of tie line shown.

The Kiglapait intrusion is critically undersaturated in silica and contains no primary Ca-poor pyroxene. Its only pyroxene trend is the augite trend, which runs at slightly increasing Ca across the diagram to the Hd-Fs sideline.

The Skaergaard trend is typical of olivine–tholeiitic-layered intrusions, of which the Bushveld complex (Atkins 1969) is another famous example. The trends for coexisting augite and Ca-poor pyroxene evidently describe the compositional traces of the pyroxene solvi in the quadrilateral. Since we know that temperature falls towards the Fe sideline and away from the Mg sideline in any such system, the decreasing Ca content of the augite limb of the solvus cannot be accounted for by

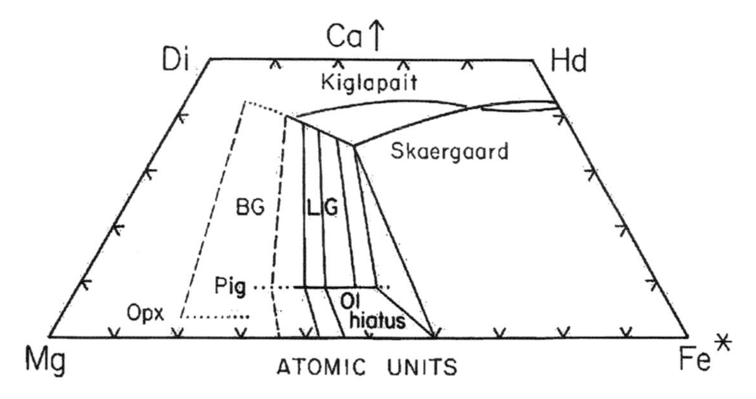

Fig. 17.11 Pyroxene quadrilateral, showing olivine and pyroxene composition trends for the Skaergaard intrusion (Wager and Brown 1968; augite trend after Chayes 1969) and the augite composition trend for the Kiglapait intrusion (Morse and Ross 2004). Di and Hd refer to diopside, $CaMgSi_2O_6$, and hedenbergite, $CaFeSi_2O_6$, respectively. Fe* refers to Fe^{2+} + Mn. BG, border group; LG, layered group; Pig, pigeonite; Opx, orthopyroxene; Ol, olivine. Olivine compositions lie on the base of the quadrilateral

falling temperature on a solvus with an isothermal crest line. It can be ascribed only to a solvus that is shaped like a plunging anticline, with axial trace running from about $Di_{25}En_{75}$ on the Mg sideline to about $Hd_{20}Fs_{80}$ on the Fe sideline. The augite trend follows a path of decreasing Ca until Ca-poor pyroxene disappears, after which it follows a path of rising Ca until it reaches the Fe sideline. The reason for this is clear: both augite and Ca-poor pyroxene must be constrained to the solvus as long as they coexist; when the Ca-poor pyroxene disappears, augite is no longer constrained to the solvus but simply follows a path dictated by the liquid. This path eventually merges, almost exactly, with the Kiglapait path.

The question of why Ca-poor pyroxene disappears has been addressed at length by Lindsley and Munoz (1969), who studied the sideline Hd-Fs. These authors deduced from several lines of evidence that the polythermal traces of the two-pyroxene solvus for silica-saturated rocks would continue along extensions of the Skaergaard paths shown in Fig. 17.11. This deduction was amply reinforced by later discovery of lunar pyroxenes falling along or near such trends, although the case of lunar pyroxenes is complicated by metastable effects, among other things (see, for example, Bence and Papike 1972). Among the evidence used by Lindsley and Munoz was the occurrence in Quebec and in Labrador of quartz monzonites with pyroxene pairs lying essentially on the extensions of the Ca-poor and Ca-rich solvus trends shown in Fig. 17.11. It is to be emphasized that these pyroxenes occur in company with abundant quartz, hence at silica saturation, whereas the Skaergaard and Bushveld intrusions have lost Ca-poor pyroxene long before their magmas become saturated in silica. Primary quartz is lacking in these intrusions except at very late stages of crystallization, where Mg is essentially absent. Lindsley and Munoz therefore reasoned that the disappearance of Ca-poor pyroxene could not be the result of a reaction relation among pyroxene, liquid, and a silica mineral (as on curve OSPL in MgO-FeO-Fe_2O_3-SiO_2) because Ca-poor pyroxene disappeared before the field boundary with a silica mineral was reached.

The stability of Ca-poor pyroxene relative to olivine is dependent on the silica content of the magma. In the absence of sufficient silica, the pyroxene will tend to decompose to olivine and a more silica-rich liquid. Such a reaction cannot be seen in a liquidus phase diagram, which deals only with liquid saturation surfaces, but it can easily be represented by a chemical equation, namely:

$$\underset{\text{(in pyroxene)}}{MgSiO_3} = \underset{\text{(in olivine)}}{MgSi_{0.5}O_2} + \underset{\text{(in liquid)}}{0.5\ SiO_2}. \tag{17.1}$$

By LeChatelier's principle, an increase in the concentration of silica would drive the above reaction to left; hence, the maximum persistence of Ca-poor pyroxene will occur when the system is saturated with silica in the form of primary quartz, as in the Quebec and Labrador quartz monzonites. Conversely, a decrease in the concentration of silica would drive the reaction to the right, causing the disappearance of Ca-poor pyroxene in favour of olivine. It therefore appears that the limiting composition of Ca-poor pyroxene in tholeiitic intrusions is in some way related to the concentration of silica in the magma. Extending the argument further, if the concentration of silica remains sufficiently low throughout crystallization, no Ca-poor pyroxene will occur, as in the case of the Kiglapait intrusion. We may therefore conclude that the pyroxene trends of layered intrusions are sensitive indicators of silica concentration.

It is preferable to deal in such cases with silica *activity* rather than concentration. For any component i, the activity a_i is related to the mole fraction X_i by $a_i = X_i \gamma_i$, where γ_i is the activity coefficient, to be determined from experience. Ideal solutions are defined as those for which $\gamma_i = 1.0$, hence for which $a_i = X_i$. Activity coefficients for complex magmas are poorly known, but can in principle be extracted through application of the law of mass action to cases where mineral compositions and liquid compositions are known. The *activity* is a thermodynamic property related to the chemical potential by

$$\mu_i = \mu_i^o + RT \ln a_i, \tag{17.2}$$

where μ_i is the chemical potential or partial molar Gibbs free energy of i, μ_i^o is the chemical potential at some standard state (such as the pure component i at the temperature and pressure of interest), R is the ideal gas constant, and T is temperature in Kelvins.

For reaction (17.1), we can write the equilibrium constant K for a given temperature and pressure as

$$K = \frac{a_{MgSi_{0.5}O_2} \cdot a^{0.5}{}_{SiO_2}}{a_{MgSiO_3}} \tag{17.3}$$

and taking the log of both sides, we get

$$\log K = \log a_{MgSi_{0.5}O_2} + 0.5 \log a_{SiO_2} - \log a_{MgSiO_3}. \quad (17.4)$$

By rearranging and multiplying to eliminate the fraction, we get

$$\log a_{SiO_2} = 2 \log a_{MgSiO_3} - 2 \log a_{MgSi_{0.5}O_2} + 2 \log K. \quad (17.5)$$

Log K is equal to $-\Delta G_r/RT$, where ΔG_r is the Gibbs free energy change for the reaction. This free energy change can in principle be calculated for any temperature of interest from tables of free energy such as those of Robie, Hemingway, and Fisher (1978). However, the results are very sensitive to small errors in the free energies of formation for forsterite and enstatite.

Equation (17.5) is a useful expression of the activity of silica for cases where the olivine and pyroxene have intermediate values of the Mg/Fe ratio. In such cases, the values of $a_{MgSi_{0.5}O_2}$ and a_{MgSiO_3} can be estimated from the mole fractions of these components in the olivine and pyroxene, respectively, by use of experimentally determined activity coefficients as for example given by Williams (1972, 1971a). Measurement of the mineral compositions, combined with knowledge of the free energy of reaction and hence log K for a given temperature, thus leads directly to an estimate of a_{SiO_2} in the liquid, provided that olivine and Ca-poor pyroxene co-existed with liquid and that the temperature of crystallization can be estimated. An internally consistent set of constants is given in the paper by Morse (1979), where procedures of the calculation method are also described. Calculations of silica activity in the Skaergaard intrusion are given by Morse, Lindsley, and Williams (1980). The parallel relation between silica activity and oxygen fugacity (which is also an activity; note the identity of the definitions in Eqs. (16.3) and (17.2)) in basic magmas is also discussed in these papers.

In the latter regard, we have already seen some basic principles about the parallel relation of the oxygen and silica activities, starting with the FMQ reaction in Chap. 16 (Eq. 16.2) and the role of oxidation in magmas identified by Kennedy (1955), Osborn (1959), and others. It is clear from all these studies that oxidation and silication go hand in hand, for example, in the oxidation of olivine to form magnetite and silica. It therefore appears that the pyroxene trends of Fig. 17.11 should say something about the relative oxidation states of the Skaergaard and Kiglapait intrusions, namely that the more silicated Skaergaard magma should have been more oxidized than the silica-undersaturated Kiglapait magma. Studies of the Fe-Ti oxide minerals in the two intrusions bear out this theoretical prediction (Morse 1980).

Examples of the use of silica activities, and other calculations based on equilibrium constants, abound in the book by Carmichael, Turner, and Verhoogen (1972) and are also given in the papers of Lindsley, Brown, and Muir (1969) and Williams (1971b). Such calculations find increasing use in igneous petrology and form an important extension of the experimental base provided by phase diagrams.

We may close this chapter with three applications of these principles to actual rock systems.

Silica Activity in the Skaergaard Intrusion

We wish to find the activity of silica in the Skaergaard magma when Ca-poor pyroxene ceased to crystallize, liberating the augite composition from the pyroxene solvus. A magma saturated with a silica mineral (tridymite in this case) has $a_{SiO_2} = 1.0$ by definition, and under such a condition, an increase in the concentration of SiO_2 would only serve to make more tridymite. Since we ascribe the loss of pigeonite to a low activity of silica, we will expect to find $a_{SiO_2} < 1.0$. For use of Eq. (17.5), we take as data the conditions at the 1580 m level of the Skaergaard intrusion, corresponding approximately to the three-phase triangle shown in Fig. 17.11. Data are taken from Wager and Brown (1968), Lindsley, Brown, and Muir (1969), and Morse, Lindsley, and Williams (1980). The temperature is taken as 1364 K (1091°C) and the pressure as 830 bars for this level of the intrusion. The term 2 log K for the pure end member reaction En = Fo + silica therefore works out to a value of −0.214, using the reaction constants given by Morse (1979) for the equation $\log K = A/T + B + C(P - 1)/T$. (The constants are $A = -364$, $B = +0.169$, $C = -0.0193$).

The mineral compositions are olivine Fo_{40} and pigeonite En_{49}. To correct from pigeonite to hypersthene, we add 2 to obtain En_{51}, by reasoning explained in Morse (1979). The corresponding activities of components Fo and En in the crystalline solutions, taken from the activity–composition relations given by Williams (1972, 1971a), are 0.473 and 0.576, respectively. Equation (17.5) now reduces to

$$\begin{aligned}\log a_{SiO_2} &= 2(-0.240) - 2(-0.326) - 0.214 \\ &= -0.480 + 0.652 - 0.214 \\ &= -0.042\end{aligned} \tag{17.6}$$

and $a_{SiO_2} = 0.9$. We therefore have verified, from the mineral compositions and estimated crystallization conditions, that $a_{SiO_2} < 1.0$, and that the magma was not saturated with tridymite, a fact observed in the rocks. But more than this, we have obtained a specific value for the activity of silica when pigeonite disappeared from the Skaergaard magma, and we can say that a_{SiO_2} must have been less than 0.9 (assuming similar T, P conditions) at a similar stage in the Kiglapait intrusion, in order for pigeonite not to have crystallized, and must have been greater than 0.9 in the Quebec and Labrador examples where Ca-poor pyroxene continued to crystallize. By such exercises, we gradually learn to attach meaningful numbers to the intensive parameters of state of crystallizing magmas.

The Expanded Pyroxene Quadrilateral and Augite–Olivine–Liquid Equilibria in the Kiglapait Intrusion

The pyroxene quadrilateral shown in Fig. 17.11 served as a specific example of a practical application. In this discussion, the main goal was to understand the constraints on augite–orthopyroxene–olivine equilibria in the Skaergaard intrusion. At the same time, the contrasting array of the Kiglapait augites showed the trend of lower silica activity when no Ca-poor pyroxene existed at the liquidus.

The augite-opx tie lines in this figure clearly outline a solvus that tends to close with lowering temperature: The tie lines become shorter as the minerals get richer in iron. What would a complete solvus look like in this quadrilateral? The answer resides in Fig. 17.12. Here, the space is contoured with isotherms, and their nodes illustrate the shape of the thermal "anticline" with its nose running to the iron-rich right. The three-phase triangles denote the coexistence of augite (C2/*c*), orthopyroxene (Pbca), and pigeonite, which is the low-Ca monoclinic pyroxene with $P2_1/c$ symmetry. The "forbidden zone" denoted by a small-dashed curve denotes the region in which the three-pyroxene triangles are metastable with respect to augite–olivine–liquid equilibria. The diagram shown here is for a pressure of 1 atm, which might be close to appropriate for the Skaergaard intrusion, so from Fig. 17.11 the last three-phase triangle (here with olivine) would fall at approximately 950°C, all other things being equal. For the Kiglapait intrusion, the rather wavy line shown in Fig. 17.11 is due to the presence of five samples near the Mg-end of the array that were not accompanied by a full fraction of olivine. Since they were not constrained by olivine saturation to a Ca-poor augite–olivine equilibrium, their Ca-rich positions influenced any attempt to draw a composite trend for the augite

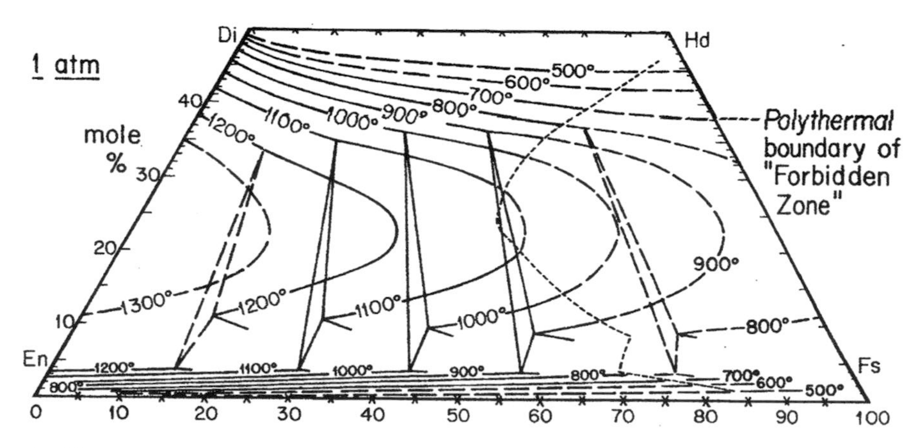

Fig. 17.12 Pyroxene quadrilateral with temperature contours at 1 atm (from Lindsley and Andersen 1983). The original text prescribes the methods of projections that must be followed to make this a valid thermometer, and it *strictly forbids* shortcuts that avoid this procedure. The original text also provides equations that correct for pressure. Similar diagrams at several pressures are given in the presidential address by Lindsley (1983)

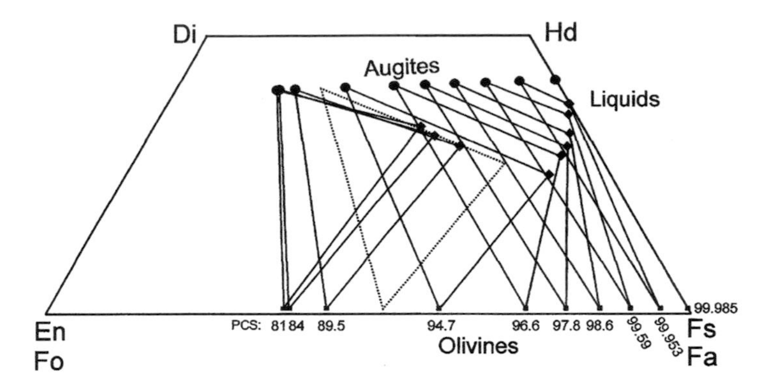

Fig. 17.13 Augite and olivine pairs with liquid compositions in the Kiglapait intrusion, Labrador. The augite and olivine compositions are from Figure 14 in Morse and Ross (2004), and this diagram is from Fig. 15 in the same paper. The liquid compositions are from Morse (1981) except in the left-most triangle, which is experimental from Morse et al. (2004). The stratigraphic position of selected samples is listed in terms of PCS, the volume percent of the magma crystallized. The diagram shows the interesting negative slope of the liquid Wo content until magnetite begins to crystallize in quantity at 93 PCS, after which the Wo content in the liquid rises sharply to the Mg-free end point. The absence of low-Ca pyroxene in this low-silica intrusion contrasts with the Skaergaard case shown in Fig. 17.11. (Diagram from Morse and Ross 2004, with permission)

compositions in the intrusion. The remaining olivine-saturated augites, however, form a statistically straight line that becomes slightly more Ca-rich with Fe content, as shown in Fig. 17.13. Here, we find a suite of three-phase triangles that hold some interesting information about fractional crystallization in a complex silicate system.

In this figure, the augites are tied to coexisting olivines and then to liquids. In one case, the liquid is experimental at 5 kbar; the rest are interpreted from bulk chemical compositions and calculated liquids from Morse (1981). The most interesting feature of the diagram is the rapid depletion of Ca (i.e. augite component) in the liquid caused by the crystallization of abundant augite in the region 84–94 PCS in the intrusion. This depletion trend terminates abruptly at about 93 PCS when Fe-Ti oxides begin to precipitate in abundance to form the massive “main ore band” of titanomagnetite (about 80 cm thick in the northern exposure), so by 94.7 PCS the trend has already begin to swing towards the final end point at Mg-free ferrohedenbergite. Here, the modifier “ferro” is classically used to call attention to the ferrosilite component (Fs in the figure), which defines a eutectic at the Fe termination of the liquid path in the figure.

Silica Activity in the Kiglapait Intrusion

In the discussion above, we found the silica activity for the Skaergaard intrusion at the point where pigeonite ceased to crystallize (an actually ill-defined place in the actual stratigraphy, where pigeonite seems to come and go over a range of stratigraphic height). We avoided any overall trend of silica activity, but in fact that was calculated in some detail for the whole intrusion by Morse et al. (1980), along with the oxygen fugacity. However, the Kiglapait intrusion presents a challenge, because for about 84% of its crystallization history, the only liquidus Fe-bearing phase is olivine, and with only Ol + liquid, one has no handle on the activity of silica. This problem is resolved by considering the fictive equilibrium among orthopyroxene, olivine, and liquid, and then erasing the Opx part by a subtle manoeuvre. For this exercise, we go way back to the En-Fo-Sil equilibria of Morse (1979). (We may be sure that this source of wisdom is outdated, and yet not quite convinced that it is useless; the results are subject to test.) For a more detailed exploration of this exercise, the reader is referred to Morse (2014), about which yet more anon.

Briefly, then, we assume lower zone equilibria olivine–orthopyroxene–liquid–(plagioclase) and write the silication EFS reaction:

$$\underset{\text{(enstatite)}}{MgSiO_3} = \underset{\text{(forsterite)}}{MgSi_{0.5}O_2} + 0.5\ \underset{\text{(silica)}}{SiO_2}. \quad (17.7)$$

and also the FMQ reaction:

$$6\ \underset{\text{(fayalite)}}{FeSi_{0.5}O_2} + \underset{\text{(oxygen)}}{O_2} = 2\ \underset{\text{(magnetite)}}{Fe_3O_4} + 3\ \underset{\text{(quartz)}}{SiO_2}. \quad (17.8)$$

and find reaction constants for these as listed in Morse (1980) and again in Supplementary Material for Morse (2014), where all of this study in silica activity is developed in full.

When the condensed phases coexist in equilibrium at P and T, they define the state of the system in terms of silica and oxygen activities, which apply also to the coexisting liquid. The equilibrium constant K for a given temperature can be written

$$\log K = \log a_{MgSi_{0.5}O_2} + 0.5 \log a_{SiO_2} - \log a_{MgSiO_3}. \quad (17.9)$$

Rearranging and multiplying to eliminate the fraction, we have

$$\log a_{SiO_2} = 2 \log a_{MgSiO_3} - 2 \log a_{MgSi_{0.5}O_2} + 2 \log K. \quad (17.10)$$

As before, Log K is equal to $-\Delta G_r/RT$ where ΔG_r is the Gibbs free energy change for the reaction. Empirical constants for the reaction are given in the form $A/T + B + C(P - 1)/T$ in Table 6 of Morse (1980) and in the 2014 Supplementary Material cited above.

Recalling the warning that this development invokes the presence of orthopyroxene that does *not* occur in the Kiglapait Lower Zone liquidus assemblage, we are sure that the estimate of the silica activity calculated here will be a maximum, from which we will need to adjust downwards to find a relevant result. First, though, we wish to plot calculated silica activities for OPX equilibria against temperature.

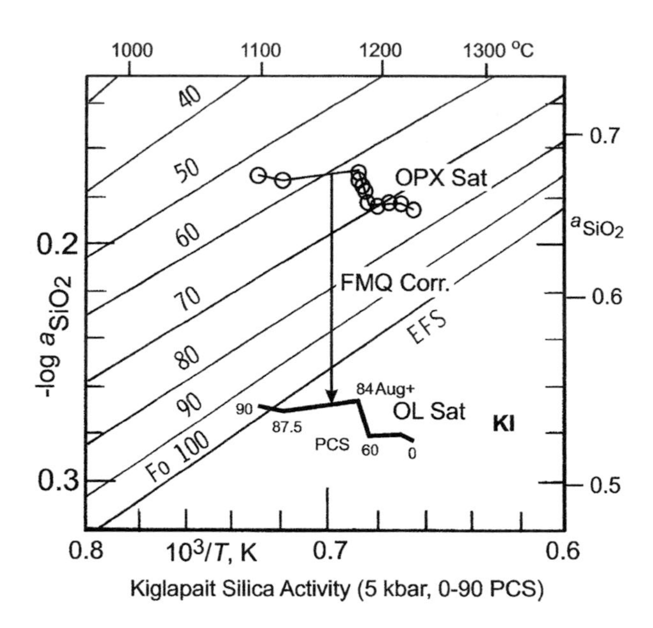

Fig. 17.14 Contoured silica activity diagram for 5 kbar plotted against Kelvin temperature. The contours are from the EFS reaction and Fig. 1b of Morse (1979a). The contours are for olivines in nominal equilibrium with Ca-poor pyroxene ("OPX Sat"). Temperatures are experimental, as discussed in Morse (2014). The array is then translated downward to the Opx-absent condition of the Kiglapait Lower Zone using a correction from FMQ equilibrium (Morse 1980), to find the nominal silica activity for this region of the intrusion. The labels on the lower array indicate the percent-solidified (PCS) values used to describe the relative volumes of rocks below the samples. The diagram is used to evaluate the silica activity from 0 to 90 PCS in the intrusion where the local (plagioclase + olivine) mineralogy by itself cannot yield an estimate of silica activity. (Diagram from Morse 2014, Fig. 7, with permission)

Using an old but adequate algorithm, we can convert observed Lower Zone olivine compositions to comparable orthopyroxene compositions: Let $X_{\mathrm{En}} \approx 0.85X_{\mathrm{Fo}} + 0.15$. Then, we apply the results to Eq. (17.10), with temperatures determined experimentally at 5 kbar by Morse et al. (2004). The results are plotted here in the upper part of Fig. 17.14. This diagram is from Morse (1979), and it shows contours of olivine composition as a function of temperature using the EFS algorithm. We also have one calculation for the actual silica activity in the Kiglapait intrusion from a study of the Fe-Ti oxide minerals by Morse (1980), so we can use this to adjust the OPX array downward to represent olivine saturation, "OL Sat" in the lower part of Fig. 17.14. The data are consistent with $-\log a_{\mathrm{SiO_2}} 0.26 \sim 0.28$, $a_{\mathrm{SiO_2}} \sim 0.53$.

Is this result in any good? Well, it can be tested by reference to direct calculations of silica activity from multicomponent cumulates from the Upper Zone, combined with calculations from the QUILF formulation of Andersen et al. (1993) and calculated by D. H. Lindsley, shown in Fig. 17.15. The components involved in these equilibria are fictive quartz (Q), ulvöspinel (U), ilmenite (I), liquid (L), and fayalite (F). This algorithm is used to calculate oxygen and silica activities from

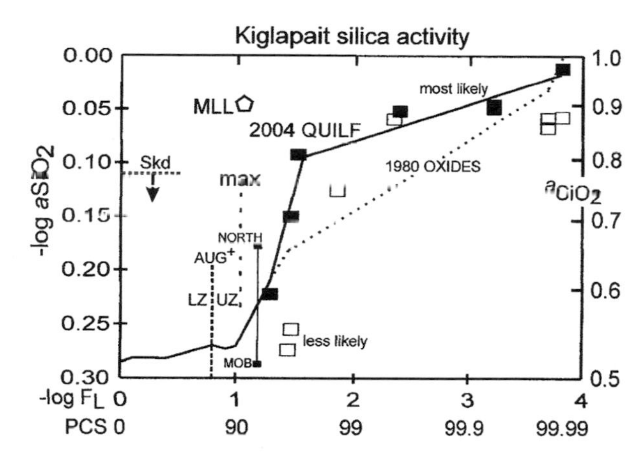

Fig. 17.15 Activity of silica in the Kiglapait intrusion. The data for the Lower Zone and early Upper Zone are taken from Fig. 17.14 where they are calculated from the EFS reaction. Here the trend is reversed along with the temperature scale. Data shown in filled rectangles for the Upper Zone refer to coexisting pairs of augite and olivine at high temperature, from the QUILF equilibrium of Andersen et al. (1993) calculated by D. H. Lindsley (Pers. Comm. 2004). The original calculations by Morse (1980) for the Upper Zone beyond the Main Ore Band (MOB) gave the open boxes and lower, dotted line, but the subsequent calculations from QUILF (black rectangles) have superseded the earlier results. The sharp increase in silica activity beyond the Main Ore Band arises from the sudden deposition of Fe-Ti oxide minerals, mostly titanomagnetite, and the resulting silication of the main magma body. Figure modified from Fig. 16 of Morse and Ross (2004). The polygon labelled "MLL" shows the silica activity for the tholeiitic Makaopuhi lava lake (Wright and Weiblen 1967) at an olivine composition of Fo_{49}, corresponding to 91 PCS in the Kiglapait intrusion. The MLL liquid is reported to have phases Ol, Opx, and Mt at 1293°C. The dotted line at 0.115 on the *Y* axis labelled "Skd" is a recalculation for the base of the Skaergaard intrusion using $T = 1168$°C and olivine Fo_{72} (Thy et al. 2009, 2013) and the EFS Fig. 1a of Morse (1979a) for conditions at 1-atm, tridymite base, pigeonite at the liquidus. This is a maximum value because pigeonite is not an early liquidus phase; accordingly, a small arrow suggests a range toward a lower value. (Diagram from Morse 2014, with permission)

minimized Gibbs free energies. As the long caption shows, the values for the Kiglapait silica activity are reasonably secure and consistent with petrographic observations. Several other systems of interest are also plotted, most especially the estimate for the maximum silica activity in the Skaergaard intrusion, somewhat below our early rough estimate of 0.9.

Chapter 18
The Effects of High Pressure

Introduction

Pressure increases with depth in the earth, to a maximum at the centre. The rate of increase depends on the material present along the path followed; it is lower, for example, in sea water than in rocks. But a useful generalization, adequate for most petrologic purposes to moderate depth, is that the pressure increases by about 3 kbar for every 10 km. Inverting this, depth increases at a rate of about 3.3 km/kbar. A *bar* is defined as 10^5 pascals and is about equal to an atmosphere of pressure[1]. The bar and its multiples are therefore convenient units for Earth science, ranging from millibars (mb) in the atmosphere through bars in the oceans, kilobars in the crust and upper mantle, and megabar (Mbar) in the deep mantle and core. The core–mantle boundary lies at a pressure of about 1.27 Mbar (127 GPa) and a temperature about 3500 K. Despite Earthly convenience, the uniform code for scientific publication is based on multiples of Pascals. Some Earthly journals still acccpt kbar for upper-crustal pressures.

As we have seen from the Clausius–Clapeyron Equation (3.1), all materials that have a positive volume change in melting will have positive melting slopes, dT_m/dP. This is true of all the solid forms of H_2O except ice-1 and of most known silicate materials at low-to-moderate pressure. The *rates* at which the melting temperatures increase with pressure vary from quite small, e.g. about 3°C/kbar for anorthite, to rather large, e.g. about 15°C/kbar for diopside. This means that phase diagrams will tend to change with pressure, with eutectics and cotectic shifting away from the phase with the higher dT_m/dP. Some familiar phase diagrams will change only a little, and others will change greatly, even so as to yield completely different relationships, as with the change to congruent melting for enstatite mentioned in

[1]Experiments or observations said to have been made at 1 bar are falsely described unless made under strict controls. Experiments made in ambient conditions and elevations above sea level are best described as made at 1 atm. It is imprecise but far more honest than 1 bar.

S. A. Morse, *Basalts and Phase Diagrams*,
https://doi.org/10.1007/978-3-030-97882-2_18

the last chapter. New solid phases appear at high pressure in many systems, and these may affect the melting relations profoundly.

The source regions of basaltic magma quite surely lie in the range 5–30 kbar and very commonly near 10 kbar (Presnall et al. 1979). We shall therefore be concerned, in this book, with melting relations at moderate pressures, usually less than 30 kbar.

In this chapter, we review in sequence the high-pressure melting relations of most of the systems previously discussed in this book. We shall be concerned chiefly with the geometrical relations of phase diagrams, both in *P–T* space and in isobaric *T–X* sections through *P–T–X* space. For ternary systems, we shall need to map thermal information onto the *X–X* diagram at constant pressure, just as we did at 1-atm pressure. The treatment in this chapter is introductory. We begin with a brief introduction to the methods of high-pressure research.

High-Pressure Apparatus

The conduct of experiments at high pressure is now a routine matter in many laboratories, and it is a highly advanced practice in select institutions. We shall undertake here only a very brief account of the main instruments in use today. The art of high-pressure research owes much to the pioneering work of the Nobel Prize-winning physicist Percy W. Bridgman[2] at Harvard University in the early part of the twentieth century. Two of the instruments in use today for petrologic research were designed by geologists who were acquainted with his laboratory.

The most commonly used apparatus today for high pressures in the range 5–60 kbar is the so-called single stage or piston–cylinder or solid-media apparatus (Boyd and England 1960; see also Ulmer 1971, for a review of this and other types of apparatus). If you take a short, thick-walled pipe, insert it in two tight-fitting opposed pistons that protrude beyond the ends of the pipe and squeeze the pistons together with a hydraulic jack, you have a piston–cylinder apparatus. In practice, the lower piston is usually a static plug of tungsten carbide. The sample is placed between the pistons and surrounded by a graphite cylinder which acts as a furnace when an electric current is applied through the piston, across the graphite, and out through the carbide plug. Temperature is measured by inserting a thermocouple next to the sample and guiding the leads through the piston, so they do not get sheared off. Pressure is measured by means of a gauge in the hydraulic line and calculated for the sample chamber by calibration against known phase transitions.

[2]The pioneering role of Percy Bridgman in the study of high-pressure mineralogy was finally recognized in the 2014 naming of bridgmanite, the major mineral stable in the lower mantle of the Earth to the core–mantle boundary (CMB). When this "post-perovskite" phase was found by diamond cell experiments to exist at the extreme pressures of the mantle, the name was reserved until a natural occurrence could be found. Lacking specimens from the CMB, this was a tall order, finally achieved by the patient search among shocked stony meteorites (Tschauner et al. 2014). It is rightly called the most abundant mineral in the Earth.

The catch in this description is the misleading scale connoted by use of the words "pipe" and "cylinder". The pistons in common use are roughly the size of your forefinger, but the cylinder may be about a foot in diameter and perhaps a third of a foot thick, with a piston-sized hole drilled along its cylindrical axis. To give the cylinder added strength, it is made up of several concentric rings, each with an interference taper, swaged into each other with great force. This scheme pre-stresses the piston chamber. Great efforts are made to reduce the friction between piston and cylinder and to reduce the shearing of the sample by directed stress. These ends are often accomplished by using pyrex glass or rock salt (which soften at high T and P) in the sample chamber.

Samples to be used in the piston–cylinder apparatus are often contained in graphite capsules. These have the advantage that a graphite rod can be cut short and drilled out with three holes distributed in a scalene triangle (so that they can be identified as 1, 2, and 3 after recovery). They have a second advantage of buffering the silica and oxygen activity to below FMQ and near WM, suitable for mafic bulk compositions that warrant study under those conditions. Other types of sample holder, such as crimped and welded noble metal tubing, can also be used. The amount of sample used is typically only a few milligrams. Water, if desired, is introduced to the tube before the sample by means of a hypodermic needle or fine syringe. Carbon dioxide may be introduced as well by adding crystals such as oxalic acid ($C_2H_2O_4$). After the apparatus is assembled with the sample in position, the run may be brought to P and T in a matter of 5–10 min, and when the experiment has run the desired length of time, the charge is quenched in less than a minute by simply turning off the power to the furnace assembly. Samples are then retrieved with a postmortem dissection process so that they stay intact and then placed in an epoxy mould and cured overnight before polishing.

The advantages of the piston–cylinder apparatus are that it is quick and easy to use, relatively inexpensive to maintain, capable of temperature limited only by the materials used (around 1650°C for platinum capsules, but much higher, 2100°C, for graphite and other materials), and capable of routine pressures to 50 or 60 kbar. Its disadvantages are related mainly to the friction problem, which introduces errors particularly at low pressures, below 5 kbar, and some limitations on the precision of temperature control.

For pressures up to 5–10 kbar, 300–700°C, an externally heated cold seal apparatus is often used, with water or argon as the pressure medium. This useful device was invented by O. F. Tuttle. The reaction vessel is a steel oblong cylinder, typically 3 cm in diameter and perhaps 30 cm long, along which a 6-mm hole is drilled to within several centimetres of one end. The sample, welded into a noble metal or silver tube, is pushed into the closed end with a filler rod, and the open end is sealed with a cone-in-cone seal held tight by a large nut threaded onto the open end of the reaction vessel. Water or gas is pumped under pressure into the vessel through an orifice in the cone-in-cone seal and maintained at pressure by closing a valve in the pressure line. Pressure is monitored by a Bourdon tube gauge in line with the delivery system. Heat is applied to the reaction vessel by putting a cylindrical furnace around it near the closed end, so that the seal area remains relatively cool.

Temperature is measured by inserting a thermocouple into a well drilled off-centre in the closed end of the vessel, so that the thermocouple junction lies beside the sample, separated from it by a few millimetres of steel. The run is brought part way to pressure by pumping the pressure medium, then heated by applying the furnace, and thus brought to *P* and *T* in a matter of some 10–15 min. It is quenched, at pressure, by removing the furnace and directing a jet of air against the vessel, followed eventually by a jet of water.

The advantages of the cold seal apparatus are its small expense, ease of operation, and suitability to accurate control of low pressures in the range 1–10 kbar. Its disadvantages are its limitation to low temperatures, beyond which the reaction vessel may fail by rupture, and some imprecision in temperature control and measurement, due in part to large thermal gradients.

An early king of instruments for high-pressure research with relatively large (>50 mg) samples was the internally heated gas medium apparatus or so-called gas apparatus. This was developed originally by Bridgman and modified for routine petrologic work by Yoder (1950). Its key feature is the Bridgman "unsupported area seal", which gets tighter the higher the pressure. The pressure vessel is a large steel alloy cylinder with a small axial hole into which is introduced the furnace assembly with its sample. The hole for the furnace assembly is of the order of 2 cm in diameter, and the enclosing cylinder is typically 20 cm in diameter by 50 cm long, wrapped with copper tubing through which water flows to cool the vessel. Each run typically contains four samples in welded platinum tubes, arranged around a thermocouple junction, embedded in granular alumina within a 30-g, heavy-walled, platinum cylinder called a "block" (because it provides a high thermal inertia to the sample area). The block, typically 8 mm in outside diameter, rests inside a ceramic tube about 15 cm long, onto which is wound a platinum resistance wire to act as a furnace. This furnace array is cemented into a cylindrical nickel support housing, which in turn is connected at one end to a "head" about 4 cm in diameter, containing stacked washers of steel, copper, and leather. When the furnace assembly is put into the pressure vessel, one end of the head fetches up against a shoulder, and the other end is forced against the washers, and these in turn against the shoulder, by means of an enormous nut (really a hollow bolt), tightened with a 3-foot wrench. The longitudinal force against the flat sides of the washers makes them extrude radially into the walls of the sealing chambers of the pressure vessel, and when the pressure medium (argon) is pumped into the vessel through an orifice in the closed end, the washers are compressed even more, to tighten the seal. This is the unsupported area seal of Bridgman.

The gas apparatus is brought to pressure, after loading, by pumping argon into the pressure vessel and simultaneously into an intensifier consisting of a hydraulic ram with its chamber expanded. When the pumping limit (2–5 kbar) is reached, the ram is advanced past the pumping orifice and the pressure in the entire system is thus further intensified (to a conventional limit of 10 kbar) by the consequent reduction in volume. The furnace is then turned on, being attached by welds to permanent pressure-tight leads through the head and thence through an axial hole in the nut. Pressure is monitored externally in the line to the intensifier by a manganin cell, whose emf is a nicely known function of P. The run is rapidly quenched by turning off the power to the furnace.

The advantages of the gas apparatus lie in its precision of T and P measurement and control, to 2°C and a few tens of bars; in its hydrostatic application of pressure; and in its high-temperature limit (near 1700°C) at 10 kbar. Its disadvantages are in the somewhat lengthy preparation needed for each run, although on good days one can quench and reload in the space of an hour; in its practical limitation to 10 kbar; and in the shielding and precautions needed for safety. A gas apparatus at pressure is not unlike the chamber of a good artillery piece being fired, and any failure is likely to be explosive. Actually, only one real humdinger of a failure is known to me, and its cause was simply wear and tear, not altogether unanticipated. The nut stripped its threads and flew out, redesigning a substantial wall and overturning a tall glassed-in bookcase in the next room: Yoder's office! The gas apparatus is under-used today.

A more spectacular, large-scale high-pressure apparatus is the opposed multi-anvil device invented in 1958 and developed at Cambridge University by David Walker and at Carnegie by Dean Presnall among others. This instrument is particularly useful for the study of upper mantle phase equilibria, with pressure limits nominally at 10–20 GPa (and 3000°C, hence the Transition Zone) but now nearly 100 GPa at high T, hence getting close to the CMB! It is based on either six anvils to compress a cube, or eight anvils to compress an octahedron. The sixfold rig has a horizontal ring of three and a vertical pair of three anvils. The hydraulic press is likely to be 6–8-ft high and extremely strong. The sample assembly lies at the junction of all the anvils, and the thermocouple wire is led out in a channel between two anvils. Experiments can be run "live" with X-ray diffraction at the sample or, more likely in modern studies, with synchrotron radiation for finer details of phase changes or characterization of run products. With this capability, the mineralogy and phase changes in almost the entire mantle have been studied.

The most versatile and productive instrument for high-pressure research is unquestionably the opposed diamond anvil apparatus, developed at the U.S. National Bureau of Standards and taken to a level of high refinement by Ho-Kwan Mao and his colleagues (including his daughter) at the Geophysical Laboratory, Carnegie Institution of Science. The instrument is ridiculously simple in principle, consisting of two brilliant-cut diamonds with their points flattened off parallel to the tables and supported, point to point, in a small thumbscrew vise. The sample resides between the flattened points, surrounded by a gasket. The entire pressure cell is mounted in the optical path of a microscope so that experiments can be observed in progress with transmitted light, and it can be demounted for X-ray diffraction or spectroscopic studies. The catch is that the instrument must be beautifully designed and built to hold the alignment of the diamond and the diamonds must be perfect and masterfully dressed. Pressure is ingeniously measured by throwing a few grains of ruby in with the sample and observing the wavelength shift of a beam of laser light, which passes through a ruby crystal and is affected by the compression of the crystal. With this instrument, Mao and Bell attained pressures near that of the core–mantle boundary, in a cell so perfect that the diamond failed by plastic flow rather than by cracking. As development continued, the Carnegie team found that the use of natural diamonds limited the depths in the Earth that could be studied, because almost all natural diamonds have flaws or strain effects or inclusions that weaken them. Now, these teams grow their own strain-free diamonds and extend their experimental capacity into the outer core. Justly heralded as a window into the deep Earth, the diamond cell is an exciting vehicle for discovery about the Earth and about the laws of physics. With the multi-anvil apparatus, it brought us the mineral we now can call bridgmanite.

The instruments described above are all static devices, but it is well known in the physics of high pressure that controlled extreme pressures can be reached by dynamic methods. A highly productive and rather exciting shock-wave instrument was devised at Caltech by the late Thomas Ahrens and is now supervised by his successor, Paul Asimow. The target area is supported by a massive stainless steel recovery chamber with a cavity into which the target material is loaded. The target itself may be a single solid composition or an assembly of layers. The target area is monitored by ultra-high sensitive optical devices that record the shock history in nanoseconds. A long, evacuated steel tube extends to a firing source connected to an air compressor. The projectile is a flat-plate flyer disc made typically of Ta 2 mm thick attached to the face of a polycarbonate sabot. When fired at ultrasonic speed, this projectile impacts the sample, which is then recovered and studied by an assortment of instruments needed to characterize the composition and structure of the target material. The muzzle velocity of the impactor is typically ~900 m/s, about the same as that of a 0.243 calibre hunting rifle. The shock event may last ~500 ns depending on the complexity of the sample. The shock pressure attainable in this apparatus is at least that of half the depth of the Earth, well into the outer core. It has recently been used to synthesize quasicrystals of Al–Cu–Fe alloys with fivefold symmetry that mimic those found in the Khatyrka meteorite, demonstrating evidence of asteroid collisions in the early history of the solar system (Asimow et al. 2016). The first natural quasicrystal to be described is named icosahedrite (Bindi et al. 2011).

We now return to the matter at hand, the results of terrestrial experiments at high pressures.

Representations of Melting Curves

Figure 18.1 shows three ways of representing the same melting curve $S = L$. All three are commonly used in geology. The third, $-P$ versus T, is very sensible from the standpoint of viewing the earth in cross section, but it can be confusing because the positive slope looks negative until one looks closely at the pressure axis. The first two types of diagram are used most commonly, and we shall follow common

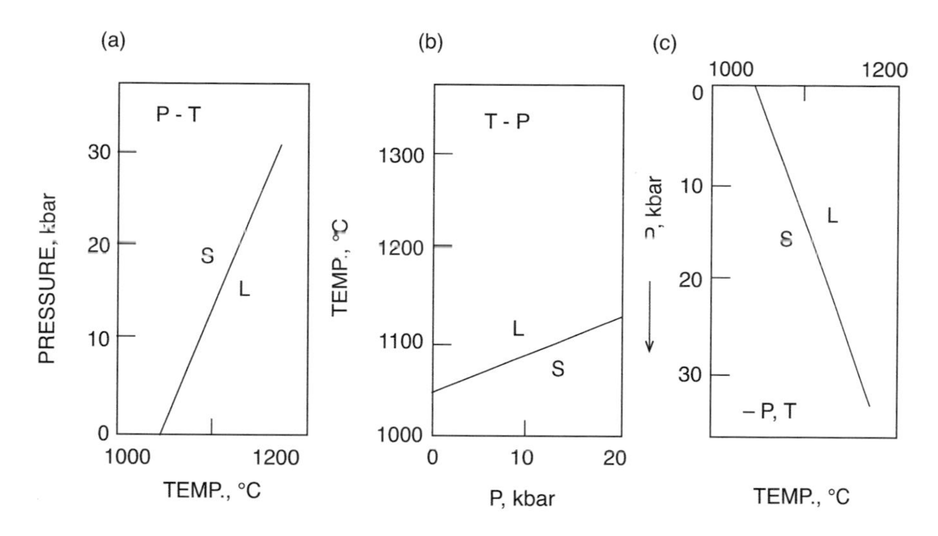

Fig. 18.1 Three ways of showing the same melting curve as a function of pressure. The third diagram most closely resembles a vertical section into the earth, but the inversion of the pressure axis makes the positive slope look negative

practice by mixing them rather indiscriminately. This is annoying at first, but one gets used to it.

Dealing with Melting Curves

The sources of melting curves will usually be referenced in the captions rather than in the text. The melting curves themselves are characteristically described by a $P = T$ equation and that can be useful in calculating approximate crystallization histories. In a previous era (e.g. the first edition of this book), the P–T slope was characterized, if at all, by a Simon equation, which has the form:

$$P - P_o = A[(T/T_o)^c - 1] \tag{18.1}$$

where P is pressure, P_o is the intercept pressure, commonly 1 atm, A and c are constants to be identified by experience, T is the Kelvin temperature, and T_o is the temperature at the pressure intercept. The solution of this equation was done by trial and error. This bit of history is included here only because this calculation may be found in older literature. The Simon equation has no physical significance and has long been superseded by the simple second-order polynomial in a spreadsheet, operating on four data points or more. It is quite common to decorate a P–T diagram with a polynomial that connects the temperature with the pressure.

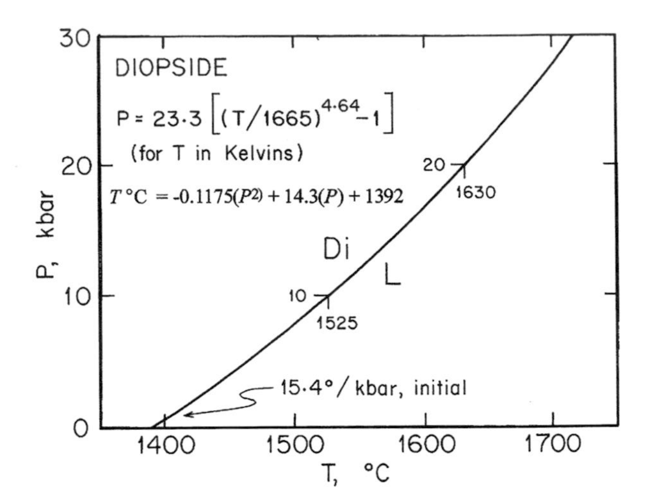

Fig. 18.2 Melting curve for diopside at high pressure, after Boyd and England (1963). The upper equation is a Simon equation, retained for historical interest. The lower equation is the second-order polynomial

Diopside

The melting curve for diopside to 30 kbar is shown in Fig. 18.2. The initial slope of the melting curve equals 15.4°C/kbar. The curve shown is the liquidus trace; the solidus trace lies somewhat lower because of incongruent melting (see Fig. 12.2 and Kushiro 1972). Two generations of equations describing the melting curve are shown in the figure.

Augite

The common monoclinic pyroxene of mafic igneous rocks is augite rather than diopside, which means that it resides within the pyroxene quadrilateral. The equilibria described in Chap. 17 reflect moderate pressures for Kiglapait and low pressures for Skaergaard. It is frequently assumed that augite gains aluminium by gaining a feldspar component or its CaAl cousin CaTs. However, at high pressure (13 kbar) and a graphite buffer, augite saturated with plagioclase and spinel becomes aluminous at a constant ratio of Ca/Al (McIntosh 2009), in effect forming a solution with garnet. For 12 oxygen, this result can be described in principle as diopside + garnet to form a mixture $CaM_{2.5}AlSi_{3.5}O_{12}$ where M is a divalent mix of Mg and Fe. In the actual experiments, the array is shifted towards olivine so as to be anchored at augite rather than diopside.

Anorthite

The melting of anorthite is much more complex than that of diopside at high pressure. The behaviour followed, however, is a very common one among silicate minerals, and we shall see many further examples of it. The phenomena involve the generation of a singular point (S_{An}) where the composition of the liquid exactly equals $CaAl_2Si_2O_8$. We have already seen singular points of this type in the system Fe–SiO_2–O, as for example S_m and S_i in Fig. 16.19.

In the anorthite system, S_{An} occurs at 1568°C and 9 kbar, where it is generated by the common tangency of two equilibria, An + *L* and An + Cor + *L* (see inset to Fig. 18.3). Corundum (Cor) occurs as an incongruent melting product of An above 9 kbar, and it forms a eutectic with An below 9 kbar. The curve An + Cor + *L* is a continuous equilibrium passing through the singular point with its metastable and stable extensions coincident; hence, the curve is a continuous curve, and the only thing that happens as it passes through S_{An} is that the sign of the reaction changes from even (below S_{An}) to odd (above S_{An}). The curve An + *L* lies everywhere to the right of the equilibrium An + Cor + *L* except at the singular point, where the two curves unite. The metastable segment of An = *L* lies on the same side of An + Cor + *L* as the stable segment.

The explanation for all this complexity can be seen in Fig. 18.4. Part (A) of the figure shows the system CAS with the locations of An, the low-pressure liquid L_1, and the high-pressure liquid L_2. Part (B) of the figure repeats the *P–T* inset of Fig. 18.3 and shows the isobaric thin section lines C_1 and C_3 corresponding to the first and third sketches in part C_1 of the figure. Part C represents the compositions as projected from CaO. At a pressure C_1, less than 9 kbar, we see in part C that the eutectic reaction Cor + An = *L* occurs at a lower temperature than the reaction An = *L*. As the pressure is raised, this eutectic migrates towards An. At 9 kbar, the pressure of

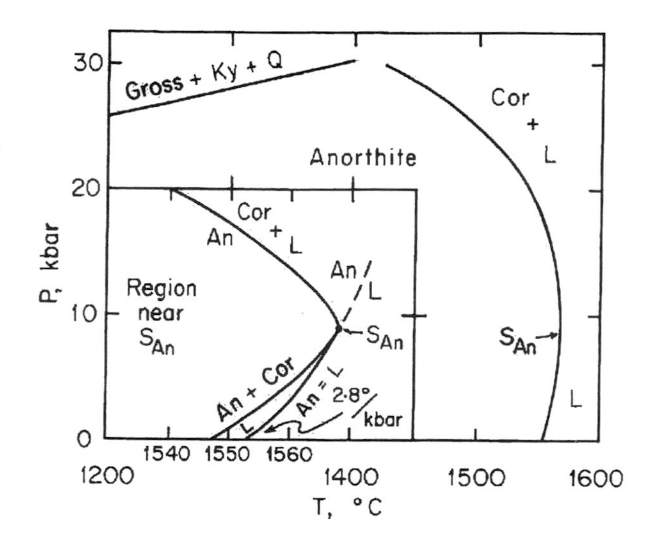

Fig. 18.3 Melting and breakdown relations of anorthite at high pressure, after Goldsmith (1980). S_{An} is a singular point where the melting of anorthite becomes incongruent (see Fig. 18.4). *Gross*, grossular; *Ky*, kyanite; *Q*, quartz; *Cor*, corundum; *An*, anorthite; *L*, liquid. For relations near the intersection of the subsolidus breakdown and the melting curve, see Lindsley (1968) and Wood (1978)

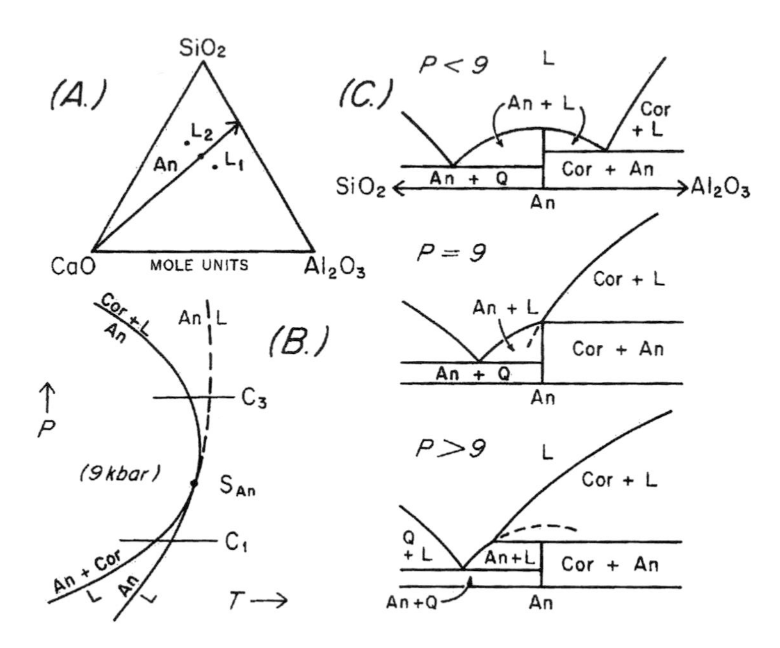

Fig. 18.4 Explanation of the singular point S_{An}. (**a**) Compositional space showing An projected from CaO onto side line SiO_2–Al_2O_3. (**b**) *P–T* diagram showing the continuous equilibrium An + Cor + *L* and the stable and metastable parts of the equilibrium An + *L*. (C_1) and (C_3) refer to the first and third diagrams to the right. (**c**) *T–X* diagrams showing melting relations at a series of pressures, including the pressure of S_{An} (middle diagram)

the singular point and the liquid in equilibrium with corundum and anorthite have reached the composition of An. This generates the special *T–X* configuration shown in the middle diagram of Fig. 18.4c. At pressure C_3, the Cor + *L* field overlaps the composition of An and the melting relationship is incongruent, or odd. Note that in the last sketch of Fig. 18.4c, the metastable melting point of An at the top of the dashed curve lies higher than the incongruent melting point, as the arrangement of the curves in part (B) of the figure also shows. The *T–X* diagram C_3, in fact, shows *why* the metastable extension of An = *L* must lie to the high-temperature side of An + Cor + *L*.

As shown in Fig. 18.4b, the singular point S_{An} does not lie at the thermal maximum on An + Cor + *L* except by chance. By chance, in fact, it does lie indistinguishably close to the maximum, as shown in Fig. 18.3. This figure also shows that at high pressures, An breaks down to grossular plus kyanite plus quartz by the subsolidus reaction:

$$\underset{(3An)}{3CaAl_2Si_2O_8} = \underset{(Gross)}{Ca_3Al_2Si_3O_{12}} + \underset{(2Ky)}{2Al_2SiO_5} + \underset{(Q)}{SiO_2}. \tag{18.2}$$

The intersection of this reaction with the melting reaction is discussed by Wood (1978); it is complicated by the presence of CaTs, $CaAl_2SiO_6$.

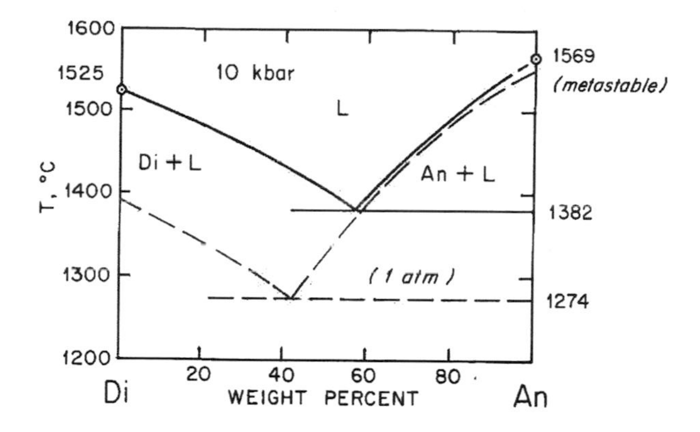

Fig. 18.5 System Di–An at 1 atm (dashed) and 10 kbar (solid lines). At 10 kbar, the melting of pure An is metastable relative to Cor + *L*. After Presnall et al. (1978)

Diopside–Anorthite

Because the melting point of diopside rises much faster with pressure than that of anorthite, we shall expect to find that the diopside field is enlarged with pressure at the expense of the anorthite field. Figure 18.5, an isobaric *T–X* section at 10 kbar, confirms this expectation. At 10 kbar, the eutectic-like point Di + An + *L* has moved from 42% An to about 57% An, and the cotectic temperature has risen from 1274 to 1382°C. The melting point for An, 1569°C, is the metastable melting point estimated from Fig. 18.3. The small field of Cor + *L* near An is omitted from the diagram.

An interesting feature of Fig. 18.5 is that the An liquidus moves hardly at all with pressure, and the shift of the cotectic is caused almost entirely by the diopside liquidus climbing up the anorthite liquidus.

The subsolidus relations of the system Di–An are even more complicated by entry of CaTs into the pyroxene than at 1 atm and are ignored here.

The change in the cotectic temperature for Di + An = *L* averages about 11°C/kbar in the first 10 kbar, but the rate of change diminishes with pressure, as shown in Fig. 18.6. The *P–X* part of Fig. 18.6 shows the sensibly linear change in the "eutectic" liquid composition with pressure, to 20 kbar.

Albite

The initial melting slope for albite in *P–T* space is about 18°C/kbar, and the melting curve is given by the polynomial equation shown in Fig. 18.7a. It is interesting to note that the pressure effect here for albite at 18°C/kbar is very much greater than that for anorthite, 2.8°C/kbar (Fig. 18.3). This relationship must imply that the plagioclase loop becomes narrower at pressure, as we shall see. At pressures near 30 kbar, albite breaks down to jadeite plus quartz according to the reaction:

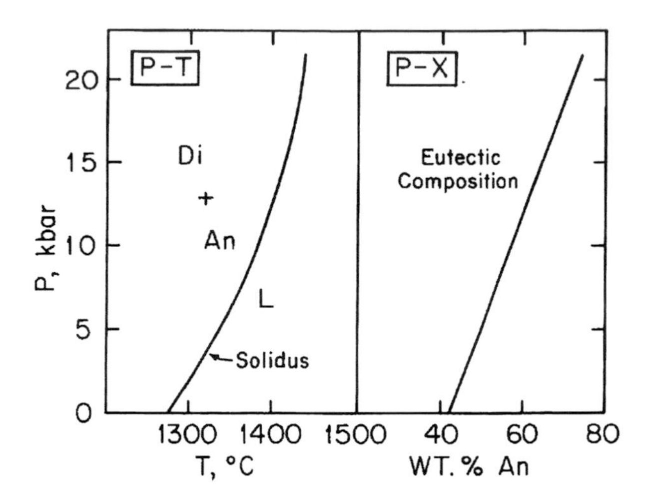

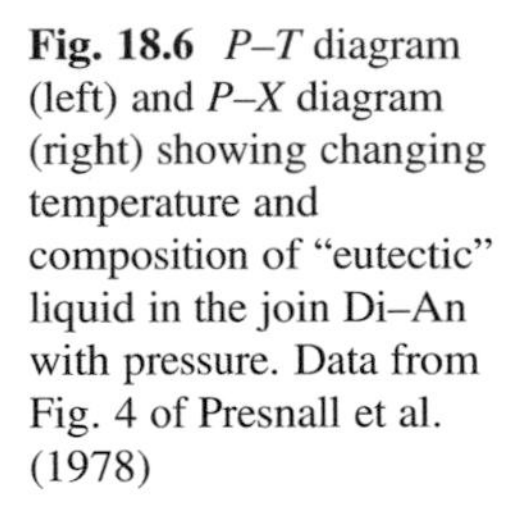
Fig. 18.6 *P–T* diagram (left) and *P–X* diagram (right) showing changing temperature and composition of "eutectic" liquid in the join Di–An with pressure. Data from Fig. 4 of Presnall et al. (1978)

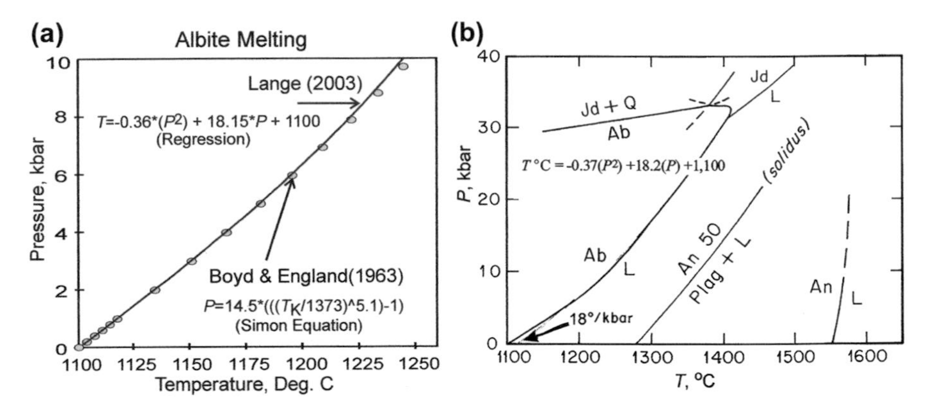

Fig. 18.7 (**a**) The melting curve of Ab at lower pressures based on Lange (2003). (**b**) The melting curve for albite (Ab) at higher pressures, from Boyd and England (1963), with albite breakdown reaction and other jadeite (Jd) relations from Bell and Roseboom (1969). Shown also, for comparison, are the melting curve for An, from Fig. 18.3, and the solidus curve for An_{50} plagioclase, drawn from data in Fig. 18.8

$$\underset{(\text{Ab})}{\text{NaAlSi}_3\text{O}_8} = \underset{(\text{Jd})}{\text{NaAlSi}_2\text{O}_6} + \underset{(\text{Q})}{\text{SiO}_2}, \tag{18.3}$$

and complications arise in the melting relations. We shall return to these later.

Plagioclase

The melting relations of plagioclase at 10 and 20 kbar are shown in Fig. 18.8. The 1-atm diagram is sketched with the 10 kbar diagram for comparison. The essential features of the plagioclase melting loop are unchanged with pressure except for the

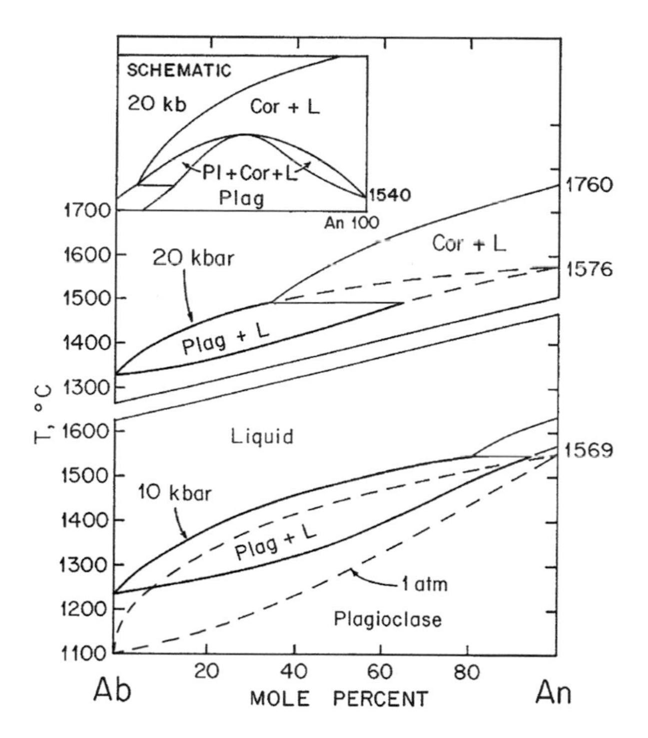

Fig. 18.8 Melting relations of plagioclase at 10 and 20 kbar compared with the 1-atm diagram (dashed). Compiled from data of Lindsley (1968) and Figs. 18.3 and 18.7. Where the plagioclase loop is dashed, it is metastable relative to Plag + Cor + Liq. Inset shows schematically the stable relations at 20 kbar, from An_{30} to An_{100}. Further details are given by Lindsley (1968)

higher temperatures of melting and the narrowing of the loop. However, equilibria with Cor + *L* begin to intrude at 9 kbar (see Fig. 18.3) and expand into the diagram as the pressure rises. At 20 kbar, only intermediate (to An_{64}) and sodic plagioclase are stable with liquid of plagioclase composition, and calcic plagioclase coexists with corundum and a silica-rich liquid. The inset to the 20 kbar figure shows, schematically, the shape of the Plag + Cor + *L* loops, which unite at a maximum. The dashed part of the plagioclase melting loop at 20 kbar is metastable. Binary equilibria are denoted by heavy lines, and ternary (incongruent) equilibria by light lines in the figure.

It is clear from Figs. 18.3, 18.7, and 18.8 that we cannot expect to find plagioclase at any great depth in the mantle. Intermediate plagioclase might be found at 20 kbar, given a suitable container, but by 30 kbar it would be represented instead by garnet, jadeitic pyroxene, kyanite, and quartz. Even more stringent restrictions occur, as we shall see, when plagioclase occurs with olivine.

Diopside–Anorthite–Albite

The "diopside" present at high pressures in equilibrium with plagioclase will be a complex solution including some CaTs and Jd. The high-pressure phase relations are poorly known, but the most interesting feature of the ternary system is the location of

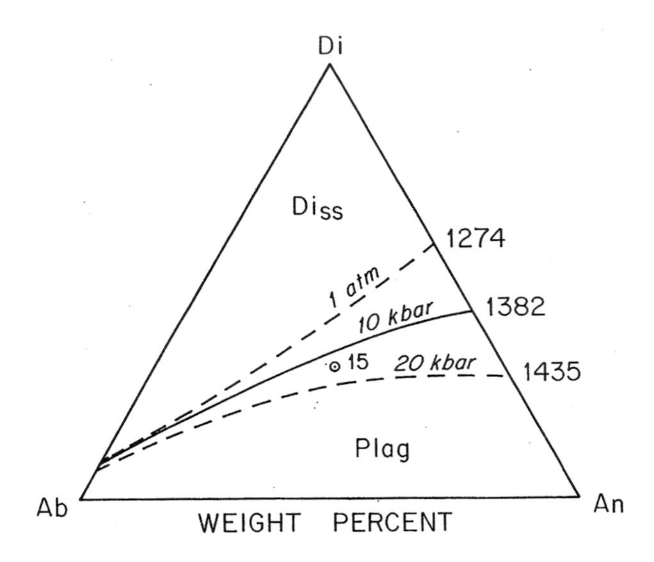

Fig. 18.9 System diopside (Di)–anorthite (An)–albite (Ab) at 10 and 20 kbar compared to 1 atm. Di–An side line after Presnall et al. (1978), and cotectic traces based on 15 kbar point determined by Lindsley and Emslie (1968)

the plagioclase–pyroxene cotectic. This is sketched in Fig. 18.9, with data for the Di–An sideline picked off Fig. 18.6. The positions of the ternary cotectic for 10 and 20 kbar are estimated from those determined by Lindsley and Emslie (1969) at 15 kbar.

The figure shows that high pressure shifts the cotectic towards plagioclase, but not enough to account for very plagioclase-rich anorthosites. It should also be noted that entry of iron into the pyroxene would reduce the shift in the cotectic. An important principle emerges, nevertheless: a cotectic liquid generated at high pressure will, on rising into the crust and cooling at lower pressure, lie in the plagioclase field. It may crystallize plagioclase for a considerable time before it again reaches saturation with pyroxene. If the plagioclase should be separated mechanically, as by floating (Grout 1928; Morse 1968; Kushiro and Fujii 1977), it could easily form a crystalline mass of anorthosite. Even after saturation is regained, plagioclase should tend to float while pyroxene sinks, and if cooling is slow enough to allow the crystals to move far enough, a bimodal rock suite should result, i.e. anorthosite overlying pyroxenite. Such a process is by no means the whole answer to the anorthosite problem, because there is evidence that very plagioclase-rich magma existed (Wiebe 1979), but some means of plagioclase concentration (and loss of mafics) must be assumed for those anorthosites containing only a trace of mafic minerals (Morse 2006).

Forsterite

The melting curve of Fo is linear, within experimental error, to 30 kbar. As shown in Fig. 18.10, the slope is about 4.8°C/kbar. A similar value was deduced by Bowen

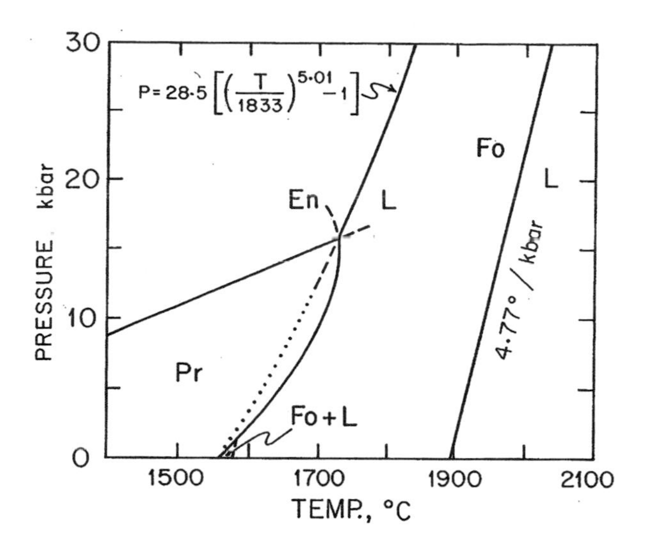

Fig. 18.10 Melting curves of forsterite (Fo; Davis and England 1964) and enstatite (Pr, En; Boyd et al. 1964; Chen and Presnall 1975). The Fo curve is linear. A Simon equation is given for the En curve above 15 kbar. The equivalent polynomial is given by $P = 0.00011994(T^2) - 0.304(T) + 182.89$, for T in degrees C

and Schairer (1935), but is subject to some error, as discussed by Davis and England (1964).

The melting curve for enstatite composition is also shown in Fig. 18.10 for comparison; this will be discussed later in connection with the system Fo–SiO_2.

Forsterite–Diopside–Anorthite

Isobaric sections at 1 atm, 5, 10, and 20 kbar are shown in Fig. 18.11. A corundum + L field intrudes near the An corner at about 9 kbar. High pressure enlarges the fields of pyroxene and spinel at the expense of forsterite and anorthite. The 5 kbar section accidentally contains the singular point where all four phases coexist with liquid at 1292°C. The effects of high pressure preclude the coexistence of olivine + plagioclase + liquid if the system composition is rich in Mg and Ca, so allivalites (basic troctolites) must be products of relatively low pressure crystallization.

None of the intersections in Fig. 18.11 are truly isobaric invariant points, but rather piercing points, because of solid solution in pyroxene and consequent non-ternary behaviour. Presnall et al. (1978) have determined the compositions of the important piercing point liquids by electron probe analysis. They found that at 1 atm, the liquid at 1270°C (Geophysical Laboratory temperature scale) is silica-poor relative to the ternary join, so the thermal maximum for the equilibrium L(Di, Fo, An) lies on the *silica-rich* side of the plane. Crystallization of pyroxene tends to drive liquids towards silica undersaturation, and the join Fo–Di–An is not a thermal barrier at 1 atm, although it closely approximates such a barrier. At about 5 kbar, the liquid lies in the ternary join, and at higher pressures, it becomes silica-rich as the fields of Fo and An move apart. Presnall et al. (1978), generalizing from the join

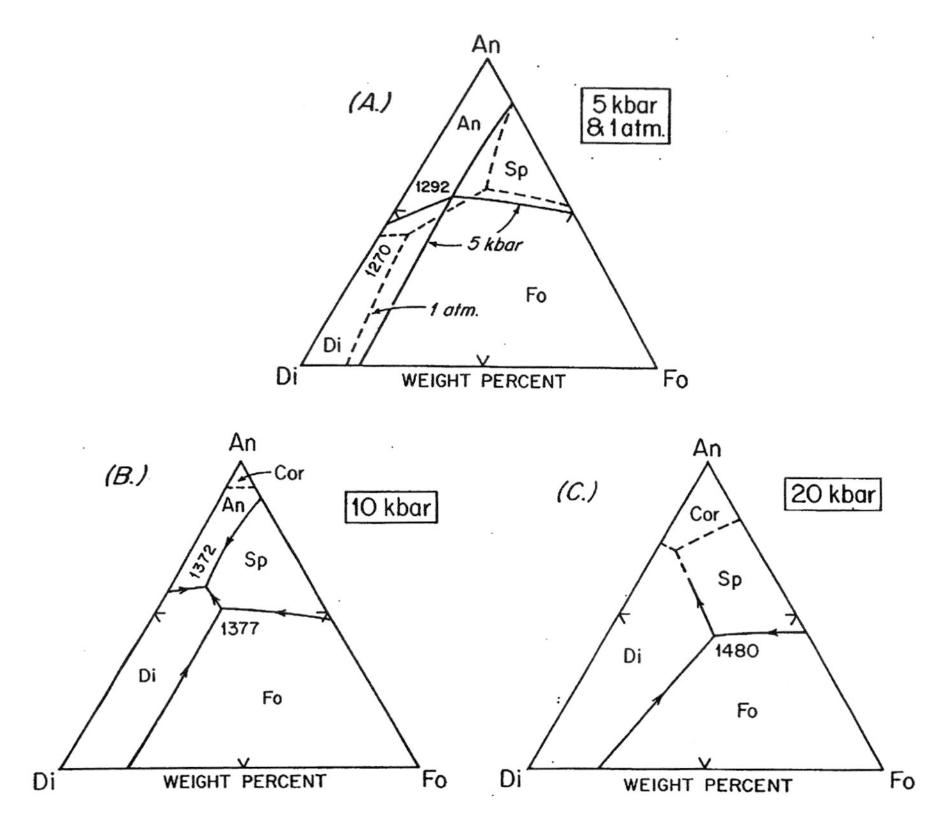

Fig. 18.11 System Fo–Di–An at 1 atm and 5, 10, and 20 kbar, after Presnall et al. (1978). Temperatures are referred to the Geophysical Laboratory temperature scale. *Sp*, spinel; *Cor*, corundum

Fo–Di–An, concluded that a thermal divide between alkali basalt and tholeiite exists to a pressure of about 4 kbar (except that crystallization of spinel would cause alkali basalt to fractionate to tholeiite). At higher pressure, to 12 kbar, they found that alkali basalt could fractionate to olivine tholeiite, but not vice versa.

From Fig. 18.11, it will be seen that initial melting reactions at high pressure will involve the production of spinel and a silica-rich liquid. In olivine-rich compositions, anorthite will be excluded from such events by a subsolidus reaction of olivine plus anorthite to pyroxene plus spinel. The beginning of melting for mantle compositions is therefore not usefully shown in the liquidus diagrams.

The point labelled 1377 in the 10 kbar diagram is a reaction point at which Fo + L = Di + Sp. The point labelled 1372 is a piercing point where Di + Sp + An crystallize together from a silica-rich liquid that does not lie in the join.

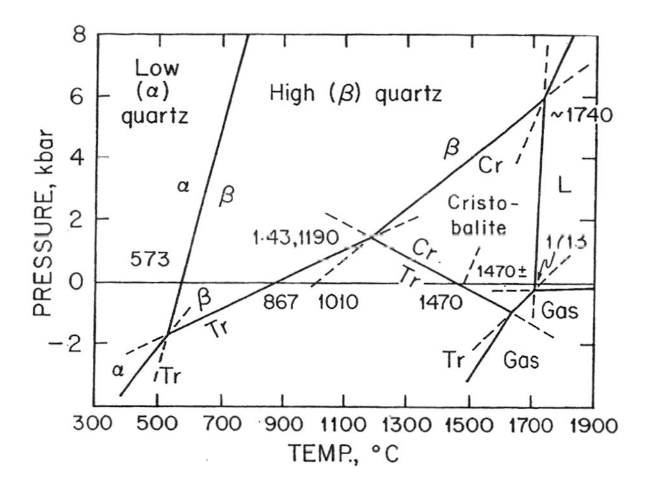

Fig. 18.12 *P–T* relations for the system SiO_2 at low pressure. Relations below zero pressure are extrapolated. *Tr*, tridymite; *Cr*, cristobalite; *L*, liquid. Data from Yoder (1950), Cohen and Klement (1967), Kracek (1953), Ostrovsky (1967), and Kracek and Clark (1966)

Silica

The subsolidus and melting relations of SiO_2 at low pressures are shown in Fig. 18.12. Relations at negative pressures are extrapolated from higher pressures and correspond to the phenomena that would occur in vacuum at varying degrees of evacuation. A region of gas is shown in the lower right-hand corner of the diagram, indicating that the vapour pressure of silica is very low. This is why we generally refer to silicate systems as condensed systems, because a gas phase of silicate composition never occurs at 1 bar pressure and at ordinary temperatures near the melting point. Every silicate, in principle, has a *S–L–V* triple point like that shown for Cr + *L* + Gas in the diagram, but at a pressure below that of interest for terrestrial geology.

All the intersections shown in Fig. 18.12 are derived with Schreinemakers' rules.

The diagram shows that tridymite is unstable above 1.4 kbar and cristobalite is unstable above 6 kbar. High quartz will be the polymorph appearing in most phase diagrams at high pressures.

Figure 18.13a shows the phase diagram over an expanded pressure range, to illustrate the melting curve and its intersection with the high-pressure polymorphs coesite and stishovite. Coesite is named for Loring Coes, Jr. of the Norton Company, who in 1953 first synthesized this polymorph, and stishovite is named for the physicist Sergey M. Stishov, who in 1961 first synthesized this silica oxide with the rutile structure. Both natural minerals were first found in impact glasses at Barringer Meteor Crater, Arizona, by Ed Chao (in 1960 and 1962, respectively). The reader is invited to compare this figure with Fig. 3.2, showing the *P–T* polymorphs of water, to discover why mineralogists are intrigued by the similarity of silica and ice.

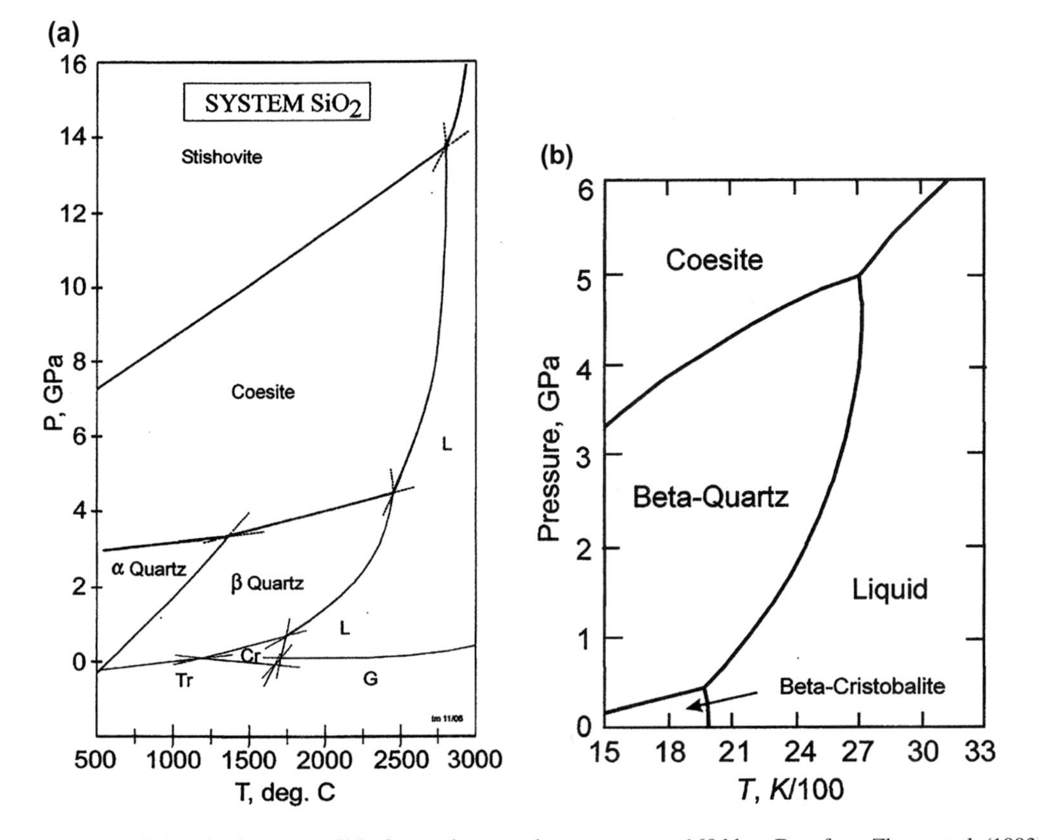

Fig. 18.13 (**a**) High-pressure P–T relations in the system SiO_2 from sub-atmosphere pressure to 160 kbar. Data from Zhang et al. (1993). (**b**) Melting of silica phases to 60 kbar from thermodynamic studies by Hudon et al. (2002)

Figure 18.13b shows a more detailed melting diagram of quartz and coesite from the thermodynamic studies of Hudon et al. (2002). Note that the temperature is in Kelvin, not Celsius degrees.

Forsterite–Silica

Figure 18.10 contained a sneak preview of the enstatite melting curve, and we now examine the relations in more detail with the aid of Figs. 18.14 and 18.15. We seek the pressure at which the melting becomes congruent. The *T–X* part of Fig. 18.14 shows the 1-atm melting relations of protoenstatite, Pr. The metastable melting point of Pr–*L* can be estimated to lie at about 1560°C at 1 atm. Chen and Presnall (1975) found an initial slope of about 17°C/kbar for the melting reaction Pr = *L*. Assuming this slope to be constant, we can draw the curve Pr = *L* originating at 1560°C (metastably), as in the *P–T* section of Fig. 18.14. The liquidus temperature for En composition at 1 atm is 1577°C, as shown in the left-hand diagram. If we assume that this temperature increases with pressure at the same rate as the melting point of Fo, 4.77°C/kbar, we obtain an intersection at 1.4 kbar, 1584°C, as shown in the *P–T* section. This intersection must denote the singularity where Pr for the first time melts to a liquid of its own composition. The geometry is essentially the same as previously encountered with anorthite, except that in this case the incongruent melting occurs at low pressure instead of high. The singular point in Fig. 18.14 lies where the continuous equilibrium changes sign from odd to even.

Returning briefly to Fig. 18.10, it appears that Boyd et al. (1964) actually determined the equilibrium En = *L* in both its stable and metastable regions (see dotted part of the curve in the figure). The main difference between their work and the later work of Chen and Presnall (1975) was in the much longer run durations used by the latter authors, who confirmed the earlier work at pressures above 15 kbar, but determined the reaction Pr = *L* at lower pressures. Chen and Presnall also determined the melting reactions involving quartz and forsterite with $MgSiO_3$; these curves are omitted here for clarity. An interesting feature of Fig. 18.10 is the rather strong curvature of the curve Pr = *L* as it approaches the invariant point *L*(Pr, En). The subsolidus reaction Pr = En generates a cusp on the melting curve, a feature we shall encounter repeatedly in other systems. Such cusps are required, in principle, by Schreinemakers' rules. The curvature is a kind of premonitory phenomenon, probably reflecting the reorganization of the melt to a more enstatite-like structure as the pressure is increased. Protoenstatite and liquid must have very similar densities near 15 kbar, where the melting curve is nearly isothermal.

Figure 18.15 shows the *T–X* geometry with increasing pressure. Although the En composition is rapidly uncovered with pressure, the shift of the eutectic *L*(Fo, En) towards Fo is very small, so that the liquid remains near the En composition even at 20 kbar.

The disappearance of the incongruent melting of enstatite composition at the low pressure of 1.4 kbar means that the reaction relation of olivine to orthopyroxene must

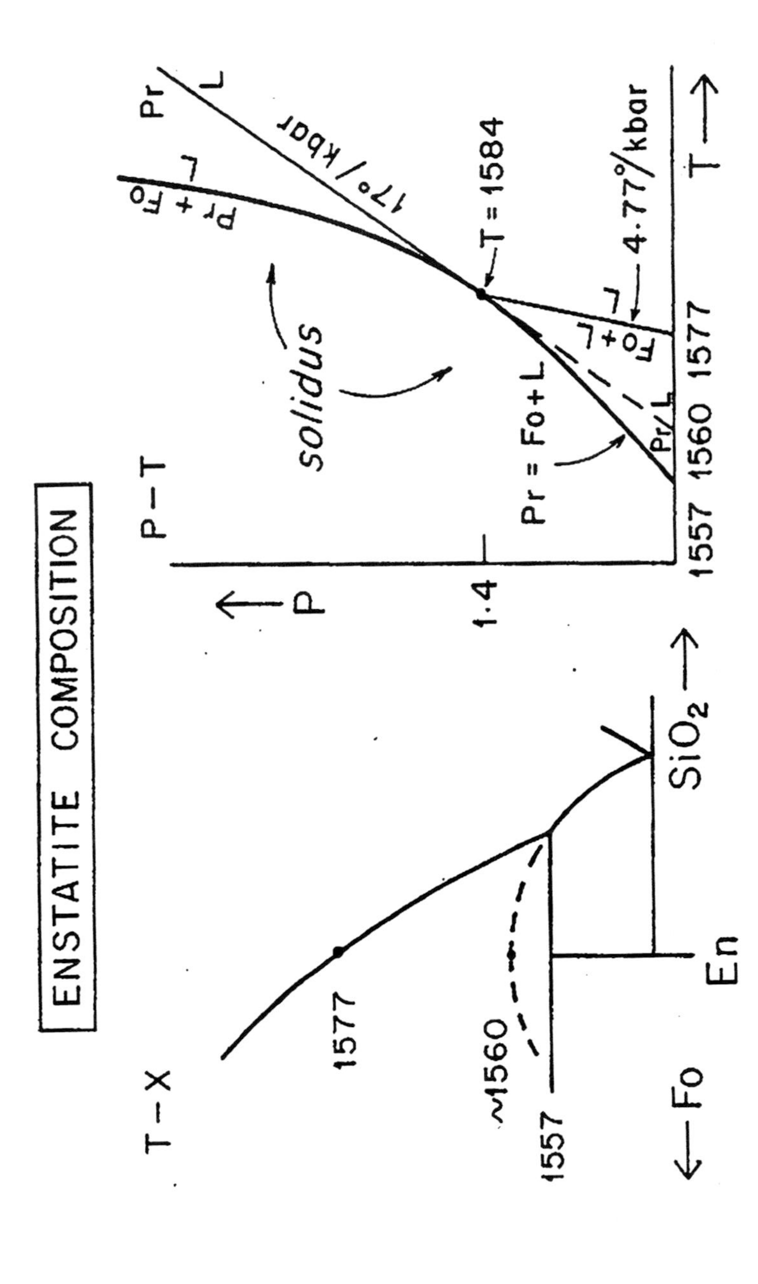

Fig. 18.14 Melting relations of enstatite composition. Left, *T–X* diagram at 1 atm showing metastable melting point of Pr and the liquidus of En composition. Right, *P–T* diagram showing a singular point at 1.4 kbar where the continuous equilibrium Fo + Pr + *L* touches the curve Pr + *L*. See Fig. 18.4 for similar geometry about S_{An}

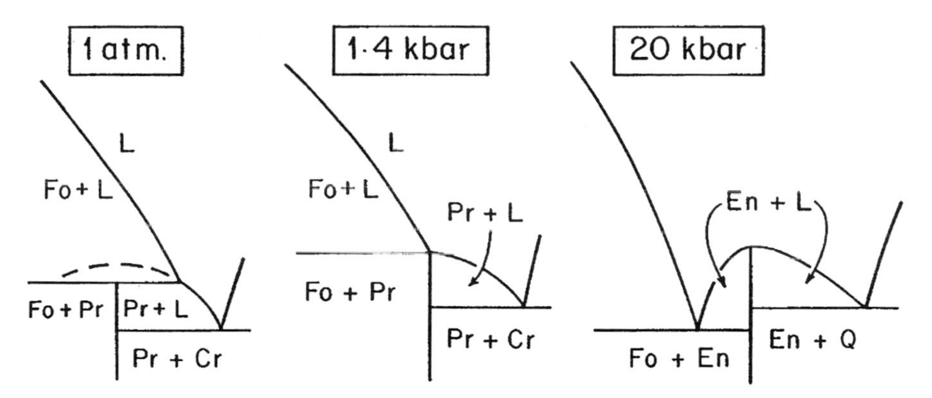

Fig. 18.15 Isobaric (schematic) *T–X* sections showing the change from incongruent to congruent melting of protoenstatite. The Pr = En curve is shown in Fig. 18.10

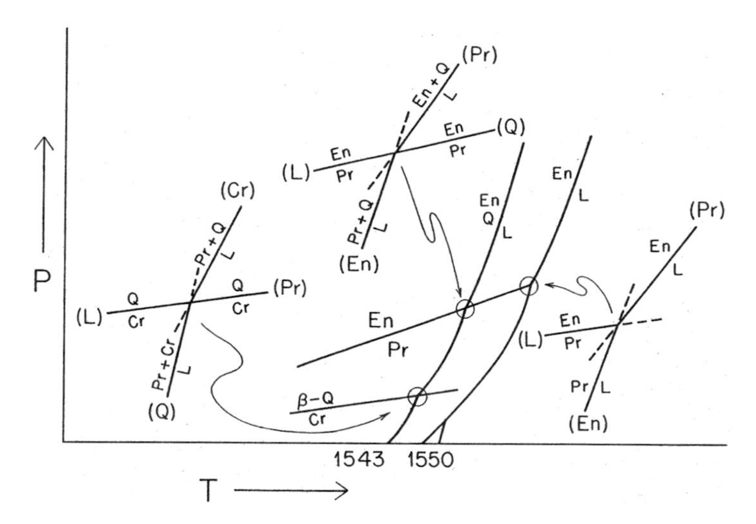

Fig. 18.16 *P–T* diagram showing the geometrical relations along the curves En + *Q* = *L* and En = *L* where they are intersected by the phase transformations Pr = En and Cr = *β* quartz

be limited to low pressures. Similarly, the crystallization together of olivine and orthopyroxene in sills and layered intrusions can be accounted for by modest pressures (as well as iron content; Chap. 17) during crystallization. However, Presnall et al. (1979) show that for liquids approaching basaltic rather than ultramafic composition, the thermal barrier originating at En does not exist at any pressure up to 20 kbar.

The melting curves of enstatite composition with and without SiO_2 encounter the polymorphic transitions Pr = En and Cr = *β* quartz at high pressures. The geometry of these intersections is shown schematically in Fig. 18.16, where it will be seen that cusps are generated on all the melting curves.

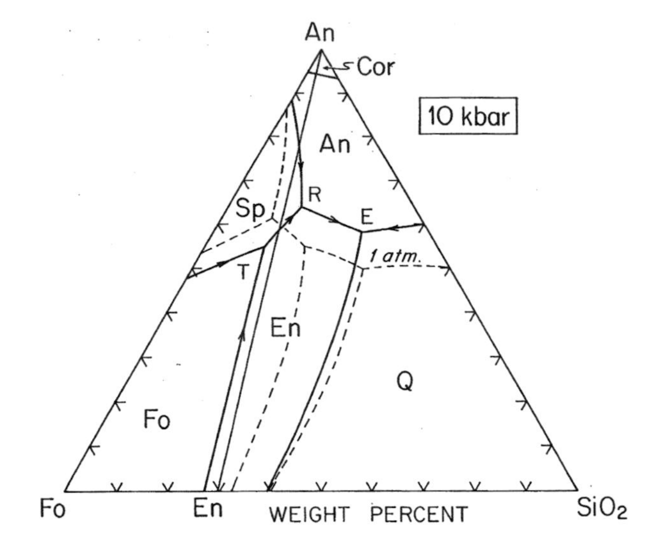

Fig. 18.17 System Fo–An–SiO_2 at 1 atm and at 10 kbar, after Presnall et al. (1979)

Forsterite–Anorthite–Silica

The system at 10 kbar is shown in Fig. 18.17. As in the system Fo–Di–An, Fo and An are prohibited from coexisting, this time by the coexistence of En + Sp. This news increases our suspicion that troctolites must crystallize at relatively low pressure and that plagioclase-bearing facies of the mantle must be restricted to modest pressures.

Forsterite and enstatite crystallize along a cotectic far into the diagram (Fig. 18.17), but their stable association with liquid terminates at the peritectic point *T*, where the reaction Fo + *L* = En + Sp sets in. Point *R* is a peritectic point involving Sp + *L* = En + An. Point *E* is a simple ternary eutectic in appearance, although the entry of Al into pyroxene at 10 kbar will cause liquid to lie out of the join.

The loss of olivine at point *T* illustrates the conclusion of Presnall et al. (1979) that enstatite cannot generate a thermal barrier for liquids rich enough in plagioclase component to be called basaltic. However, the addition of iron (see Chap. 17) may complicate the issue. It is significant that norites, rocks composed almost solely of plagioclase (about 75%) and orthopyroxene (25%), occur in abundance among anorthositic rocks and tend to contain very little of either residual quartz or olivine. No phase diagram yet studied at moderate pressure adequately explains the existence of voluminous liquids of norite composition except by accidental restriction to the plagioclase–orthopyroxene join.

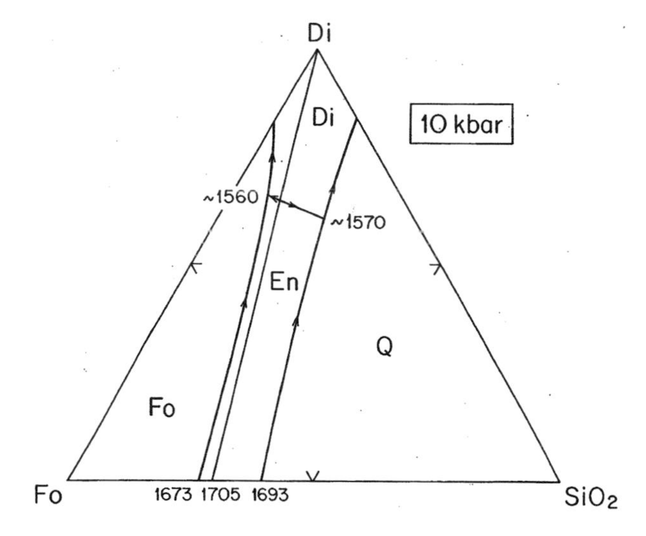

Fig. 18.18 System Fo–Di–SiO_2 at 10 kbar, after Kushiro (1969) and Presnall et al. (1979)

Forsterite–Diopside–Silica

The system at 10 kbar is sketched in Fig. 18.18. The pigeonite field has (probably) disappeared, the pyroxenes are Di-rich and En-rich solid solutions, and Fo and En crystallize together. The pyroxene join forms a thermal barrier shedding liquids to either side. The olivine reaction relation is absent, since liquids in equilibrium with olivine are silica-poor. The first melt of a peridotite will be *L*(Di, Fo, En), and if the peridotite is diopside-poor, diopside will be used up first and the liquid will travel along *L*(Fo, En) with further melting.

Nepheline–Silica

The role of albite as a thermal barrier at 1 atm was stressed in Chap. 14. This barrier disappears at high pressure as albite melts incongruently to Jd + *L* and eventually becomes metastable altogether relative to Jd + Q. The sequence in which this happens is illustrated schematically in Figs. 18.19 and 18.20. Here, we find a continuous equilibrium Ab + Jd + *L* containing two singular points representing the degenerate reactions Jd = *L* and Ab = *L*. At relatively lower pressures, jadeite melts incongruently to Ab + *L* (P_1 in Figs. 18.19 and 18.20), and the metastable melting point of Jd lies at a higher temperature. At the singular point S_{Jd}, the liquid composition has backed off exactly to the jadeite composition, and melting can be described completely in terms of the one-component (degenerate) system $NaAlSi_2O_6$. At a slightly higher pressure P_2, jadeite forms a thermal barrier flanked by two eutectics *L*(Jd, Ne) and *L*(Jd, Ab). With further increases in pressure, jadeite becomes stable to higher and higher temperatures, raising its liquidus eventually to

Fig. 18.19 *T–P* relations showing the melting behaviour of jadeite (Jd) and albite (Ab) in the system Ne–SiO_2, after Bell and Roseboom (1969)

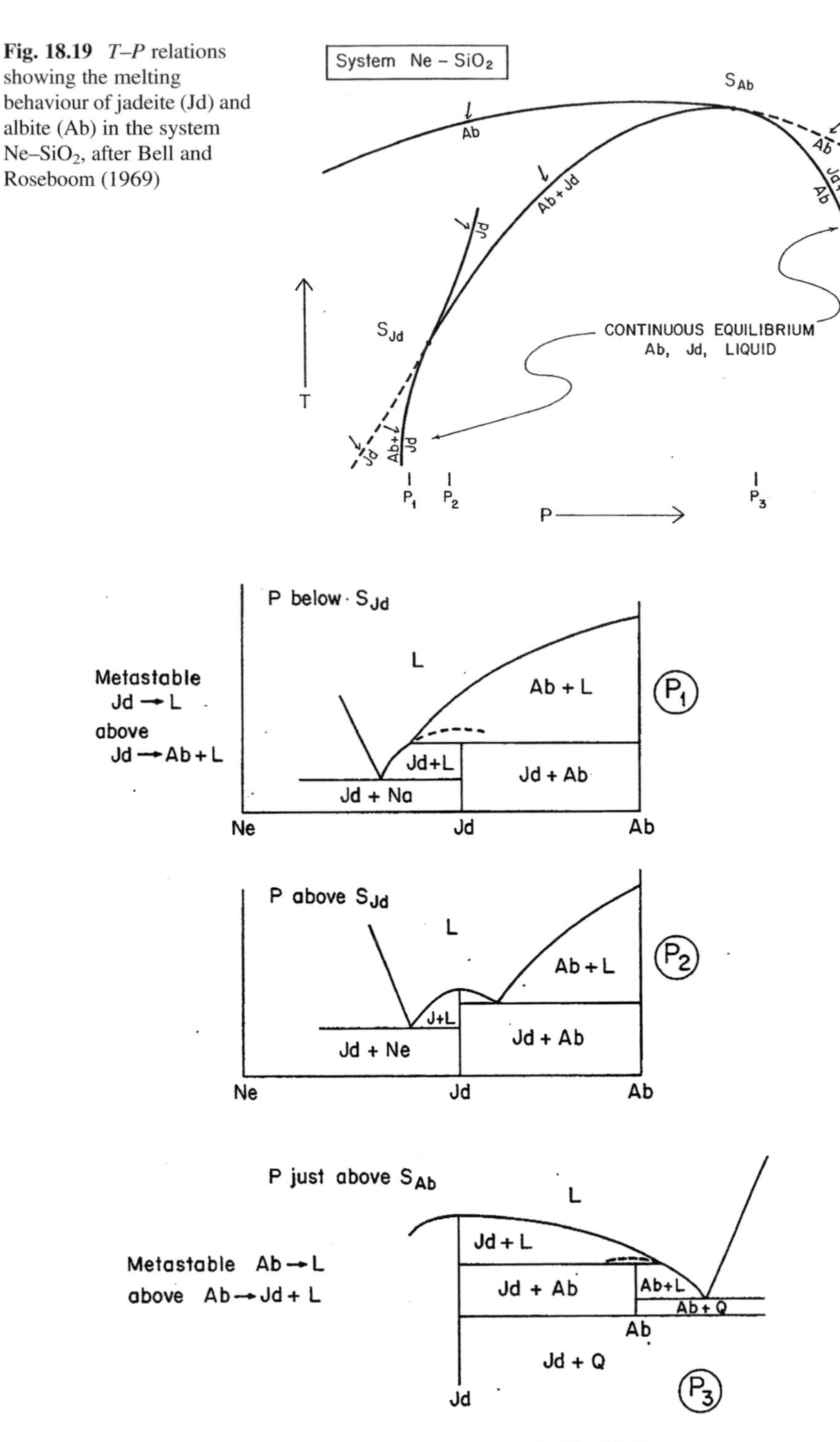

Fig. 18.20 *T–X* sections at three pressures shown as P_1–P_3 in Fig. 18.19

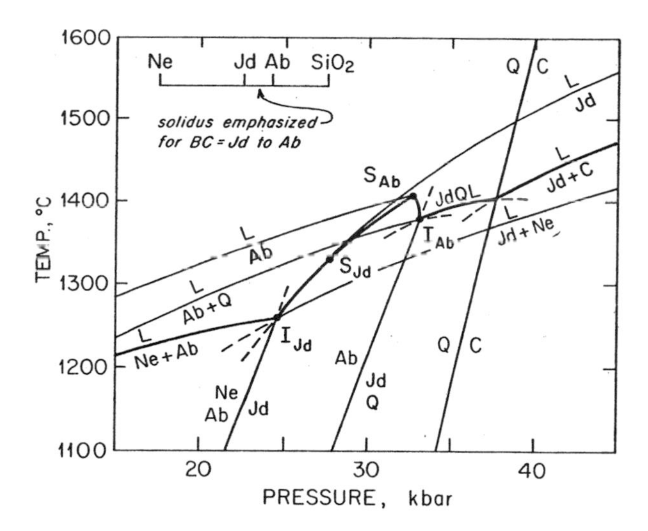

Fig. 18.21 Scale-true *T–P* diagram showing high-pressure reactions and melting relations in the system Ne–SiO_2, after Bell and Roseboom (1969). Singular points *S* are shown in Fig. 18.19. Invariant points are labelled *I*. *C* = coesite. The solidus for composition from Jd to Ab is shown as a heavy line (after Presnall et al. 1979)

reach and then cover the albite composition. At S_{Ab}, the system is again degenerate (to $NaAlSi_3O_8$) as the liquid reaches albite composition, and just above the pressure of S_{Ab} at P_3, albite melts incongruently to Jd + *L*. The geometrical relations of the curves and the changes of reaction signs from odd to even at the singular points are in principle the same as with the anorthite diagram, Fig. 18.4, and require no further comment.

Melting and subsolidus relations in Ne–SiO_2 are shown to scale in the *T–P* diagram of Fig. 18.21, where the solidus is emphasized for bulk compositions between and including Jd and Ab. The liquidus for Ab has previously been seen in Fig. 18.7. In Fig. 18.21, the two singular points S_{Jd} and S_{Ab} are flanked by two truly invariant points, I_{Jd}, and I_{Ab}. The first of these marks the lowermost *P–T* stability of Jd with liquid, and the second marks the upper limit of Ab with liquid.

Proceeding from the left along the solidus, we note that for all bulk compositions slightly less silica-rich than Ab, the melting begins with production of liquid at the *L*(Ne, Ab) eutectic. This eutectic has a smaller slope dT_m/dP than the Ab = *L* curve, and the eutectic reaction terminates at about 24.5 kbar, at I_{Jd} where it intersects the subsolidus reaction Ne + Ab = Jd. This invariant point is generated by the intersection of four univariant curves involving the phases Ne, Ab, Jd, and *L*. The labels of the curves, proceeding clockwise from the subsolidus reaction, are (*L*), (Jd), (Ne; see Fig. 18.19), and (Ab). The solidus for the complete system Ne–SiO_2 now runs along the (Ab) curve, the eutectic reaction Jd + Ne = *L*, but we are more interested in the restricted, mantle-like compositions between Jd and Ab. For these, the solidus is the reaction (Ne), namely Jd = Ab + *L*; see Figs. 18.19 and 18.20, section P_1. This curve has a greater slope than the curve Ab = *L*.

We now pause to note a very important feature of the melting curves around I_{Jd}. The *T–P* slope of the solidus changes dramatically when a new phase appears, and the solidus forms a *cusp* at I_{Jd} (and at I_{Ab} and at the point defined by the equilibrium Jd + *Q* + *C* + *L*). If we imagine a mantle composed of Jd + Ab, and slowly raise a geotherm, having a slope like Ab = L, through the subsolidus region, it will first touch the solidus at I_{Jd}. The solid assemblage will now begin to melt to a liquid poorer in silica than the Jd composition, hence an alkalic liquid (P_1 in Fig. 18.20). The melting process will absorb calories and may stabilize our rising geotherm to a fixed position. However, if the temperature continues to rise, the liquid composition will change *away* from Ne in either of two ways. First, suppose that jadeite is completely used up (see P_1 in Fig. 18.20); then, the liquid produced on further melting will move towards Ab by simple *dilution*. Second, suppose that Jd persists as a solid phase. Then, according to P_2 in Fig. 18.20, the stable liquid composition will migrate through S_{Jd} to a eutectic composition between Jd and Ab, and this composition will continue to migrate towards (and beyond) Ab with rising T and P.

Although our assumed mantle composition is slightly ridiculous, the principles mentioned in the above paragraph are quite appropriate to the real earth and are very important. The first principle, discussed in detail by Presnall et al. (1979), is that a cusp on a solidus should tend to lock or stabilize a geotherm, acting as a *thermal buffer*. We return to this principle in a more appropriate context in Chap. 20. The second principle is that *first melts* may tend to be *alkalic*, but as the proportion of melt increases, they become less alkalic and more silicic. Countless observations and deductions support this principle. Among these are experimental determinations of liquid compositions produced at high pressure from natural and synthetic peridotites (e.g. Mysen and Kushiro 1977), the abundant theoretical evidence and empirical evidence that "large ion lithophile" (LIL) elements, including alkalies, are happier in silicate melts than in coexisting mantle-type minerals, and geochemical observations on natural lavas, combined with theory (e.g. Gast 1968). We shall see other examples of this principle, again in a more appropriate context, in Chap. 20.

The remaining features of the heavy solidus in Fig. 18.21 can readily be followed by reference to the preceding figure for details. The negative melting slope of the reaction Ab = Jd + L is an interesting curiosity; it arises from the fact that the assemblage Jd + L is denser than Ab, because Jd is so very much denser than Ab.

Kalsilite–Silica

The high-pressure melting behaviour of sanidine composition, $KAlSi_3O_8$, is shown in Fig. 18.22. High pressure reduces the stability of the open framework silicate leucite (Lc), and at a pressure of around 19 kbar, sanidine begins to melt congruently. The curve San = L is continuous through the singular point S_1, and its metastable (low pressure) segment runs to the metastable melting point at 1200°C, 1 atm, given by Waldbaum and Thompson (1969). The complete geometry of the melting relations is shown schematically in Fig. 18.23. The figure is distorted for clarity; in reality, S_1 lies at a higher temperature than S_2. The correct locations are S_1, 19.5 kbar, 1440°C; S_2, 28 kbar, 1415°C; and I, 30 kbar, 1400°C, as shown by Lindsley (1966a). The geometry of the singular points and the invariant point can be understood by reference to previous examples of the type, and to Fig. 18.23. It

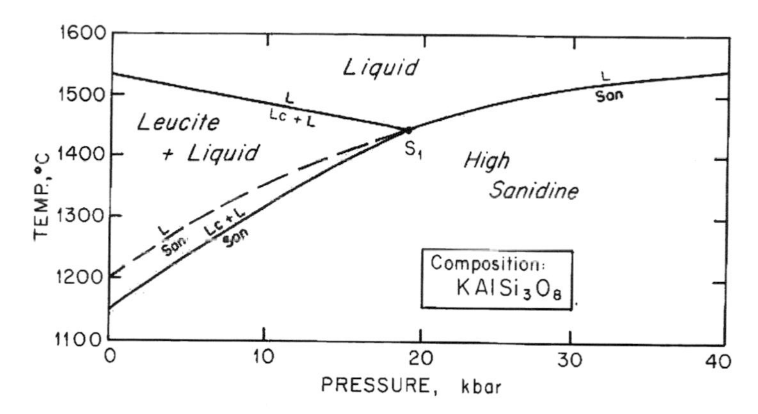

Fig. 18.22 Melting relations of the composition $KAlSi_3O_8$, after Lindsley (1966a). Metastable 1-atm termination of San = *L* is from Waldbaum and Thompson (1969). Other 1-atm relations from Schairer and Bowen (1955). The curve Lc + *L* = *L* is univariant by restriction to composition $KAlSi_3O_8$. The other curves are univariant in the system *Ks*–SiO_2. S_1 is a singular point shown also in the next figure

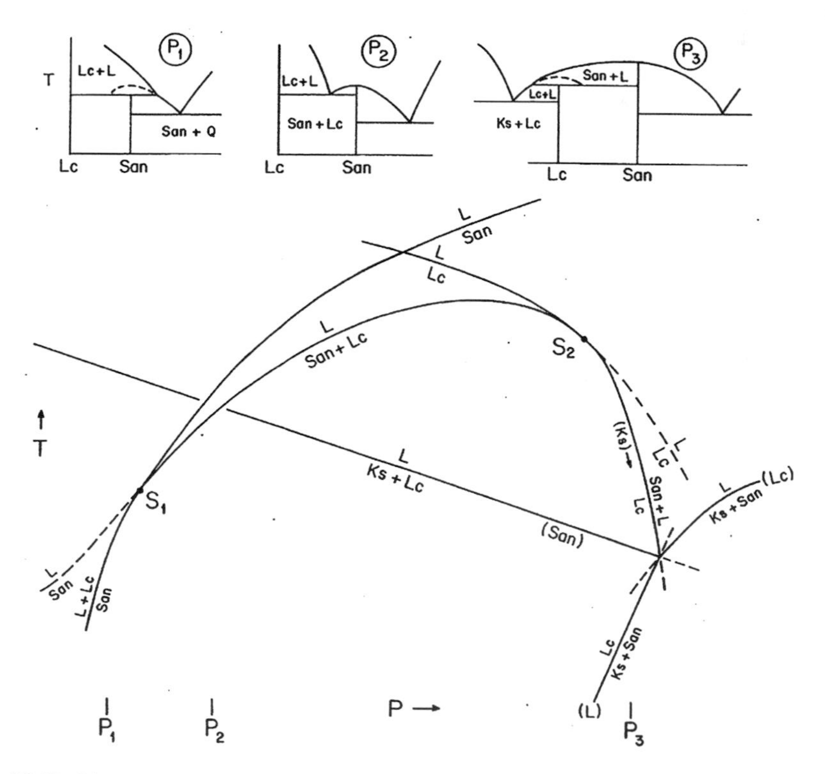

Fig. 18.23 Distorted, schematic *T–P* diagram showing the continuous equilibrium San + Lc + *L* and the two singular points lying on it. *T–X* diagrams at top show isobaric relations at three selected pressures. After Lindsley (1966a). The curve Lc = *L* is a rare example of a silicate with a negative melting curve

should be noted that the curve Lc + $L = L$ in Fig. 18.22 is simply the liquidus for the restricted composition $KAlSi_3O_8$ and is not a true univariant curve in the binary system.

From the isobaric *T–X* sketches in Fig. 18.23, it may appear that the previously announced important principle about the alkalic nature of first melts is violated. Indeed, it would be if the Earth were rich in potassium relative to sodium, but it is not. The first liquid obtained on melting sanidine is silica-rich, and only the higher pressure liquids are alkalic- and silica-undersaturated. But the opposite effect found in Ne–SiO_2 dominates in nature because Na is more abundant than K in the Earth. The case might be otherwise on another planet.

Alkali Feldspar

The alkali feldspar system is illustrated at 1 atm in Fig. 15.4. From this beginning, we can make a few statements about the system at high pressure. The initial melting slope for Ab is 18°C/kbar (Fig. 18.7), and the corresponding metastable slope for Or is also 18°C/kbar (Fig. 18.22). The average slopes to 20 kbar are 10.8 and 12.5°C/kbar, respectively. Since San = *L* moves faster than Ab = *L*, the minimum composition in the alkali feldspar system should move somewhat towards Ab with pressure. The minimum lies near 1/3 Or, 2/3 Ab, so the rise in the minimum temperature should be something like 1/3 (12.5) + 2/3 (10.5) = 11.4°C/kbar. The calculated minimum temperature at 20 kbar would thus be 1063 + 227 = 1280°C. At this pressure, no leucite field would be present, and the melting diagram would be of the simple minimum type terminating at Ab = *L*, 1316°C and San = *L*, 1449°C.

The alkali feldspar solvus also rises with pressure, at a rate variously estimated at 13–20°C/kbar. Assuming that the solvus crest (critical temperature, T_c) lies at about 650°C at 1 atm, and assuming the largest value of dT_c/dP, the solvus and the solidus would intersect only somewhere near 60 kbar, where we know that Ab is grossly unstable relative to Jd + *Q*, so there is no prospect that two alkali feldspars will crystallize directly from the melt at any pressure in the system Ab–Or. There is therefore no need to draw the alkali feldspar diagram at 20 kbar: if we did, it would look essentially like Fig. 15.26 after removal of the leucite field and some adjustment of the temperatures and the minimum composition.

Despite all this characterization, the natural alkali feldspar system at pressure becomes very complicated in the presence of water and very different in the presence of aluminous clinopyroxene and fayalite.

Fayalite and Olivine Solutions

The melting curve of fayalite has an initial slope of about 8°C/kbar. The curve shown in Fig. 18.24 is a hand-calculated solution to the Simon equation using the

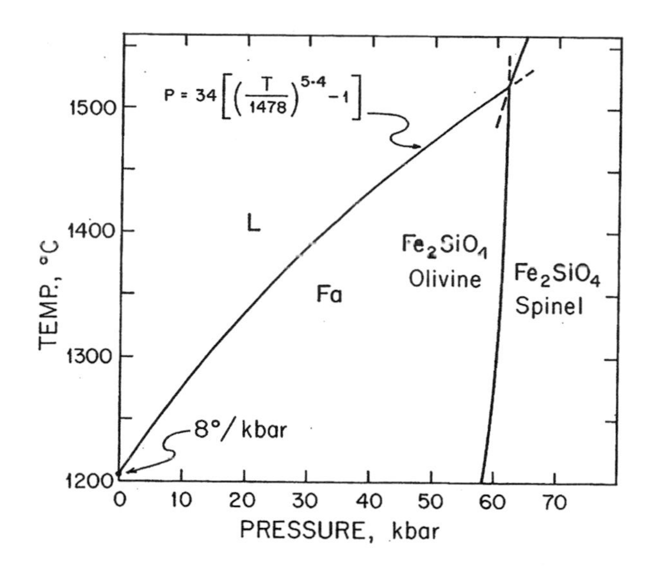

Fig. 18.24 High-pressure melting and breakdown relations of fayalite (Fa), after Lindsley (1966b) and Akimoto et al. (1967). The Simon equation represents an approximate fit to both sets of data and differs slightly from that given by Akimoto et al. The corresponding polynomial is $y = -0.059x^2 + 7.79x + 1205.3$, where $y = T$ °C and $x = P$, kbar

combined data of Lindsley (1966b) and Akimoto et al. (1967). The position of the high-pressure phase transformation to the spinel structure of Fe_2SiO_4 is taken from the latter authors. Note the characteristic cusp in the melting curve where it is intersected by the polymorphic transition.

From the melting curves for forsterite and fayalite, and the assumption that little change occurs in the form of the melting loop at high pressures up to 50 kbar, the approximate form of the olivine diagram can be sketched as in Fig. 18.25. The melting points for the end members are indicated at 10 kbar intervals. Melting loops are sketched for 1 atm, 20 kbar, and 50 kbar. At 20 kbar, a liquid of composition Fo_{50} would be in equilibrium with crystals of composition Fo_{42} in weight per cent. The equivalent mole per cent values are Fo_{59} and Fo_{87}, and these are reasonably similar to typical values for basaltic liquids and peridotite, respectively.

Fe–SiO₂–O

Pressure stabilizes the pyroxene ferrosilite, $FeSiO_3$, relative to Fa + *Q*, and the resulting P–T–f_{O_2} relations have been thoroughly expounded by Lindsley et al. (1968). The present treatment takes up where we left off in Chap. 16. We omit all equilibria involving wüstite and focus at first on equilibria among olivine, silica, liquid, magnetite, and iron. In Fig. 16.20, the l-atm (isobaric) invariant point *osml* lies at higher f_{O_2} and lower T than the invariant point *osil*. These isobaric invariant

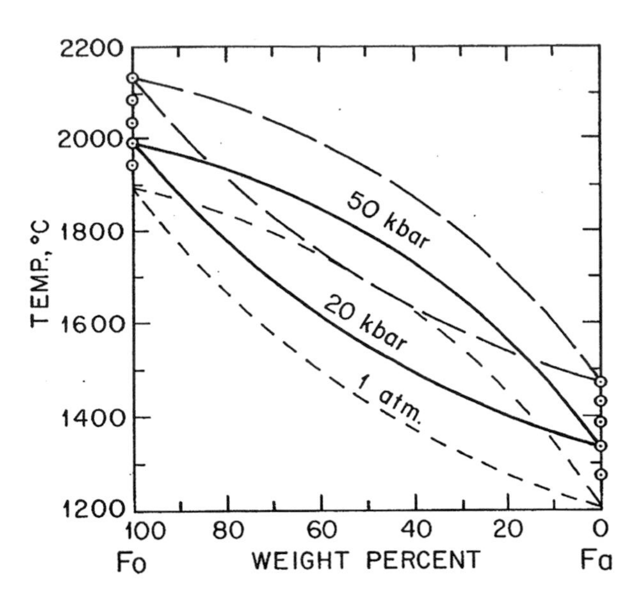

Fig. 18.25 Approximate high-pressure melting loop for olivine, compared with the 1-atm diagram. Deduced from Figs. 18.10 for Fo and 18.25 for Fa. End member melting points are plotted for each 10–50 kbar. It should be noted that these simple loops are not likely to describe olivine–liquid relations in complex basaltic bulk compositions

points are univariant *curves* when the pressure restriction is removed, i.e. when $W_p = 0$, $W = 1$. We can speak of a two-dimensional, 3-coordinate *intensive variable space* such as P–T–f_{O_2} space. The isobaric 1-atm diagram (Fig. 16.20) is a section through that space, normal to the pressure axis, at 1 atm. We now wish to turn around and look at the P–T relationships, but it would not be useful to take a constant f_{O_2} *section* through the P–T–f_{O_2} space, because arbitrary restriction to a constant f_{O_2} is neither experimentally convenient nor conceptually informative. Instead, we shall use a projection parallel to the f_{O_2} axis onto the T–P plane. Such a *projection* preserves all of the geometrical properties of the curves and invariant points, and we only need to remember that any given equilibrium represents a variety of f_{O_2} values along its curve; i.e., f_{O_2} is variable and unspecified in the diagram. Figure 18.26 is such a *T–P projection* along the f_{O_2} axis.

In the figure, we see at the left two pyroxene-absent curves labelled (p). The high-temperature curve is the equilibrium *osil*, now univariant in T–P projection. The low-temperature curve is the equilibrium *osml*. Both curves originate at the corresponding 1-atm points in Fig. 16.20 (but of course the equilibria pass continuously through the 1-atm points to conditions of vacuum, so they do not cease to exist at pressures below 1 atm). The pyroxene-absent equilibria encounter the pyroxene-forming reaction $o + s = p$ shown in the figure by a straight line. The intersections so formed generate two truly invariant points, *ospil* and *ospml*. The sequence of curves around each of these points is obtainable from Schreinemakers' rules (or Richardson's rule) with the help of the chemographic diagrams. Samples of the compatibility diagrams are shown in the figure.

Because pyroxene lies on the line o–s, the reaction $o + s = p$ is degenerate and is truly univariant even though only three phases suffice to define the reaction. The

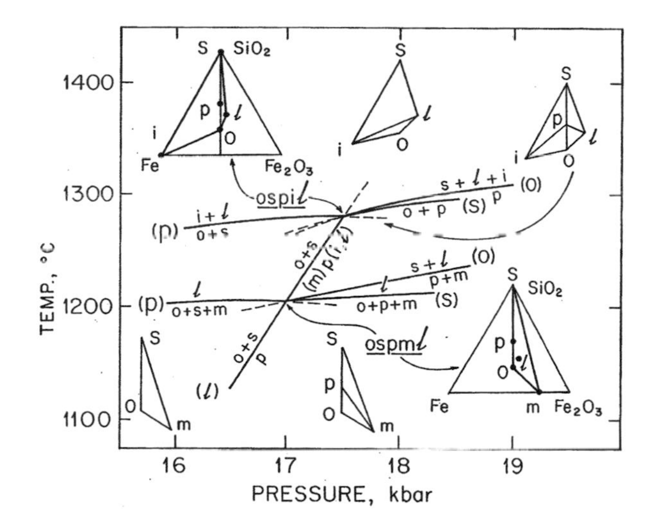

Fig. 18.26 *T–P* projection for the system Fe–SiO_2–O, after Lindsley et al. (1968). Abbreviations as in Chap. 16 except for *p* pyroxene, $FeSiO_3$. Relations involving wüstite are omitted

assemblage *osp* itself does not begin to melt at 17 kbar, 1205°C (at *ospml*), but liquid is generated there in the presence of magnetite. Just above *ospml*, the equilibrium *osp* is the magnetite-absent reaction (*m*). The reaction then becomes both (*i*) and (*l*) near the invariant point *ospil*. The equilibrium is metastable above this invariant point, because now the tie line iron + liquid cuts the line olivine + pyroxene. The point *ospil* therefore represents the maximum stability of fayalite plus ferrosilite at any f_{O_2}. Curve *osp* represents the minimum stability of ferrosilite and the maximum stability of fayalite + quartz. The high pressures encountered along this curve make it fairly obvious why fayalite + quartz are found in granites, but pure ferrosilite is not. As we shall see, however, addition of Mg changes the story.

We now return to f_{O_2}–T space to recover a picture analogous to Fig. 16.20. But now we shall again *project* the equilibria, this time along the pressure axis onto the f_{O_2}–T plane. In this way, we shall be able to look at f_{O_2}–T relations in complete detail, without concern for the pressure at which they occur. Figure 18.27 is such a projection. The same two invariant points occur in this projection as in the *T–P* projection, and their arrangement is not unlike that of the points *osml* and *osil* in Fig. 16.20. Indeed, the equilibria (*p*) lead directly to the 1-atm invariant points, and these are shown by dots in the figure. The curves are extended beyond the dots for convenience in labelling, and this means that parts of the curves (*p*) refer to pressures less than 1 atm.

The curves (*o*) and (*s*) in Fig. 18.27 have been corrected from those in the original paper, thanks to a timely communication (1979) from D. H. Lindsley. The liquid-absent (*l*) curves in Fig. 18.27 separate the regions where *p* + *s* or *s* + *i* are stable relative to olivine. Connecting these two subsolidus curves is a *ruled surface* labelled *osp*. This surface is the locus of the reaction *o* + *s* = *p*, and it is everywhere rigorously parallel to the f_{O_2} axis. This means that the reaction *o* + *s* = *p* is independent of f_{O_2}. The surface is inclined to the pressure axis, and each ruling

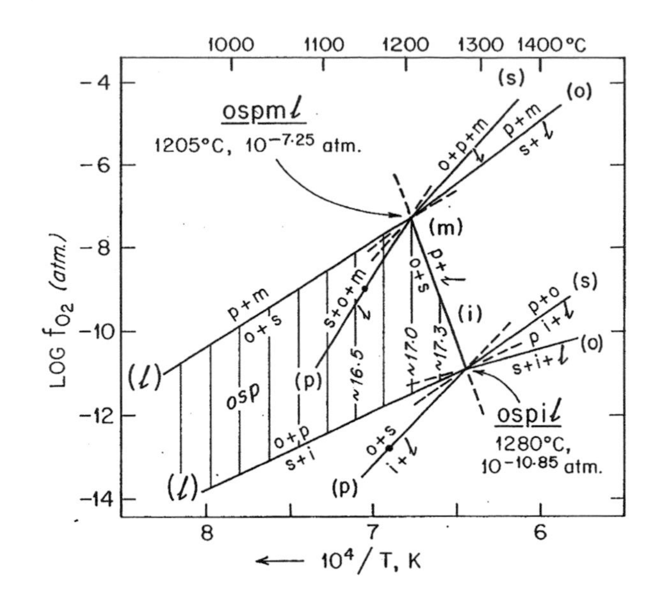

Fig. 18.27 Projection of part of the system Fe–SiO_2–O parallel to the pressure axis onto the T–f_{O_2} plane, after Lindsley et al. (1968). The ruled surface labelled *osp* is a divariant surface rigorously parallel to the f_{O_2} axis but varying in pressure: each ruling represents the intersection of an isobaric plane with the surface *osp* in P–T–f_{O_2} space. Several isobars are labelled with pressures corresponding to the isobaric sections shown in the next figure. Abbreviations as in the previous figure. Symbols in parentheses denote the missing phase of a truly univariant assemblage. Heavy dots on curves *osml* and *osil* show the location of the 1-atm isobar on those curves

represents the trace of an isobaric plane intersecting the ruled surface. The trace of the ruled surface is shown in T–P space by the curve *osp* in Fig. 18.26. Several isobars, including the 17 kbar isobar intersecting *ospml*, are labelled on the ruled surface in Fig. 18.27. We can make use of the labelled isobars to show examples of isobaric T–f_{O_2} sections through P–T–f_{O_2} space, as shown in Fig. 18.28.

Diagram (a) of Fig. 18.28 illustrates the isobaric T–f_{O_2} relations at 16.5 kbar and is valid in principle for any pressure in the range 10–17 kbar. The *osp* curve, rigorously parallel to the f_{O_2} axis, is the 16.5 isobar in Fig. 18.27. Pyroxene is stable on the low-temperature side and fayalite + quartz on the high-temperature side of the curve. The reaction is isobarically univariant because of degeneracy. It terminates at the isobaric invariant points *ospm* and *ospi*, (univariant lines in Fig. 18.27), where it intersects the curves $o = s + m$ and $o = s + i$ inherited from the 1-atm diagram (Fig. 16.20). New equilibria involving pyroxene are generated at the low-T sides of the invariant points. At the hi-T sides, the curves *osm* and *osi* terminate in the familiar invariant points *osml* and *osil*, which have the same surrounding geometry as in Fig. 16.20. The singular points and the curve $o = l$, as well as all wüstite equilibria, are omitted from Fig. 18.28, but can be seen in Fig. 16.20.

As the pressure rises from 16.5 kbar, the reaction $o + s = p$ moves to higher temperatures, rapidly compared to the melting reactions involving $o + s$; see

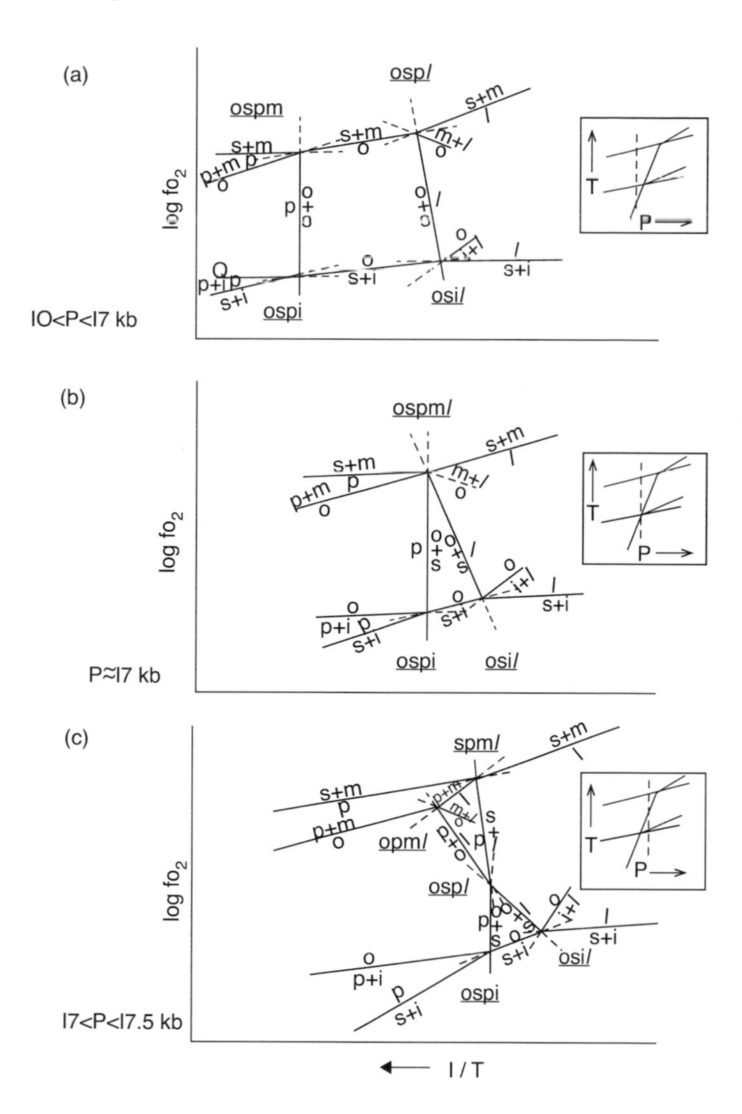

Fig. 18.28 Schematic isobaric sections through P–T–f_{O_2} space for the system Fe–SiO_2–O at several high pressures. Insets show where each section is taken relative to the P–T projection in Fig. 18.26. The singular points S_i and S_m are omitted; they would appear on the curves *oil* and *oml* and would be connected by the curve $o = l$ (see Fig. 16.20). (**a**) Section at about 16.5 kbar. Points *osml* and *osil* are present at 1 atm. Points *ospm* and *ospi* reflect the stability of ferrosilite at high pressures. The curve *osp* is rigorously parallel to the f_{O_2} axis and is equivalent to the 16.5 kbar isobaric ruling in this figure. Points *ospm* and *ospi* reflect the intersections of curve *osp* (Fig. 18.27) at two different values of f_{O_2}. (**b**) Section at about 17 kbar, and the pressure of the invariant point *ospml*. (**c**) Section at about l7.3 kbar. From Lindsley, Speidel, and Nafziger (1968), reproduced by permission

Fig. 18.26. Thus, at 17 kbar, the curve *osp* has reduced the curve *osm* to a point and intersected *osml* to form the truly invariant point *ospml* (Fig. 18.28b). This point, in the isobaric section, has "too many" (6) curves emanating from it, because these curves are really isobaric sections through divariant surfaces in P–T–f_{O_2} space. There could conceivably be as many as 10 such curves representing the 10 divariant surfaces *pml*, *sml*, *spl*, *spm*, *oml*, *opl*, *opm*, *oil*, *osm*, and *osp*. However, if both the limiting curves of a divariant surface lie at lower or higher pressure than the isobaric section, that surface will not cut the section and will not appear as a line. In the present case, six divariant surfaces have limiting curves that straddle the pressure 17 kbar, so six curves appear. Curves *osm* (at lower *P*) and *spl*, *opi*, and *pml* (at higher *P*) do not appear in the section. The sequence of curves about the truly invariant point *opml* in such an "accidental" isobaric section cutting the point cannot be determined directly by Schreinemakers' rules, but can be deduced, as done here, by merging the isobaric invariant points, each carrying its own configuration, into the truly invariant point. This discussion is amplified in somewhat more detail by Lindsley et al. (1968).

As the pressure rises above 17 kbar, curve *osp* moves through the truly invariant point and begins to cut the curve *osl*, thus generating the isobaric invariant point *ospl* (Fig. 18.28c). Two other new isobaric invariant points also appear, *opml* and *spml*; these are points on the univariant curves of the same names in Fig. 18.26, the *T*–*P* projection. The geometry of all these new points is straightforward and can be derived, using care, from the usual rules.

With further rise in pressure, the *osp* curve intersects the other truly invariant point, *ospil,* and eventually passes through it. This process eliminates the equilibrium *osi* from the isobaric sections and produces instead a new equilibrium, *psi*, and two new isobaric invariant points, *opil* and *spil.* Such isobaric sections are omitted from this book, but the reader may care to try her luck with them (taken at 17.5 kbar and a higher pressure) as a useful exercise. Those who do so may be rewarded by seeing the original versions, for comparison, in Lindsley et al. (1968).

It may be noted in Fig. 18.28c that ferrosilite melts incongruently to quartz plus liquid (curve *psl*).

MgO–FeO–SiO₂

We will not explore this system in any detail at high pressures; an exposition is given by Speidel and Nafziger (1968), and it contains quantitative information on the compositions of coexisting olivine and pyroxene as a function of *P*, *T*, and f_{O_2}. Oxidation drives both mineral compositions towards Mg, as Fe is used up as ferric iron in oxide minerals or liquid.

A few general remarks are pertinent. It may not have escaped the reader's notice that the invariant points *ospil* and *ospml* in Figs. 18.26 and 18.27 have identical counterparts at 1 atm in the system MgO–FeO–Fe_2O_3–SiO_2, where they are named with capital letters (OSPIL and OSPML in Fig. 17.4). They are, of course, the same equilibria. As Mg is added to $FeSiO_2$–O, the pressure required to stabilize pyroxene decreases, until it reaches 1 atm for pyroxene compositions near $En_{15}Fs_{85}$. The composition of pyroxene in equilibrium with olivine and silica is therefore a sensitive indicator of pressure in some crustal rocks. Wheeler (1965) recognized the breakdown of iron-rich orthopyroxene to olivine plus quartz in a Labrador adamellite, and Smith (1971) studied the reaction experimentally. Berg (1977) made use of this and similar assemblages in metamorphosed iron formation to estimate pressures in the contact metamorphic aureole of the anorthositic Nain Plutonic Suite, Labrador.

Summary

We have now ripped through the essential features of our earlier 1-atm phase diagrams at high pressures, omitting petrogeny's residua system and ternary feldspars. The main messages of high pressure are contained in the generally positive melting slopes, the frequent appearance of singular points as liquid compositions move past crystal compositions, the occurrence of cusps where solidi intersect solid reactions, and the geometrical analysis of reactions in intensive variable space. Many phase diagrams are incompletely known at high pressure, but are sufficiently known for estimating behaviour in the mantle to a first approximation. Our information on temperatures is incomplete, but is gradually being improved by the many experimental studies recently made on selected compositions approaching natural peridotite.

The analysis of crystallization and melting paths proceeds at a given high pressure just as it does in isobaric diagrams at 1 atm, so it has not been discussed here.

Chapter 19
Effects of Volatiles at High Pressure

Introduction

Melting temperatures are lowered dramatically when H_2O is added under pressure to a system of anhydrous silicate crystals. This is because a silicate melt can readily dissolve H_2O, thereby reducing the volume of H_2O from that occupied by a gas to that occupied by a liquid. The silicate crystals can dissolve only minute amounts of H_2O. A system composed of an anhydrous phase plus H_2O gas has a large volume just below the melting point, and the formation of hydrous melt will lead to a smaller volume above the melting point, so high pressure favours the formation of melt.

At low-to-moderate pressures, carbon dioxide is much less soluble in silicate and melts than H_2O, so it has a less pronounced effect on the lowering of melting temperatures. Water and carbon dioxide are the chief volatiles of interest for the Earth's mantle. Both are observed in substantial amounts in volcanic eruptions, but this cannot be taken as evidence that they originate with the magma, because the isotopic ratios of H, C, and O show that most of the volatiles come from meteoric (recycled) waters. The search for juvenile water, long frustrated by this overwhelming flux of meteoric water, has now been rewarded by recognition of the $^3He/^4He$ ratio as the important tracer. This ratio can demonstrate the existence of unrecycled mantle water in volcanic emanations. But the primary evidence for water in the mantle comes from phenocrysts of the hydrous mineral phlogopite in kimberlites and from primary hydrous amphibole in mantle xenoliths (although both these minerals probably contain a large oxy-component rather than their full complement of hydroxyl). Similarly, the primary (and impressive) evidence for CO_2 in the mantle comes from the carbonate-rich composition of kimberlites. The composition of fluid inclusions in mantle minerals also shows the existence of H_2O and CO_2. Whether these are important components in the genesis of basalt magma depends on their concentration. They are important if present in sufficient amounts, as for example in subduction zones. Moreover, the presence of the hydrosphere and atmosphere

S. A. Morse, *Basalts and Phase Diagrams*,
https://doi.org/10.1007/978-3-030-97882-2_19

requires de-gassing of the planet through time, so the effect of volatiles in magmas plays an important role in planetary evolution.

If water is added to albite and the system is heated at 1 atm, the water will soon turn to steam and escape, so no effect will be noted in the melting relations. In order for water or any volatile to have an observable effect on melting, the experimental system must be closed, and pressure must be applied to prevent the escape of steam. If albite plus 1% H_2O is contained at a pressure of 5 kbar at, say, 500°C, the system consists of crystals plus supercritical gas, and it is common parlance to say that the water pressure equals the total pressure ($P_{H_2O} = P_T$), because the gas always fills the container and therefore bears the load, hydrostatically, with the crystals. Purists regularly and insistently decry the term "water pressure" or "P_{H_2O}" in such cases, and they are, of course, right to do so, because the gas at these conditions is no longer composed of H_2O only, but is a solution of H_2O and some of the silicate material (mostly silica, as it turns out). Therefore the pressure is sustained not by pure H_2O gas but by an H_2O–silicate gaseous solution. If this is so, the partial pressure of the H_2O part of the gas is less than the total pressure, the rest being made up by the silicate component of the gas.

Numerous dodges have been resorted to in the interests of accuracy and clarity when talking about hydrous systems with a gas phase under pressure. Perhaps the best is to use the term "steam" and "steam pressure", for nobody knows or cares what steam is composed of, and in this way, we tell no lies. Another helpful device is to use P_{aq}, meaning "the pressure of an aqueous fluid", likewise a guiltless phrase. The most common dodge, in some ways the most appealing, goes something like this: purists are tedious people, and everybody knows very well that the stuff is a complex solution, so let us just call it water pressure and assume that everybody will read this as a euphemism for whatever it really is.

Returning to the albite with 1% H_2O, if we now take the system to 800°C at 5 kbar, it will consist of a single phase, melt. The melt will consist of 99% $NaAlSi_3O_8$ and 1% H_2O (as components, not as molecular species) in solution. Now, no aqueous gas phase exists to bear the pressure load, so purists and dodgers alike are sure that the partial pressure of water vapour is less than the total pressure, i.e. $P_{H_2O} < P_T$. The criterion of *vapour* (gas) *saturation* is of crucial interest in petrology, and in common parlance, saturation is meant by the notation $P_{H_2O} = P_T$ and undersaturation is meant by the notation $P_{H_2O} < P_T$. In the first case, a gas *phase* is present; in the second case, it is absent. For the case of Ab–H_2O, a gas phase would be present with melt at 800°C, 5 kbar if the system composition were 20% H_2O, 80% Ab, because only about 9% H_2O is needed to saturate the liquid with gas at 800°C, 5 kbar (see Burnham and Davis 1974).

This chapter contains a selective review of the melting relations of common silicate systems with water and a very limited discussion of the role of carbon dioxide in such systems. The subject of volatiles in magma genesis is vast, and we can attempt here no more than an introduction. For an appreciation of the scope of the subject, interested readers may wish to consult the excellent (and somewhat divergent) reviews by Wyllie (Presidential Address 1979), Eggler and Holloway (1977) and Mysen (1977).

Forsterite–H_2O

Figure 19.1 shows by means of a *P–T* projection the dramatic effect on the melting point of forsterite when water is added at high pressures. The diagram shows the dry melting curve Fo = *L* on the right, and the H_2O-saturated curve Fo + *G* = *L* + *G* running out to the left, to lower temperatures.

Figure 19.2 is an isobaric *T–X* section at 10 kbar showing the eutectic-like melting of the system Fo–H_2O. (It is not truly eutectic because the gas phase contains much more silica than magnesia, so neither the gas nor the liquid lie in the Fo–H_2O join, as shown by Nakamura and Kushiro 1974.) The assemblage Fo + *G* melts at 1520°C, 418°C lower than pure Fo at 10 kbar. The coexistence of Fo + *G* is

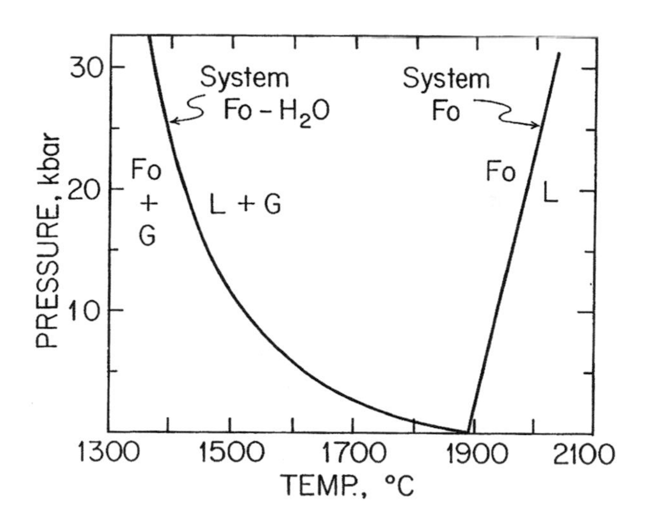

Fig. 19.1 *P–T* projections showing the melting of forsterite, dry and with H_2O in excess. Dry curve from Fig. 18.10; wet curve from Kushiro and Yoder (1968)

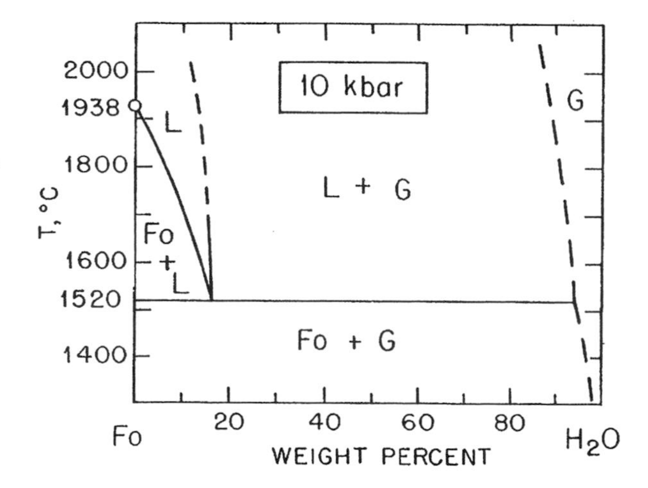

Fig. 19.2 Isobaric 10 kbar *T–X* section showing melting relations in the join Fo–H_2O. Melting is not strictly eutectic because the gas and liquid compositions lie off the join. Temperatures from Fig. 19.1; liquid composition after Hodges (1974)

interrupted by liquid at all higher temperatures. The "eutectic" liquid contains about 18% H_2O in solution.

The Fo liquidus describes the composition of liquid in equilibrium with forsterite crystals at higher temperatures; this curve shows that the H_2O content of the liquid decreases to zero at the dry melting point, 1938°C at 10 kbar. A liquid field in the shape of a gore shows the range of bulk compositions that occur as gas-undersaturated liquid. This field is bounded to the left by the liquidus and to the right by a saturation curve denoting saturation of the liquid with an aqueous gas phase. This curve is part of a solvus whose crest is beyond temperatures of interest and whose right-hand limb is seen in the right part of the diagram. The solvus curve delimits the two-phase $L + G$ field from the L field on the left and the G field on the right. The gas phase contains a nontrivial amount of silicate component.

The fields L and Fo + L in Fig. 19.2 are those for which P_{H_2O} is clearly less than the total pressure, 10 kbar, because a gas phase is absent from those fields. Elsewhere in the diagram, $P_{aq} = P_T$.

The relation between Figs. 19.1 and 19.2 is shown in the perspective sketch of the P–T–X prism in Fig. 19.3. This figure shows the relationship between a *projection* and a *section*. Two isobaric T–X sections are shown, at 10 and 20 kbar; these resemble Fig. 19.2 seen from the back of the page. The isobaric sections show the T–X relations at single values of pressure. The actual trace of the first liquid composition with varying T and P is shown by the dashed curve within the prism; this trace connects the first liquids formed at all pressures. It is this trace, *projected* along the composition axis, which appears as the curve Fo + G = L + G in Fig. 19.1

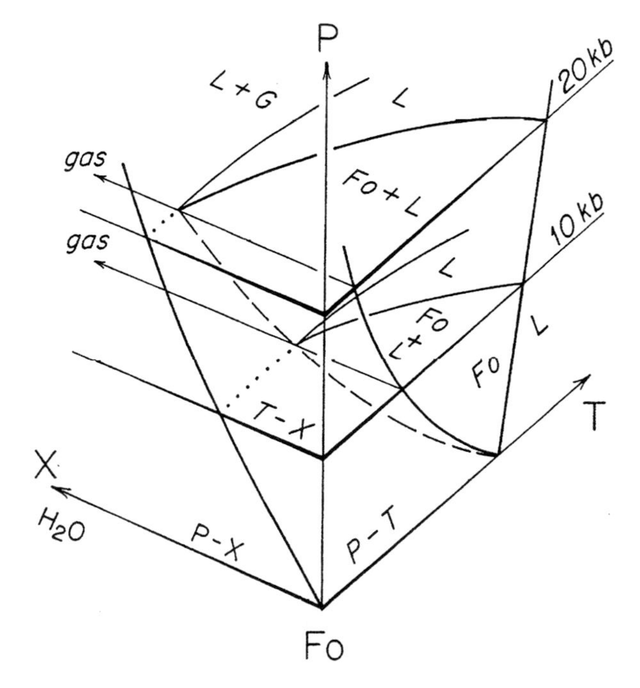

Fig. 19.3 P–T–X prism for the system Fo–H_2O, showing isobaric sections at 10 and 20 kbar. Only the H_2O-poor compositions are shown; similar relations could be sketched for H_2O-rich compositions. The dashed line is the trace, within the prism, of the "eutectic" liquid composition. This trace is projected along X onto the P–T plane and along T onto the P–X plane

and in the *P–T* face of the prism in Fig. 19.3. Any isocompositional *P–T* section could be used to collect this projected trace, but it is most convenient to use the anhydrous face of the prism for this purpose, since that face truly contains the dry melting curve. The same dashed curve tracing the liquid composition in equilibrium with Fo + *G* is projected, in Fig. 19.3, onto the *P–X* face to give a *P–X* projection. This projection shows the composition of the liquid, rather than the temperature, as a function of pressure.

SiO_2–H_2O

The wet and dry melting curves for silica are shown in *P–T* projection in Fig. 19.4. The point labelled *K* is a *critical end point*, to be described. The same relations are shown schematically in Fig. 19.5. The dashed curve in this figure is the critical curve $L = G$ running from $K_{S\text{-}H}$, the critical end point for SiO_2–H_2O, to K_S, the critical point for SiO_2. The wet melting curve generates invariant points at its intersections with the tridymite field, and it terminates *without metastable extension* at the triple point for SiO_2. The absence of a metastable extension here is due to the fact that the hydrous melting reaction is undefined in the anhydrous system SiO_2. That the wet melting curve originates at the triple point for SiO_2 can be deduced from the fact that gases of compositions H_2O and SiO_2 are completely miscible in all proportions, and

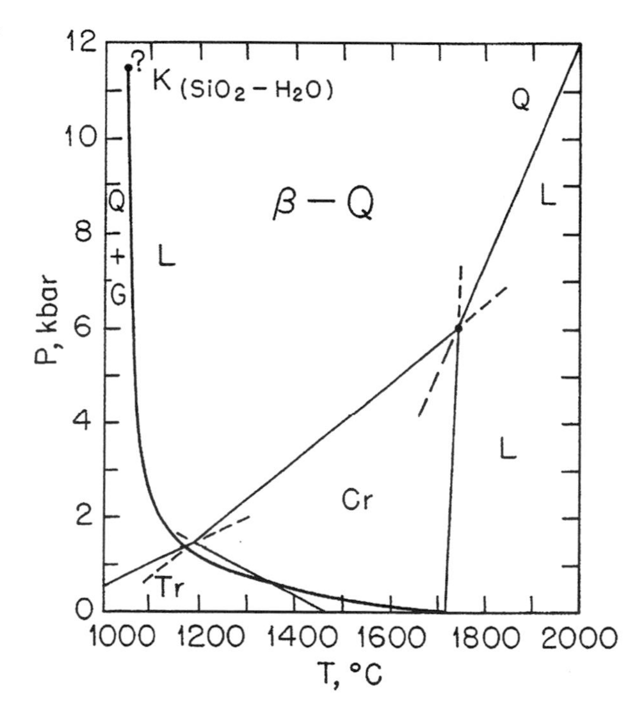

Fig. 19.4 *P–T* projection of melting relations in the system SiO_2–H_2O. Point *K* is a critical end point. Data for SiO_2 from Fig. 18.13 and for SiO_2–H_2O from Kennedy et al. (1962) and Stewart (1967)

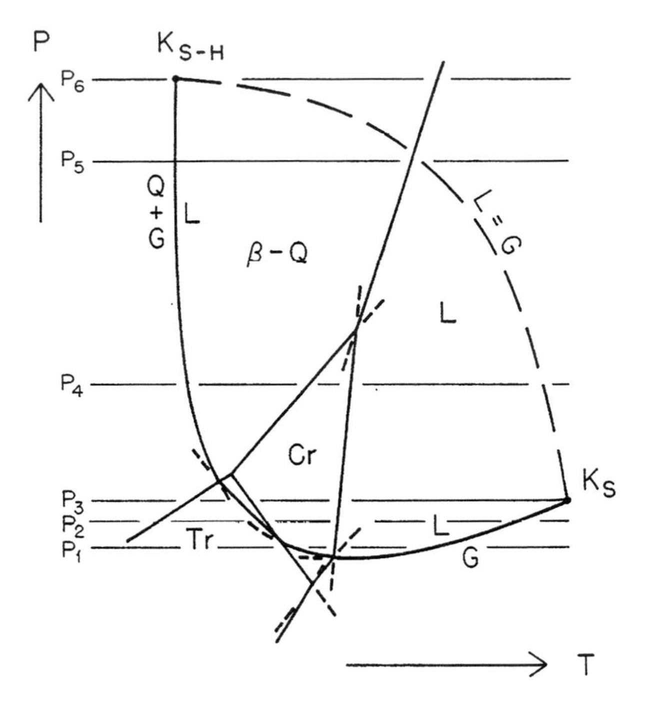

Fig. 19.5 Schematic P–T projection for SiO_2–H_2O showing locations of isobaric sections P_1 through P_6 and the critical curve $L = G$ connecting the two critical (end) points $K_{S\text{-}H}$ (for the binary system) and K_S (for the unary system SiO_2). The wet melting curve runs to the triple point in the system SiO_2 and ends there without metastable extension

the resulting fact that the G field in the limiting system SiO_2 must be continuous with the G field in SiO_2–H_2O. All wet melting curves for silicates likewise originate at the triple point for the anhydrous silicate rather than at the 1-atm melting point, because of course there is nothing special about the pressure of 1 atm except that we love it. The wet and dry melting curves are commonly drawn to meet at 1 atm simply because the difference between this case and the real one is experimentally undetectable in ordinary work.

A series of isobaric T–X sections, corresponding to the pressures P_1 through P_6 in Fig. 19.5, is shown in Fig. 19.6 to illustrate the melting and critical relations for SiO_2–H_2O. The diagrams are valid in principle for all such systems, but are drawn in the book only for this one, in which enough data are available to make the relations qualitatively meaningful. The story told by these T–X sections requires little elaboration. Section P_3 shows the nature of the critical point in the system SiO_2, where the distinction between gas and liquid vanishes and the state of the system is described as that of a supercritical fluid. The following sections show the continued diminution of the $L + G$ field in the binary system SiO_2–H_2O as the pressure is raised. Section P_6 shows the nature of the critical end point $K_{S\text{-}H}$ in the binary system. Figure 19.7 shows the supercritical condition at 15 kbar in the binary system; this figure is based on actual data obtained by Nakamura (1974).

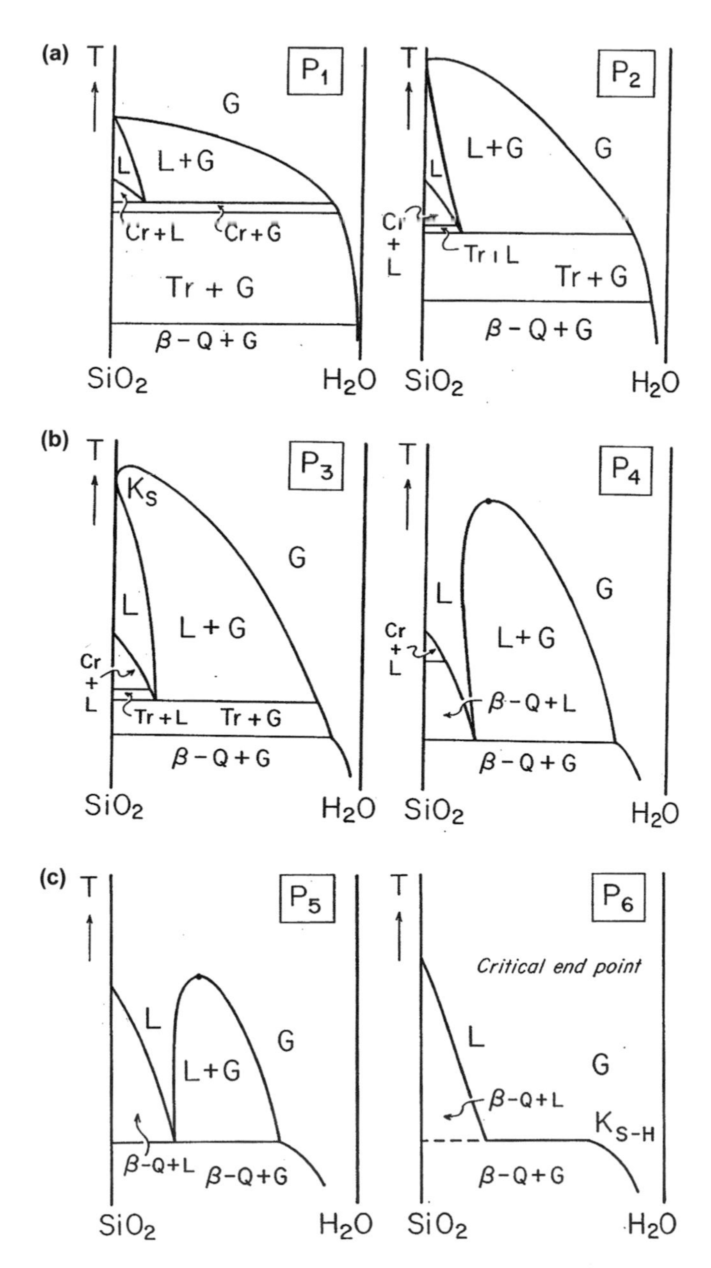

Fig. 19.6 Schematic isobaric *T*–*X* sections P_1 through P_6 for SiO_2–H_2O. Symbols: *β*–*Q*, high quartz; Tr, tridymite; Cr, cristobalite; *L*, liquid; *G*, gas; K_S, critical point; $K_{S\text{-}H}$, critical end point

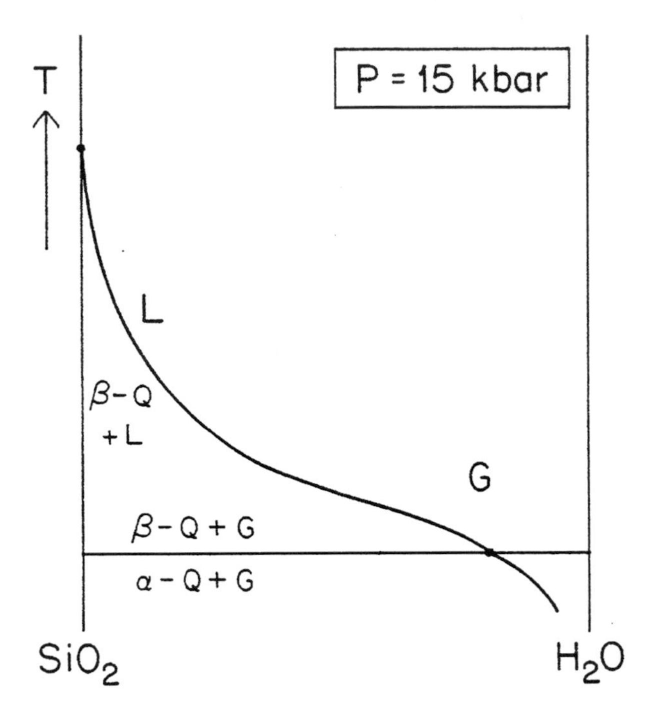

Fig. 19.7 Isobaric *T–X* section at 15 kbar for the system SiO_2–H_2O. After Nakamura (1974)

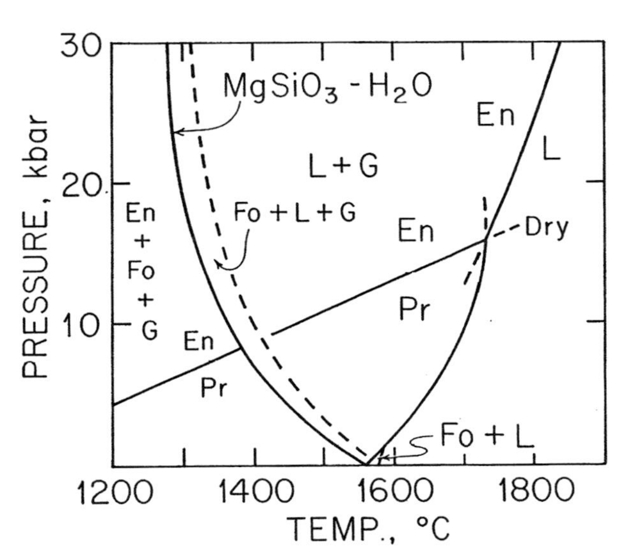

Fig. 19.8 *P–T* projection of the melting relations of $MgSiO_3$–H_2O. After Kushiro et al. (1968) and Chen and Presnall (1975)

Enstatite–H_2O

The *P–T* projection for $MgSiO_3$–H_2O is shown in Fig. 19.8, and the melting behaviour is further illustrated in Figs. 19.9 and 19.10. Figure 19.10 shows that enstatite dissolves incongruently with H_2O at 10 kbar to Fo + *G*. It also melts

Fig. 19.9 Isobaric 10 kbar T–X section along the join $MgSiO_3$–H_2O, after Kushiro et al. (1968)

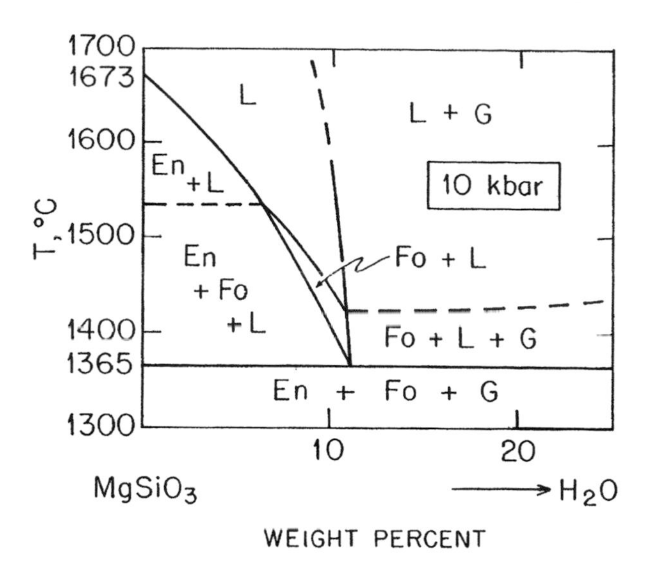

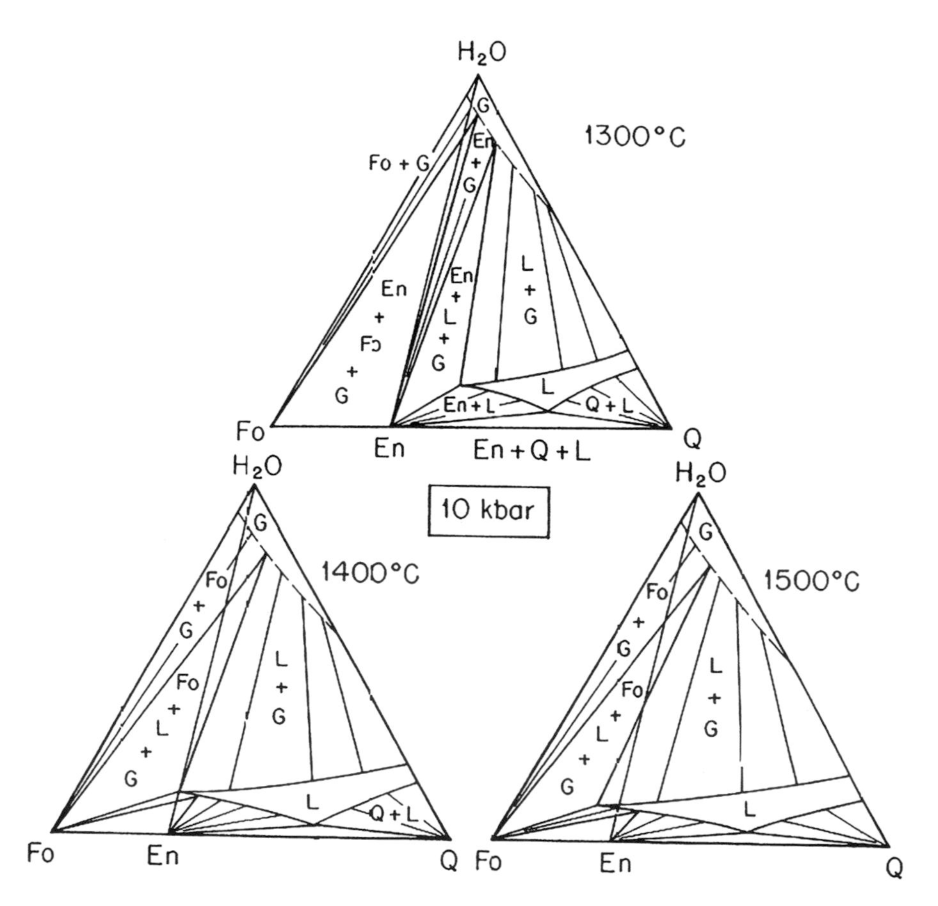

Fig. 19.10 Isobaric (10 kbar)–isothermal sections for the system Fo–Q–H_2O, after Kushiro et al. (1968)

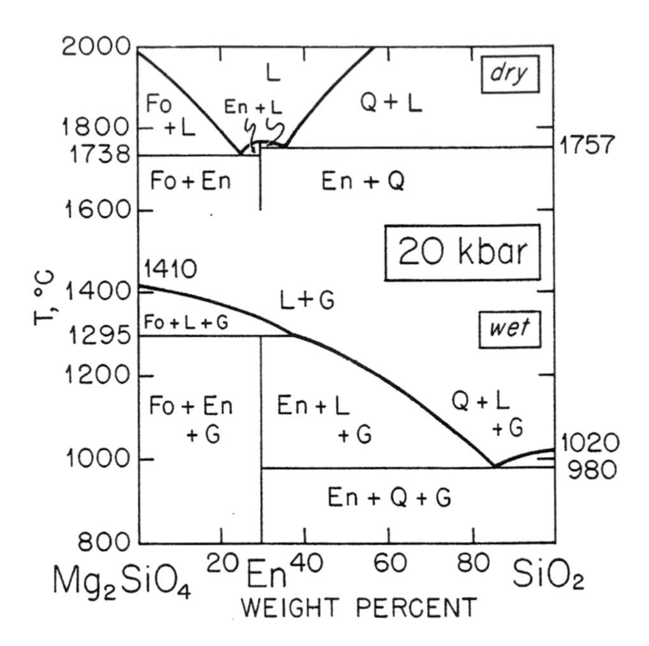

Fig. 19.11 Dry and wet melting relations of enstatite at 20 kbar. Dry diagram at top is an isobaric *T–X* section; wet diagram is a projection from H_2O. After Kushiro (1969) and Chen and Presnall (1975)

incongruently to Fo + *L* + *G*, as shown by the transition from the 1300°C to the 1400°C isothermal, isobaric section in Fig. 19.10. The *T–X* section of Fig. 19.9 can be understood by reference to the ternary sections of Fig. 19.10.

The wet melting of enstatite at 20 kbar is compared with the dry melting behaviour, at the same pressure, in Fig. 19.11. The 20 kbar diagram emphasizes the fact that addition of H_2O at high pressure changes the melting behaviour of enstatite from congruent to incongruent, thus destroying the thermal barrier present in the dry system. Ultramafic sills that show cotectic crystallization of olivine and bronzite must have crystallized from relatively dry liquids.

Forsterite–Plagioclase–Silica–H_2O

The liquidus relations at 15 kbar, projected from H_2O onto the anhydrous plane Fo–An_{50}–SiO_2, are shown in Fig. 19.12. Olivine and plagioclase do not coexist with liquid under these conditions, but react to form a pargasitic amphibole instead. This evidence can be used to argue that troctolites crystallize from H_2O-undersaturated or relatively dry melts. Enstatite and plagioclase do coexist with liquid, so high water pressure is, in principle, no bar to the crystallization of norites.

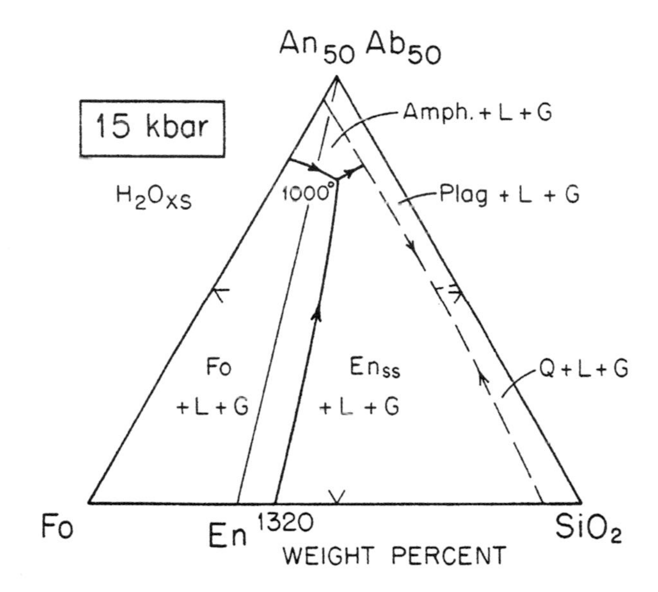

Fig. 19.12 Gas-saturated liquidus relations in the system Fo–An_{50}–SiO_2 at 15 kbar, after Kushiro (1974). "Amph." is pargasitic amphibole

Forsterite–Diopside–Silica–H_2O

Figure 19.13 shows that the incongruent melting of enstatite–H_2O extends far into the system Fo–Di–SiO_2–H_2O. Mantle-like compositions in the triangle Fo–Di_{SS}–En_{SS} will begin to melt, with H_2O, to a liquid at the peritectic composition *X*, 1220°C. Such a liquid may be likened to a basaltic andesite; it is clearly silica-saturated in terms of its normative composition.

The shift of quartz-saturated equilibria towards the SiO_2 corner is remarkable. Yoder (1973) argued that the liquid composition *Y* was an adequate proxy for rhyolite and used the 20 kbar diagram to explain the contemporaneous eruption of rhyolite and basalt from the same vent. The analysis requires a mantle source composed of two pyroxenes plus quartz. Such a composition would melt, with water, to a rhyolitic liquid *Y* at 960°C. If the melt were fractionally removed, or removed in a batch when the TSC reached the join En–Di, no further melting would occur until the temperature rose to 1220°C, supposing a new supply of H_2O to be added to replace that removed in the rhyolitic liquid. The basaltic liquid would be generated at 1220°C until the TSC reached the tie line En_{SS}–Fo, at which point melting would cease unless the temperature rose above 1220°C.

Figure 19.13 shows a fractional fusion exercise for a bulk composition BC in the field En_{SS} + Di_{SS} + *Q* + H_2O. The TSC moves from BC to *C* with extraction of the first melt (ILC) of composition *Y*. The TSC then moves from *C* to *D*, away from *X*, while the TLC moves on the mixing line *KY*, as shown by the dotted line and the sample lever at F_L 0.38. When the total solid composition has reached *D*, diopside is used up, and the liquids move along *XW*, the peritectic boundary. The total liquid composition moves towards this collection of liquids *XW*, sweeping rapidly towards

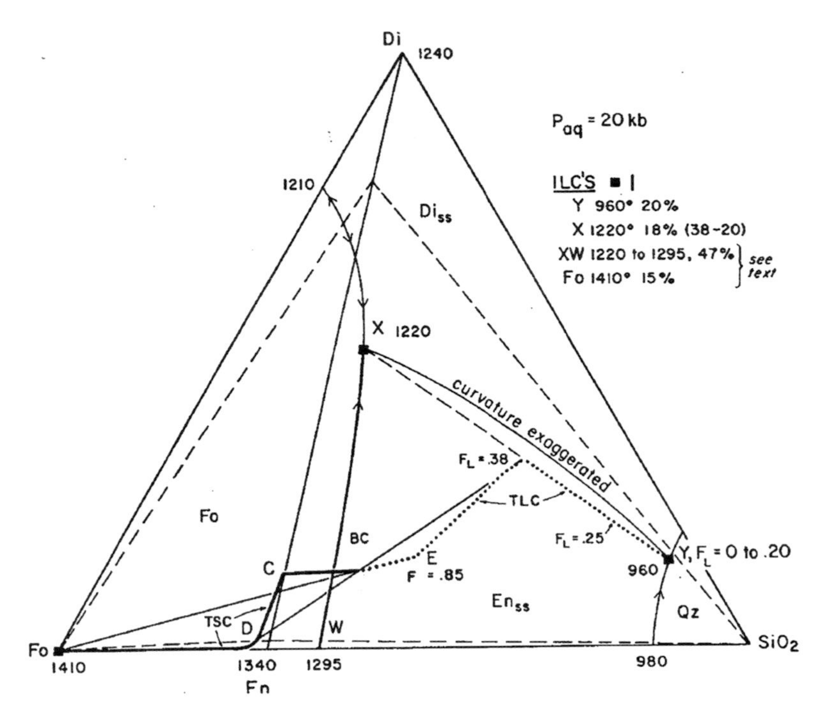

Fig. 19.13 Liquidus diagram of the system Fo–Di–SiO_2–H_2O at 20 kbar, after Kushiro (1969). The boundary *XY* has an exaggerated curvature to show the mixing line used for an analysis of fractional fusion (Yoder 1973; Morse 1976). Dashed lines show the assumed limits of pyroxene solid solution, after Yoder (1973). (Reproduced from Morse (1976) by permission)

W. The TLC path to *E* is constructed as follows. First, point *E* is located as the TLC when all En is exhausted and the TSC has reached Fo; point *E* is therefore a point on the line Fo–BC, extended. The location of *E* on this line is found to sufficient accuracy by determining F_L when the TSC has just reached Fo, as follows. Point *D* and the line *XW* are projected from diopside onto the base of the triangle, and the lever relation using *D* as a fulcrum gives 76% liquid of composition $\overline{XW}$ and 24% Fo crystals. The total amount of the system previously present as a TSC composition *D* was 62%, i.e. 1–0.38 = 0.62 as a fraction, and so the new fraction of solid forsterite is 0.62 (0.24) = 0.15, and $F_L = 0.38 + 0.47 = 0.85$. These values locate point *E* as the last TLC before pure forsterite begins to melt. The TLC path to *E* is now constructed with the constraint that it must lie everywhere below the extension of the line *W–E*, which is the leading tangent to point *E*.

From this analysis, it quickly becomes apparent that the TLC vector towards *X* and related liquids is microscopically short and that the liquids sweep rapidly towards *W*. This is reflected in the TSC path, which quickly approaches the sideline Fo–En.

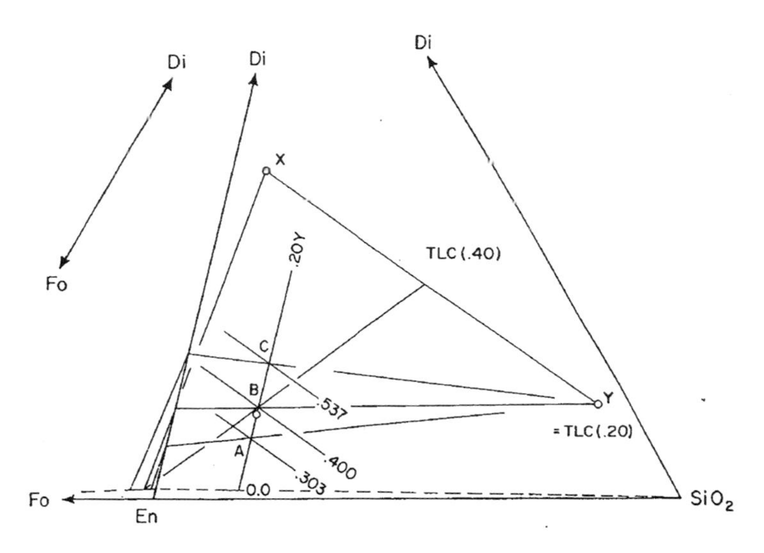

Fig. 19.14 Explicit graphical solution for equal amounts of liquids *Y* and *X* on fractional melting. The line labelled 0.20*Y* is parallel to En–Di, and the lines labelled 0.303–0.537 are parallel to *XY*. Both sets are lines of constant F_L. Point *B* is the desired bulk composition for $X = 0.20$ and $Y = 0.20$. The circle is point BC of Fig. 19.13. (From Morse (1976), with permission)

This exercise is done with Yoder's bulk composition, which yields 20% liquid *Y* and 18% liquid *X*. The amount of liquid *X* is obtained by noting that $F_L = 0.38$ when diopside is exhausted and that this fraction represents the sum of liquids *X* and *Y*. Since $F_Y = 0.20$, $F_X = 0.38–0.20 = 0.18$. An explicit solution for *equal* amounts of liquids *X* and *Y* is shown in Fig. 19.14. Lines of constant yield of liquid *Y* are rigorously parallel to the En–Di join as can easily be shown by similar triangles. Such a line for a yield of 20% liquid *Y* is plotted in Fig. 19.14 and marked 0.20*Y*. The desired bulk composition, supposing an initial liquid yield of 20% is desired, lies somewhere on this line. For an equal yield of liquid *X*, we seek a bulk composition such that F_L (as TLC) $= 0.40$ when Di is exhausted in the solids. For any such bulk composition, all lines of constant F_L lie parallel to the mixing line *XY* as can be shown with similar triangles. The desired line can be found accurately enough by choosing two random bulk compositions near the desired one on the line 0.20*Y* and calculating F_L for the case when each of these has just lost all Di to liquid *X*. The examples chosen, *A* and *C* in Fig. 19.14, gave $F_L = 0.303$ and 0.537, respectively. Plot the distances from *XY* versus F_L on a graph and find by interpolation the distance required for $F_L = 0.40$. A line parallel to *XY* at this distance intersects 0.20*Y* in the desired bulk composition, point *B* in the figure. The lever through *B* from the TSC on Fo–En_{SS} gives $F_L = 0.40$, as desired. Yoder's bulk composition, yielding 18% *X*, is shown as a circle near point *B* in Fig. 19.14. It lies very close to the bulk composition yielding equal amounts of liquids *X* and *Y*, thus showing that the relative amounts of liquid are quite sensitive to small changes in the bulk composition. This is so because liquids *X* and *Y* are both quite remote from the bulk composition.

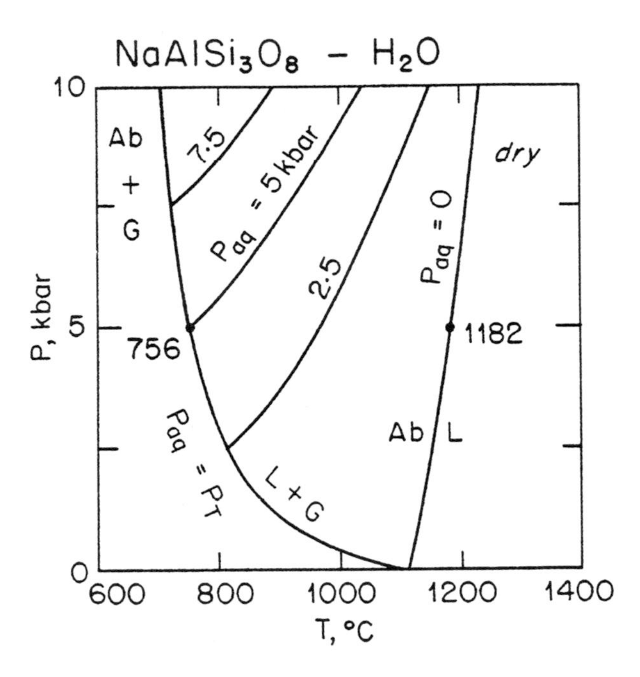

Fig. 19.15 Melting of albite with water at high pressure, shown in a *P–T* projection. Dry melting curve from Fig. 18.7; wet melting curves after Burnham and Davis (1974). Curves labelled with $P_{aq} = P_T$ show liquidus temperature in the absence of a gas phase. The curve Ab + *G* + *L* is the trace of the liquidus; the actual solidus lies at lower temperatures because of incongruent melting in Ab–H_2O

Albite–H_2O

The *P–T* projection for the melting of Ab–H_2O is shown in Fig. 19.15. The figure also shows three liquidus curves for the cases where insufficient H_2O is present to saturate the liquid at all pressures; these are curves for $P_{aq} = P_T$.

The curve Ab + *G* + *L* is, in principle, the solidus curve for all compositions; it is also the liquidus curve for vapour-saturated conditions. (The actual solidus curve lies somewhat below the curve drawn, because Ab dissolves incongruently in H_2O and melts over an interval of temperature.) Suppose the system contains just enough H_2O to saturate the liquid with gas at 5 kbar. Then, the system will melt completely at 756°C. Now, transfer this liquid isothermally to 7.5 kbar. Crystals of Ab will now form until the liquid is again just saturated with gas. At 7.5 kbar, the temperature will have to be raised to 920°C in order to melt the new Ab crystals. At 920°C, the liquid will again be undersaturated with gas, P_{aq} being only 5 kbar, while P_T is 7.5 kbar. If the liquid is now returned to 5 kbar, 756°C, it will be just saturated with gas again.

Albite–Orthoclase–H_2O

The entry of H_2O into the liquid at high pressures so reduces the solidus temperature that it cuts into the Ab–Or solvus at 5 kbar, as shown in Fig. 19.16. The reaction Ab + Sa + *G* = *L* occurs at 701°C at 5 kbar. The solvus crest rises with pressure, at a

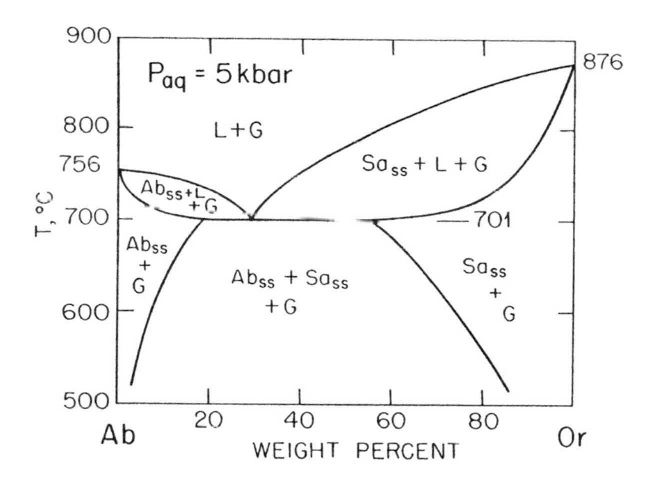

Fig. 19.16 Melting of alkali feldspars at 5 kbar, shown in an isobaric T–X projection from H_2O onto the Ab–Or join. After Yoder et al. (1957) and Morse (1970), solvus is that of Waldbaum and Thompson (1969) translated upward at a rate of 18°C/kbar

rate we shall take to be 18°C/kbar. The intersection of the solidus (trace of the minimum temperature for Ab–Or–H_2O) and the solvus *critical line* (trace of T_c with pressure) occurs at about 4.2 kbar, 715°C (Morse 1970). However, this is an *indifferent crossing* because nothing happens there. The reason for this may be seen with the aid of Figs. 19.17 and 19.18. The critical composition X_c on the solvus and the minimum composition X_m are not required to be the same. This means that when T_c and T_m are the same, at the indifferent crossing, the melting loop still lies above the solvus crest X_c and the solidus and solvus are not in contact. The actual contact occurs, in the general case, at some pressure S_1, where the solidus just kisses the solvus (Fig. 19.18a). This generates the reaction Ab + G = Sa + L, using simplified notation in which Ab and Sa are solid solutions near or at the solvus. The pressure S_1 marks the first coexistence of two feldspars with liquid. The singular point S_1, caused by the collinearity of Ab, Sa, and L in projection, lies at a minimum pressure on the continuous equilibrium curve Ab + Sa + L (Fig. 19.17). At a pressure slightly higher than S_1, a field of Sa + L occurs in the T–X projection (Fig. 19.18b). This field is bounded downward in temperature by the incongruent melting reaction Ab = Sa + L and upward by the same reaction in reverse, i.e. Sa + L at a lower temperature going to Ab at a higher temperature. This geometry requires the curve Ab + Sa + L to be concave upward, as in Fig. 19.17. The composition of sanidine in equilibrium with liquid, without Ab, is not constrained to the solvus and therefore lies on a new sanidine-rich solidus outside the solvus (Fig. 19.18b).

When the pressure is increased to that of point A in Fig. 19.17, the Ab solidus and the Sa solidus have just united at the critical point on the solvus. At a slightly higher pressure (Fig. 19.18c), the solidus becomes totally detached from the solvus except at one point and rapidly begins to change shape, moving away from the solvus. This change is shown schematically by two new lower bounds to the Sa + L loop in Fig. 19.18c. That the change in solidus shape is continuous and correct can be shown by constructing a series of G–X diagrams for the sequence shown in Fig. 19.18.

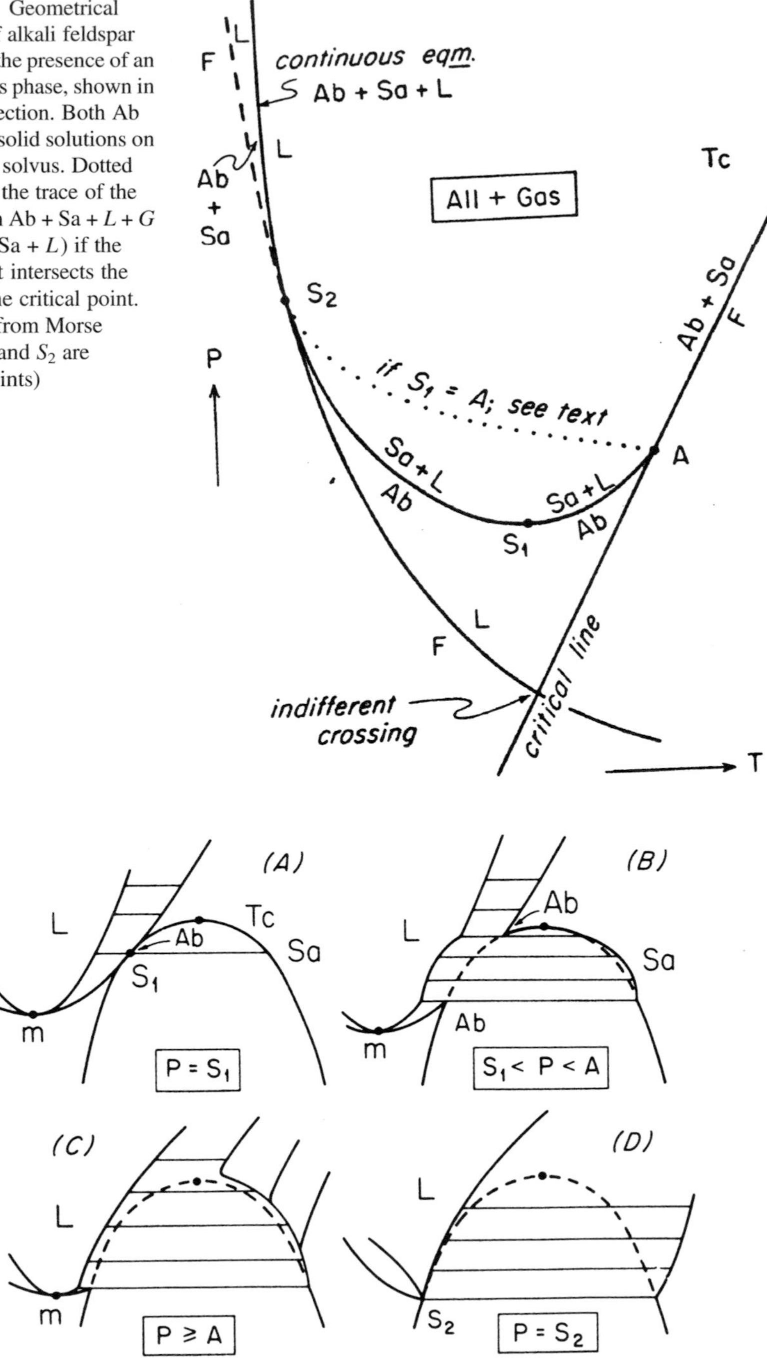

Fig. 19.17 Geometrical relations of alkali feldspar melting in the presence of an aqueous gas phase, shown in a *P–T* projection. Both Ab and Sa are solid solutions on or near the solvus. Dotted line shows the trace of the equilibrium Ab + Sa + *L* + *G* (i.e. Ab = Sa + *L*) if the solidus first intersects the solvus at the critical point. (Modified from Morse (1970). S_1 and S_2 are singular points)

Fig. 19.18 Schematic isobaric *T–X* projections from H_2O onto the Ab–Or join, showing melting relations corresponding to those displayed in Fig. 19.17

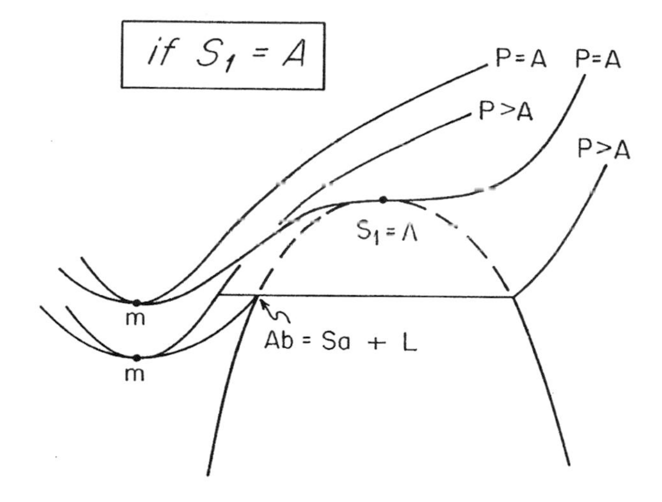

Fig. 19.19 Polybaric *T–X* projection from H_2O showing the case where the solidus first touches the solvus at the critical point. Motion of the solvus with pressure is ignored

Eventually, the solidus becomes so deeply embedded in the solvus that the minimum composition X_m touches the Ab limb of the solvus, generating the singular point S_2 (Figs. 19.17 and 19.18d). Below the pressure of S_2, all melting involving two feldspars is peritectic, and above S_2 it is eutectic. S_2 is therefore a familiar type of singular point occurring where a reaction changes from odd to even. This singular point marks the upper limit in P of the minimum melting reaction $F = L$ (where F stands for feldspar solid solution of minimum composition).

The above analysis has been made completely general, so that it applies to any system in which the projected trace of a solidus cuts a solvus at some point away from the crest. In the alkali feldspar system, however, X_c and X_m are very close together, and therefore, the solidus must have a slope near zero in *T–X* projection near the critical composition. Moreover, it can be argued from experience and ideas about the structure of highly polymerized liquids that the solidus undergoes a premonitory flattening as it approaches the solvus. In this case, $dT/dX = 0$ for the solidus and the solvus simultaneously, and the first point of contact occurs at A in Fig. 19.17, on the critical line. In such a case, S_1 is a singular point by identity rather than by collinearity of composition, and it lies at A. The continuous equilibrium Ab + Sa + L then rises from A to S_2, as shown by the dotted line in Fig. 19.17. This is the geometry deduced by Morse (1970). It is illustrated here in a *T–X* sketch (Fig. 19.19). The same geometry has also been found experimentally in feldspar and melt compositions from the Mg-free syenitic rocks of the Kiglapait intrusion (Morse 2016), thus giving new evidence for the structural features of the melt and crystals discussed above.

Diopside–H_2O

The *P–T* projection for Di–H_2O is shown in Fig. 19.20. The wet melting curve is unusual in showing a minimum near 20 kbar, implying that at higher pressures, the volume change on melting is positive.

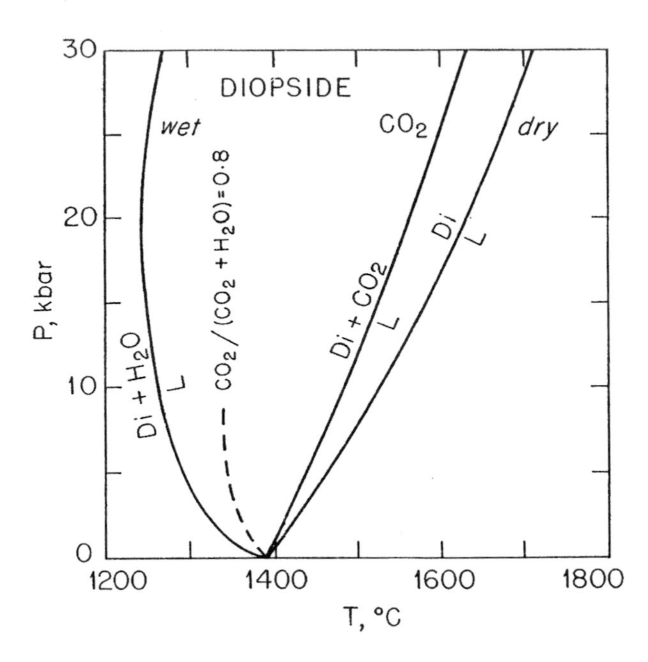

Fig. 19.20 Melting of diopside dry, with CO_2, and with water, shown in a *P–T* projection. Dry curve from Fig. 18.2. Wet curve from Yoder (1965), Eggler (1973), and Hodges (1974). CO_2 curves from Eggler (1973) and Rosenhauer and Eggler (1975)

Diopside–CO_2–H_2O

At low pressures, carbon dioxide enters silicate liquids only in small quantity; it thus lowers the melting temperatures only moderately, as shown by the Di–CO_2 curve in Fig. 19.20. When CO_2 and H_2O are mixed in a molar ratio $CO_2/(CO_2+H_2O) = 0.8$ and added to diopside, the solidus lies as shown by the dashed line in Fig. 19.20. The effect of H_2O, even in small amounts, clearly dominates the slope of the melting curve. The solubilities of H_2O and CO_2 in silicate melts are reviewed by Mysen (1977) and with other volatiles in primitive basalt and the Earth's upper mantle by Saal et al. (2002).

Phlogopite–H_2O

When a hydrous phase is heated under pressure, it will tend to decompose by dehydration to an assemblage of anhydrous crystals plus gas. The gas so released will tend to cause melting, leading to a water-poor melt, which absorbs all the released H_2O and is therefore far from being saturated with a gas phase. The case of the magnesian mica phlogopite furnished the first demonstration of this important principle (Yoder and Kushiro 1969). An isobaric 10 kbar *T–X* projection of the relevant system is shown in Fig. 19.21. In the absence of gas, the first liquid is generated at 1220°C from an assemblage of phlogopite with forsterite, leucite, and kalsilite. This liquid contains only about 2.5% H_2O by weight. Moreover, the

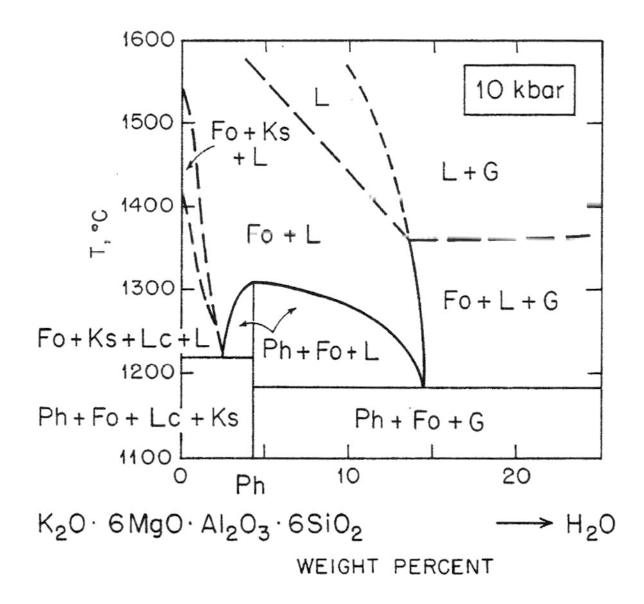

Fig. 19.21 Melting of phlogopite, $KMg_3AlSi_3O_{10}(OH)_2$, at 10 kbar total pressure. After Yoder and Kushiro (1969)

diagram shows that upon crystallization from melts poorer in H_2O than the phlogopite composition, the phlogopite extracts H_2O from the melt, causing the latter to become even poorer in H_2O. The initiation of melting in a phlogopite-bearing mantle would presumably be like that on the H_2O-poor side of phlogopite in the diagram. It is interesting to note that the H_2O-poor melting temperature, 1220°C, is not very much higher than that of the gas-saturated melting reaction at 1185°C. These two temperatures, in fact, shift past one another as the pressure varies, as will now be shown.

General *P–T–X* Relations for the Model System A–H_2O

The melting behaviour of phlogopite illustrates the general class of melting relations in a system A–H_2O containing an anhydrous phase (or phases) *A* and a hydrous phase *H*. The generalized *P–T–X* relations for such a system have been reviewed in detail by Eggler (1973b) and Eggler and Holloway (1977). Figure 19.22 illustrates such an analysis. Figure 19.22a is a *P–T* projection containing two singular points *S* and two invariant points *I*. The *P* and *T* scales are arbitrary. Figure 19.22b consists of a series of isobaric *T–X* sections, arranged in sequence from the upper left corner. In the figure, L_1 denotes the water-poor liquid, unsaturated with a gas phase, and L_2 denotes the gas-saturated liquid. The conventional missing phase notation is used in the *P–T* projection to identify the reaction assemblage.

The two invariant points are each generated by the intersection of a dehydration curve (*L*) and a solidus curve (*H*). The two singular points are each generated by the coincidence of liquid composition with the hydrous phase composition. Below the

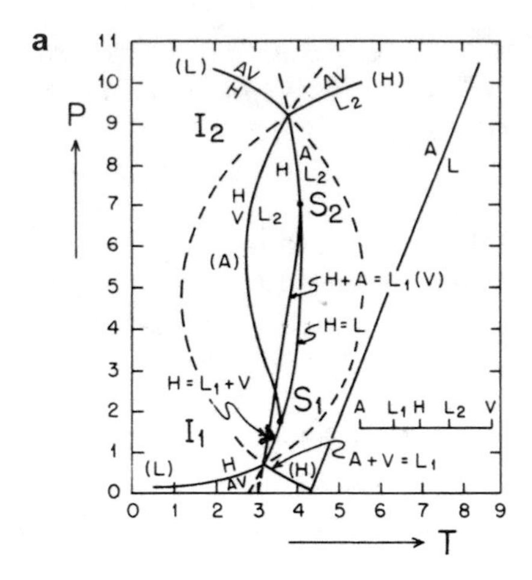

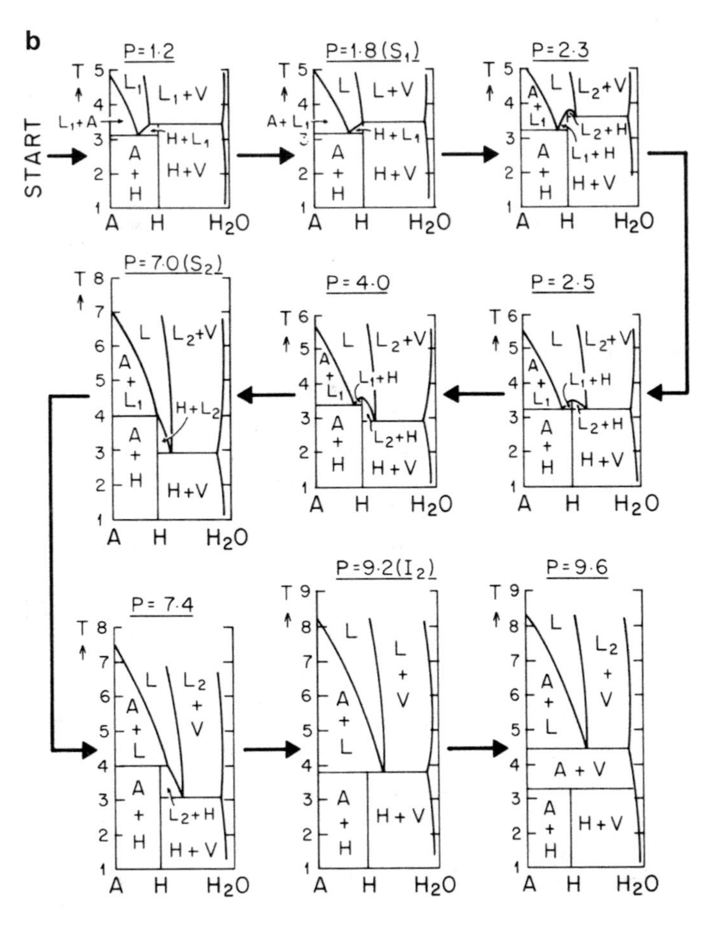

Fig. 19.22 Phase relations in a model system A–H_2O containing an anhydrous phase *A* and a hydrous crystalline phase *H*. The first diagram is a *P–T* projection; others are isobaric *T–X* sections, arranged in the sequence shown by the arrows. All scales are in arbitrary units. (Adapted from Eggler and Holloway (1977))

pressure of I_1, the hydrous phase dehydrates to $A + V$ (*vapour*), and at a higher temperature $A + V$ melts to a water-poor liquid L. These two reactions merge at I_1, and at a pressure just above I_1, the relations shown in Fig. 19.22a occur. At this pressure, the hydrous phase melts incongruently to liquid plus vapour, while a eutectic $A + H = L_1$ is formed at a lower temperature. At a pressure $P = 1.8$, the first singularity occurs when the liquid composition reaches H. At a higher pressure 2.3, the hydrous phase melts congruently and two eutectic liquids occur. Both liquids become progressively richer in H_2O as the pressure rises, and the solidus reaction switches to liquid L_2 at $P = 2.5$. The shift towards more H_2O-rich compositions finally causes L_1 to reach H, as in the section labelled $P = 7.0$, the pressure of singular point S_2. Above this pressure, the hydrous phase again melts incongruently, now to $A + L$. Eventually, this incongruent melting reaction occurs at the same temperature as the reaction $H + V = L$, generating the invariant point I_2 (section labelled $P = 9.2$). At higher pressures than I_2, the hydrous phase again dehydrates to $A + V$ and the melting reaction is $A + V = L_2$. At such high pressures, the anhydrous phase A would crystallize directly from L_2 on cooling, with release of gas.

The upper limit of stability of the hydrous phase H is given, in the P–T projection, by the congruent melting curve $H = L$. This curve mimics the observed stability of amphibole, which recurves sharply to lower temperatures in the range 20–30 kbar (see Wyllie 1979, for summary).

The relations illustrated in Fig. 19.22 have been extended to the ternary model system A–B–H_2O and the model system A–H_2O–CO_2 by Eggler and Holloway (1977), whose work should be consulted for a more complete appreciation of the principles involved. The application of dehydration melting principles to crustal anatexis involving pelitic and granitic rocks is discussed in detail by Thompson and Algor (1977).

Summary

The component H_2O drastically reduces the beginning of melting (solidus) temperature of silicate minerals if it is present in sufficient quantity to generate an aqueous gas phase. Even the merest trace of gas will cause melting at the solidus. The amount of liquid generated can be predicted from the lever rule: for example if the system contains 1% H_2O and the first liquid contains 10% H_2O, then 10% liquid will be generated at the isobaric invariant point involving *S–L–G*. This is generally accounted sufficient to cause magma separation by intergranular flow. If 0.1% H_2O is present in such a system, 1% melt will be generated. This is generally accounted insufficient for magma separation, but sufficient for seismic attenuation to yield the Low Velocity Zone (LVZ, asthenosphere) of the upper mantle (roughly 100–200 km deep). This interpretation of the LVZ must be viewed with caution, however, because the required grain-boundary relaxation may be an intrinsic property of unmelted mantle material at these depths (Anderson and Minster 1979).

Studies of the role of H_2O and CO_2 in magma genesis have reached a high degree of sophistication, including such tricks as β-track mapping of the CO_2 content and distribution in quenched silicate melts (Mysen and Seitz 1975). There has simultaneously arisen an increased understanding of silicate melt structure and the thermodynamic mixing properties of silicate melts (e.g. Burnham 1979). The resulting payoff in terms of our theoretical and pragmatic understanding of magma generation has been enormous, even if large questions about the real processes in the Earth remain.

G–X Diagrams for Ab–Or–H_2O

The system Ab–Or–H_2O furnishes a useful example of *G–X* diagrams for a ternary system, projected onto a binary join. The crystalline phases lie rigorously (we assume) in the join Ab–Or, but the liquid composition is ternary, and must be projected from H_2O. Figure 19.23 shows that the result is similar to that obtained with a simple binary system, as previously shown in Figs. 12.25 and 15.23. The dashed lines in Fig. 19.23 refer to ternary equilibria in which the liquid composition is projected onto Ab–Or, and the solid lines refer to compositions in the join Ab–Or. The diagrams are schematic, and the relative positions of the minimum composition (or eutectic composition) and the critical composition on the solvus are deliberately offset to avoid confusion.

Figure 19.23a shows, schematically, the case for a total pressure near 2 kbar; four *G–X* diagrams are superimposed on a *T–X* diagram, and all are isobaric projections from H_2O. With falling temperature, the solvus limbs theoretically reach the pure end members at 0° K. At temperature T_4, the *G–X* curve for the crystalline solution is composed of three segments, the outer two of which are concave up and the middle of which is concave down. A common tangent is drawn to the two nodes on this *G–X* curve. This tangent, where it touches the *G–X* curve, defines the stably coexisting compositions on the solvus; clearly, at this temperature, a mechanical mixture of the two nodal compositions is minimized in *G* relative to homogeneous solid solution. At the higher temperature T_3, the picture is much the same, but the nodes have drawn closer together. At T_2, the critical temperature, the nodes have united at the critical composition X_c, and the solid solution is now described in *G–X* space by a single concave-up curve. In Fig. 19.23a, it is assumed that steam (water vapour) is present at 2 kbar. Under these conditions, a minimum melting relationship obtains for alkali feldspars, as suggested by the dashed loops in the *T–X* section. At temperature T_1, the system is at the minimum melting point, and the *G–X* diagram has the form seen in Fig. 12.26. Note that the minimum composition X_m and the critical composition X_c. are not, in general, the same.

Figure 19.23b illustrates the case when a solidus is embedded in a solvus. This corresponds to the case of alkali feldspars in the presence of steam at about 6 kbar. Note that the *G–X* curves for the solvus are defined completely, as before, despite the presence of liquid. At temperature T_3, liquid has a high *G* relative to any solids and is

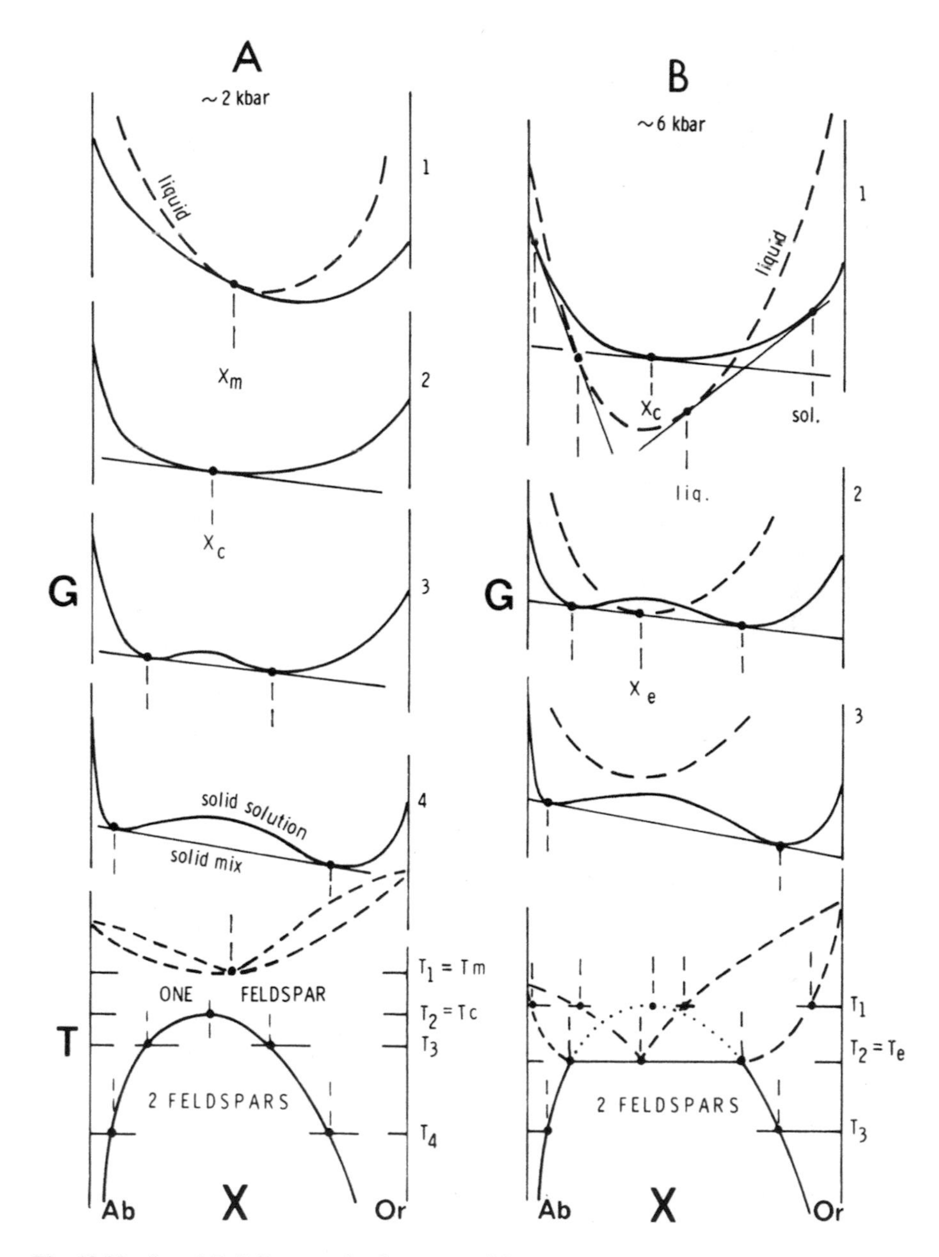

Fig. 19.23 *G*- and *T*–*X* diagrams for the system albite–orthoclase at high pressures near 2 kbar in (**a**) and near 6 kbar in (**b**). Melting (dashed lines) refers to a steam-saturated system

therefore metastable: the system is solid. At $T_2 = T_e$, the node of the liquid curve just touches the binodal tangent connecting the stable feldspar compositions, and this is the eutectic condition. At any temperature slightly above this, the binodal tangent is interrupted by two crystal–liquid tangents, which are minimized in G relative to feldspars on the solvus. The solvus is therefore metastable above T_2. It continues to be metastable up to the critical temperature, above which it does not exist. Section 1 depicts the G–X relations at the critical temperature. Above this, the G–X relations are essentially those of any eutectic diagram.

Chapter 20
Leftovers

"Leftovers" in this treatise are brief descriptions of rocks whose origin and nature illustrate interestingly one or more principles (or mysteries) of crystallization in Earth history. Brief references are given.

Anorthosites

These rocks deserve first place because they are historically enigmatic and contain surprises that are instructive. The literature is abundant; we shall be picky. Nothing in petrology is more astounding than some of the rocks in the Nain Anorthosite suite of Labrador. Here, Robert Wiebe (1988) expected to find magma mixing—unusual mixtures of rocks—and hit the jackpot with enormous, metre-scale pillows of quenched mafic rocks—essentially troctolites—within what was obviously molten anorthosite at depth. The principle of rejected solute, emphasized in this book, is illustrated in these rocks. The melting point of the mafic rocks was clearly much greater that than of the anorthosite because it was quenched into it. Yet molten plagioclase is very hot stuff as well. Massif anorthosite itself is another bit of rejected solute, emplaced into a shallower depth over time and, in this case, dragging some of the molten mafic residue up with it.

More curiosities occur in the Labrador and Norwegian anorthosites, such as the Susie Brook Slab in Labrador and many other very mafic bodies that have been coughed up from mantle depths. These high-temperature bodies contain high-alumina orthopyroxene megacrysts (HAOMs) and very mafic bodies that contain up to six mineral species that were once part of the whole as melts and crystals but have since exsolved in crystallographically oriented slices. Megacryst crystal sizes range from 0.2 to 1 m, and all contain exsolution lamellae of plagioclase. The inferred depth of these megacrysts and hence the origin of anorthosites themselves are that of the Precambrian MOHO (Bybee et al. 2014). This is a combined geochemical study of three anorthosite massifs—two in Labrador and one in

S. A. Morse, *Basalts and Phase Diagrams*,
https://doi.org/10.1007/978-3-030-97882-2_20

Norway—and shows that the mafic *megacrysts* are consistently about 100 million years *older* than the anorthosites. The implication of this result is that the low-density anorthosites are the rejected solute from the crystallization of the orthopyroxene megacrysts, the whole process originating at the MOHO. The solute found its way into the ~5-kbar crust, eventually to be uncovered after glaciation. It left behind very dense bodies of molten mafics that stayed hot until they were injected into the same places as the now-crystalline anorthosites, using routes prepared for them earlier.

The Kunene Anorthosite Complex in Western Africa is possibly the largest well-exposed and best-dated massif anorthosite in the world. It occupies some 18,000 square kilometres in Angola and Namibia. It is loaded with rejected solutes (Ashwal and Bybee 2017).

How the Study of the Earth Advances

The above tale is one example of such advance. But at this moment of our science, the insights into the evolution of the Earth have become so intensely interesting that a brief list of the illuminations must capture our admiring attention.

We may back off a bit to get to the starting point, even though repeating ourselves. Percy Bridgman was a Professor of Physics at Harvard University in Massachusetts. His main attention was the study of the deep Earth, using tools of his own design. Hatten S. Yoder (Jr) was a graduate student at the nearby Massachusetts Institute of Technology (MIT) who also wished to have a career making experiments that helped to explain the Earth. So he began to design what became the first internally heated pressure vessel. Seeking advice, he went to Bridgman's office and rang the bell, as instructed by a label near the door. After several minutes, the door opened and the Professor said, "You have three minutes". Clearly, he was making experiments that needed his attention every five minutes. Yoder loved to tell this tale.

In recent times, the geophysical community has made stunning advances in the understanding of the deep Earth composition, density, volume, and mineralogy. The discoverers have been rewarded by their names being put to their discoveries. So after Olivine at ~250 km, we have Wadsleyite at 410 km, Ringwoodite at 520 km, and until 2019 no name for all the rest down to the Core. It was the agreement of the entire bulk Earth community that the biggest chunk of the Earth must be named for Bridgman, the founder of bulk earth physics and predictor of its deepest rocks.

But how do you sample something so deep in the Earth? You don't. Instead, you find something at the same level of pressure that contains Orthorhombic (Mg,Fe) SiO_3 in the perovskite form, which is the knowable constitution of that deep part of the Earth. The choices are: find a deep Earth diamond (very hard to find) or a meteorite. Even the meteorite will be hard to find, but Oliver Tschauner at the University of Nevada at Las Vegas overcame severe difficulties to sample a highly shocked meteorite, as reported in Tschauner (2019).

Now, the mineral Bridgmanite occupies all of the Earth's Lower Mantle, from 660 km to the Core–Mantle Boundary at 2891 km.

Now, we come to the problem of ice_{VII} that Barclay Kamb, with Briant Davis, gave us as the densest possible form of water. If you raised the pressure above the experimental pressure, there was simply a re-arrangement of the critical atoms that increased the interatomic distances instead of collapsing them (NAS publication 1964 communicated by Member Linus Pauling, Kamb's PhD supervisor and father-in-law).

So how do you find a natural sample of ice_{VII}? In a high-pressure diamond, of course, as described by Tschauner (2019, p. 1707). A colleague among the authors, George Rossman, was kind enough to resolve some queries about the enterprise. He writes (personal communication):

> The discovery of ice_{VII} was pure luck. This came about with two primary methods. We examined diamonds with infrared spectroscopy (fast and easy) to find ones with CO_2 and H_2O inclusions. Our search was for CO_2 not ice. Samples that have interesting IR patterns go to the synchrotron (that is what Oliver T. did). There, the ice_{VII} diffraction pattern was obvious. The synchrotron really pinned it down.

(The synchrotron is the Advanced Photon Source at the Argonne National Laboratory.)

The text of the paper mentions ice_{VII} found in diamonds from both Africa and China. Their Figure 1 is a beautiful X-ray diffraction pattern with six major peaks. Other figures show multiple observations at pressures from 2 to 26 Gpa (which is at the lower mantle).

Mantle Minerals

The original edition of this book on phase diagrams closed with a long tutorial on applications to basalt magma genesis. There are countless papers on that subject and descriptions a-plenty. Here, instead it will be more profitable to mention a few of the floods of information on phase transitions in $(Mg,Fe)_2SiO_4$ and its neighbours, for which I acknowledge the excellent compilations of our colleague David Snoyenbos, and then to expand on the new methods of studying phase equilibria.

We may begin by discussing phase transitions in the relatively familiar simple chemistry of $(Mg,Fe)_2SiO_4$, in which crystal forms depend on depth and the Mantle geotherm.

By *depth in the Earth* from the top down, we have the *upper mantle* in which ferropericlase is stable into the upper mantle where it is silica-undersaturated.

Now, we have the familiar orthosilicate *olivine* $(Mg,Fe)_2SiO_4$, orthorhombic, from near surface to 410 km.

Then, the spinelloid *wadsleyite*, also $(Mg,Fe)_2SiO_4$, orthorhombic, to 520 km.

Then, *ringwoodite*, again $(Mg,Fe)_2SiO_4$, but now cubic *spinelloid*, to 660 km.

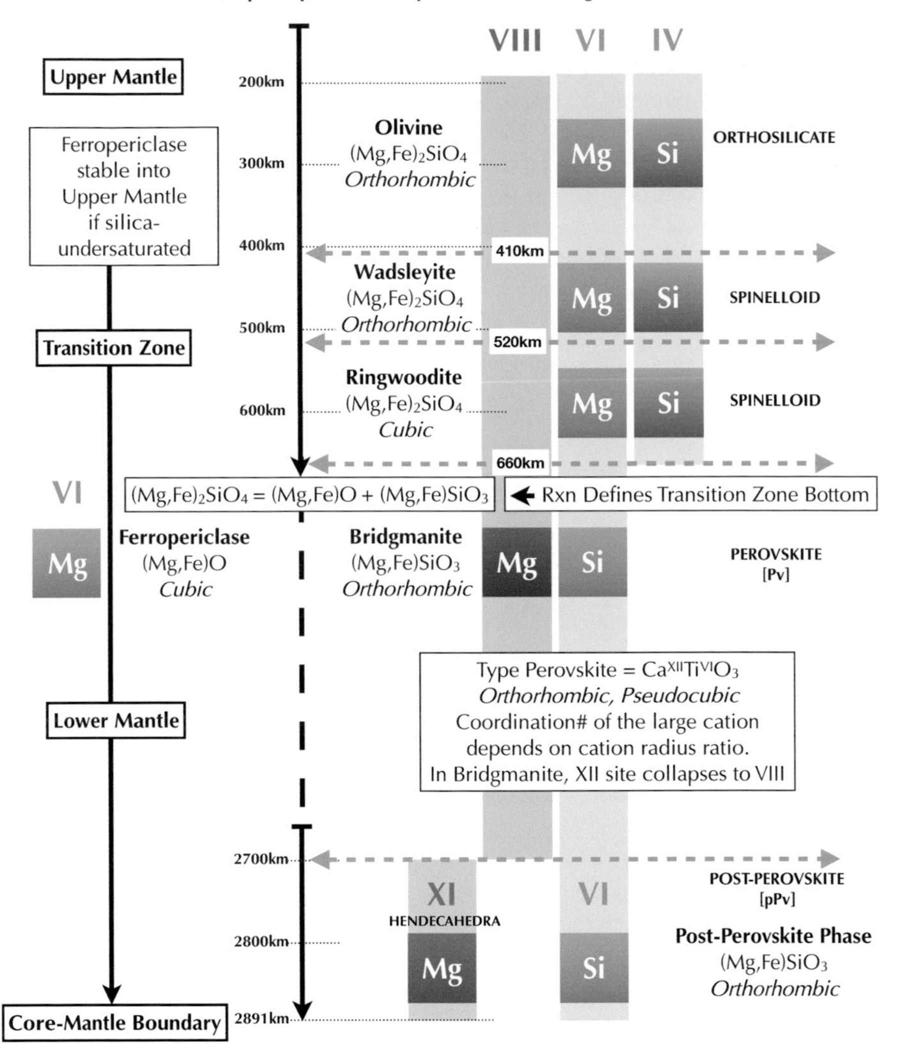

Volume fractions of lower mantle minerals for Pyrolite and MORB compositions. Mineral abbreviations are: *Bdg*, bridgmanite; *Per*, (ferro)periclase; *Maj*, majorite; *Rwd*, ringwoodite; *Sti*, stishovite; *Sft*, seifertite; *CaPv*, calcium–silicate perovskite; *CF*, calcium ferrite-type phase; *NAL*, new aluminous phase; *pPv*, post-perovskite. (Adapted from Wicks and Duffy (2016))

And now, the reaction $(Mg,Fe)_2SiO_4 = (Mg,Fe)SiO_3 + (Mg,Fe)O$, the reaction that defines the *bottom of the transition zone*.

Now, there is VI Mg in *ferropericlase* (Mg,Fe)O, cubic, and more importantly the *perovskite* (Pv) *bridgmanite* $(Mg,Fe)SiO_3$, orthorhombic, all the way to 2700 km.

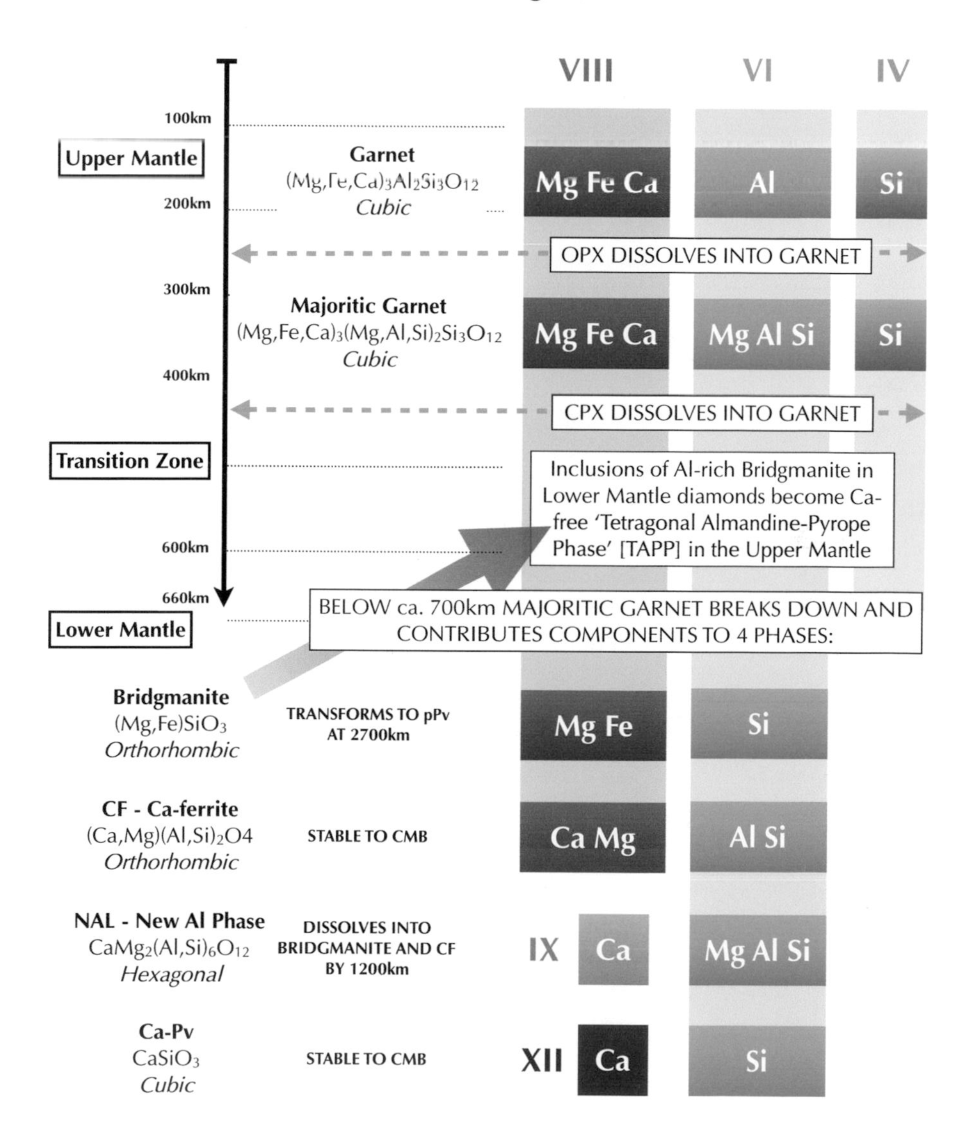

After this, there is a *post-perovskite phase* $(Mg,Fe)SiO_3$, orthorhombic to the core–mantle boundary at 2891 km.

(Here is an auspicious place to mention that diligent seismology has, at least once traced what we might call an eruption from the core–mantle boundary all the way, somewhat twisting, to somewhere near the Colorado Plateau in the USA.)

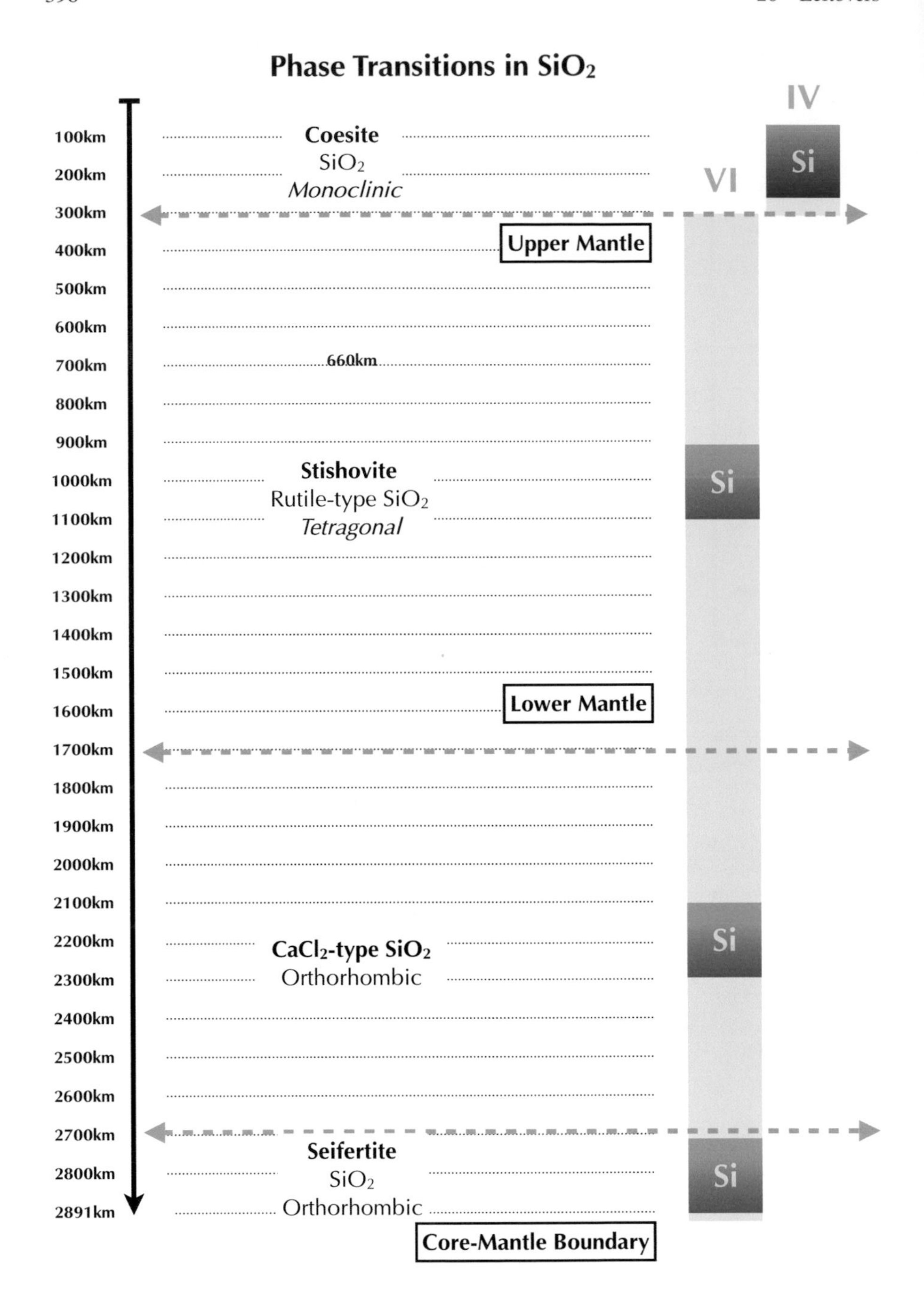

Phase Transitions in SiO_2
IV
VI
Si
100km
200km
300km
400km
500km
600km
700km
800km
900km
1000km
1100km
1200km
1300km
1400km
1500km
1600km
1700km
1800km
1900km
2000km
2100km
2200km
2300km
2400km
2500km
2600km
2700km
2800km
2891km
Coesite
SiO_2
Monoclinic
Upper Mantle
660km
Stishovite
Rutile-type SiO_2
Tetragonal
Lower Mantle
$CaCl_2$-type SiO_2
Orthorhombic
Seifertite
SiO_2
Orthorhombic
Core-Mantle Boundary

There are similar phase transitions in the more complicated chemistry (Mg,Fe, Ca)$_3$Al$_2$Si$_3$O$_{12}$ in which we may find *garnet*, where at ~ 250 km OPX dissolves into garnet, followed by majoritic garnet, beyond which CPX dissolves into garnet.

In the lower mantle below 700 km, majoritic garnet breaks down and contributes components to four more complex phases which need not bother us here.

There are phase transitions in SiO_2 that will be familiar: at 100 km, *coesite* becomes present as monoclinic SiO_2, which in turn at 300 km becomes *stishovite*, the tetragonal SiO_2, all the way to the lower mantle at 1700 km. Beyond that, there is a $CaCl_2$-type orthorhombic SiO_2 to 2700 km and then, finally, the orthorhombic *seifertite* SiO_2 to the core–mantle boundary.

Water in Magmas

We have gone to some lengths in this treatment to avoid mentioning water in magmas. The reason is partly historical: in the early to middle decades of the past century, there was a monumental argument, mostly between Great Britain and America, about whether magmas in general may be wet or dry. This discussion included conferences and lectures in which some famous moments included Bowen's remark that the opposition must have its drinks "like an old soak seeking the penitent's bench." Bowen did not think much of water in rocks.

That was before it was clear that most of the magma on Earth is in the middle of the oceans, most of which are not very dry. Although it is true that many rocks, especially older ones, may represent very dry magmas, it is clear that many igneous rocks have evolved in the presence of water and in some cases even a vapour phase. Water has a profound effect on phase equilibria and especially temperature and hence especially on evolution from mafic to felsic. As it turns out, there is a strong linear relationship between water content and the reaction sequence basalt—basaltic-andesite—andesite—dacite—rhyolite, from < 0.5 to 2%.

Water in the Earth

It is easy to assume that the interior of the Earth is essentially dry with respect to water, the oceans having long ago taken most of it in tow. On the other hand, what comes up must eventually go down if it is not a mountain, and the subduction of lithosphere must certainly feed some lighter elements back into the mantle. A paper from Shirey et al. (2020) describes this circuit with the ability to compare slab pressure–temperature seismic paths within the lower mid-mantle with cold slabs that can transport water to depths within the transition zone with seismicity between 500 and 700 km. Such a depth "also coincides with the *P*/*T* conditions at which oceanic crust in cold slabs is expected to intersect the carbonate-bearing basalt solidus to produce carbonatitic melts." Both types of fluids are well represented by sublithospheric diamonds whose inclusions record the existence of melts, fluids, or supercritical liquids derived from hydrated or carbonate-bearing slabs at depths (~300–700 km) generally coincident with deep focus earthquakes.

Phase Equilibria

The mineral sapphirine occurs in quantity on a ridge in the Wilson Lake region of Labrador, and it occurs also in Peekskill, New York, and in Antarctica and many other places in the world. It is a magnesium–aluminium silicate that may occur with quartz, hypersthene, sillimanite, kyanite, and spinel. These are high-temperature, high-pressure minerals, and one might wonder where and why they occur. That would take a good bit of research, but it is already done in phase diagrams using THERMOCALC, which calculates phase diagrams for rocks. THERMOCALC is the product of Tim Holland in Cambridge, UK, and Roger Powell in Melbourne, Australia. Fortunately, they move from time to time and have been seen together at meetings. One of the calculated diagrams has the assemblage sapphirine–Opx–Sil–Qtz beginning at 970°C, 7.4 kbar, and extending to 1070°C and 9.4 kbar. Here, we have metamorphic rocks beginning to encroach on the territory of magmas.

New tools for calculating the Gibbs free energy minimization derive from the work of Holland and Jennings on the one hand and Mark Ghiorso and colleagues on the other hand, and these have now been used by Stolper et al. (2020) to study in detail the effects of solid–solid phase equilibria on the fO_2 of a closed system, subsolidus peridotite. This paper, a product of the Roebling Medal of the Mineralogical Society of America given to Professor Stolper, is a beautiful treatment of phase equilibria which, among other virtues, puts the two major results together in coloured zones side by side in plots of temperature against pressure. The first two panels plot the log of the oxygen fugacity between the "pMELTS" and "Perple_X (JH15)" results. The two results are fundamentally similar but differ particularly in temperature.

The next row in the figure records the variation of the standard FMQ buffer between the two models, and the third row contains two separate *P*–*T* diagrams plotting the various lherzolites (Plagioclase, Spinel, Garnet).

The further text and new figures deal with details of chemistry as they vary with *T* and *P* between the two treatments.

The paper moves on with detailed discussions of the phase equilibria, comparisons with previous work, a graphical analysis of the Plag - lherzolite facies, and new pages of phase triangles giving schematic subsolidus phase equilibria in a five-component system. The article goes on to discuss the effects of spinel, iron in garnet, and other considerations for several more pages. Further outputs of the paper and its two models can be found in the endnote.

In short, at 26 pages, this masterful treatise on magmas and phase equilibria of the upper mantle is among the giants of phase equilibria in fertile application to the genesis of magmas.

The hard work of hot rocks is carried out through a combination of fieldwork and experiments. The literature is rocking in projects like these, to the advantage of all.

Appendix

Kiglapait Intrusion, Labrador Coast: A Record of Discovery

The 1.307 Ga old, 560 km^2 area-exposed Kiglapait Intrusion on the coast of Labrador, quite visible from space, is the first major layered intrusion to have been studied initially by a graduate student. I was given two summers for the study, with a 1-week lift of a Bell helicopter and pilot, Company Director, Provincial Geologist, Company Field Officer, Field Assistant, and cook (whose kit was off-limits to the Director). The project led to my PhD thesis, which was the first description of a layered intrusion conducted within the framework of the Igneous Cumulate concept of Wager, Brown, and Wadsworth (*Journal of Petrology* 1, issue 3).[1] In the end, the description of Kiglapait by Malcolm Brown in their subsequent volume on layered igneous rocks was admirably condensed from my PhD thesis.

How did all this come about? Bear with me: it is a bit of a story.

As a freshman at Dartmouth College, I expected to spend my life in forestry. This changed when I was recruited to be field assistant to an able veteran of the US Navy, the sociologist Elmer Harp, in southern Labrador and Newfoundland. We were based on the hundred-foot schooner *Blue Dolphin*, commanded by its owner, the oceanographer David Clark Nutt. He in turn had spent five summers on the schooner *Effie M. Morrissey* (now *Ernestina-Morrissey*) under the direction of its owner Captain Robert Bartlett, who had taken Peary to attempt the North Pole.

Once bitten, I spent two more seasons on coastal Labrador doing oceanography and any other tasks at hand, including navigation, off-shore work, sailing, steering, and the work of assistant engineer for an 8/10 four cylinder diesel engine. Immediately upon my return to the USA after my two year stint in the Army,[2] I was drafted

[1] It also led to a broken leg and, as an indirect result of that, to the best possible wife.

[2] While stationed in Germany I had the opportunity to attend Language School in Oberammergau, partake in downhill ski races in Garmisch-Partenkirchen, and make satisfying friendships in a new culture.

S. A. Morse, *Basalts and Phase Diagrams*,
https://doi.org/10.1007/978-3-030-97882-2

to once again help the *Blue Dolphin* crew, this time studying late-season hydrology in two fjords—the northern Hebron Fjord and the mid-southern Nain Bay. I was provided with winches, reels and messengers, bottles, etc., and with warm clothes, and taken to the US Armed Services Air Base in Western Massachusetts to be flown to Goose Bay, Labrador. From there I traveled by helicopter in successive hops to Saglek Bay, where the Moravian missionary, Fred Grubb, picked me up in the Mission's boat. During an ensuing spell of bad weather I stayed with the Grubbs, translating from German to English the diary entries made by the missionary to Hebron during the influenza epidemic of 1918. Fred's wife was in Goose Bay having a baby, while Fred took care of their girls. (In due course, the baby was born and named Anthony after me.) When the weather cleared I carried out the research for which I had brought all those supplies.

Having conducted the Hebron Bay portion of the research, I traveled by a series of mail and local boats to Nain, where I was given the Mission's boat to work with. We completed our study (though not without dragging anchor for three days in a poorly sheltered outer bay), and I returned with gear and samples as I had come, by helicopters and Air Force aircraft.[3]

Once in graduate school at McGill University I was fortunate to become part of a lively Geology Department enhanced by students from around the world, and to spend a summer field season in coastal Quebec under the supervision of great teachers and in the company of good students.

Having fractured a leg skiing in New Hampshire, which put me out of the running for far-flung field work that year, I was taken on as a lab technician at Dartmouth, in my hometown of Hanover, NH. My tutor and friend was Professor John Lyons, whose teaching of thermodynamics and phase equilibria had earlier captured my undergraduate attention. During this summer's work we collected Standing Pond Amphibolite in Vermont and New Hampshire; this work was the topic of my master's thesis, supervised in part by Lyons.

In 1957, wondering how to get back to Labrador, I visited the Montreal office of BRINCO, the British-Newfoundland Exploration Company, to see if they could accommodate me as a PhD student working on anorthosites in the Nain region. "With whom did you wish to speak?" asked the clerk. "Anybody in geology," I answered. After a minute or two, "Dr. Bevan will see you." The General Manager and Chief Geologist became a lifelong friend. After mentioning my summers and early winter in Labrador, I expressed my interest in the Nain anorthosites. Dr. Bevan asked if I had any specific places in mind. "Oh no, I'm just starting to think about that." He then suggested that it might be worth looking at rocks having "banding" reported by Dr. Wheeler at Kiglapait Harbor.[4] I agreed, and mentioned that I had

[3]This was in the days when a primary source of national science support was the Office of Naval Research (ONR), just before the welcome advent of the National Science Foundation.

[4]Dr. Everett Pepperell Wheeler was a devoted student of Labrador rocks over a long career; our family became deeply and happily acquainted with both him and his wife, Eleanor.

actually been ashore at Kiglapait Harbor during a storm. I then asked, "What do I do now?" and was told, "Study hard, get good grades, and come back in June."

In just 15 minutes, I had been directed to the most glorious layered intrusion anybody could want, without a clue to how huge it was until we laid out that 3 × 4 foot photo mosaic in the Company headquarters at Northwest River, Labrador, the following June.

In the spring of 1960, I was called to the office of my supervisor, Dr. E. H. Krank, where I found, to my astonishment, Professor Laurence Richard Wager of Oxford. He handed me a Contents page from the third issue of the *Journal of Petrology* containing the title "Types of igneous cumulates." I inquired, "What the hell is a cumulate?" to which he, unfazed, replied, "You will find out." We then had a conversation about things that accumulate, even in the sky, and eventually Dr. Krank suggested that we might walk down town and have lunch while discussing layered intrusions.

It turned out that Wager and his student Malcolm Brown were writing a book for which they wanted to include as much information on Kiglapait as feasible. While preparing my thesis as Memoir 112 for the Geological Society of America, I fed much of it to Malcolm, who ably developed it into a chapter for their *Layered Igneous Rocks* (1969).

For most of my life thereafter, I conceived of Wager's appearance in Montreal as a pleasant happenstance. At about the age of 85, it finally dawned on me that the Kiglapait Intrusion and my thesis were the purpose of his visit. He wanted to be sure I got it right. Sadly, he livedonly five more years, dying at 61.

Malcom Brown subsequently spent a year at the Geophysical Laboratory in Washington. By that time I was at Franklin and Marshall College, and we brought him to Lancaster, PA, to serve as an external examiner for a master's degree student there. Eventually Malcolm became Head of the Department at Durham. During a sabbatical year in Oslo we took our three daughters around Britain, arriving in Durham in a VW bus to enjoy Malcolm's hospitality, including his Scotch. Malcolm was later knighted for his work as the Director of the British Geological Survey, retired to Oxford, and died suddenly and far too young, a great loss.

This Appendix concludes with a recent image of the Kiglapait Intrusion in winter, showing the curves of the Kiglapait Mountains to the North against the frozen North bays, and the ridges that fill the intrusion and tend to follow the igneous layering.

Doc Searls from Santa Barbara, USA (https://commons.wikimedia.org/wiki/File:Kiglapait_Mountains,_north_coast_of_Labrador.jpg), "Kiglapait Mountains, north coast of Labrador", https://creativecommons.org/licenses/by/2.0/legalcode

Bibliography

Akimoto, S.I., Komada, E., and Kushiro, I. (1967): Effect of pressure on the melting of olivine and spinel polymorphs of Fe_2SiO_4. *J. Geophys. Res.,* vol. 72, pp. 679–686.

Andersen, O. (1915): The system anorthite-forsterite-silica. *Am. J. Sci.,* 4th ser., vol. **39,** pp. 407–454.

Anderson, A.T. (1976): Magma mixing: petrological process and volcanological tool. *J. Volcan. Geotherm. Res., vol.* **1,** pp. 3–33.

Anderson, D.L. (1979): The upper mantle transition region: eclogite? *Geophys. Res. Letters,* **vol. 6,** pp. 433–436.

Anderson, D.L. and Minster, J.B. (1979): (abs) The physical mechanism of subsolidus creep and its relation to seismic wave attenuation. *EOS Trans. Am. Geophys. Union,* **vol. 60,** pp. 378.

Anderson, D.J., Lindsley, D. H., Davidson, P. M. (1993) QUILF: a Pascal program to assess equilibria among Fe-Mg-Mn-Ti oxides, pyroxenes, olivine, and quartz. *Comput. Geosci.* Vol. 9: pp. 1333–1350.

Anovitz, L.M., and Blencoe, J.G. (1999): Dry melting of albite. *American Mineralogist*, vol. 84, pp. 1830–1842.

Appleman, D.E. (1966): Crystal chemistry of the pyroxenes in *Short Course Notes on Chain Silicates,* American Geological Institute, Palo Alto, Calif., p. 10.

Ashwal, L. D. and Bybee, G. M. (2017): Crustal evolution and the temporality of anorthosites. Earth-Science Reviews 173, 307–330

Asimow P.D. & Ghiorso M.S. (1998) Algorithmic modifications extending MELTS to calculate subsolidus phase relations, American Mineralogist 83,1127–1132.

Asimow, P.D., Lin, C., Bindi, L., Ma, C., Tschauner, O., Hollister, L. S. and Steinhardt, P. J. (2016) Shock synthesis of quasicrystals with implications for their origin in asteroid collisions. Proc. National Academy of Sciences **113 (26)** 7077–7081.

Atkins, F.B. (1968): Pyroxenes of the Bushveld intrusion, South Africa. *J. Petrol., vol.* **10,** pp. 222–249.

Bachinski, S.W., and Muller, G. (1971): Experimental determinations of the microcline-low albite solvus. *J. Petrol.,* **vol. 12,** pp. 329–356.

Bailey, E.B., Clough, C.T., Wright, W.B., Richey, J.E. and Wilson, G.V. (1924): Tertiary and post-tertiary geology of Mull, Loch Aline, and Oban. *Mem. Geol. Surv. Scot.*

Barth, T.F.W. (1962): *Theoretical petrology.* Wiley, New York.

Bell, P.M., and Roseboom, E.H. Jr. (1969): Melting relationships of jadeite and albite to 45 kilobars with comments on melting diagrams of binary systems at high pressures. *Mineral. Soc. Am., Spec. Pap.* 2, pp. 151–162.

S. A. Morse, *Basalts and Phase Diagrams*,
https://doi.org/10.1007/978-3-030-97882-2

Bence, A.E., and Papike, J.J. (1972): Pyroxenes as recorders of lunar basalt petrogenesis: chemical trends due to crystal-liquid interaction. *Proc. Third. Lunar Sci. Conf., Geochim. Cosmochim. Acta.* Supp. 3, vol. **1,** pp. 431–469.

Berg, J.H. (1977): Regional geobarometry in the contact aureoles of the anorthositic Nain complex, Labrador. *J. Petrol.,* vol. 18, pp. 399–430.

Bindi, L., Steinhardt, P. J., Yao, N., and Lu, P. J. (2011): Icosahedrite, $Al_{63}Cu_{24}Fe_{13}$, the first natural quasicrystal. American Mineralogist **96**, 928–931.

Bollmann, W., and Nissen, H.-U. (1968): A study of optimal phase boundaries: the case of exsolved alkali feldspars. *Acta Crystallogr.,* vol.A24, pp. 546–557.

Bottinga, Y., and Richet, P. (1986): Thermochemical properties of silicate glasses and liquids: A review. *Reviews of Geophysics*, vol. 24, pp. 1–25.

Bowen, N. L. (1913): The melting phenomena of the plagioclase feldspars. *Am. J. Sci.,* vol. 35, pp. 577–599.

Bowen, N. L. (1914): The ternary system diopside-forsterite-silica. *Am. J. Sci.,* 4th ser., vol. 33, pp. 551–573.

Bowen, N. L. (1915): The crystallization of haplobasaltic, haplodioritic, and related magmas. *Am. J. Sci.,* vol. 40, pp. 161–185.

Bowen, N. L. (1928): *The evolution of the igneous rocks.* Princeton University Press, Princeton, N.J.

Bowen, N. L. (1941): Certain singular points on crystallization curves of solid solutions. *Proc. N.A. S.,* vol. 27, pp. 301–309.

Bowen, N. L. (1945): Phase equilibria bearing on the origin and differentiation of alkaline rocks. *Am. J. Sci.,* vol. 243A, pp. 75–89.

Bowen, N.L., and Andersen, O. (1914): The binary system $MgO\text{-}SiO_2$. *Am. J. Sci.,* 4th ser., vol. 37, pp. 487–500.

Bowen, N.L:, and Schairer, J.F. (1932): The system $FeO\text{-}SiO_2$. *Am. J. Sci.,* 5th Ser., vol. 24, pp. 177–213.

Bowen, N.L:, and Schairer, J.F. (1935): The system, $MgO\text{-}FeO\text{-}SiO_2$. *Am. J. Sci.,* 5th Ser., vol. 29, pp. 151–217.

Bowen, N.L., and Schairer, J.F. (1938): Crystallization equilibrium in nepheline-albite-silica mixtures with fayalite. *Journal of Geology,* vol. 46, pp. 397–411.

Boyd, F.R. (1973): A pyroxene geotherm. *Geochim. Cosmochim. Acta,* vol. 37, pp. 2533–2546.

Boyd, F.R., and England, J.L. (1963): Effect of pressure on the melting of diopside, $CaMgSi_2O_6$, and albite, $NaAlSi_3O_8$, in the range up to 50 kilobars. *J. Geophys. Res.,* **vol. 68,** pp. 311–323.

Boyd, F.R., England, J.L., and Davis, B.T.C. (1964): Effects of pressure on the melting and polymorphism of enstatite, $MgSiO_3$. *J. Geophys. Res.,* **vol. 69,** pp. 2101–2109.

Bradley, R.S. (1962): Thermodynamic calculations on phase equilibria involving fused salts. Part II. Solid solutions and application to the olivines. *American Journal of Science*, vol. 260, pp. 550–554.

Brooks, K. (2018): *Over eighty years at the core of Petrological Research. The Skaergaard Intrusion. The history of research, its environment and annotated bibliography.* Kopenhagen, GEUS, 150 pp.

Burnham, C.W. (1979): The importance of volatile constituents, *in* Yoder, H.S. Jr., ed.; *The evolution of the igneous rocks,* Princeton University Press, Princeton, N.J., pp. 439–482.

Burnham, C.W., and Davis, N.F. (1974): The role of H_2O in silicate melts: II. Thermodynamic and phase relations in the system $NaAlSi_3O_8\text{-}H_2O$ to 10 kilobars, 700° to 1100°C. *Am. J. Sci.,* vol. 274, pp. 902–940.

Bybee, G. M., Ashwal, L. D., Shirey, S. B., Horan, M., Mock, T., and Anderson, T.B. (2014): Pyroxene megacrysts in Proterozoic anorthosites: Implications for tectonic setting, magma source, and magmatic processes art the Moho. Earth and Planetary Science Letters 389 (2014) p. 77–85.

Carmichael, I.S.E., Turner, F.J., and Verhoogen, J. (1972): *Igneous petrology.* McGraw-Hill, New York.

Cawthorn, R.G., Editor (1996): *Layered Intrusions*. Elsevier, Developments in Petrology, vol. 15, 531 pp.

Chaplin, M. (2016): Phase diagram of water. https://web.archive.org/web/20200311043944/http://www1.lsbu.ac.uk/water/water_phase_diagram.html (accessed 19 September 2023).

Charlier, B., Namur, O., Latypov, R., and Tegner, C., Editors (2015): *Layered Intrusions*. Springer, 748 pp.

Chayes, F. (1963): On pyroxene molecules in the CIPW norm. *Geol. Mag., vol. 100,* pp. 7–10.

Chayes, F. (1969): Curve fitting in the ternary diagram. *Carnegie Inst. Washington Yearb.,* vol. 67, pp. 236–239.

Chayes, F. (1972): Silica saturation in Cenozoic basalt. *Philos. Trans. R. Soc. London.* A, vol. 271, pp. 285–296. (Data extracted from computer output for this paper, with permission.)

Chayes, F., and Metais, D. (1964): On the relation between suites of CIPW and Barth-Niggli norms. *Carnegie Inst. Washington Yearb.,* vol. 63, pp. 193–195.

Chen, C.H., and Presnall, D.C. (1975): The system Mg_2SiO_4 - SiO_2 at pressures up to 25 kilobars. *Am. Mineral.,* vol. 60, pp. 398–406.

Clark, S.P. Jr. (1959): Effect of pressure on the melting points of eight alkali halides. *J. Chem. Phys.,* vol. 31, pp. 1526–1531.

Clark, S.P. Jr. (ed.) (1966): *Handbook of physical constants. Geol. Soc. Am. Mem., vol.* 97.

Clark, S.P. Jr., Schairer, J.F., and de Neufville, J. (1962): Phase relations in the system $CaMgSi_2O_6$ - $CaAl_2SiO_6$ - SiO_2 at low and high pressure. *Carnegie Inst. Washington Yearb.* vol. 61, pp. 59–68.

Cohen, L.H., and Klement, W. Jr. (1967): High-low quartz inversion: determination to 35 kilobars. *J. Geophys. Res.,* vol. 72, pp. 4245–4251.

Cross, W., Iddings, J.P., Pirsson, L.V., and Washington, H.S. (1903): *Quantitative classification of igneous rocks.* Univ. Chicago Press, Chicago.

Czamanske, G.K., and Wones, D.R. (1973): Oxidation during magmatic differentiation, Finnmarka complex, Oslo area, Norway: part 2, the mafic silicates. *J. Petrol., vol. 14,* pp. 349–380.

Davis, B.T.C., and England, J.L. (1964): The melting of forsterite up to 50 kilobars. *J. Geophys. Res.,* vol. 69, pp. 1113–1116.

Day, A.L., and Allen, E.T. (1905): *The isomorphism and thermal properties of the feldspars. Carnegie Inst. Washington Publ.,* vol 31.

Deer, W.A., Howie, R.A. and Zussman, J. (2001) Rock-forming Minerals. Volume 4A. Second Edition. Framework silicates: Feldspars. London (The Geological Society) 972 pp. ISBN 1-86239-081-91.

DeVries, R.C. and Osborn, E.F. *J. Am. Ceram. Soc.* vol. 40, p. 9.

Eggler, D.H. (1973a): Role of CO_2 in melting processes in the mantle. *Carnegie Inst. Washington Yearb.,* vol. 72, pp. 457–467.

Eggler, D.H. (1973b): Principles of melting of hydrous phases in silicate melt. *Carnegie Inst. Washington Yearb.,* vol. 72, pp. 491–495.

Eggler, D.H. (1974): Effect of CO_2 on the melting of peridotite. *Carnegie Inst. Washington Yearb.,* vol. 73, pp. 215–224.

Eggler, D.H. (1975): CO_2 as a volatile component of the mantle: the system Mg_2SiO_4-SiO_2-H_2O-CO_2. *Phys. Chem. Earth,* vol. 9, pp. 869–881.

Eggler, D.H. (1976): Does CO_2 cause partial melting in the low-velocity layer of the mantle? *Geology,* vol. 4, pp. 69–72.

Eggler, D.H. (1977): The principle of the zone of invariant vapor composition: an example in the system CaO-MgO-SiO_2-CO_2-H_2O and implications for the mantle solidus. *Carnegie Inst. Washington Yearb.,* vol. 76, pp. 428–435.

Eggler, D.H. (1978): The effect of CO_2, upon partial melting of peridotite in the system Na_2O-CaO-Al_2O_3-MgO-SiO_2-CO_2 to 35 kb, with an analysis of melting in a peridotite-H_2O-CO_2 system. *Am. J. Sci., vol.* **278,** pp. 305–343.

Eggler, D.H., and Holloway, J.R. (1977): Partial melting of peridotite in the presence of H_2O and CO_2: Principles and review, *in* Dick, H.J.B., ed., *Magma genesis. Oreg. Dep. Geol. Miner. Ind. Bull.* **96,** pp. 15–36.

Ellis, D.E., and Wyllie, P.J. (1979): Carbonation, hydration, and melting relations in the system MgO-H_2O-CO_2 at pressures up to 100 kbar. *Am. Mineral.,* **vol. 64,** pp. 32–40.

Emslie, R.F. (1971): Liquidus relations and subsolidus reactions in some plagioclase-bearing systems. *Carnegie Inst. Washington Yearb.,* **vol. 69,** pp. 148–155.

Engel, A.E.J., Engel, C.G., and Havens, R.G. (1965): Chemical characteristics of oceanic basalts and the upper mantle. *Geol. Soc. Am. Bull.,* **vol. 76,** pp. 719–734.

Eskola, Pentti (1954): A proposal for the presentation of rock analyses in ionic percentage. *Ann. Acad. Sci. Fenn.,* Ser. A, III, vol. **38,** pp. 2–15.

Eugster, H.P. (1959): Reduction and oxidation in metamorphism: *in* Abelson, P.H., ed., *Researches in geochemistry,* Wiley, New York, pp. 397–426.

Eugster, H.P., and Wones, D.R. (1962): Stability relations of the ferruginous biotite, annite. *J. Petrol., vol.* **3,** pp, 82–125.

Fenner, C.N. (1929): The crystallization of basalts. *Am. J. Sci.,* 5th Ser., vol. **18,** pp. 225–253.

Fermor, L.L. (1913): Preliminary note on garnet as a geological barometer and on an infra-plutonic zone in the earth's crust. *Rec. Geol. Surv. India, vol.* **43,** pt. 1, pp. 41–47.

Ford, W.E. (1932): *Dana's textbook of mineralogy.* Wiley, New York, p. 559.

Franco, R.R., and Schairer, J.F. (1951): Liquidus temperatures in mixtures of the feldspars of soda, potash, and lime. *J. Geol., vol.* **59,** pp. 259–267.

Fudali, R.F. (1963): Experimental studies bearing on the origin of pseudoleucite and associated problems of alkalic rock systems. *Geol. Soc. Am. Bull., vol.* **74,** pp. 1101–1126.

Gast, P.W. (1965): Terrestrial ratio of potassium to rubidium and the composition of the earth's mantle. *Science, vol.* **147,** pp. 858–860.

Gast, P.W. (1968): Trace element fractionation and the origin of tholeiitic and alkaline magma types. *Geochim. Cosmochim. Acta.,* vol. 32, pp. 1057–1080.

Ghiorso, M.S. (1997) Thermodynamic models of igneous processes. *Annual Review of Earth and Planetary Sciences,* 25, 221–241.

Ghiorso, M. S. & Sack, R. O. (1995): Chemical mass transfer in magmatic processes IV. A revised and internally consistent thermodynamic model for the interpolation and extrapolation of liquid - solid equilibria in magmatic systems at elevated temperatures and pressures. Contributions to Mineralogy and Petrology **119**, 197–212.

Goldsmith, J.R. (1980) Melting and breakdown reactions of anorthite at high pressures and temperatures. *Am. Mineral.,* vol. 65, pp. 272–284.

Greenland, L.P. (1970): An equation for trace element distribution during magmatic crystallization. *Am. Mineral.,* **vol. 55,** pp. 455–465.

Greig, J.W. (1927): Liquid immiscibility in silicate melts. *Am. J. Sci.,* 5th ser., vol. **13,** p. 15.

Greig, J.W., and Barth, T.F.W. (1938): The system $Na_2O{\cdot}Al_2O_3{\cdot}2SiO_2$ (nephelite, carnegieite)-$Na_2O{\cdot}Al_2O_3{\cdot}6SiO_2$ (albite). *Am J. Sci.,* 5th ser., vol. **35A,** pp. 93–112.

Grout, F.F. (1918): The lopolith: an igneous form exemplified by the Duluth Gabbro: *American Journal of Science*, vol. 46, pp. 516–522.

Grout, F.F. (1928): Anorthosites and granite as differentiates of a diabase sill from Pigeon Point, Minnesota. *Geol. Soc. Am. Bull.,* vol. 39, pp. 555–578.

Guthrie, Frederick (1884): On eutexia. *Philos. Mag. 17,* vol. 5, pp. 462–482.

Hess, H.H. (1939): Extreme fractional crystallization of a basaltic magma; the Stillwater igneous complex. *Trans. Am. Geophys. Union,* Pt. 3, pp. 430–432.

Hess, H.H. (1960): Stillwater igneous complex, Montana: a quantitative mineralogical study. *Geol. Soc. Am. Mem.,* vol. 80.

Hodges, F.N. (1974): The solubility of H_2O in silicate melts. *Carnegie Inst. Washington Yearb.,* vol. 73, pp. 251–255.

Holness, M.B., Nielsen, T.F.D., and Tegner, C. (2007): Textural maturity of cumulates: a record of chamber filling, liquidus assemblage, cooling rate and large-scale convection in mafic layered intrusions. *Journal of Petrology,* Vol 48, pp. 141–157.

Holness, M. B., Tegner, C., Nielsen, T.F.D., Stripp, G., and Morse, S. A. (2007) A textural record of solidification and cooling in the Skaergaard intrusion, East Greenland. *Journal of Petrology,* Vol 48, pp. 2359–2377.

Hudon, P., Jung, I., and Baker, D. (2002): Melting of β-quartz up to 2.0 GPa and thermodynamic optimization of the silica liquidus up to 6.0 GPa. *Physics of the Earth and Planetary Interiors*, vol. 130, pp. 159–174.

Hytonen, Kai, and Schairer, J.F. (1961): The plane enstatite-anorthite-diopside and its relation to basalts. *Carnegie Inst. Washington Yearb.,* vol. 60, pp. 125–134.

Irvine, T.N. (1970): Crystallization sequences in the Muskox intrusion and other layered intrusions. *Geol. Soc. S. Afr., Spec. Publ. 1,* pp. 441–476.

Irvine, T.N. (1975): Olivine-pyroxene-plagioclase relations in the system Mg_2SiO,-$CaAl_2Si_2O_8$-$KAlSi_3O_2$-SiO_2 and their bearing on the differentiation of stratiform intrusions. *Carnegie Inst. Washington Yearb.,* vol. 74, pp. 492–500.

Irvine, T.N., and Smith, D.H. (1967): The ultramafic rocks of the Muskox intrusion, Northwest Territories, Canada, *in* Wyllie, P.J., ed., *Ultramafic and related rocks,* Wiley, New York, pp. 38–49.

Jackson, E.D. (1961): Primary textures and mineral associations in the ultramafic zone of the Stillwater Complex, Montana. *U.S. Geol. Surv., Prof. Pap.,* vol. 358.

Jackson, E.D. (1967): Ultramafic cumulates in the Stillwater, Great Dyke, and Bushveld intrusions. *in* Wyllie, P.J., ed., *Ultramafic and related rocks,* Wiley, New York, pp. 20–38.

Jackson, E.D. (1970): The cyclic unit in layered intrusions—a comparison of repetitive stratigraphy in *the* ultramafic parts of the Stillwater, Muskox, Great Dyke, and Bushveld complexes. *Geol. Soc. S. Afr., Spec. Publ. 1,* pp. 391–424.

Johannes, W., and Holtz, F. (1996). The Haplogranite System Qz-Ab-Or. In: Petrogenesis and Experimental Petrology of Granitic Rocks. Minerals and Rocks, vol 22. Springer, Berlin, Heidelberg. https://doi.org/10.1007/978-3-642-61049-3_2

Kennedy, G.C. (1955): Some aspects of the role of water in rock melts, *in* Poldervaart, A., ed., Crust of the Earth. *Geol. Soc. Am., Spec. Pap. 62,* pp. 489–503.

Kennedy, W.Q., and Anderson, E.M. (1938): Crustal layers and the origin of magmas. *Bull. Volcanol.,* 2nd ser., vol. **3,** pp. 23–82.

Kennedy, G.C., Wasserburg, G.J., Heard, H.C., and Newton, R.C. (1962): The upper three-phase region in the system SiO_2-H_2O. *Am. J. Sci.,* **vol. 260**, pp. 501–521.

Kracek, F.C. (1953): Polymorphism: *Encyclopedia Britannica.*

Kracek, F.C. and Clark, S.P. Jr. (1966): Melting and transformation points in oxide and silicate systems at low pressure, *in* Clark, S.P. Jr., ed., *Handbook of physical constants, Geol. Soc. Am. Mem.,* **vol. 97,** pp. 301–322.

Kuno, H. (1959): Origin of Cenozoic petrographic provinces of Japan and surrounding areas. *Bull. Volcanol.,* 2nd ser., vol. **20,** pp. 37–76.

Kuno, H. (1960): High-alumina basalt. *J. Petrol., vol.* **1,** pp. 121–145.

Kushiro, I. (1968): Composition of magma formed by partial zone melting of the earth's upper mantle. *J. Geophys. Res.,* **vol. 73,** pp. 619–634.

Kushiro, I. (1969): The system forsterite-diopside-silica with and without water at high pressures. *Am. J. Sci.,* vol. 267-A (Schairer Volume), pp. 269–294.

Kushiro, I. (1972a): Determination of liquidus relations in synthetic silicate systems with electron probe analysis: the system forsterite-diopside-silica at 1 atmosphere. *Am. Mineral., vol. 57,* pp. 1260–1271.

Kushiro, I. (1972b): Effect of water on the composition of magmas formed at high pressures. *J. Petrol., vol.* **13,** pp. 311–334.

Kushiro, I. (1973): The system diopside-anorthite-albite: determination of compositions of coexisting phases. *Carnegie Inst. Washington Yearb.,* vol. 72, pp. 502–507.

Kushiro, I. (1974): Melting of hydrous upper mantle and possible generation of andesitic magma: an approach from synthetic systems. *Earth Planet. Sci. Letters, vol.* **22,** pp. 294–299.

Kushiro, I., and Fujii, T. (1977): Flotation of plagioclase in magmas at high pressures and its bearing on the origin of anorthosite. *Proc. Jpn. Acad., vol.* **53,** Ser. **B,** pp. 262–266.

Kushiro, I., and Schairer, J.F. (1963): New data on the system diopside-forsterite-silica. *Carnegie Inst. Washington Yearb., vol.* **62,** pp. 95–103.

Kushiro, I., and Schairer, J.F. (1970): Diopside solid solutions in the system diopside-anorthite-albite at 1 atm. and at high pressures. *Carnegie Inst. Washington Yearb., vol.* **68,** pp. 222–226.

Kushiro, I., and Walter, M.J. (1998): Mg-Fe partitioning between olivine and mafic-ultramafic melts. *Geophysical Research Letters*, vol. 25, pp. 2337–2340.

Kushiro, I., and Yoder, H.S. Jr. (1966): Anorthite-forsterite and anorthite-enstatite reactions and their bearing on the basalt-eclogite transformation. *J. Petrol., vol.* 7, pp. 337–362.

Kushiro, I., and Yoder, H.S. Jr. (1969): Melting of forsterite and enstatite at high pressures under hydrous conditions. *Carnegie Inst. Washington Yearb.,* **vol. 67,** pp. 153–158.

Kushiro, I., and Yoder, H.S. Jr. (1970): Stability of field of iron-free pigeonite in the system $MgSiO_3$-$CaMgSi_2O_6$. *Carnegie Inst. Washington Yearb., vol.* **68,** pp. 226–229.

Kushiro, I., Yoder, H.S. Jr., and Nishikawa, M. (1968): Effect of water on the melting of enstatite. *Geol. Soc. Am. Bull., vol.* **79,** pp. 1685–1692.

Lange, R.A. (2003): The fusion curve of albite revisited and the compressibility of $NaAlSi_3O_8$ liquid with pressure. *American Mineralogist*, vol. 88, pp. 109–120.

Lindsley, D.H. (1966a): *P-T* projection for part of the system kalsilite-silica. *Carnegie Inst. Washington Yearb., vol.* **65,** pp. 244–247.

Lindsley, D.H. (1966b): Pressure-temperature relations in the system FeO-SiO_2. *Carnegie Inst. Washington Yearb., vol.* **65,** pp. 226–230.

Lindsley, D.H. (1968): Melting relations of plagioclase at high pressure, *in* Isachsen, Y.W., ed., *Origin of anorthosite and related rocks, N.Y. State Mus. Sci. Serv. Mem. vol.* **18,** pp. 39–46.

Lindsley, D. H. (1983): Pyroxene thermometry. [Presidential Address] American Mineralogist **68,** 477–493.

Lindsley, D. H. (2016): personal communication.

Lindsley, D. H. & Andersen, D. J. (1983): A two-pyroxene thermometer. Proceedings Thirteenth Lunar Planetary Science Conference. Part 2. *Journal of Geophysical Research*, **88**, Supplement, A887–A906.

Lindsley, D.H., and Emslie, R.F. (1968): Effect of pressure on the boundary curve in the system diopside-albite-anorthite. *Carnegie Inst. Washington Yearb., vol.* **66,** pp. 479–480.

Lindsley, D.H., and Munoz, J.L. (1969): Subsolidus relations along the join hedenbergite-ferrosilite. *Am. J. Sci.,* vol. 267-A (Schairer Volume), pp. 295–324.

Lindsley, D.H., and Smith, D. (1971): Chemical variations in the feldspars. *Carnegie Inst. Washington Yearb.,* vol. 69, pp. 274–278.

Lindsley, D.H., Speidel, D.H., and Nafziger, R.H. (1968): P-T-f_{02} relations for the system Fe-O-SiO_2. *Am. J. Sci.,* vol. 226, pp. 342–360.

Lindsley, D.H., Brown, G.M., and Muir, I.D. (1969): Conditions of the ferrowollastonite-ferrohedenbergite inversion in the Skaergaard intrusion, East Greenland. *Mineral. Soc. Am. Spec. Pap. 2,* pp. 193–201.

Lipin, B.R. (1978): The system Mg_2SiO_4-Fe_2SiO_4-$CaAl_2Si_2O_8$-SiO_2 and the origin of Fra Mauro basalts. *Am. Mineral.,* vol. 63, pp. 350–364.

Longhi, J., and Hays, J.F. (1979): Phase equilibria and solid solution along the join $CaAl_2Si_2O_8$-SiO_2. *Am. J. Sci.,* vol. 279, pp. 876–890.

Macdonald, G.A. (1949): Petrography of the Island of Hawaii. *U.S. Geol. Surv., Prof. Pap.,* vol. 214D, pp. 51–96.

Macdonald, G.A., and Katsura, T. (1964): Chemical composition of Hawaiian lavas. *J. Petrol., vol.* **5,** pp. 82–133.

Maun, A., and Osborn, E.F. (1956): Phase equilibria at liquidus temperatures in the system MgO-FeO-Fe_2O_3-SiO_2. *J. Am. Ceram. Soc., vol.* **39,** pp. 121–140.

McBirney, A. R. and Naslund, H. R. (1990): The differentiation of the Skaergaard intrusion. A discussion of Hunter and Sparks (*Contributions to Mineralogy and Petrology*, vol. 95, pp. 451–461), *Contributions to Mineralogy and Petrology*, vol. 104, pp. 235–240.

McIntosh, D.C.B. (2009): High pressure liquidus studies of the inferred magma composition of the Kiglapait layered intrusion, Labrador, Canada. M.S Thesis, Department of Geosciences, University of Massachusetts, 91 pp.

Michot, Jean (1960): La palingenese basique. *Acad. R. Belg. Bull. 1. sci. 5, vol.* **46,** pp. 257–268.

Morey, G.W. (1964): Phase-equilibrium relations of the common rock-forming oxides except water. *U.S. Geol. Surv., Prof. Pap.* 440-L.

Morse, S.A. (1961): The geology of the Kiglapait layered intrusion, coast of Labrador, Canada. Unpub. Ph.D. thesis, McGill Univ.

Morse, S.A. (1968): Layered intrusions and anorthosite genesis, *in* Isachsen, Y.W., ed., *Origin of anorthosite and related rocks, N.Y. State Mus. Sci. Serv. Mem., vol.* **18,** pp. 175–187.

Morse, S.A. (1969): *The Kiglapait layered intrusion, Labrador. Geol. Soc. Am. Mem., vol.* **112.**

Morse, S.A. (1969a): Syenites. *Carnegie Inst. Washington Yearb., vol.* **67,** pp. 112–120.

Morse, S.A. (1969b): Ternary feldspars. *Carnegie Inst. Washington Yearb.,* vol. 67, pp. 124–126.

Morse, S.A. (1970): Alkali feldspars with water at 5 kb pressure. *J. Petrol., vol.* **11,** pp. 221–251.

Morse, S.A. (1976): The lever rule with fractional crystallization and fusion. *Am. J. Sci.,* vol. 276, pp. 330–346.

Morse, S. A. (1979a): Reaction constants for En-Fo-Sil equilibria: an adjustment and some applications. American Journal of Science **279,** 1060–1069.

Morse, S.A. (1979b): Kiglapait Geochemistry I, (1979c) MG. II: J. Petrol. *20,* 555–624.

Morse, S. A. (1979d) Kiglapait geochemistry 1: systematics, sampling and density. *J Petrol.* Vol. 20, pp. 555–590.

Morse, S. A. (1980): Kiglapait mineralogy II: Fe-Ti oxide minerals and the activities of oxygen and silica. J. Petrology **21,** 685–719.

Morse, S. A. (1981): Kiglapait geochemistry IV: The major elements. Geochimica et Cosmochimica Acta **45,** 461–479.

Morse, S. A. (1986) Convection in aid of adcumulus growth. *J Petrol.* Vol. 27, pp. 1183–1215.

Morse, S.A. (1994): *Basalts and Phase Diagrams.* Krieger, Melbourne, FL.

Morse, S.A. (1996) Kiglapait minerialogy III. Olivine compositions and Rayleigh fractionation models. *Journal of Petrology* 37, 1037–1061.

Morse S. A. (1997): Binary solutions and the lever rule revisited: *Journal of Geology*, vol. 105, pp. 471–482.

Morse, S.A. (2000): Linear partitioning in binary solutions. *Geochimica et Cosmochimica Acta,* vol. 64, pp. 2309–2319.

Morse, S. A. (2006): Labrador massif anorthosites: Chasing the liquids and their sources. Lithos **89,** 202–221.

Morse, S. A. (2008) Toward a thermal model for the Skaergaard Liquidus. *Am. MM.* Vol 93, pp. 248–251.

Morse, S. A. (2011) The fractional latent heat of crystallizing magmas. *Am. Min.* Vol. 96, pp. 682–689.

Morse, S. A. (2012) Plagioclase An range and residual porosity in igneous cumulates of the Kiglapait Intrusion. *J Petrol.* Vol. 53, pp. 891–918.

Morse, S. A. (2013) Solidification of trapped liquid in rocks and crystals. *Am. MM.* Vol. 98: 888–896.

Morse, S. A. (2014): Plagioclase fractionation in troctolitic magma. J. Petrology **55**, 2403–2418.

Morse, S. A. (2015a): Linear partitioning in binary solutions: A review with a novel partitioning array. American Mineralogist **100**, 1021–1032.

Morse, S. A. (2015b): Kiglapait Intrusion, Labrador. *In* Charlier et al. (eds), *Layered Intrusions* Springer-Dordrecht, 589–648.

Morse, S. A. (2017) Kiglapait mineralogy V: Feldspars in a hot, dry magma. American Mineralogist 102, 2084–2095.

Morse, S. A. & Brady, J. B. (2017) Thermal history of the Upper Zone of the Kiglapait intrusion. Journal of Petrology 58, 1319–1332.

Morse, S. A. & Ross, Malcolm (2004): Kiglapait mineralogy IV: The augite series. American Mineralogist 89, 1380-1395.

Morse, S. A., Lindsley, D.H. & Williams R. J. (1980): Concerning intensive parameters in the Skaergaard intrusion. American Journal of Science **280-A** (Jackson Vol.), 159–170.

Morse, S. A., Brady, J. B. & Sporleder, B. A. (2004): Experimental petrology of the Kiglapait intrusion: Cotectic trace for the Lower Zone at 5 kb in graphite. *Journal of Petrology* **45,** 2225–2259.

Morse, S.A., and Brady, J.B., and Banks, D.C. (2020): 87 million years of recorded history in Labrador: birth, life, and sleep of the Kiglapait *intrusion. Canadian Mineralogist*, vol. 58, pp. 461–475.

Murphy, W.M. (1977): An experimental study of solid-liquid equilibria in the albite-anorthite-diopside system. Unpublished MS thesis, Univ. of Oregon, Eugene.

Mysen, B.O. (1977): Solubility of volatiles in silicate melts under the pressure and temperature conditions of partial melting in the upper mantle, *in* Dick, H.J.B., ed., *Magma genesis. Oreg. Dep. Geol. Miner. Ind. Bull.,* **96,** pp. 1–14.

Mysen B. (2007): Partitioning of calcium, magnesium, and transition metals between olivine and melt governed by the structure of the silicate melt at ambient pressure. *American Mineralogist*, vol. 92, pp. 844–862.

Mysen, B.O., and Kushiro, **I.** (1977): Compositional variations of coexisting phases with degree of partial melting of peridotite in the upper mantle. *Am. Mineral.,* vol. 62, pp. 843–865.

Mysen, B., and Richet, P. (2018): *Silicate Glasses and Melts*, Second Edition. Elsevier Science. 720p.

Mysen, B.O., and Seitz, M.G. (1975): Trace element partitioning determined by beta-track mapping—an experimental study using carbon and samarium as examples. *J. Geophys. Res.,* vol. 80, pp. 2627–2635.

Mysen, B.O., Virgo, D., and Seifert, F.A. (1982): The structure of silicate melts: Implications for chemical and physical properties of natural magma. *Reviews of Geophysics*, vol. 20, pp. 353–383. https://doi.org/10.1029/RG020i003p00353

Nakamura, Y., (1974): The system SiO_2-H_2O-H_2 at 15 kbar. *Carnegie Inst. Washington Yearb.; vol.* **73,** pp. 259–262.

Nakamura, Y., and Kushiro, I. (1974): Composition of the gas phase in Mg_2SiO_4-SiO_2-H_2O at 15 kbar. *Carnegie Inst. Washington Yearb., vol.* **73,** pp. 255–258.

Navrotsky, A., Hon, R., Weill, D.F., and Henry, D.J. (1980): Thermochemistry of glasses and liquids in the systems $CaMgSi_2O_6$-$CaAl_2Si_2O_8$-$NaAlSi_3O_8$, SiO_2-$CaAl_2Si_2O_8$-$NaAlSi_3O_8$ and SiO_2-Al_2O_3-CaO-Na_2O. *Geochimica et Cosmochimica Acta*, vol. 44., pp. 1409–1423. https://doi.org/10.1016/0016-7037 (80)90107-6

Neuman, H., Mead, J., and Vitaliano, C.J. (1954): Trace element variation during fractional crystallization as calculated from the distribution law: *Geochim. Cosmochim. Acta.,* vol. 6, pp. 90–100.

Nielsen, T.F.D, Rudashevsky, N.S., Rudashevsky, V.N., Weatherley, S.M., and Andersen, J.C. Ø. (2019) Elemental distributions and mineral parageneses of the Skaergaard PGE–Au mineralization: Consequences of accumulation, redistribution, and equilibration in an upward-migrating mush zone. *Journal of Petrology*, vol. 60, pp. 1903–1934. https://doi.org/10.1093/petrology/egz057

Niggli, P. (1936): Molekularnormen zu Gesteinsberechnung. *Schweiz. Mineral. Petrogr. Mitt., vol.* **16,** p. 295.

O'Hara, M.J. (1977): Geochemical evolution during fractional crystallization of a periodically refilled magma chamber. *Nature, vol.* **266,** pp. 503–507.

Osborn, E.F. (1942): The system $CaSiO_3$-diopside-anorthite. *Am. J. Sci.,* **vol. 240,** pp. 751–788.

Osborn, E.F. (1959): Role of oxygen partial pressure in the crystallization and differentiation of basaltic magma. *Am. J. Sci., vol.* **257,** pp. 609–647.

Osborn, E.F., and Schairer, J.F. (1941): The ternary system pseudowollastonite-akermanite-gehlenite. *Am. J. Sci., vol.* **239,** pp. 715–63.

Osborn, E.F., and Tait, D.B. (1952): The system diopside-forsterite-anorthite. *Am. J. Sci.,* Bowen vol., pp. 413–433.

Ostrovsky, I.A. (1967): On some sources of errors in phase-equilibria investigations at ultra-high pressures; phase diagram of silica. *Geol. J., vol. 5,* pp. 321–328.

Parsons, Ian, Fitz Gerald, J.D., and Lee, M.R. (2015) Routine characterization and interpretation of complex alkali feldspar intergrowths. American Mineralogist 100, 1277–1303.

Paster, T.P., Schauwecker, D.S., and Haskin, L.A. (1974): The behavior of some trace elements during solidification of the Skaergaard layered series. *Geochim. et Cosmochim. Acta,* **vol. 38,** pp. 1549–1577.

Peck, D.L., Wright, T.L., and Moore, J.G. (1966): Crystallization of tholeiitic basalt in Alae lava lake, Hawaii. *Bull. Volcano!.,* **vol. 29,** pp. 629–656.

Presnall, D.C. (1966): The join forsterite-diopside-iron oxide and its bearing on the crystallization of basaltic and ultramafic magmas. *Am. J. Sci., vol.* **264,** pp. 753–809.

Presnall, D.C. (1969): The geometric analysis of partial fusion. *Am. J. Sci.,* vol. 267, pp. 1178–1194.

Presnall, D.C., Dixon, S.A., Dixon, J.R. O'Donnell, T.H., Brenner, N.L., Schrock, R.L., and Dycus, D.W. (1978): Liquidus phase relations on the join diopside-forsterite-anorthite from 1 atm to 20 kbar: their bearing on the generation and crystallization of basaltic magma. *Contr. Mineral. Petrol.,* **vol.** 66, pp. 203–220.

Presnall, D.C., Dixon, J.R., O'Donnell, T.H., and Dixon, S.A. (1979): Generation of mid-ocean ridge tholeiites. *J. Petrol.,* **vol. 20,** pp. 3–36.

Rankin, G.A., and Wright, F.E. (1915): The ternary system CaO-$A1_2O_3$-SiO_2. *Am. J. Sci.,* vol. 39, pp. 1–79.

Rayleigh, J.W.S. (1896): Theoretical considerations respecting the separation of gases by diffusion and similar processes. *Phil. Mag.,* vol. 42, pp. 493–498.

Rhodes, J.M., Dungan, M.A., Blanchard, D.P., and Long, P.E. (1979): Magma mixing at mid-ocean ridges: evidence from basalts drilled near 22°N on the Mid-Atlantic ridge. *Tectonophysics,* vol. 55, pp. 35–61.

Robie. R.A.; Hemingway, B.S., and Fischer, J.R. (1978): Thermodynamic properties of minerals d related substances at 298.15 K and 1 bar (10^5 pascals) pressure and temperatures. *US Geol. Surv. Bull.* 1452.

Robinson, P. et al. 1971, Am. Min. *56,* 909–939.

Roeder, P.L. (1974): Paths of crystallization and fusion in systems showing ternary solid solution. *Am. J. Sci., vol.* **224,** pp. 48–60.

Roeder, P.L., and Emslie, R.F. (1970): Olivine-liquid equilibria. *Contrib. Mineral. Petrol.,* vol. 29, pp. 275–289.

Ronov, A.B., and Yaroshevsky, A.A. (1969): Chemical composition of the earth's crust, *in The earth's crust and upper mantle,* P.J. Hart, ed., *Am. Geoph. Union Monograph, vol.* **13,** pp. 37–57.

Rosenhauer, M., and Eggler, D.H. (1975): Solution of H_2O and CO_2 in diopside melt. *Carnegie Inst. Washington Yearb.,* **vol. 74,** pp. 474–479.

Rudnick, R.L., and Gao, S. (2005) Composition of the continental crust. in R.L. Rudnick, ed., The Crust, Treatise on Geochemistry, Volume 3, Elsevier, 1–64.

Ryan, B (1990): Geological map of the Nain Plutonic Suite and surrounding rocks (Nain–Nutak, NTS 14SW). Newfoundland Department of Mines and Energy, Geological Survey Branch, Map 90-44, scale 1:500 000.

Saal, A. E., Hauri, E., Langmuir, C. H., & Perfit, M. R. (2002): Vapour undersaturation in primitive mid-ocean-ridge basalt and the volatile content of the Earth's upper mantle. Nature **419**, 451–455.

Schairer, J.F. (1950): The alkali feldspar join in the system $NaAlSiO_4$-$KAlSiO_4$-SiO_2. *J. Geol., vol.* **58,** pp. 512–517.

Schairer, J.F. (1951): Phase transformations in polycomponent silicate systems, *in Phase transformations in solids,* R. Smoluchowski, ed., Wiley, New York, pp. 278–295.

Schairer, J.F. (1954): The system K_2O-MgO-Al_2O_3-SiO_2. I. Results of quenching experiments on four joins in the tetrahedron cordierite-forsterite-leucite-silica and on the join cordierite-mullite-potash feldspar. *Journal of the American Ceramic Society*, vol. 37, pp. 501–533.

Schairer, J.F., and Bowen, N.L (1947): Melting relations in the systems Na_2O-Al_2O_3-SiO_2 and K_2O-Al_2O_3-SiO_2. *American Journal of Science*, vol. 245, pp. 193–204.

Schairer, J.F., and Bowen, N.L. (1955): The system K_2O-$A1_2O_3$-SiO_2. *Am. J. Sci., vol.* **253,** pp. 681–746.

Schairer, J.F., and Bowen, N.L. (1956): The system Na_2O-$A1_2O_3$-SiO_2. *Am. J. Sci.,* **vol. 254, pp.** 129–195.

Schairer, J.F., and Kushiro, I. (1964): The join diopside-silica. *Carnegie Inst. Washington Yearb., vol.* **63,** pp. 130–132.

Schairer, J.F., and Yoder, H.S., Jr. (1958): The quaternary system Na_2O-MgO-Al_2O_3-SiO_2. *Carnegie Institution of Washington Yearbook*, vol. 57, pp. 210–212.

Schairer, J.F., and Yoder, H.S. Jr. (1960): The nature of residual liquids from crystallization, with data on the system nepheline-diopside-silica. *Am. J. Sci.,* **vol. 258A,** pp. 273–283.

Schairer, J.F., and Yoder, H.S. Jr. (1962): The system diopside-enstatite-silica. *Carnegie Inst. Washington Yearb.,* **vol. 61,** pp. 75–82.

Schairer, J.F., and Yoder, H.S. Jr. (1971): The joins $Na_2O{\cdot}5MgO{\cdot}12SiO_2$-sodium disilicate and $2Na_2O{\cdot}3MgO{\cdot}5SiO_2$-sodium disilicate in the system Na_2O-MgO-SiO_2. *Carnegie Inst. Washington Yearb.,* **vol. 69,** pp. 157–160.

Schreinemakers, F.A.H. (1915–1925): In-mono- and di-variant equilibria: 29 papers in vols. 18–28 of *Proc. K. Akad. Wet.,* Amsterdam, and reproduced by the College of Earth and Mineral Science, Penn. State Univ., University Park, Pa.

Seck, H.A. (1971): Koexistierende Alkalifeldspate and Plagioclase im System $NaAlSi_3O_8$-$KAlSi_3O_8$-$CaAl_2Si_2O_8$-H_2O bei Temperaturem von 650°C bis 900°C. *Neues Jahrb. Mineral. Abh.,* **vol. 115,** pp. 315–345.

Shirey, S. B., Wagner, L. S., Walter, M. J., Pearson, D. G., and van Keken, P. E. (2020); Sublithospheric diamonds and deep earthquakes demonstrate an arc-avoiding subduction pathway for C-O-H-N-S volatiles. Science Advances DOI:https://doi.org/10.1126/sciadv.abe9773.

Smith, C.H. (1962): Notes on the Muskox intrusion, Coppermine River area, District of MacKenzie. *Geol. Surv. Can., Pap.* 61-25.

Smith, D. (1971): Stability of the assemblage iron-rich orthopyroxene-olivine-quartz. *Am. J. Sci., vol.* **271,** pp. 370–382.

Smith, J. V. and Brown, W. L. (1988) Feldspar Minerals: Volume 1: Crystal Structures, Physical, Chemical, and Microtextural Properties. Springer-Verlag, Berlin-Heidelberg, 828 pp.

Smith, C.H., and Kapp, H.E. (1963): The Muskox intrusion, a recently discovered layered intrusion in the Coppermine River area, Northwest Territories, Canada. *Miner. Soc. Amer., Spec. Pap. 1,* pp. 30–35.

Smith, C.H., Irvine, T.N., and Findlay, D.C. (1966): Geologic maps of the Muskox intrusion. *Geol. Surv. Can.* Maps 1213-A and 1214-A.

Smith, E. M., Ni, Peng, Shirey, S. B., Richardson, S. H., Wany, W., and Shahar, A. (2012: Heavy iron in sublithospheric diamonds reveals deep serpentine subduction. (Abs) Goldschmidt 2021 online.

Solomon, S.C. (1976): Geophysical constraints on radial and lateral temperature variations in the upper mantle. *Am. Mineral.,* **vol. 61,** pp. 788–803.

Sosman, R.B. (1952): Temperature scales and silicate research. *Am. J. Sci.,* Bowen volume, pp. 517–528.

Speidel, D.H., and Nafziger, R.H. (1968): P-T-f_{02} relations in the system Fe-O-MgO-SiO_2. *Am. J. Sci., vol.* **266,** pp. 361–379.

Stern, C.R., Huang, W.-L., and Wyllie, P.J. (1975): Basalt-andesite-rhyolite H_2O: crystallization intervals with excess H_2O and H_2O-undersaturated liquidus surfaces to 35 kilobars, with implications for magma genesis. *Earth Planet. Sci. Lett., vol.* 28, pp. 189–196.

Stewart, D.B. (1967): Four-phase curve in the system $CaAl_2Si_2O_8$-SiO_2-H_2O between 1 and 10 kilobars. *Schweiz. Mineral. Petrogr. Mitt., vol.* **47,** pp. 35–59.

Stewart, D.B., and Roseboom, E.H. (1962): Lower temperature terminations of the three-phase region plagioclase-alkali feldspar-liquid. *J. Petrol.,* **vol. 3,** pp. 280–315

Stolper, E. M., Oliver Shorttle, Paula M. Antoshechkina, and Paul D. Asimow, (2020): The effects of solid-solid phase equilibria on the oxygen fugacity of the upper mantle. (Roebling Medal Paper) American Mineralogist Vol. 105, p 1445–1471.

Stone, E., Lindsley, D.H., Piggot, V., Harbottle, G., & Ford, M.T. (1998): From shifting silt to solid stone: the manufacture of synthetic "basalt" in ancient Mesopotamia. Science **280**, 2091–2093.

Tammann, G. (1903): *Kristallisieren and Schmelzen:* J.A. Barth, Leipzig. English translation *in States of aggregation* (1925) D. Van Nostrand.

Thompson, A.B., and Algor, J.R. (1977): Model systems for anatexis of pelitic rocks I. Theory of melting reactions in the, system $KAlO_2$-$NaAlO_2$-Al_2O_3-SiO_2-H_2O. *Contr. Mineral. Petrol.,* **vol. 63,** pp. 247–269.

Thy, P., Lesher, C. E., and Tegner, C. (2009): The Skaergaard liquid line of descent revisited. *Contributions to Mineralogy and Petrology*, vol. 157, pp. 735–747.

Thy, P., Lesher, C. E., and Tegner, C. (2013): Further work on experimental plagioclase equilibria and the Skaergaard liquidus temperature. *American Mineralogist*, vol. 98, pp. 1360–1367.

Tiller, W.A. (1991): The science of crystallization: microscopic interfacial phenomena. Cambridge University Press, U.K., 391 p.

Tilley, C.E. (1950): Some aspects of magmatic evolution. *Quart. J. Geol. Soc.,* Lond., vol. **106,** pp. 37–61.

Tilley, C.E., Yoder, H.S. Jr., and Schairer, J.F. (1963): Melting relations of basalts. *Carnegie Inst. Washington Yearb.,* vol. 62, pp. 77–84.

Tomita, T. (1935): On the chemical compositions of the Cenozoic alkaline suite of the circum-Japan sea region. *J. Shanghai Sci. Inst.,* sect. IIi, pp. 227.

Toplis M. J. (2005): The thermodynamics of iron and magnesium partitioning between olivine and liquid: criteria for assessing and predicting equilibrium in natural and experimental systems. *Contributions to Mineralogy and Petrology*, vol. 149, pp. 22–39.

Toplis, M.J., Brown, W.L., and Pupier, E. (2007): Plagioclase in the Skaergaard intrusion. Part 1: Core and rim compositions in the layered series. *Contributions to Mineralogy and Petrology*, vol. 155, pp. 329-340. DOI: 10.1007/s00410-007-0245-1

Tschauner O (2019) High-pressure minerals. American Mineralogist **104**, 1701–1731.

Tschauner, O., Ma, C. Beckett, J. R., Prescher, C., Prakapenka, V. B., & Rossman, G. R. (2014): Discovery of bridgmanite, the most abundant mineral in Earth, in a shocked meteorite. Science **346**, 1100–1102.

Turcotte, D. L. and Schubert, G. (2002): Geodynamics, 2nd Ed. Cambridge University Press, U.K., 456 p.

Tuttle, O.F., and Bowen, N.L. (1950): High temperature albite and contiguous feldspars. *J. Geol.,* **vol. 58,** pp. 572–583.

Tuttle, O.F., and Bowen, N.L. (1958): Origin of granite in the light of experimental studies in the system $NaAlSi_3O_8$-$KAlSi_3O_8$-SiO_2-H_2O. *Geol. Soc. Am. Mem.,* **vol. 74**.

Tuttle, O.F., and Smith, J.V. (1958): The nepheline-kalsilite system II. Phase relations. *Am. J. Sci.,* **vol. 256,** pp. 571–589.

Ubbelohde, A.R. (1965): *Melting and crystal structure.* Clarendon Press, Oxford.

Ubbelohde, A.R. (1979): *The molten state of matter. Melting and crystal structure.* Wiley-Interscience, New York.

Ulmer, G.C., ed. (1971): *Research techniques for high pressure and high temperature.* Springer-Verlag, New York.

van Alkemade, A.C. van Rijn (1893): Graphische Behandlung einiger thermodyn-amischen Probleme fiber Gleichgewichtszustande von Salzlosungen mit festen Phasen. *Z. Phys. Chem., vol. 11,* pp. 289–327.

Wager, L.R. (1960): The major element variation of the layered series of the Skaergaard intrusion and a reestimation of the average composition of the hidden layered series and of the successive residual magmas. *J. Petrol., vol.* **1,** pp. 364–398.

Wager, L.R., and Brown, G.M. (1968): *Layered igneous rocks.* W.H. Freeman and Co., San Francisco.

Wager, L.R., and Deer, W.A. (1939): Geological investigations in East Greenland, Pt. III. The petrology of the Skaergaard intrusion, Kangerdlugssauq, East Greenland. *Medd. Groenl.,* vol. 105, No. 4, pp. 1–352.

Wager, L.R., Brown, G.M., and Wadsworth, W.J. (1960): Types of igneous cumulates. *J. Petrol., vol.* **1,** pp. 73–85.

Waldbaum, D.R. (1969): Some observations on preparing glass for compositions on the alkali feldspar join. *Am. J. Sci.,* vol. 267, pp. 1249–1253.

Waldbaum, D.R., and Thompson, J.B. Jr. (1969): Mixing properties of sanidine crystalline solutions. IV. Phase diagrams from equations of state: *Am. Mineral.,* vol. 54, pp. 1274–1298.

Weaver, J.S., Chipman, D.W., and Takahashi, T. (1979): Comparison between thermochemical and phase stability data for the quartz-coesite-stishovite transformations. *Am. Mineral.,* vol. 64, pp. 604–614.

Wheeler, E.P. 2nd (1965): Fayalitic olivine in northern Newfoundland-Labrador. *Canad. Mineral., vol.* **8,** pp. 339–346.

Whittaker, E.J.W., and Muntus, R. (1970): Ionic radii for use in geochemistry. *Geochim. et Cosmochim. Acta.,* **vol. 34,** pp. 945–956.

Wicks, J. and Duffy, T.S. (2016): Crystal structures of minerals in the lower mantle. In Deep Earth: Physics and Chemistry of the Lower Mantle and Core. American Geophysical Union 68–88.

Wiebe, R.A. (1979): Fractionation and liquid immiscibility in an anorthosite pluton of the Nain complex, Labrador. *J. Petrol., vol.* **20,** pp. 239–270.

Wiebe, R. A. (1988) Structural and magmatic evolution of a magma chamber: The Newark Island Layered Intrusion, Nain, Labrador. Journal of Petrology Vol. 29, pt. 2, p 383–411.

Williams, R.J. (1971a): Reaction constants in the system Fe-MgO-SiO_2-O_2: I. Experimental results. *Am. J. Sci.,* **vol. 270,** pp. 334–360.

Williams, R.J. (1971b): Reaction constants in the system Fe-Mg-SiO_2-O_2: intensive parameters in the Skaergaard intrusion, East Greenland. *Am. J. Sci.,* **vol. 271,** pp. 132–146.

Williams, R.J. (1972): Activity-composition relations in the fayalite-forsterite solid solution between 900° and 1300°C at low pressure. *Earth Planet. Sci. Lett., vol.* **15,** pp. 296–300.

Willie, P.J. (1963): Effects of the changes in slope occurring on liquidus and solidus paths in the system diopside-anorthite-albite. *In* Mineralogical Society of America, Special Paper 1, pp. 204–212

Windom, K.E., and Boettcher, A.L. (1977): (abs) Melting of albite at high pressure: a redetermination. *EOS Trans. Am. Geophys. Union, vol.* 58, p. 1243.

Wood, B.J. (1978): Reactions involving anorthite and $CaAl_2SiO_6$ pyroxene at high pressures and temperatures. *Am. J. Sci.,* **vol. 278,** pp. 930–942.

Wright, T. L., and Weiblen, P.W. (1967): Mineral composition and paragenesis in tholeiitic basalt, 1965 Makaopuhi lava lake, Hawaii. *Geological Society of America, Abstracts with Programs,* pp. 242–243.

Wyllie, P.J. (1963): Effects of the changes in slope occurring on liquidus and solidus paths in the system diopside-anorthite-albite. *In* Mineralogical Society of America, Special Paper 1, pp. 204–212.

Wyllie, P.J. (1978): Mantle fluid compositions buffered in peridotite-CO_2-H_2O by carbonates, amphibole, and phlogopite. *J. Geol., vol.* **86,** pp. 687–713.

Wyllie, P.J. (1979): Magmas and volatile components. *Am. Mineral., vol.* **64,** pp. 469–500.

Wyllie, P.J., and Huang, W.L. (1975): Peridotite, kimberlite, and carbonatite explained in the system CaO-MgO-SiO_2-CO_2. *Geology,* **vol. 3,** pp. 621–624.

Wyllie, P.J., and Huang, W.L. (1976): Carbonation and melting reactions in the system CaO-MgO-SiO_2-CO_2 at mantle pressures with geophysical and petrological applications. *Contr. Mineral. Petrol.,* vol. 54, pp. 79–107.

Yoder, H.S. Jr. (1950): High-low quartz inversion up to 10,000 bars. *Trans. Am. Geophys. Union,* vol. 31, pp. 827–835.

Yoder, H.S., Jr. (1965): Diopside-anorthite-water at five and ten kilobars and its bearing on explosive volcanism. *Carnegie Institution of Washington Yearbook*, vol. 64, pp. 82–89.

Yoder, H.S. Jr. (1973): Contemporaneous basaltic and rhyolitic magmas. *Am. Mineral.,* vol. 58, pp. 153–171.

Yoder, H.S. Jr. (1976): *Generation of basaltic magma:* National Acad. Sci., Washington, D.C.

Yoder, H.S. Jr., and Kushiro, I. (1969): Melting of a hydrous phase; phlogopite: *Am. J. Sci.,* vol. 267-A (Schairer Volume), pp. 558–582.

Yoder, H.S. Jr. and Tilley, C.E. (1957): Basalt magmas. *Carnegie Inst. Washington Yearb.,* **vol. 56,** pp. 156–161.

Yoder, H.S. Jr. and Tilley, C.E. (1962): Origin of basalt magmas: an experimental study of natural and synthetic rock systems. *J. Petrol.,* vol. 3, pp. 342–532.

Yoder, H.S. Jr., Stewart, D.B., and Smith, J.V. (1957): Ternary feldspars. *Carnegie Inst. Washington Yearb.,* **vol. 56,** pp. 206–214.

Yund, R.A. (1975): Microstructure, kinetics, and mechanisms of alkali feldspar exsolution. Ch. 6 *in* Ribbe, P.H. (ed.) *Feldspar Mineralogy,* M.S.A. Short Course Notes *2,* Mineral. Soc. Am., Washington, D.C.

Zavarizskii, A.N. and Sobolev, V.S. (1964): *The physiochemical principles of igneous petrology.* Israel Program for Scientific Translations, Jerusalem.

Zen, E-an. (1963): Components, phases, and criteria of chemical equilibrium in rocks. *Am. J. Sci.,* **vol. 261,** pp. 929–942.

Zen, E-an. (1966): Construction of pressure-temperature diagrams for multi-component systems after the method of Schreinemakers—a geometric approach. *U.S. Geol. Sum, Bull. 1225.*

Zen, E-an. (1977): The phase-equilibrium calorimeter, the petrogenetic grid, and a tyranny of numbers. *Am. Mineral.,* **vol. 62,** pp. 189–204.

Zhang, J., Liebermann, R.C., Gasparik, T., Herzberg, C.T., & Fei, Y. (1993): Melting and subsolidus relations of SiO_2 at 9–14 GPa. Journal of Geophysical Research Solid Earth 98, pp. 19785–197893.

Printed in the United States
by Baker & Taylor Publisher Services